Methods in Molecular Biology

For further volumes:
http://www.springer.com/series/7651

For over 35 years, biological scientists have come to rely on the research protocols and methodologies in the critically acclaimed *Methods in Molecular Biology* series. The series was the first to introduce the step-by-step protocols approach that has become the standard in all biomedical protocol publishing. Each protocol is provided in readily-reproducible step-by-step fashion, opening with an introductory overview, a list of the materials and reagents needed to complete the experiment, and followed by a detailed procedure that is supported with a helpful notes section offering tips and tricks of the trade as well as troubleshooting advice. These hallmark features were introduced by series editor Dr. John Walker and constitute the key ingredient in each and every volume of the *Methods in Molecular Biology* series. Tested and trusted, comprehensive and reliable, all protocols from the series are indexed in PubMed.

Computer Optimized Microscopy

Methods and Protocols

Edited by

Elena Rebollo

Molecular Imaging Platform, Molecular Biology Institute of Barcelona (CSIC), Barcelona, Spain

Manel Bosch

Scientific and Technological Centers (CCiTUB), Universitat de Barcelona, Barcelona, Spain

Humana Press

Editors
Elena Rebollo
Molecular Imaging Platform
Molecular Biology Institute
of Barcelona (CSIC)
Barcelona, Spain

Manel Bosch
Scientific and Technological Centers (CCiTUB)
Universitat de Barcelona
Barcelona, Spain

ISSN 1064-3745 ISSN 1940-6029 (electronic)
Methods in Molecular Biology
ISBN 978-1-4939-9685-8 ISBN 978-1-4939-9686-5 (eBook)
https://doi.org/10.1007/978-1-4939-9686-5

This Humana imprint is published by the registered company Springer Science+Business Media, LLC, part of Springer Nature.
The registered company address is: 233 Spring Street, New York, NY 10013, U.S.A.

Preface

The significant advances in the microscopy field during the last decades have placed image quantification at the heart of many experimental routines in biomedical research. Concurrently, the evolution of open-source tools has greatly encouraged the development of image processing and analysis workflows specifically tailored to solve particular biomedical problems, together with wider solutions for more general imaging problematics. Even so, the lack of systematic training in digital imaging during regular degrees imposes biomedical researchers the need to search for proper instruction and know-how to be able to cope with the basics of digital image quantification and automation. *Computer Optimized Microscopy: Methods and Protocols* provides a comprehensive collection of open-source-based image analysis workflows covering relevant image analysis problematics, such as colocalization, particle counting, 3D structural analysis, ratio imaging and FRET quantification, particle tracking, high-content screening, or machine learning. Each workflow, based on a particular biological application, explains in detail the main processing steps required to succeed and provides image sample sets along with macroinstructions to help the reader get a better understanding of the protocol and also develop the skills to further customize solutions in a wide range of biomedical applications. We would like to acknowledge all the imaging experts who, over the last 10 years, have contributed with their knowledge to the Computer Optimized Microscopy Course, in which this volume is inspired.

Barcelona, Spain

Elena Rebollo
Manel Bosch

Contents

Preface . . . . *v*
Contributors . . . . *ix*

PART I OVERVIEW

1 Main Steps in Image Processing and Quantification: The Analysis Workflow 3
José F. Pertusa and Jose M. Morante-Redolat

2 Open Source Tools for Biological Image Analysis 23
Romain Guiet, Olivier Burri, and Arne Seitz

PART II METHODS BASED ON IMAGEJ MACRO PROGRAMMING

3 Proximity Ligation Assay Image Analysis Protocol: Addressing Receptor-Receptor Interactions 41
Marc López-Cano, Víctor Fernández-Dueñas, and Francisco Ciruela

4 Introduction to ImageJ Macro Language in a Particle Counting Analysis: Automation Matters 51
Manel Bosch

5 Automated Macro Approach to Quantify Synapse Density in 2D Confocal Images from Fixed Immunolabeled Neural Tissue Sections 71
Elena Rebollo, Jaume Boix-Fabrés, and Maria L. Arbones

6 Automated Quantitative Analysis of Mitochondrial Morphology 99
Anna Bosch and Maria Calvo

7 Structure and Fluorescence Intensity Measurements in Biofilms 117
Bertrand Cinquin and Filipa Lopes

8 3D + Time Imaging and Image Reconstruction of Pectoral Fin During Zebrafish Embryogenesis 135
Hanh Nguyen, Jaume Boix-Fabrés, Nadine Peyriéras, and Elena Kardash

9 Automated Macro Approach to Remove Vitelline Membrane Autofluorescence in *Drosophila* Embryo 4D Movies 155
Jaume Boix-Fabrés, Katerina Karkali, Enrique Martín-Blanco, and Elena Rebollo

10 Which Elements to Build Co-localization Workflows? From Metrology to Analysis 177
Patrice Mascalchi and Fabrice P. Cordelières

11 Triple-Colocalization Approach to Assess Traffic Patterns and Their Modulation 215
Daniel Sastre, Irene Estadella, Manel Bosch, and Antonio Felipe

12 Photobleaching and Sensitized Emission-Based Methods for the Detection of Förster Resonance Energy Transfer 235
Timo Zimmermann

13 In Vivo Quantification of Intramolecular FRET Using RacFRET Biosensors 275
Manel Bosch and Elena Kardash

14 Cell Proliferation High-Content Screening on Adherent Cell Cultures 299
Pau Carrillo-Barberà, Jose M. Morante-Redolat, and José F. Pertusa

15 HCS Methodology for Helping in Lab Scale Image-Based Assays 331
Joaquim Soriano, Gadea Mata, and Diego Megias

Part III Methods Based on ImageJ Plugin Development

16 Filopodia Quantification Using FiloQuant 359
Guillaume Jacquemet, Hellyeh Hamidi, and Johanna Ivaska

17 Coincidence Analysis of Molecular Dynamics by Raster Image Correlation Spectroscopy 375
David F. Moreno and Martí Aldea

18 3D Tracking of Migrating Cells from Live Microscopy Time-Lapses 385
Sébastien Tosi and Kyra Campbell

Part IV Methods Based on Machine Learning

19 A Cell Segmentation/Tracking Tool Based on Machine Learning............ 399
Heather S. Deter, Marta Dies, Courtney C. Cameron, Nicholas C. Butzin, and Javier Buceta

20 2D + Time Object Tracking Using *Fiji* and *ilastik* 423
Andrea Urru, Miguel Angel González Ballester, and Chong Zhang

21 Machine Learning: Advanced Image Segmentation Using *ilastik* 449
Anna Kreshuk and Chong Zhang

Index *465*

Contributors

MARTÍ ALDEA • *Molecular Biology Institute of Barcelona, IBMB-CSIC, Barcelona, Catalonia, Spain; Department of Basic Sciences, Universitat Internacional de Catalunya, Barcelona, Spain*

MARIA L. ARBONES • *Molecular Biology Institute of Barcelona (IBMB-CSIC), Barcelona, Spain*

JAUME BOIX-FABRÉS • *Molecular Imaging Platform, Molecular Biology Institute of Barcelona, Spanish Research Council (CSIC), Barcelona, Spain*

ANNA BOSCH • *Scientific and Technological Centers (CCiTUB), Universitat de Barcelona, Barcelona, Spain*

MANEL BOSCH • *Scientific and Technological Centers (CCiTUB), Universitat de Barcelona, Barcelona, Spain*

JAVIER BUCETA • *Chemical and Biomolecular Engineering Department, Lehigh University, Bethlehem, PA, USA; Bioengineering Department, Lehigh University, Bethlehem, PA, USA*

OLIVIER BURRI • *Faculty of Life Sciences (SV), BioImaging and Optics Platform (BIOP), Ecole Polytechnique Fédérale de Lausanne (EPFL), Lausanne, Switzerland*

NICHOLAS C. BUTZIN • *Biology and Microbiology Department, South Dakota State University, Brookings, SD, USA*

MARIA CALVO • *Advanced Optical Microscopy Facility, Scientific and Technological Centers of University of Barcelona, University of Barcelona, Barcelona, Spain*

COURTNEY C. CAMERON • *Biology and Microbiology Department, South Dakota State University, Brookings, SD, USA*

KYRA CAMPBELL • *Department of Biomedical Science, Firth Court, University of Sheffield, Sheffield, UK; Bateson Centre, Firth Court, University of Sheffield, Sheffield, UK*

PAU CARRILLO-BARBERÀ • *Departamento de Biología Celular, Biología Funcional y Antropología Física, Universitat de València, Burjassot, Spain; Estructura de Recerca Interdisciplinar en Biotecnologia i Biomedicina (ERI BIOTECMED), Universitat de València, Burjassot, Spain; Centro de Investigación Biomédica en Red sobre Enfermedades Neurodegenerativas (CIBERNED), Madrid, Spain*

BERTRAND CINQUIN • *Laboratoire de Biologie et Pharmaceutique Appliquée, ENS Saclay, Cachan, France*

FRANCISCO CIRUELA • *Unitat de Farmacologia, Departament Patologia i Terapèutica Experimental, Facultat de Medicina i Ciències de la Salut, IDIBELL, Universitat de Barcelona, L'Hospitalet de Llobregat, Barcelona, Spain; Institut de Neurociències, Universitat de Barcelona, Barcelona, Spain*

FABRICE P. CORDELIÈRES • *Bordeaux Imaging Center, UMS 3420 CNRS—Université de Bordeaux—US4 INSERM, Pôle d'imagerie photonique, Centre Broca Nouvelle-Aquitaine, Bordeaux, France*

HEATHER S. DETER • *Biology and Microbiology Department, South Dakota State University, Brookings, SD, USA*

MARTA DIES • *Chemical and Biomolecular Engineering Department, Lehigh University, Bethlehem, PA, USA*

IRENE ESTADELLA • *Molecular Physiology Laboratory, Departament de Bioquímica i Biomedicina Molecular, Institut de Biomedicina (IBUB), Universitat de Barcelona, Barcelona, Spain*
ANTONIO FELIPE • *Molecular Physiology Laboratory, Departament de Bioquímica i Biomedicina Molecular, Institut de Biomedicina (IBUB), Universitat de Barcelona, Barcelona, Spain*
VÍCTOR FERNÁNDEZ-DUEÑAS • *Unitat de Farmacologia, Departament Patologia i Terapèutica Experimental, Facultat de Medicina i Ciències de la Salut, IDIBELL, Universitat de Barcelona, L'Hospitalet de Llobregat, Barcelona, Spain; Institut de Neurociències, Universitat de Barcelona, Barcelona, Spain*
MIGUEL ANGEL GONZÁLEZ BALLESTER • *BCN-MedTech, DTIC, Universitat Pompeu Fabra, Barcelona, Spain; ICREA, Barcelona, Spain*
ROMAIN GUIET • *Faculty of Life Sciences (SV), BioImaging and Optics Platform (BIOP), Ecole Polytechnique Fédérale de Lausanne (EPFL), Lausanne, Switzerland*
HELLYEH HAMIDI • *Turku Bioscience Centre, University of Turku and Åbo Akademi University, Turku, Finland*
JOHANNA IVASKA • *Turku Bioscience Centre, University of Turku and Åbo Akademi University, Turku, Finland*
GUILLAUME JACQUEMET • *Turku Bioscience Centre, University of Turku and Åbo Akademi University, Turku, Finland; Faculty of Science and Engineering, Cell Biology, Åbo Akademi University, Turku, Finland*
ELENA KARDASH • *BioEmergences Laboratory (USR 3695), CNRS, University Paris-Saclay, Gif-sur-Yvette, France*
KATERINA KARKALI • *Molecular Biology Institute of Barcelona (IBMB-CSIC), Barcelona, Spain*
ANNA KRESHUK • *EMBL, Heidelberg, Germany*
FILIPA LOPES • *Laboratoire Génie des Procédés et Matériaux CentraleSupélec, Gif-sur-Yvette, France*
MARC LÓPEZ-CANO • *Unitat de Farmacologia, Departament Patologia i Terapèutica Experimental, Facultat de Medicina i Ciències de la Salut, IDIBELL, Universitat de Barcelona, L'Hospitalet de Llobregat, Barcelona, Spain; Institut de Neurociències, Universitat de Barcelona, Barcelona, Spain*
ENRIQUE MARTÍN-BLANCO • *Molecular Biology Institute of Barcelona (IBMB-CSIC), Barcelona, Spain*
PATRICE MASCALCHI • *Bordeaux Imaging Center, UMS 3420 CNRS—Université de Bordeaux—US4 INSERM, Pôle d'imagerie photonique, Centre Broca Nouvelle-Aquitaine, Bordeaux, France*
GADEA MATA • *Confocal Microscopy Unit, Spanish National Cancer Research Centre-CNIO, Madrid, Spain*
DIEGO MEGIAS • *Confocal Microscopy Unit, Spanish National Cancer Research Centre-CNIO, Madrid, Spain*
JOSE M. MORANTE-REDOLAT • *Departamento de Biología Celular, Biología Funcional y Antropología Física, Universitat de València, Burjassot, Spain; Estructura de Recerca Interdisciplinar en Biotecnologia i Biomedicina (ERI BIOTECMED), Universitat de València, Burjassot, Spain; Centro de Investigación Biomédica en Red sobre Enfermedades Neurodegenerativas (CIBERNED), Madrid, Spain*
DAVID F. MORENO • *Molecular Biology Institute of Barcelona, IBMB-CSIC, Barcelona, Catalonia, Spain*

HANH NGUYEN • *BioEmergences Laboratory (USR 3695), CNRS, University Paris-Saclay, Gif-sur-Yvette, France*
JOSÉ F. PERTUSA • *Departamento de Biología Celular, Biología Funcional y Antropología Física, Universitat de València, Burjassot, Spain*
NADINE PEYRIÉRAS • *BioEmergences Laboratory (USR 3695), CNRS, University Paris-Saclay, Gif-sur-Yvette, France*
ELENA REBOLLO • *Molecular Imaging Platform, Molecular Biology Institute of Barcelona (CSIC), Barcelona, Spain*
DANIEL SASTRE • *Molecular Physiology Laboratory, Departament de Bioquímica i Biomedicina Molecular, Institut de Biomedicina (IBUB), Universitat de Barcelona, Barcelona, Spain*
ARNE SEITZ • *Faculty of Life Sciences (SV), BioImaging and Optics Platform (BIOP), Ecole Polytechnique Fédérale de Lausanne (EPFL), Lausanne, Switzerland*
JOAQUIM SORIANO • *Confocal Microscopy Unit, Spanish National Cancer Research Centre-CNIO, Madrid, Spain*
SÉBASTIEN TOSI • *Advanced Digital Microscopy Core Facility (ADMCF), Institute for Research in Biomedicine (IRB Barcelona), The Barcelona Institute of Science and Technology, Barcelona, Spain*
ANDREA URRU • *BCN-MedTech, DTIC, Universitat Pompeu Fabra, Barcelona, Spain*
CHONG ZHANG • *BCN-MedTech, DTIC, Universitat Pompeu Fabra, Barcelona, Spain*
TIMO ZIMMERMANN • *Advanced Light Microscopy Unit, Centre for Genomic Regulation, Barcelona, Spain*

Part I

Overview

Chapter 1

Main Steps in Image Processing and Quantification: The Analysis Workflow

José F. Pertusa and Jose M. Morante-Redolat

Abstract

In the last decades, the variety of programs, algorithms, and strategies that researchers have at their disposal to process and analyze image files has grown extensively. However, these are only pointless tools if not applied with the careful planning required to achieve a succesful image analysis. In order to do so, the analyst must establish a meaningful and effective sequence of orderly operations that is able to (1) overcome all the problems derived from the image manipulation and (2) successfully resolve the question that was originally posed. In this chapter, the authors suggest a set of strategies and present a reflection on the main milestones that compose the image processing workflow, to help guide the way to obtaining unbiased quantitative data.

Key words Image processing, Image analysis, Image quantification, Main step's analysis

1 Introduction

The commercial image analysis devices available in the market in the 1980s of the twentieth century, such as the powerful IBAS2000 Kontron Bildanalyse, were based on Intel 8086 processors whose image dedicated memory rarely reached over 2 MB. Despite being expandable up to 16 MB, such upgrade was so expensive at that time that it was unaffordable to most labs. This original configuration greatly limited the number of images that could be processed simultaneously. A machine such as the IBAS2000 supported up to six images of 512 × 512 pixels, which could be converted into four of 768 × 512 or twelve of 256 × 256 pixels, assuming that images had 256 gray levels plus an extra bit in overlay mode, which implied a maximum of 9-bit depth (*see* **Note 1**). Therefore, the image analysis protocol had to be designed assuming that most of the time, the intermediate images generated by each operation would be progressively overwritten and unavoidably lost along the process. The limited memory, thus, forced to design the analysis sequence in a carefully ordered fashion.

Elena Rebollo and Manel Bosch (eds.), *Computer Optimized Microscopy: Methods and Protocols*, Methods in Molecular Biology, vol. 2040, https://doi.org/10.1007/978-1-4939-9686-5_1, © Springer Science+Business Media, LLC, part of Springer Nature 2019

On the other hand, we must recognize the influence of Niklaus Wirth and his wisely entitled manual *Algorithms + Data Structures = Programs* [1] on those who were learning computer programming at that time. Contravening the popular programming languages of the moment, such as *FORTRAN* or *BASIC*, Wirth's programming style, instead of including labels for unconditional jumps within the program, consisted in sequences that started with a "BEGIN" and progressed, line by line, until reaching the command "END" in the final line. These two circumstances might well justify why we present in this chapter a structured and ordered workflow for image analysis and encourage the readers to design their own in the same manner. However, we might find more reasons that have led the scientific community to design their image analysis protocols as algorithms. If we think in the tremendous technological advances befallen in the last 30 years, which have made possible to obtain hundreds of high-quality images in a single experiment, we will easily understand this requisite. Processing such number of images becomes a titanic task unless we are able to automate it, but an automated routine necessarily requires an ordered and structured sequence of steps that, ideally, could be executed by a computer in an unsupervised mode.

2 Image Analysis Sequences in the Form of Mathematical Algorithms: A Few Examples

Be as it may, it is an advisable practice to design our analysis routines following the principles of structured programming, which will enable us to track and keep control of the images resulting of each operation, make decisions about the result, and establish checkpoints to verify if we are achieving the expected goals. Actually, we can find these notions in the *Encyclopaedia Britannica*: "Algorithm is a systematic procedure that produces—in a finite number of steps—the answer to a question or the solution of a problem" [2]. From a mathematical point of view, Cormen and collaborators [3] state that an algorithm is "...any well-defined computational procedure that takes some value, or set of values, as input and produces some value, or set of values, as output. An algorithm is thus a sequence of computational steps that transform the input into the output."

However, we think that in order to prevent ambiguities, this definition should somehow include the concept of "order," that is, that the sequence of procedures must be, apart from finite, arranged in a specific order. In fact, provided that the output data from one operation will serve as input for the next one, each intermediate transformation must be carefully planned and executed in the right order. Moreover, and despite their close relationship, a tiny subtlety still differentiates a program from an algorithm, being the program a specific implementation of the algorithm, subjected to the limitations imposed by the input data and the software [4].

In summary, these principles can be translated into two simple rules. First, it has to be clearly determined what is to be evaluated; this would entail establishing which are the objects and/or regions of interest, how they can be found in the image, and, consequently, which are the parameters required to that purpose. Second, omitting for the moment the quality of our image, the work sequence necessary to perform the measurements has to be defined; this conforms the authentic sequence of instructions—as stated in the aforementioned definition of algorithm—that will eventually become an executable routine.

We will use two examples to illustrate such workflow. On the first example (Fig. 1), there is an image of the ganglion cell layer of a reptilian retina taken with the aim of determining the distribution pattern of its neurons, which appear on the image as dark dots scattered among the surface. However, we notice that the background (*see* **Note 2**) is not homogeneous, and the presence of dark regions might difficult the detection of the objects of interest. Provided that they can be identified and separated from the surrounding background, the position of each neuron could be determined by its centroid, and therefore we could calculate the distance between each pair of neurons. However, these estimated parameters themselves do not provide a straightforward answer to whether the distribution pattern is random or, conversely, neurons tend to aggregate in certain areas, but it can be easily deduced by mathematically working with the obtained measures.

On the second example (Fig. 2), we find two optical sections corresponding to two different mouse zygotes at metaphase of the

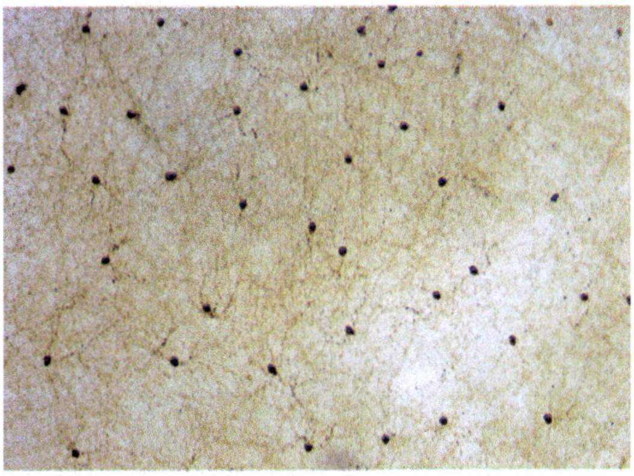

Fig. 1 A reptilian retina with stained ganglion neurons extended over a slide. The ganglion cells seem to be homogenously distributed among the tissue, with a fairly constant average distance between them

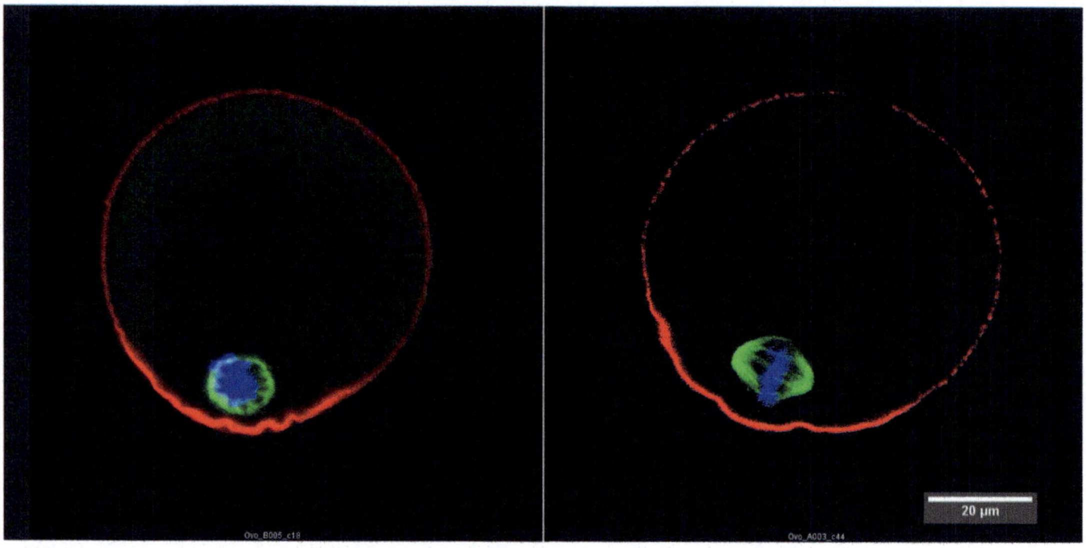

Fig. 2 Two murine zygotes during metaphase of the first mitosis. Images correspond to Z planes extracted from the original stack, showing the chromosome metaphase plate

first cell division. The researcher is interested in studying whether handling or treatments applied to the oocyte prior to in vitro fertilization might result in morphological changes in their spindle size or shape. To detect such differences, the object of interest must be retrieved after a 3D reconstruction, since a single section may not reflect the full spindle shape due to the stochastic positioning of the zygote on the glass slide. Therefore, the image requires manipulation to properly orientate the objects before taking any measurements.

Once determined the subject of the analyses, an algorithm needs to be created to perform the measurements. On the first example image (Fig. 1), the algorithm must be a sequence designed to avoid the problem of background spots, identify the neurons, make sure that the performed operations keep at least the pixel occupied by the centroid of each object, and, finally, calculate their coordinates in the plane and/or measure the distance to each of their neighbors. This imposes several conditions that must be included in the analysis sequence. First, the image has to be processed to correct the uneven illumination and reduce or eliminate persistent noise elements (image restoration). Then, the objects need to be distinguished, either by thresholding (*see* **Note 3**) or by searching for minimum local values in the image (*see* **Note 4**). As a result, a binary image is obtained (*see* **Note 5**), which needs to be further corrected for errors by extra particle elimination, filling of holes, separation of connected objects, or elimination of objects touching the image border to avoid bias in the final calculations. Such operations are sequentially performed until at least the pixel in the centroid of each object remains. Next, the obtained binary mask needs to be verified to make sure it contains the expected objects to be measured. Finally, data retrieval and storage in the form of tables are required for the subsequent analysis.

Creating an algorithm for our second example, the case of the mouse zygotes, must include the manipulation of the images in the three dimensions, so that the achromatic spindles can be fully measured over their most representative projection. Object selection in this case would be greatly facilitated by the obvious difference between the fluorescent staining and the black background, which enables an unequivocal threshold selection, although the possible presence of unspecific noise in the binary image might require further treatment.

In summary, although in practice the researcher usually develops an intuitive perception of what each specific problem requires, a proper workflow must nevertheless be established allowing an optimized image analysis that results in the correct measurements. Figure 3 depicts what we propose as a proper workflow,

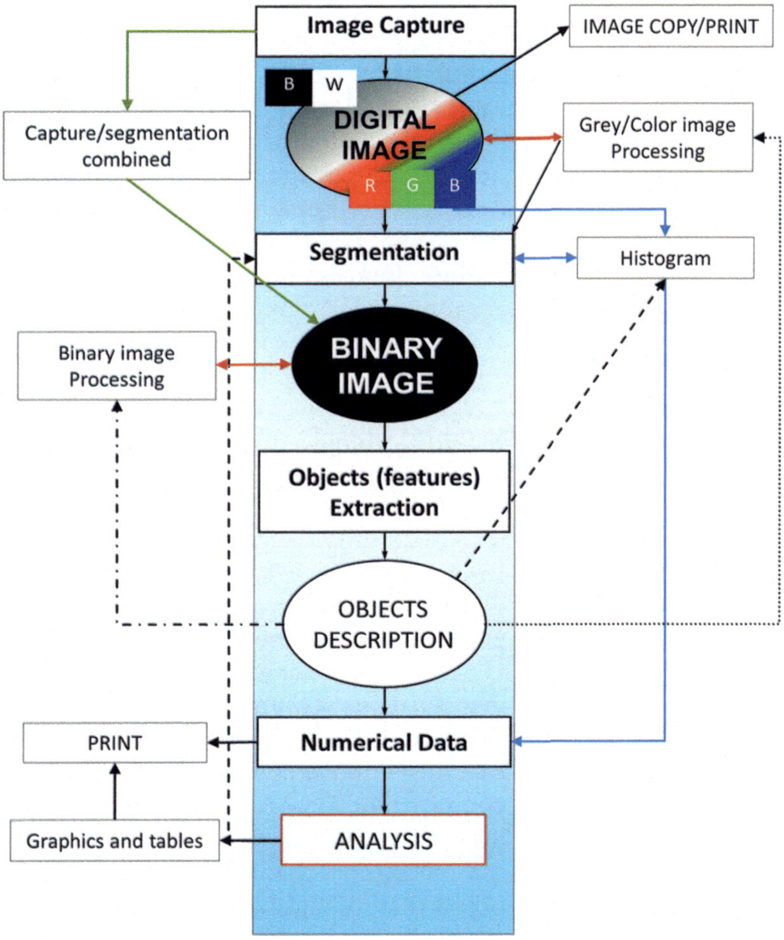

Fig. 3 Schematic representation of the general workflow for image analysis

schematized so as to include each step and the corresponding quality checkpoints. Nonetheless, its value is essentially educational since, in practice, each particular problem will require a specific pipeline. In the following lines, we will elaborate the proposed scheme and justify the need for each particular step.

3 The Digital Image: Image Quality

We want to highlight that, although the workflow starts with the already acquired digital image, the process of digitalization is decisive to determine whether the analysis will be straightforward or would require an extra effort: in other words, we are talking about how image quality has an impact on the analysis of the images. Off-center microscope illumination (*see* **Note 6**) and scratches in the objectives, for instance, are two circumstances that will introduce serious perturbations during image processing. Similarly, high cell confluence or excessive tissue thickness may lead to objects appearing clumped and connected to one another, requiring intensive work on the binary image to individualize them, if even possible. Regarding illumination, inadequate light intensity, either by excess or by defect, may distort, for instance, the densitometry values (*see* **Note 7**), and uneven illumination between different images may prevent a proper comparison of the obtained results. Additionally, image resolution (*see* **Note 8**) can be critical when dealing with measurements that approach the diffraction limit of the system, i.e., the minimum distance this system can resolve. To acquire images at optimal resolution, we suggest taking into account the Whittaker-Nyquist-Kotelnikov-Shannon sampling theorem, commonly known as the Nyquist-Shannon theorem [5, 6].

4 Knowing What to Measure: Objects and Background

Our experience as biological observers enables us to immediately recognize those parts in the image that are relevant to our study. Intuitively, we call "objects" to those distinguishable and independent elements that clearly stand out from the surrounding background, which contains no relevant information.

Actually, we are able to recognize a set of structures and elements having a common nature (color, shape, texture). To facilitate the nomenclature, the term "phase" is commonly used to refer to those parts in the image that share certain properties, e.g., an image composed of some objects perceived over the background contains in fact two phases. Conversely, an image with just a single phase cannot be more than either a bare background or just one part of an object that, because of its size, has been divided into several images.

Thus, we perceive the information contained in an image due to the visual contrast between the different phases present.

Obvious as it may seem, the definition of object is not that simple from an academic point of view. Many authors consider that the here presented term "object" is inadequate, for the reasons stated below. On the one hand, in computerized image analysis, objects are only identified after image segmentation, an operation that divides the pixels in the image in two groups assigning them only two possible values: pixels belonging to objects become "1" (or "0"), regardless of their original gray value, whereas those being part of the background become "0" (or "1"). The image is consequently presented in a new form, the binary image. This processing reduces the original information to "islands" of uniform elements scattered in an "ocean" of background, hence, a flat image. On the other hand, these "islands of information" might not be independent objects but parts of the same real object. This may, for instance, occur when confocal optical sections are obtained from three-dimensional samples. For this reason, instead of "objects," many authors prefer to use the term "features," reserving "objects" exclusively to designate 3D structures. In some cases, both concepts will concur, as happens in macrophotography, electrophoretic images, or those taken through a dissecting microscope. Conversely, hardly ever will a histological section display real objects, but features.

Ultimately, from the image analysis point of view, an object contained in an image will be characterized by its numerical properties, i.e., the parameters that define the object. Therefore, when working with flat surfaces, either from projections or very thin sections, we can directly use morphometric parameters (*see* **Note 9**) to numerically characterize the objects.

5 Enhancement and Restoration

Although the researcher always tries to image the samples in the best possible conditions, the resulting images may nevertheless display certain deficiencies. It is not uncommon to find errors derived from uneven microscope illumination or noise introduced by the capture devices themselves or by particles of dust accumulated somewhere along the optical axis. In most occasions, poor image quality derives from the intrinsic characteristics of the sample itself and from the imaging technique of choice. Light diffraction accounts for most of the blurring effect observed on the image, an effect named convolution. Because of the nature of light, a single illuminated point does not behave as a uniform light surface but a series of intensity waves that spread from the central point. This pattern defines what we call point spread function (PSF) and is perceived as concentric rings (known as Airy rings). This diffraction phenomenon is even more pronounced in fluorescence microscopy. In the last 20 years, a great variety of methods intended to solve or circumvent convolution have been developed. In fact, confocal microscopy itself is a very effective practical solution to improve this problem: by means of a pinhole, a small diaphragm located at the end of the optical axis greatly eliminates the blurring effect that we are describing. Nonetheless, confocal microscopy might not be always an available option so the development of specific programs for the correction of convolutions (deconvolution programs) and their application to the optical microscopy field has helped enormously to overcome these light diffraction problems [7, 8].

Regarding those other alterations of the image, in the form of noise or illumination defects, the digital nature of the image allows the application of simple operations that can attenuate, correct, or eliminate most of the problems we have indicated, a process known as "image enhancement and restoration." It includes any procedure aimed to enhance some parts of the image or restore it by removing the artifacts or improving its quality. In some cases, this process starts during the image acquisition stage. For instance, it is a deep-seated habit, especially among microscopy experts, to correct the overall illumination by subtracting a reference image [9] (mainly used in wide-field systems) or by applying a Kalman filter during the digitalization process [10] (mainly used in confocal microscopy). This a priori correction is intended to prevent the presence of the noise that comes from illumination, the optical lenses, or the quantic noise. It is important to highlight that, far from being exclusively an aesthetic transformation, the purpose of the "enhancement and restoration" process is to facilitate the subsequent identification and selection of the objects of interest. In fact, it is not an obligated step in the image analysis workflow, because it might not be required to perform the subsequent operations. On the contrary, quite often manipulating the image might be counterproductive and could lead to inaccuracies or omissions affecting the reliability of the results. For this reason, we have placed enhancement and restoration out of the main flow in the diagram depicted in Fig. 3.

There are four main types of techniques that can be applied to the image once it has been digitized: adjustment of the histogram, arithmetic operations between two images, application of filters based in matrices (*see* **Note 10**), and operations in the frequency domain. The histogram adjustment operations result in an enhancement of the image contrast. They are so popular that most image analysis programs include a specific implemented function, named *normalize*, that stretches the histogram transforming the image gray levels to bring them to the maximum possible range, i.e., 0 to 255 in 8-bit images. The *normalize* function does not produce significant alterations in morphometric parameters but might dramatically alter the result of densitometry measurements (gray-level optical density in particular). A common consequence of normalization is that reassigned pixels accumulate in the extreme values, leading to a loss of image information.

The arithmetic operations between two images are frequently applied for the correction of uneven illumination or to eliminate defects that appear as a constant in the image. Most of the time, image imperfections caused by the illumination of the sample during the digitalization go unnoticed or cannot be avoided. This is because they appear as a consequence of the dynamic adaptation of the digital camera sensors to the light intensity or by the perturbations that the beam of light suffers as it crosses the microscope lenses. As a result, images are not uniformly illuminated which interferes with the subsequent treatments, especially with segmentation operations.

In order to remove the described defects, we require a reference image, containing just the undesirable background noise, to be used as the second operand, which is mathematically subtracted from the original image we want to correct. The result is usually very effective, but we must keep in mind that, like what happens with normalization, the operation changes the original gray-level value of each pixel, hence altering the densitometry parameters of the image.

Operations with matrix filters are probably the most difficult to apply because the profound transformations that they can originate in the image might not be easily foreseen. However, when properly used, they become powerful tools to enhance or eliminate specific features on the image. Interestingly, the most frequently used filter, the Gaussian filter, is intended to produce a blurring of the image, which might seem completely counterproductive and contrary to the concept of restoration that we have been presenting. However, its widespread use comes from the fact that it effectively attenuates the quantum noise of the image. The Gaussian filter works by averaging groups of contiguous pixels (neighbors), smoothing their gray levels. Therefore, it can be used to artificially generate blurred background images that might serve as reference in an arithmetic operation, as described before. Another filter used for the elimination of noise is the median filter. This filter removes the noise of the image provided that the undesired particles do not exceed the size of the kernel of the applied filter.

Regarding the risk of applying an arbitrary correction to the images, it must be kept in mind that some manipulations might distort the size of the objects and/or the gray level, and this is to be avoided in order to get reliable measurements [11].

6 Segmentation

If there is a process that we should place in the center of our quantitative analysis framework, that would undoubtedly be segmentation. The main purpose of this operation is to isolate the elements of interest, the ones that will be measured, from the rest of the image. As a result, a new image susceptible of being studied and measured by the computer is obtained, in a binary form (*see* **Note 5**). The process by which the objects are separated from the background is, essentially, a discrimination operation. This, far from being an obvious operation, is usually the most important decision we will have to make during the analysis. In fact, if the objects can be properly extracted, the remaining operations will be basic logic sequences.

The main problem is how to precisely determine the limits of the objects. To this purpose, a series of toolsets and operations have been developed specially aimed to find those limits in order to help us differentiate the objects from the background. We could always delimitate an object by manually drawing its perimeter, but it would be imprecise and time-consuming. A much practical approach requires finding a property which, being common to all the objects of interest, can be used as an argument for computer-based automatic discrimination. Such property must highlight the object upon the surrounding area, as, for instance, color or gray-level intensity do for color or monochromatic images, respectively. In short, an object appears as a discontinuity in the tapestry that we call background, and it is composed of a set of pixels, which are similar in color, luminance, or gray-level intensity with respect to the background.

From a formal point of view, the segmentation algorithms are based on two basic object properties: the discontinuity and the gray level [12]. The algorithms based on discontinuity explore the object limits, finding the line that comprises their boundary. Necessarily, the object must be closed. But, is it possible for an object not being closed? This may happen for several reasons being the most frequent that the object is touching one of the image borders. In any case, the object limit is reached as long as there are peripheral pixels whose gray level matches the background's. Thus, it is the user's responsibility to decide what gray-level value determines the threshold between objects and background and when the boundary has been reached, as well as to evaluate the possible impact of such decisions on the size of the objects. A classic procedure for boundary search is the use of filters for border detection [13, 14].

The segmentation based on object similarity uses as a criterion the inclusion of the pixel gray level within a range enclosed by two thresholds (*see* **Note 3**). This method does not take into account whether the selected pixels are actually part of an object or not. Consequently, this operation generates a certain amount of noise in the binary image in the form of isolated pixels or other structures that, despite being selected, are not objects of our interest. In other words, if we assume that a key property of any object is that all its pixels must be connected, those we find isolated or forming small groups will be rendered as background noise (we will have to set the minimum size to be considered as an object). The use of threshold segmentation clearly applies to images with at least two phases. Thus, the desired thresholds will define a range where to include all the gray levels corresponding to a specific phase. As a result, any value out of this range will be set as background. The difficulty lies precisely in determining the gray value that, set as a threshold, is able to separate the objects from the background. Considering a dark background, it would be reasonable to set one of the threshold limits in 255 (white). Similarly, for a white background, we could set a black threshold (gray level = 0). But, how is the gray level that separates the two phases found? If they were equivalent, the answer would be the average value of the gray-level histogram. However, if each phase can be represented as a Gaussian curve, the best option would then be the minimum value found between the two relative maximum values. Additionally, other image properties could be used such as gray-level entropy, the shape of the accumulated frequency curve, etc. ImageJ incorporates a collection of algorithms as configurable options of the threshold command, each of them based on a different strategy to find the proper segmentation threshold. We believe that it is important to pay special attention to this matter and we recommend reading the work of Tajima and Kato [15].

7 The Binary Image

After the segmentation operation, the resulting image might be considered as a mask of the original image, in other words, a sort of "copy in clear acetate film" containing just the regions of interest from the original image (now pixels with value = 1). Actually, this idea suits so perfectly the use we make of the binary image that the image analysis community has adopted the term "binary mask" (*see* **Note 5**) to designate it. However, as much care as we might have put during the processes of restoration, enhancement, thresholding, and binarization, the resulting binary image often retains imperfections. As aforementioned, threshold segmentation tends to yield binary masks with certain amount of background noise in the form of isolated pixels. This happens because the gray levels of those pixels, despite being part of the background, are included within the range of the established threshold. If this binary mask were to be used to get measurements from the original image, those extra pixels will irremediably be included as objects of interest, thus biasing the results. Additionally, two close objects displaying blurred, or even out-of-focus, edges in the original image could end up connected as a single object in the binary mask if the segmentation threshold included the gray level corresponding to their fuzzy boundaries. Other possibility is that segmentation does not retrieve the whole object, which may contain holes in the binary image. This happens when the threshold setting leaves out of the range some gray values that correspond to both background and object pixels. Finally, it is even possible to retrieve, especially in the corners of the image, partial objects that would surely distort the results if ever included. The solution to these problems is discussed in the next section.

8 Processing of the Binary Image: Mathematical Morphology

Mathematical morphology is a nonlinear technique that has become essential to depurate binary image imperfections. Published by Matheron and Serra after being reformulated [16], it is based on set theory, lattice theory, topology, and random functions. Although it has become very popular in digital imaging, it can also be used in graphics, solid bodies, polygonal meshes, and other spatial structures. It is nowadays possible to find good implementations of the mathematical morphology operations among the functions available in *ImageJ*, plus an extra set carefully introduced by Gabriel Landini [17].

The most commonly used mathematical morphology operations are *Open*, *Close*, *Fill Holes*, *Skeletonize*, and *Watershed*. Applied in the correct order and with the proper number of iterations, they can be used to eliminate undesired pixels, restore incomplete objects, or separate and individualize those that, despite being separated entities, appear contiguous in the image and would be interpreted as one by the analysis software. Another set of extremely useful operations are those called logic or Boolean. When properly applied to the binary image, they make possible to identify the pixels that were modified by the mathematical morphology operations along the process and also allow for the step-by-step depuration of the binary image until the proper mask where to perform the measurements is obtained.

For instance, a simple way to retrieve the inner area of a tube section, using morphology operations, could be as follows: First fill the inner empty area inside the tube wall by applying the *Fill Holes* operation; then, obtain the difference between the original binary image and the filled one. However, because of the particular properties of the binary image, we cannot perform a simple subtraction. As mentioned before, the pixels in the binary image do not admit values other than 0 and 1, so conventional arithmetic operations between binary images might yield inadmissible pixel values. For instance, subtraction would easily result in operations such as "$0 - 1 = -1$" or, a simple addition, others such as "$1 + 1 = 2$," being both results outside of the binary space. And believe us when we tell you that you do not want to end up facing a division by zero while attempting an arithmetic division between two binary images: the "undefinition" might as well cost you a lifetime exile from the binary realm.

Obviously, operations with binary images require a special type of operator that produces only binary results. Boolean algebra solves that problem because comparisons (operations) between binary images always result in a binary image [18]. In Mathematics, Boolean algebra is a branch of Algebra fundamental for the functioning of computers and many common automatons, apart from being an invaluable tool in the final phases of the image analysis process (readers interested in the mathematical concepts of Boolean algebra can consult the book of BH Arnold) [19].

Going back to the tube example, the Boolean operation required in this case is *XOR* (Exclusive-OR), which compares two binary images and produces a new image, binary as well, that contains only those elements present in one and not the other image.

The other three available Boolean operators are *AND*, *OR* (*inclusive-OR*), and *NOT*. The *AND* operation returns only the pixels with value "1" that are present in both source images, thus resulting in a new binary image containing only the common objects. The *OR* (*inclusive-OR*) operation works similarly but returning all the positive (value = 1) pixels in either image, as if we were adding both images considering that 1 + 1 = 1. Finally, the Boolean *NOT* operation, applied to a single binary image, results in its pixels interchanging their value, that is, the pixels that were off in the source image will appear on and vice versa [20].

We have already discussed the fatal consequences of applying algebraic operations to a binary image. Seemingly, applying Boolean operations to 8-bit images, either in gray level or RGB, can produce unforeseen results. Therefore, it is important to keep in mind that Boolean operations always require that their two operands be binary images. Curiously, some image analysis programs accept that one or both of them are gray-level images, but it is difficult to imagine the result of such operation and its usefulness for the subsequent analysis.

9 Object Identification

At this point in the process of image processing, we will be ready to measure and obtain the corresponding table of results. We assume that all the objects have been perfectly isolated, there is no trace of background noise and we have completely eliminated the border touching objects... but, how can we be sure?

Before taking measures, our old IBAS2000 made a simple operation that was called identification. It consisted in applying a false-color look-up table (LUT) that assigned a different color to each object. This way the observer could visually check that all the contiguities had been properly solved and all the objects appear individualized.

The growing trend to implement routines in which the computer is able to improve the processes and learn from each run (machine learning) is completely opposed to a visual identification of the objects. However, it is possible to filter the results, introducing a kind of secondary quality control, ignoring particles above or under a certain pre-established value. Therefore, the machine's learning process starts with a pre-evaluation in order to set the minimum particle size to be considered as an object.

In the current versions of *ImageJ*, it is possible to reproduce such identification: in the [Analyze > Analyze Particles...] menu, there is a drop-down list of options to *Show* being one of them *Count Masks* which, applying a gray LUT, assigns a different gray level to each of the objects found in the image. It starts with the first one and proceeds incrementing by one the gray level assigned to each new object until all of them have been processed (or until the maximum gray level, according to the image's bit depth, is reached). Later, applying a *glasbey invert* LUT, we can convert the gray image into a polychromatic one, enabling the visual discrimination we were describing (Fig. 4).

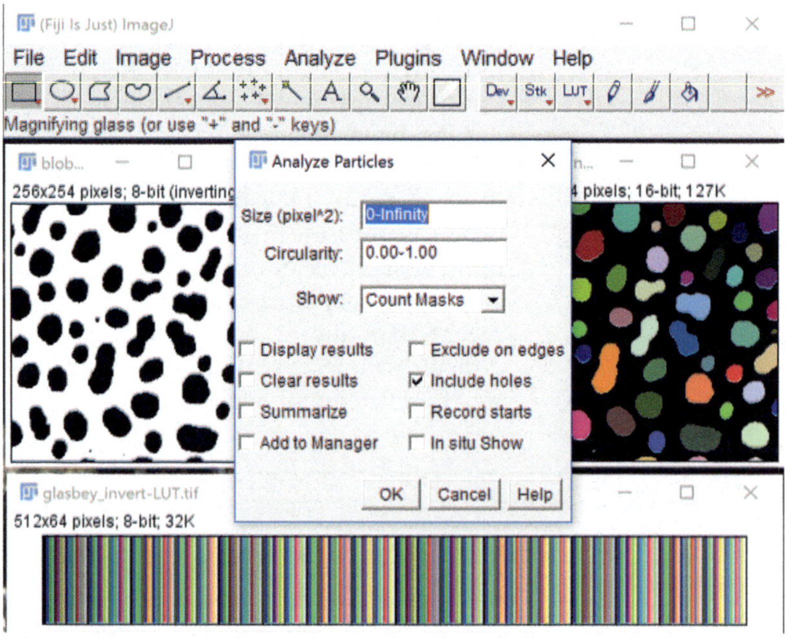

Fig. 4 Particle identification. The upper panel shows the *ImageJ* program interface (Fiji version). The middle panel contains a binary image (left), the *Analyze Particles* menu (center), and the *Count Masks* output image indexed with the *glasbey invert* LUT (right). The lower panel depicts the *glasbey invert* LUT

10 Obtaining Data

Having identified and isolated the objects of interest, finally the long-sought data are at hand. It is only a question of selecting the proper parameters and letting the analysis software measure them for us. However, choosing the right parameters might not be as easy as it might seem. We have already discussed that, from the experimenter point of view, images and image analysis are nothing but means to get to an objective, a specific question that needs to be answered. In some occasions, basic morphological (*size*, *perimeter*, *circularity*, etc.) or densitometric parameters (*mean* or *max and min gray level*) might give a straightforward answer. Other cases would require the use of less intuitive measurements, such us *center of mass*, *Feret's diameter*, or *skewness*. We recommend familiarizing with the catalogue of parameters and their mathematical nature, i.e., which object features are measured and combined to obtain them, before designing your analysis protocol [21]. In any case, as happened with the reptilian retina example that we presented above, most of the time the table of measurements we get from the segmented objects does not directly provide the answer to the problem. These raw data need to be further processed, by combining mathematical and statistical tools with a certain dose of creativity, in order to solve the mystery. This is the reason why, after "Object Description" in the pipeline we present in Fig. 3, a step called "numerical data," representing the objects direct measurements, precedes the final "analysis" step.

Finally, we would like to draw your attention to the feedback arrow that connects the analysis with the segmentation strategy in the same figure. Along with "Object Identification" (see above), numerical data might reveal that something went wrong along the image analysis protocol and the obtained results are somehow incoherent or clearly biased. If that were the case, the segmentation method or the binary image processing could be re-evaluated to correct the observed deviations.

Once we confirm that the protocol works and the obtained results can be trusted, the following step would be to convert it into a macroinstruction able to run the whole process automatically, but this would be the subject for another chapter.

11 Notes

1. *Bit depth/range of gray levels*: Also known as color depth, it refers to the number of different color (or gray) shades that a single dot in an image can display. Therefore, it measures the quantity of information that can be stored in a single pixel or, in other words, how many bits of information will be required to

store all the different color (or gray) tones that a specific image can display. Since one bit accepts only two possible values (1, 0), the maximum range of tones results from 2 raised to the power of the bit depth of the image. Henceforth, we typically use the exponent to express the bit depth. A binary image (*see* **Note 5**), therefore, has a bit depth of 1 and can only display two (2^1) tones: black and white. Consequently, an 8-bit image can display 256 (2^8) levels, a 16-bit image 65.536 (2^{16}), and a 24-bit image more than 16 million tones ($2^{24} = 16{,}777.216$).

2. *Background*: In the context of digital imaging, it refers to the set of pixels that constitute the "filling" of the image; above which it is possible to distinguish the objects of interest. It may be difficult to define what the background in a picture of a landscape would be, because all the elements contribute to create a coherent entity. However, in a portrait of two people, we would easily identify the background as the part of the image that is not occupied by people. In microscopic images, it is very common to expand the term and include the possible illumination defects, the presence of dust and dirt specks, any unspecific signal result of an immunohistochemical staining, or any other artefactual element that could be considered "background noise."
3. *Thresholding*: A type of image segmentation based on the selection of two values, among the image grayscale range, that becomes the superior and inferior limits of an interval in which all the pixels of interest (object pixels) are included. The pixels with gray values outside of the defined interval are, therefore, rendered as "background."
4. *Minimum/maximum local values*: It is a way of referring to the relative minimum/maximum in a certain area of the image, when studying the gray-level values region by region. If we draw a horizontal line in the center of an image and graphically represent the gray intensity value of each pixel on the line, we will obtain a roughly flat profile (actually a low amplitude irregular sawtooth wave) where the line crosses just background pixels. However, crossing an object, either brighter or darker than the background, would result in a necessary discontinuity in the gray level, and the graph profile will gain height (maximum) or sink (minimum), returning to the background sawtooth wave once the object has been crossed. This situation will be repeated for each of the objects intersected by the line, and each peak or valley will correspond to a local maximum or minimum, respectively. Most importantly, the search of local peaks and valleys can be used as an automatic method for finding objects in the images thus for segmentation.

5. *Binary image/binary mask*: A binary image is the result of applying a segmentation operation to a grayscale or color image. The applied segmentation seeks to only leave the pixels of interest visible, eliminating the content of those left as background. This is achieved by "switching off" the background pixels, assigning all of them a single value (typically "zero") while "illuminating" the pixels of interest with a different value (typically "one"). As a result, the original image is transformed into a flat image, a sort of carbon copy or "mask" of the objects of interest. Because the shape and distribution of the objects is maintained, the binary image can be used to obtain morphological and spatial information. Additionally, it can be redirected, acting as a "mask" to the original image to obtain specific densitometry data from the segmented objects. It is important to highlight that the mathematical nature of the binary image fits perfectly with Boole's algebra.
6. *Off-center illumination*: Even the most sophisticated microscopes are susceptible of causing illumination defects on the captured image if not properly adjusted. Besides the correct alignment of the light source with the optical axis, microscopes include a condenser and a diaphragm that, as August Köhler proposed, can be used to illuminate the sample with a field of uniform light, whose diameter is equal to the objective capture area. Any image taken without such adjustment will suffer from off-center illumination.
7. *Densitometry parameters*: Parameters that measure the light properties such as brightness, luminosity, clarity, or luminance of the objects in the image. The term comes from one of the most common light-related parameters: *optic density* (physical magnitude that measures the ability of an optic element to absorb a certain light). Another densitometry parameter is *center of mass*, calculated as the average of the coordinates of all the pixels in an object, weighted by their respective gray level.
8. *Image resolution*: A measure of the quality of a digital image. It is used analogously to the "resolution power" of a microscope as it gives an indication of the quantity of detail that can be visualized in an image. There are two ways of referring to digital resolution: (a) by size, taking into account the dimensions of the image, measured in pixels and expressed as *length*width*, and (b) by density, indicating how many pixels would be included if we draw a 1-inch segment within the image, expressed as *dots per inch* (dpi).
9. *Morphometric parameters*: Parameters that measure the dimensions and shape of the objects. One of the first used morphometric parameters was the "calibrator diameter," known as

Feret's diameter (honoring L.R. Feret). It measures the length of the object projected over a Cartesian axis; thus it consists of two values: the x-axis projection and the y-axis projection. Another important shape descriptor is *circularity*, defined as the ratio between the area and the square perimeter of the object, which would be "1" if we were measuring a perfect circle and diminishes as the shape becomes less round.

10. *Filter (kernel)*: A numerical matrix that, applied iteratively to each of the pixels in an image, transforms its value to a weighted average depending on the neighboring pixels and the numbers on the filter. Each of the coefficients in the matrix is used to multiply the value of the pixel in the corresponding position. Then, the value of the central pixel (the one that is currently being evaluated by the filter) is calculated as an average of the matrix-obtained values of itself and of the neighboring pixels. The matrix then traverses the whole image, obtaining a weighted average for all the pixels it contains. The assigned value for each pixel is dependent on the size of the matrix used, the most common sizes being 3×3, 5×5, and 7×7. It is important to highlight that the matrix cannot evaluate the pixels in the border of the image as they do not have enough neighboring pixels.

Acknowledgments

The authors thank Katherina-Theresa Mantikas for her *English touch* in our glossary definitions. We would also like to thank Dr. Enrique Lanuza for granting us free access to his histological collection.

References

1. Wirth N (1978) Algorithms + Data Structures = Programs. Prentice Hall PTR, Upper Saddle River, NJ
2. Britannica encyclopaedia. https://www.britannica.com/topic/algorithm. Accessed 15 Jan 2018
3. Cormen TH, Leiserson CE, Rivest SC (2009) Introduction to algorithms, 3rd edn. MIT Press, Cambridge, MA
4. Peña Marí R (2005) Diseño de Programas. Formalismo y abstracción. Prentice Hall, Madrid
5. Davidson MW (2016) Basic properties of digital images. In: Molecular expressions. https://micro.magnet.fsu.edu/primer/digitalimaging/digitalimagebasics.html. Accessed 8 Jan 2018
6. Davidson MW (2016) Digital image sampling frequency. In: Molecular expressions. https://micro.magnet.fsu.edu/primer/java/digitalimaging/processing/samplefrequency/index.html. Accessed 8 Jan 2018
7. Van de Sompel D, Sasportas LS, Jokerst JV, Gambhir SS (2016) Comparison of deconvolution filters for photoacoustic tomography. PLoS One 11(3):e0152597
8. Hernández Candia CN, Gutiérrez-Medina B (2014) Direct imaging of phase objects enables conventional deconvolution in bright field light microscopy. PLoS One 9(2):e89106
9. Afanasyev P, Ravelli RB, Matadeen R, De Carlo S, van Duinen G, Alewijnse B, Peters PJ, Abrahams JP, Portugal RV, Schatz M, van Heel M (2015) A posteriori correction of

camera characteristics from large image data sets. Sci Rep 5:10317

10. Kuan DT, Sawchuk AA, Strand TC, Chavel P (1985) Adaptive noise smoothing filter for images with signal-dependent noise. IEEE Transactions on Pattern Analysis and Machine Intelligence PAMI-7(2):165–177
11. Davidson MW (2016) Basic concepts in digital image processing. In: Molecular expressions. https://micro.magnet.fsu.edu/primer/digital imaging/imageprocessingintro.html. Accessed 8 Jan 2018
12. Zhang YJ (1997) Evaluation and comparison of different segmentation algorithms. Pattern Recogn Lett 18:963–974
13. Maini R, Aggarwal H (2009) Study and comparison of various image edge detection techniques. Int J Image Processing (IJIP) 3:1–12
14. Senthilkumaran N, Rajesh R (2009) Edge detection techniques for image segmentation – a survey of soft computing approaches. Int J Recent Trends Eng 1:250–254
15. Tajima R, Kato Y (2011) Comparison of threshold algorithms for automatic image processing of rice roots using freeware ImageJ. Field Crop Res 121:460–463
16. Matheron G, Serra J (2000) The birth of mathematical morphology. Dissertation Xerox Center, Palo-Alto. http://cmm.ensmp.fr/~serra/pdf/birth_of_mm.pdf. Accessed 18 Jan 2018
17. Landini G (2008) Advanced shape analysis with ImageJ. Proceedings of the Second ImageJ User and Developer Conference, Luxembourg, 6--7 November, 2008. p 116–121. ISBN 2-919941-06-2. Plugins. http://www.mecourse.com/landinig/software/software.html. Accessed 18 Jan 2018
18. Null L, Lobur J (2014) Chapter 3: Boolean Algebra and digital logic. In: The essentials of computer organization and architecture. Jones & Bartlett Publishers, Burlington, MA, pp 138–145
19. Arnold BH (2011) Logic and Boolean algebra. Dover Publications Inc., New York
20. Russ J (2011) The image processing handbook, 6th edn. CRC Press, Boca Raton
21. Ferreira TA, Rasband W. (2012). The ImageJ User Guide – Version 1.46r. http://imagej.net/docs/guide/user-guide.pdf. Accessed Apr 2018

Chapter 2

Open Source Tools for Biological Image Analysis

Romain Guiet, Olivier Burri, and Arne Seitz

Abstract

Visiting the Bio Imaging Search Engine (BISE) (Bio, BISE, Engine, http://biii.eu/, Imaging, Search) website at the time of writing this article, almost 1200 open source assets (components, workflows, collections) were found. This overwhelming range of offer difficults the fact of making a reasonable choice, especially to newcomers. In the following chapter, we briefly sketch the advantages of the open source software (OSS) particularly used for image analysis in the field of life sciences. We introduce both the general OSS idea as well as some programs used for image analysis. Even more, we outline the history of *ImageJ* as it has served as a role model for the development of more recent software packages. We focus on the programs that are, to our knowledge, the most relevant and widely used in the field of light microscopy, as well as the most commonly used within our facility. In addition, we briefly discuss recent efforts and approaches aimed to share and compare algorithms and introduce software and data sharing good practices as a promising strategy to facilitate reproducibility, software understanding, and optimal software choice for a given scientific problem in the future.

Key words Image analysis, Open source software, ImageJ, Interoperability

1 Introduction

The ultimate goal of scientific research is to be able to generate and share knowledge. An important role played by methods and analysis strategies is that other scientists may use them to further expand them, answer new questions, and push the scientific field forward. In particular, publicly funded research strongly relies on the accessibility of knowledge in terms of results (e.g., published articles) as well as on the methods used to obtain these results. However, both results and methods, especially in the field of image analysis, often make use of commercial software licenses. Monetary barriers aside, the lack of access to the underlying algorithms frequently blocks reproduction and validation of the obtained results. Therefore, the open source software idea is extremely appealing in order to cope with the aforementioned challenge. In brief, it means that a software is published along with its source code, thus allowing anyone to reuse and/or modify it for its own purposes and even use it for its

Elena Rebollo and Manel Bosch (eds.), *Computer Optimized Microscopy: Methods and Protocols*, Methods in Molecular Biology, vol. 2040, https://doi.org/10.1007/978-1-4939-9686-5_2,

own software as subject to the terms of the specific open source license that applies.

In the field of image analysis and image processing, about 30 years ago, an open source project emerged which managed to become indispensable over the years: *ImageJ* [1]. The major strengths of *ImageJ* and its offspring were their adaptability to cope with complicated tasks and their attractiveness for beginners due to the ease of use. The need for an open source software for image handling (including image inspection, image processing, and image analysis) was linked to the fact that image-based methods became crucial for life scientists in the last decades. In particular, light microscopy, having always been an invaluable tool to life scientists, has evolved dramatically within the last 30 years. Digital acquisitions have replaced hand-drawn observations, and new imaging modalities have kept constantly appearing, driven by increasing advances in electronics and optics toward more accurate, more reproducible, and always higher throughputs. It is interesting to note that the light microscopy research field has received two Nobel prizes in the last decade [2, 3]. The first one awarded the discovery, purification, expression, and application of GFP [4–6], which turned fluorescence microscopy into an invaluable tool as virtually every protein in live cells and organisms could be fluorescently labeled. The second prize awarded the development of super-resolution microscopy techniques [7–9], which stand among the more advanced techniques available in the field. These newly developed microscopy methods make it necessary to automate the analysis of the obtained images in order to achieve reliable and reproducible results. This need for automation is further motivated by the fact that nowadays it is common that the amount of images produced by a single instrument is far too large to be analyzed manually by a single scientist. To cope with the challenge of performing reproducible image analysis on light microscopy data, it is necessary to create as-automatic-as-possible image processing and analysis workflows and to report these within the context of the experiment.

In this chapter, we will first formalize the definitions of image analysis and open source software (OSS) and show how they help make science reproducible. We will then focus on some of the most common open source image analysis software available, with a particular focus on *ImageJ*. Finally, we wish to share our vision of good practices in scientific image analysis and offer a perception of what the future ecosystem could look like through the already ongoing collaborative efforts of the open source imaging analysis community.

2 What Is Image Analysis?

The most powerful and flexible image analysis machinery up to now is a part of the human brain called the visual cortex. It is able to analyze images in "real time"—meaning that it has almost immediate access to the information extracted from the visual cortex and can cope with a myriad of different objects, lighting conditions, and visual obstructions in such an efficient way that one cannot notice the difference between seeing things and recognizing them. In terms of flexibility and speed, it is still unmet. However, while the human visual system is extremely powerful, it has some disadvantages, which become especially evident in science. An example illustrating this fact is adaptation. The cells of the human visual system have a limited response range in terms of light intensities. Adaptation helps to cope with the huge range of intensities, from daylight levels of around 10^8 cd/m^2 to night luminance levels of approximately 10^{-6} cd/m^2 [10]. In addition, everything we are seeing is interpreted in a context-dependent fashion, i.e., humans are very good at detecting changes, but precise quantification capabilities are lacking.

Seeing is a matter of learning. Therefore, human analysis capabilities (or analysis routines) can only be shared with others via teaching or training (looking at images together). This is one of the reasons why pathologists need to undergo several years of training before they can reliably judge tissue sections and detect, e.g., signs of cancer. This demonstrates that, to make image analysis reproducible, scalable, and shareable, computational image analysis approaches are needed.

An image analysis workflow can be seen as a method to reproducibly extract information out of an image by the means of image processing and image analysis routines. Digital image processing became popular in 1960 and emerged as a subfield of digital signal processing. The used algorithms are well-defined mathematical procedures and thus reproducible and easily sharable. However, single algorithms are in most of the cases not sufficient for the analysis of images in life sciences. Typically, a combination of different algorithms is needed to provide quantitative and reproducible data extraction from an image.

The challenge to reliably quantify images in life sciences stems from the fact that each and every question demands a tailored analysis workflow. The quality of the images directly influences the performance of the analysis routine. In fact, it largely depends on the sample preparation and the image acquisition itself. Thus, reproducible image analysis results are only achievable if sample preparation and image acquisition are part of the analysis workflow. Image analysis routines perform well with images of a certain image quality. However, they might fail completely if, for example, the signal-to-noise ratio (SNR) or the signal-to-background ratio (SBR) drops below a certain value or threshold [11]. Quantifying the image quality and the knowledge about the factors influencing its quality is key in order to successfully perform image analysis in life sciences.

The easiest image analysis workflow consists of at least one image analysis component (IAC), applied to an image that was acquired beforehand. In most of the cases, an image processing component (IPC) is applied to the image beforehand in order to facilitate the extraction of data. A prominent example is the task to count the number of cells in a given field of view. Before detecting the cells via a simple thresholding procedure, the image is often "smoothed" with a filter (e.g., Gaussian filter) in order to account for noise during the acquisition and make the computational detection of objects more robust. Also, a background subtraction algorithm is useful and serves the same purpose. Both steps as well as the thresholding to distinguish the cell from the background will be classified as IPCs, as input and output are images. The algorithm detecting the cells is an IAC, leading to a table with at least the *xy*-coordinates of the object although, potentially, other features like size, shape intensity, etc. can also be reported. A generic workflow that leads us from the sample to the numbers is summarized in Fig. 1.

In order to obtain the same result when repeating a workflow at a later time point, it is necessary that the algorithms as well as the used parameters are known and remain unchanged. Even if the parameters fed into the algorithm are carefully documented, it becomes impossible to obtain the same results if the algorithm itself was modified. This is in particular a problem when using proprietary software packages, since the algorithm is often known to only a restricted part of the community (mostly developers of the software company). For all the others, it is a "black box," and scientific users are unable to track changes. Companies might be reluctant in publishing algorithms and their evolvement as it interferes with their intellectual property. As a consequence reproducibility, a key requirement of science, cannot be guaranteed. A solution to this problem is the usage of open source software.

3 Open Source Software (OSS) and Science

The idea to share and co-develop software packages was brought up by scientists in the early days of computing. They wanted to learn from each other and (co)evolve the field. Prominent examples following this mindset are Donald Knuth in 1979 with the *TeX* typesetting system [12] or Richard Stallman in 1983 with the *GNU* operating system [13]. However, it took until the late 1990s for the idea of open source software (OSS) to be formally postulated by

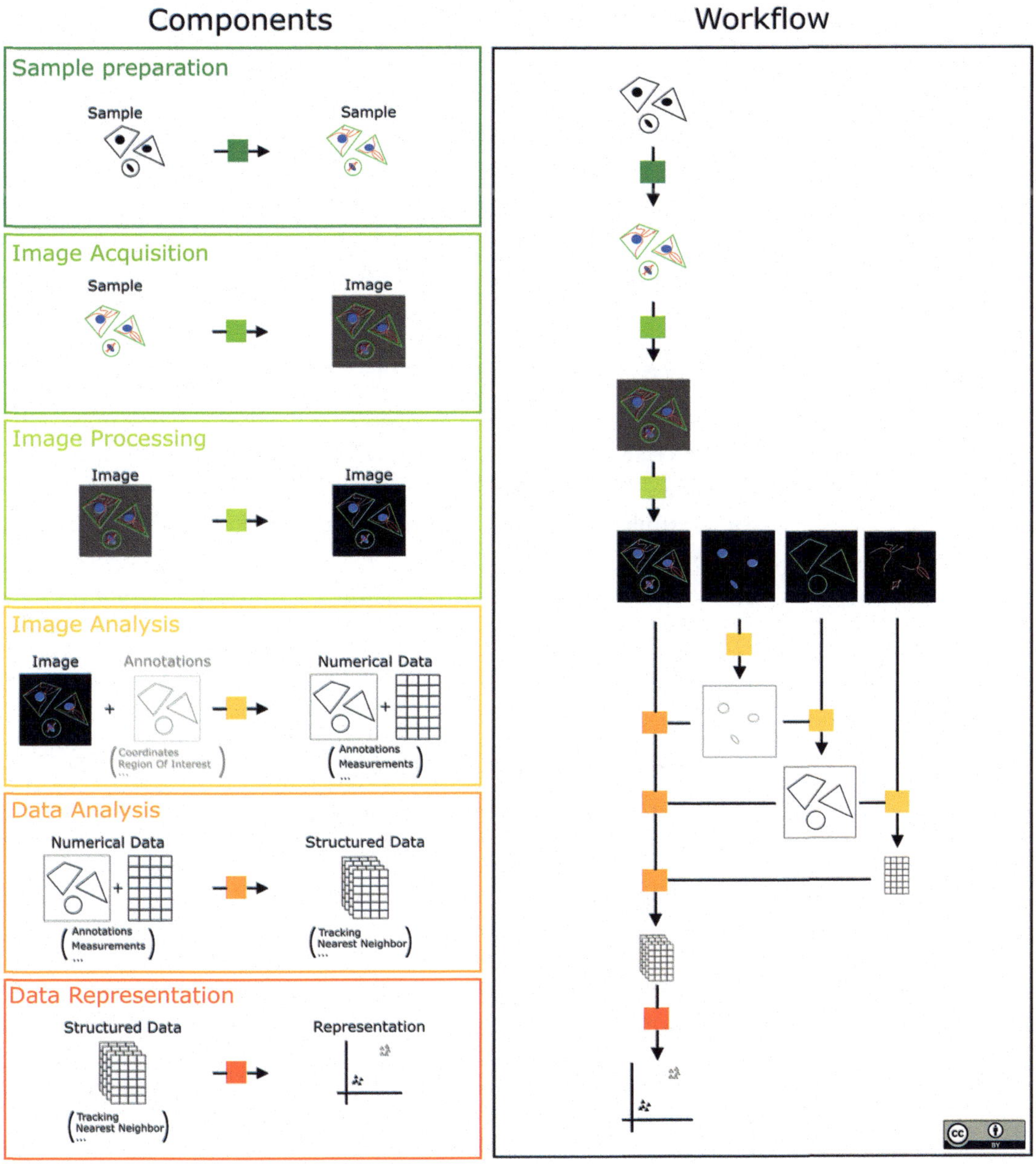

Fig. 1 The workflow of a typical experiment involving imaging is composed of several different components. It begins with "Sample Preparation" (sample to sample), where a sample is functionalized for imaging, followed by "Image Acquisition" (sample to image), where a selected imaging system is used to produce a digital image of the sample. These two components are typically not part of an image analysis workflow, however nevertheless crucial for the success in general. It is typically followed by an "Image Processing" (image to image) step, where certain aspects of the image are enhanced, removed, or restored to facilitate the subsequent "Image Analysis" (image [and optionally annotations] to numerical data). This is the crucial step that extracts number(s) from an image. Finally, "Data Analysis" (numerical data to structured data) is the step that gathers numerical data and extracts new information of it, which can then be turned into "Data Representation" (structured data to graphical representation) that visually summarizes information for human interpretation. The latter step is not strictly linked to data from microscopy. (Figure CC-BY Romain Guiet-PTBIOP available at https:/go.epfl.ch/aV8)

Eric S. Raymond, Bruce Perens, and others within the Open Source Initiative (OSI) [14]. The seven main pillars of this movement became [15]:

1. Security, meaning the possibility of having full control over what the software is doing
2. Affordability, meaning access to the software is not blocked by a monetary barrier
3. Transparency, meaning changes and modifications are apparent and directly observable from the code
4. Perpetuity, meaning there are no barriers for the code to be maintained by a new entity, should this become necessary
5. Interoperability, meaning access to the source code allows for developers to create interfaces that allow for communication between different software
6. Flexibility, meaning new improvements can be added as needed by any entity
7. Localization, meaning sourcing software locally in order to receive the most direct and adapted support or documentation

It is important to make the distinction between *open source software* and *free software* and understand their fundamental difference. Free software does not mean that the underlying code is available for scrutiny. Accordingly, open source software is built around the core principle that source code should be publicly available. The beauty of the OSI resides in how it naturally meshes with the scientific method and the peer review approach. Image analysis methodologies, algorithms, and tools, developed by the scientific community, are easily—and freely—shared, tested, verified, and built upon. The availability of the source code prevents deprecation, as other groups can take over development as needed. It promotes open discussions through its practically automatic transparency, which usually grants fast response times to implement new ideas and developments. However, traditionally open source projects were often geared toward programmers and scientists with extensive software knowledge. Within this context interfaces were scarce; most of the interaction was based on command-line scripting. So while open, it lacked the accessibility needed for the greater scientific community to comfortably adopt these tools. This made commercial software, which had a strong emphasis on user experience, the most attractive solution. Those open source projects that acknowledged the need for seamless user experience became the most popular tools of the twenty-first century, like *Ubuntu*, *LibreOffice*, *Gimp*, or *VLC*. These tools have one particular feature in common. They have integrated enterprise-level best practices to software development, including version control, bug tracking, task lists, testing tools, and package management. This became possible because all the tools needed now also exist as open source. Because of this, a point has been reached where research groups have also invested significant time in making convenient software, leveraging state-of-the-art approaches and algorithms while keeping user interfaces simple and intuitive and development rigorous.

Through access to open source tools, scientific research can continue to "stand on the shoulders of giants" and continue to foster innovation. Open access to source code and code sharing tools helps foster direct discussion between and with developers to address issues, suggest improvements, and receive advice on tool use and best practices. One of the most popular to date is GitHub [16]. The GitHub web platform leveraged the growing popularity of the GIT version control model into a seamless web interface that hosts code, discussions, and most importantly, tools to make collaboration between very geographically distinct entities nearly effortless. Projects are stored into "repositories" which work as a completely self-contained unit that can be downloaded. The general structure should always contain a Readme file that explains the purpose of the project as well as installation instructions and dependencies. Each repository has its own issue tracking (for bugs or requests) and discussion threads, which promotes interaction and subsequent improvement of the project.

All the tools around open source allow for rapid adoption and adaptation of projects. One prominent example was the development of the OpenSPIM [17], which triggered multiple laboratories around the world to build their own system based on the open design and contribute back [18]. For biological image analysis, one such tool that gained a reputation and a large user base over its 30 years of existence is *ImageJ*.

4 NIH Image-ImageJ-Fiji

ImageJ, successor to *NIH Image*, is an extremely popular open source image processing and analysis tool. Development began over 30 years ago on the Mac II and continues today with cross-institution efforts to adapt its functionality to the ever-changing panorama of scientific image acquisition. Its key to success lies in the community-driven approach its developer, Wayne Rasband, decided to use. After a first framework was built, new functionalities and bug fixes were driven according to the user needs; such external contribution adopted the form of plugins that implemented the constant community feedback. Asides from the 100+ functions built into *ImageJ*, over 700 contributed plugins are listed on the *ImageJ* website. Furthermore, due to the early adoption of the Java runtime environment, the scope of the community was no longer limited by hardware and operating system, which greatly contributed to its adoption.

ImageJ was scriptable almost from the start. Further expansion came with the creation of the *ImageJ Macro language*, which allows for users with very little to no programming knowledge to automate simple (and even complex) image processing and analysis tasks. This language also makes prototyping analysis pipelines and joining complex pipelines together simple and serves as a glue to combine user-contributed functionality together with ease. Recent efforts have now also brought the most common scripting languages to be able to work within *ImageJ* (*Beanshell*, *Clojure*, *Groovy*, *JavaScript*, *Python*, *R*, *Ruby*, *and Scala*), allowing users to use their favorite language.

A massive effort is made in providing up-to-date documentation of all the functionalities and access to the code via the *ImageJ* website [19], bringing transparency and almost automatically validation of the plugins and tools. A mailing list [20] and, more recently, a very active discussion forum [21] serve as meeting points to discuss workflows, pipelines, code, and best practices among beginners and experts. The forum became so popular in fact, that in august 2018, it was merged with the *CellProfiler* forum, and now many open source image analysis software can use it as an entry point for discussion.

While being a great advantage, community-driven contributions tend to have a less strict structure in terms of program behavior, code quality, and/or output, as these are created by different individuals with different approaches and skill levels. This situation can lead to unexpected problems that can be addressed by programmers or bioimage analysts to refurbish the code into a more streamlined version, or at least to adapt it to a new project or need. Moreover, while being constantly adapted to cope with new imaging modalities, extra data dimensions are limited to five for the user (width, height, depth, color, and time), which sometimes proves insufficient nowadays (lifetime data, spectral information, sample orientation in 3D with eventual rotation). Finally, *ImageJ* was never built to handle the large amount of data that can be produced by current state-of-the-art acquisition devices (light sheet microscopes, whole slide scanners, high-throughput microscopes) in an integrated and convenient way. Workarounds for these limitations exist, but are not built-in. These limitations and new aspects of biomedical image analysis aim to be covered by new functionalities from *ImageJ2* and *SCIFIO* (the *Scientific Image Format Input and Output*, part of the *ImageJ2* backbone), and future iterations will see the aforementioned constrains becoming a problem of the past, with new problems emerging along the way, of course.

ImageJ's initial broadness triggered the creation of multiple *ImageJ* flavors, such as *Fiji* [22], *Bio7* [23] (ecological modeling), *SalsaJ* [24], and *AstroImageJ* [25] (Astronomy), each catering to a different scientific community that found in *ImageJ* the backbone to help fill the gap in their toolboxes.

Currently, for bioimage analysis, its most prominent and active iteration is *Fiji* (*Fiji* Is Just *ImageJ*), which also serves as the *ImageJ2* showroom [1]. It is maintained by a very dynamic team of developers, which interacts through GitHub and an online forum [21] where beginners can easily seek advice or assistance. One of the developments we particularly take advantage of is the concept of "Update Sites" that eases the installation of new plugins and also keeps them updated without much effort or manual intervention.

While *ImageJ* is a big part of the open source image analysis ecosystem, it is by far not the only tool available to researchers, nor the magic bullet that can tackle all image acquisition protocols. Many open source tools exist—with their own stories and reasons for their existence—to fulfill different goals. Some are geared toward machine learning, others tailored to high-throughput analysis, and others toward performance. The tools we seek to cover here are the ones that contain user interfaces and make it possible for non-programming-savvy scientists to interact comfortably with their powerful functionalities.

5 Representative List of Open Source Software and Interoperability

Visiting the BISE [26] website at the time of writing this article, almost 1200 open source assets (components, workflows, collections) were found. The number is impressive and frightening at the same time. Especially newcomers will find it extremely difficult to understand the differences of the individual software packages. Making a reasonable choice might thus be hindered by the overwhelming offer which, at the same time, makes it almost impossible to test all the available algorithms in a reasonable amount of time.

During the last decade, many other software tools implemented modularity and scriptability in their own way. One can find an exhaustive list of existing solutions on BISE [26]; however, here we focus on some software commonly used within our core facility. This use has been motivated by several factors, not least of which (a) the size of the community using these tools to which we can access for support and discussions, (b) the responsiveness of the teams behind the software, (c) the software must be user-friendly to cater to our end users (usually biologists with little programming experience), and (d) the software must allow for batch processing in one way or another. That is, it must be able to let us move from processing single files to folders easily. And finally (e) the software must be as extensible as possible via scripting on add-ons/plugins. The tools we develop are almost always tailored to a specific experiment or biological question, so whatever software we use must have an accessible programming interface in order to create pipelines by novel combination of image processing and image analysis algorithms.

The *suite* composed of *CellProfiler* [27] and *CellProfiler Analyst* [28] offers a versatile tool usable by almost anyone. The workflow, from image loading to output and passing through segmentation, is constructed within a "pipeline" consisting of "modules." Every single module and sub-element has an exhaustive documentation attached, which provides the explanation of all parameters, rationale, examples, and references. Pipelines are saved as simple *.cpproj files, making them easily shareable. One module in particular can run *ImageJ* commands, scripts, or macros. This friendly software is optimized for screening assays (file naming, results parsing, parallel processing) but is versatile enough to be used for varied imaging experiments. Furthermore, *CellProfiler Analyst* can combine the measurements from *CellProfiler* and perform single cell classification via machine learning. Additionally, it offers tools for data visualization. Recently, independent solutions emerged to offer new data analysis and visualization interfaces [29], or an overall management of the data (from sample to imaging to image and statistical analysis [30]) by leveraging the power of yet another open source project, *R* [31].

ICY [32] provides a very complete and user-friendly environment. It is already bundled with many indispensable tools, and many plugins are available (which can be added seamlessly). A community comments and rates all the plugins, which is beneficial when deciding which one to install. *ICY* embeds *ImageJ* and enables the exchange of images (and ROIsets). Finally, great effort was poured in a visual interface that helps the user to connect plugins and create what are called "protocols." Protocols are batch-ready and easily saved and shared with other users. Finally, more advanced users can also write scripts in *JavaScript* or *Python*.

ilastik [33] is an open source implementation of image segmentation via machine learning. It offers an interface for dataset annotation, pixel classification, and finally object classification and, more recently, tracking. Pipelines can be saved and loaded, and many tutorials and examples are available. Furthermore, the pixel classification step can be called from other software (*CellProfiler*, *KNIME*) in order to create integrated interoperable workflows.

QuPath [34] is (as far as we know) the only open source solution for digital pathology that can handle large images (>50 k × 50 k pixels) and that readily integrates image analysis tools and interaction with *ImageJ*. It is a recent project, with a growing community. *QuPath* embeds its own *ImageJ*, which enables the exchange of images and ROIs. Furthermore, almost any *ImageJ* plugin can be added to enrich the built-in *ImageJ*. Finally, more advanced users can also write scripts in groovy and make full use of the *QuPath* API.

For too long the only interoperability between different software has been the ability to read common file formats, and for the last years, the Open Microscopy Environment group's Bio-Formats tool [35, 36] has been a game changer! Nowadays, as we have seen above, some tools include *ImageJ* or make it possible to run *ImageJ* commands, scripts, or macros. Other tools act as bridge between software like *MIJ* that can connect *ImageJ* and *Matlab* [37]. A convenient solution could come from *KNIME* [38, 39], a workflow creator that leverages the APIs of any software to create modules (as long as they have been initially coded by someone). These modules, which wrap the desired functionality of third party tools, can now be interconnected through the graphical user interface to create simple (or more complex) workflows. In Fig. 2 we depict a desirable future where *KNIME* helps us connect different components from different collections seamlessly.

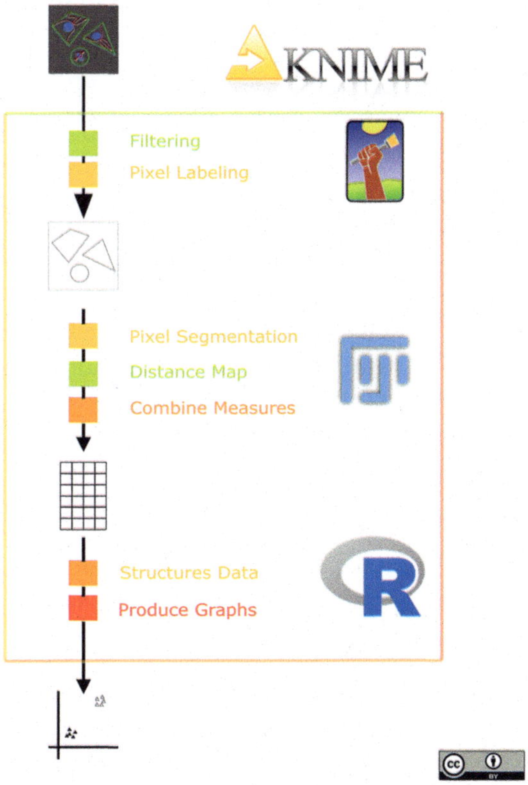

Fig. 2 The use of different software should be easy (in a close future). Using *KNIME* to create workflow that integrates machine learning capability of *ilastik*, further processing and analysis in *ImageJ*/*Fiji*, and finally *R* for data analysis and representation (Fig. CC-BY Romain Guiet-PTBIOP, available at https:/go.epfl.ch/aV7)

6 Compare and Share via Scripts

As previously mentioned, our main motivation for choosing open source software was partly due to their expandability via scripts or plugins. Sharing scientific image analysis workflows is essential for two main reasons: Numerous routine tasks (e.g., cell counting) can profit from standardization, and, more importantly, image analysis is an integral part of scientific protocols. Failure to report workflows hinders access to the scientific work by others and harms the reproducibility of the results. We sincerely believe that poor documentation or absence of proper image analysis pipeline sharing is not a malicious effort. It is the consequence of two mature and traditionally distinct fields (informatics and biology) fusing because of the advancement of instrumentation, and the growing documented complexity of living organisms. As such, it can be daunting for biologists to make sense of the field of image analysis or for computer scientists to gain an understanding of a biological phenomenon and formulate an experiment in order to study it. Moreover, the development and documentation of usable software is not well recognized or rewarded by traditional academic metrics. This chasm is being mended through international efforts such as NEUBIAS [40] and the work of image analysts, whose expertise lies in their capacity to communicate with ease with the world of biology and computer vision. Similarly, the Software Sustainability Institute [41] and the Research Software Engineer Association [42] aim at having the role of software recognized within the academia in the UK. Through efforts like these, modern good practices for sharing image analysis workflows and datasets are being taught and disseminated throughout the community. What follows is a short introduction to how workflows can be shared and tested.

One can distinguish two types of workflows, which determines the way in which they are shared. Complex yet general workflows (cell counting, object tracking, image alignment, STORM, light sheet multiview fusion) are usually shared as packages or plugins and involve an important effort (months, years) to develop. These are usually maintained and curated by active developers working alongside biologists.

Another type of workflow consists of specific combinations of IPCs and IACs constructed to answer a specific biological question, e.g., tracking cell mitosis in micropatterns [43]. These workflows are part of the data analysis that helps answer the biological question and makes use of multiple packages or plugins to achieve that goal. Scientists should always share these along with the relevant raw data in order to allow for reproduction of the analysis by their peers. The language (*ImageJ* macros, *Groovy* scripts, *Python*, *ICY* protocols, *CellProfiler* pipelines, etc.) becomes irrelevant and mainly a personal choice; however, the act of sharing allows for the validation of the analysis, its eventual extension, and other aspects that were mentioned in the open source software section. It is important to sufficiently document the workflow, so new users can quickly test and use it. This usually involves determining the dependencies needed (what software is needed to run the workflow) and the steps needed to go from raw data to final result.

However sharing the workflow is often not sufficient. As mentioned, these are highly specific workflows that function under specific experimental and image acquisition conditions and thus are heavily dependent on the data being analyzed. This motivates the need to share the data as well as the workflow. Services like *Zenodo* [44] allow for data to be linked to a DOI and publication, with whatever copyright is required by the publisher, absolutely free of cost. Sharing the raw data serves the supplemental purpose of acting as a new dataset for testing and benchmarking future pipelines or novel algorithms. Benchmarking has existed for many years in the form of image databases [45], like the *STARE* dataset for retinal blood vessel segmentation [46] and challenges [47]. These contain human-annotated ground truths that are ideal for testing novel workflows and algorithms. If being part of so-called challenges, these datasets can directly report on the performance of workflows. Similarly, in silico benchmarking, where "phantoms" (artificial images that mimic real image data with auto-generated ground-truth) [48] can also serve as a tool to assess performance of workflows. However, these databases tend to be highly specific too and cannot encompass the myriad of imaging modalities that exist in bio imaging today nor the specificity of a novel biological experiment. For these reasons, our field would greatly benefit from having datasets accompanying their corresponding image processing and analysis workflows. One attractive way of sharing code and in particular workflows in the future is *Jupyter Notebook* [49]. It is an open source web application that allows creating and sharing documents that contain live code, equations, visualizations, and narrative text. Therefore, it is optimally suited to serve as a link between biologically oriented scientists and image analysists.

7 Conclusion

In this introductory chapter, we have outlined the concept of open source software and how it relates and meshes with scientific image acquisition and image processing workflows. We outline how OSS is accessible to all, without the need to purchase expensive licenses, which allows for the critical evaluation of algorithms by all. The open nature of scientific reasoning requires extremely customizable systems to perform cutting-edge analysis through novel workflows and algorithms, for which OSS is in an ideal position to provide. By introducing the open source ecosystem available to bioimage analysis based on the ease of use, expandability, and community support, we provide an entry point for researchers new to the field. Finally, we provide references to tools for sharing scientific data and workflows as well as suggested good practices.

Acknowledgments

We would like to thank the faculty of Life Science (SV) of the EPFL for the continuous support of the bioimaging and optics platform. We are grateful to Peter Bankead and Christian Tischer for the insightful comments on the manuscript.

References

1. Schneider CA, Rasband WS, Eliceiri KW (2012) NIH image to *ImageJ*: 25 years of image analysis. Nat Methods 9:671–675
2. Betzig E, Hell SW, Moerner WE (2014) Nobel prize chemistry. https://www.nobelprize.org/nobel_prizes/chemistry/laureates/2014/
3. Shimomura O, Chalfie M, Tsien RY (2008) Nobel prize chemistry. https://www.nobelprize.org/nobel_prizes/chemistry/laureates/2008/
4. Shimomura O, Johnson FH, Saiga Y (1962) Extraction, purification and properties of aequorin, a bioluminescent protein from the luminous hydromedusan, aequorea. J Cell Comp Physiol 59:223–239
5. Chalfie M, Tu Y, Euskirchen G et al (1994) Green fluorescent protein as a marker for gene expression. Science 263:802–805
6. Heim R, Cubitt AB, Tsien RY (1995) Improved green fluorescence. Nature 373:663–664
7. Klar TA, Hell SW (1999) Subdiffraction resolution in far-field fluorescence microscopy. Opt Lett 24:954–956
8. Moerner WE, Fromm DP (2003) Methods of single-molecule fluorescence spectroscopy and microscopy. Rev Sci Instrum 74:3597–3619
9. Betzig E, Patterson GH, Sougrat R et al (2006) Imaging intracellular fluorescent proteins at nanometer resolution. Science 313:1642–1645
10. Ledda P, Santos LP, Chalmers A (2004) In: New York, NY (ed) A local model of eye adaptation for high dynamic range images. ACM Digital Library
11. Dima AA, Elliott JT, Filliben JJ et al (2011) Comparison of segmentation algorithms for fluorescence microscopy images of cells: comparison of segmentation algorithms. Cytometry A 79A:545–559
12. Gaudeul A (2007) Do open source developers respond to competition? The LATEX case study. Rev Netw Econ 6:2
13. The GNU Operating System. https://www.gnu.org/
14. History of the OSI. https://opensource.org/history
15. Casson T, Ryan PS (2006) Open standards, open source adoption in the public sector, and their relationship to Microsoft's market dominance. Social Science Research Network, Rochester, NY
16. The world's leading software development platform – GitHub. https://github.com/
17. Gualda EJ, Vale T, Almada P et al (2013) OpenSpinMicroscopy: an open-source integrated microscopy platform. Nat Methods 10:599–600
18. Who has an OpenSPIM. http://openspim.org/Who_has_an_OpenSPIM%3F
19. *ImageJ*.net Website. https://*ImageJ*.net
20. *ImageJ* Mailing List. https://*ImageJ*.nih.gov/ij/list.html
21. Image.sc Forum. https://forum.image.sc/
22. Schindelin J, Arganda-Carreras I, Frise E et al (2012) Fiji: an open-source platform for biological-image analysis. Nat Methods 9:676–682
23. Bio7. https://*ImageJ*.net/Bio7
24. SalsaJ. http://www.euhou.net/index.php/salsaj-software-mainmenu-9
25. Astro*ImageJ*. http://www.astro.louisville.edu/software/astro*ImageJ*/

26. BISE Bio Imaging Search Engine. http://biii.eu/
27. Carpenter AE, Jones TR, Lamprecht MR et al (2006) CellProfiler: image analysis software for identifying and quantifying cell phenotypes. Genome Biol 7:R100
28. Dao D, Fraser AN, Hung J et al (2016) CellProfiler analyst: interactive data exploration, analysis and classification of large biological image sets. Bioinforma Oxf Engl 32:3210–3212
29. HTM Explorer. https://github.com/tischi/HTM_Explorer
30. Shiny HTM. https://github.com/hmbotelho/shinyHTM
31. The R Project for Statistical Computing. https://www.r-project.org/
32. de Chaumont F, Dallongeville S, Chenouard N et al (2012) Icy: an open bioimage informatics platform for extended reproducible research. Nat Methods 9:690–696
33. Sommer C, Straehle C, Kothe U, et al (2011) Ilastik: interactive learning and segmentation toolkit. In: Presented at the IEEE International Symposium on Biomedical Imaging
34. Bankhead P, Loughrey MB, Fernández JA et al (2017) QuPath: open source software for digital pathology image analysis. Sci Rep 7 (1):16878
35. Linkert M, Rueden CT, Allan C et al (2010) Metadata matters: access to image data in the real world. J Cell Biol 189:777–782
36. Swedlow JR, Goldberg I, Brauner E et al (2003) Informatics and quantitative analysis in biological imaging. Science 300:100–102
37. Sage D, Prodanov D, Tinevez J-Y, et al (2012) MIJ: making interoperability between *ImageJ* and Matlab possible. In: Presented at the *ImageJ* User & Developer Conference, Luxembourg
38. Berthold MR, Cebron N, Dill F et al (2008) KNIME: the Konstanz information miner. In: Preisach C, Burkhardt H, Schmidt-Thieme L et al (eds) Data analysis, machine learning and applications. Springer Berlin Heidelberg, Berlin, Heidelberg, pp 319–326
39. Fillbrunn A, Dietz C, Pfeuffer J et al (2017) KNIME for reproducible cross-domain analysis of life science data. J Biotechnol 261:149–156
40. NEUBIAS: Network of European BioImage Analysts. http://eubias.org/NEUBIAS/
41. Software Sustainability Institute. http://software.ac.uk
42. Research Software Engineer Association. http://rse.ac.uk
43. Burri O, Wolf B, Seitz A et al (2017) TRACMIT: an effective pipeline for tracking and analyzing cells on micropatterns through mitosis. PLoS One 12:e0179752
44. Zenodo. http://about.zenodo.org/
45. Cell image library. http://www.cellimagelibrary.org/home
46. Second international challenge on 3D Deconvolution microscopy. http://bigwww.epfl.ch/deconvolution/challenge/
47. Grand Challenges in Biomedical image analysis. https://grand-challenge.org
48. Rajaram S, Pavie B, Hac NEF et al (2012) SimuCell: a flexible framework for creating synthetic microscopy images. Nat Methods 9:634–635
49. Thomas K, Benjamin R-K, Fernando P et al (2016) Jupyter notebooks – a publishing format for reproducible computational workflows. Stand alone. IOS Press, Amsterdam, pp 87–90

Part II

Methods Based on ImageJ Macro Programming

Chapter 3

Proximity Ligation Assay Image Analysis Protocol: Addressing Receptor-Receptor Interactions

Marc López-Cano, Víctor Fernández-Dueñas, and Francisco Ciruela

Abstract

Proximity ligation assay (PLA) is an antibody-based method that permits studying protein-protein interactions with high specificity and sensitivity. In brief, when a pair of specific antibodies is in close proximity, the complementary DNA strands they bear engage into a rolling circle amplification and generate, in situ, a single fluorescent signal, which indicates the presence of a protein-protein interaction. Proper image analysis methods are needed to provide accurate quantitative assessment of the obtained fluorescent signals, namely, PLA data. In this chapter, we outline basic aspects of image analysis (including software, data import, image processing functions, and analytical tools) that can be used to extract PLA data from confocal microscopy images using *ImageJ*. A step-by-step protocol to determine and quantify PLA fluorescence signals is included. Overall, the accurate capture and subsequent analysis of PLA confocal images constitutes a crucial step to properly interpret data obtained with this powerful experimental approach.

Key words Proximity ligation assay, Image analysis, G protein-coupled receptor, Protein-protein interaction, ImageJ

1 Introduction

G protein-coupled receptors (GPCRs) represent the largest family of plasma membrane receptors [1] and constitute the first-class molecular target for approved drugs [2]. The most recent classification of GPCRs is the one termed GRAFS, which is formed by five families, glutamate (G), rhodopsin (R), adhesion (A), frizzled/taste2 (F), and secretin (S), and includes more than 1000 proteins [3]. GPCRs are widely distributed throughout the body and sense a big number of extracellular signal molecules (i.e., light, local mediators, hormones, and neurotransmitters), whose information is mainly transduced into the cell through G protein-dependent pathways [4]. Indeed, the diversity of GPCR biology is not only

Electronic supplementary material: The online version of this chapter (https://doi.org/10.1007/978-1-4939-9686-5_3) contains supplementary material, which is available to authorized users.

Elena Rebollo and Manel Bosch (eds.), *Computer Optimized Microscopy: Methods and Protocols*, Methods in Molecular Biology, vol. 2040, https://doi.org/10.1007/978-1-4939-9686-5_3, © Springer Science+Business Media, LLC, part of Springer Nature 2019

dependent on the rich variety of receptors and ligands but also on the interaction with accessory proteins and/or GPCRs. In such way, GPCR oligomerization, in which a direct protein-protein interaction occurs between receptors, may determine the functioning of GPCRs. Thus, the formation of a receptor-receptor complex (i.e., oligomer) can affect receptor biosynthesis, maturation, trafficking, plasma membrane diffusion, and the pharmacology of the former receptors (for review *see* Ref. [5]). In addition, understanding the basis of the GPCR oligomerization phenomenon would also be important in the pathophysiology of many diseases and even provide novel therapeutic targets. Hence, it has been shown that changes on either the expression or the functionality of receptors forming a named oligomer can occur in pathological conditions [6], thus restoring the physiological balance could help to a better management of these diseases.

Given the relevance of GPCR oligomerization, remarkable technical advances have been made to fully demonstrate and characterize this phenomenon. From the different approaches developed to elucidate protein-protein interactions, one of the more outstanding ones is the proximity ligation assay (PLA) (Fig. 1). PLA is a method (*see* **Note 1**) that permits the visualization of the interaction between two putative-interacting proteins by means of

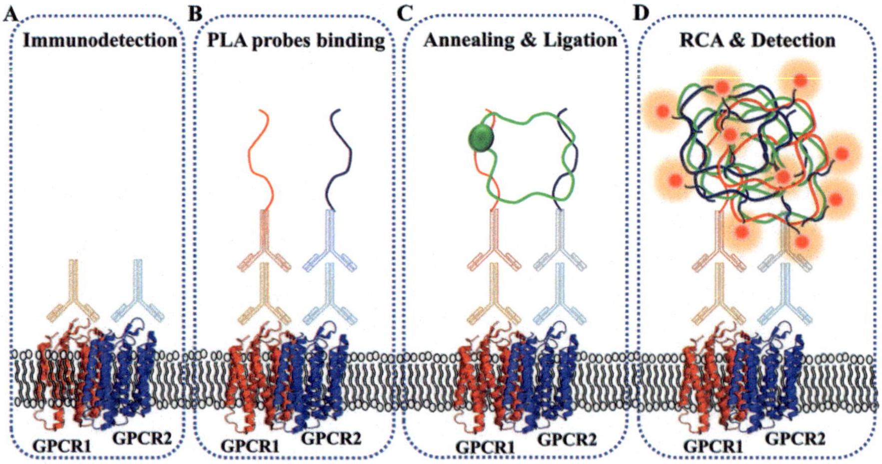

Fig. 1 Schematic representation of the proximity ligation (PLA). (**a**) Specific primary antibodies against GPCRs (i.e., GPCR1 and GPCR2) are used to specifically detect each receptor. (**b**) Secondary species-specific antibodies conjugated with oligonucleotides (PLA probe MINUS and PLA probe PLUS, illustrated as red and blue, respectively) recognize the primary antibody. (**c**) Two oligonucleotides (illustrated as red and blue curve lane) hybridize with the PLA probes. In case these PLA probes are in close proximity ($\leq$40 nm), a circular ssDNA (green curvy circle) might be formed. The open circular ssDNA could be closed by the action of the T4 DNA Ligase (illustrated as green oval). (**d**) The closed circular ssDNA may serve as a template for the phi29 DNA polymerase which extends the 3′-OH end of one of the PLA probes acting as a primer for rolling circle amplification (RCA). Finally, the generated concatemeric product is hybridized with fluorescent oligonucleotide probes (red spot)

the use of specific antibodies. Briefly, a pair of antibodies, raised in distinct species, are used to selectively recognize each protomer within the complex (Fig. 1a). Subsequently, a pair of secondary antibodies, containing single complementary short DNA strands (also called PLA probes), are used to selectively recognize the corresponding species-specific fragment of the primary antibodies (Fig. 1b). Only when the two complementary PLA probes are in close proximity (≤40 nm distance), they can hybridize (Fig. 1c). Then, in the presence of linear connector oligonucleotides, it is possible to generate an amplifiable circular DNA strand (Fig. 1c) that serves as template for a local rolling circle amplification (RCA) reaction (Fig. 1d). The product of this reaction is a single-stranded DNA molecule coupled with the antibody complex that can be detected by means of fluorescence-labeled complementary oligonucleotides (Fig. 1d).

Confocal microscopy imaging permits the visualization of the PLA fluorescent signal, which is observed with high subcellular resolution. However, once obtained the PLA images, it is necessary to properly process the information to validate and determine the presence of a direct protein-protein interaction. Here, we detail the most important and relevant aspects of the analysis of the PLA images acquired, which are processed using the free *ImageJ* software.

2 Materials

2.1 Software and Image Files

1. The *ImageJ* program can be downloaded from the *ImageJ* directory [7]. Specific details regarding installation on Macintosh and Windows are also available at the *ImageJ* directory [8]. The version used in this protocol was the 1.51v (9 March 2018). For quantitative purposes BMP, TIFF, PNG, and GIF formats are preferred as they do not miss information.
2. An example of DAPI-stained nuclei and PLA probe stacks of images can be downloaded from the Springer website. Those images where acquired as described in **Note 2**.

3 Methods

3.1 Image Processing

This section covers the preprocessing of the images to remove background and noise and adjust their brightness and contrast. These preprocessing steps facilitate the extraction of the objects of interest that will be analyzed in the next section.

1. Launch *ImageJ* and open the "DAPI.tif" and "PLA.tif" stacks at [File > Open...] or by drag and drop the files on the *ImageJ* menu bar.

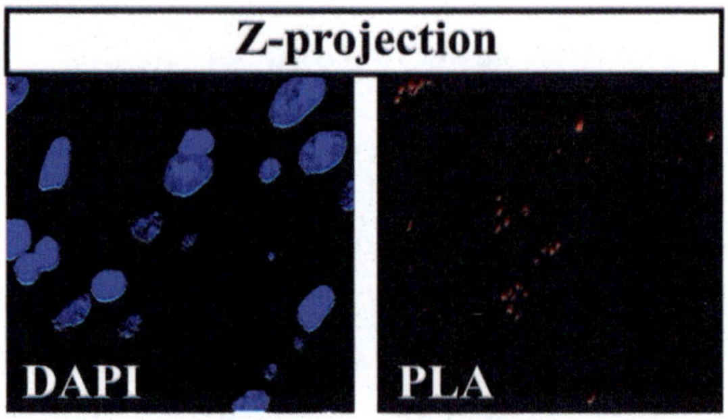

Fig. 2 Maximum intensity projection images of DAPI and PLA stacks

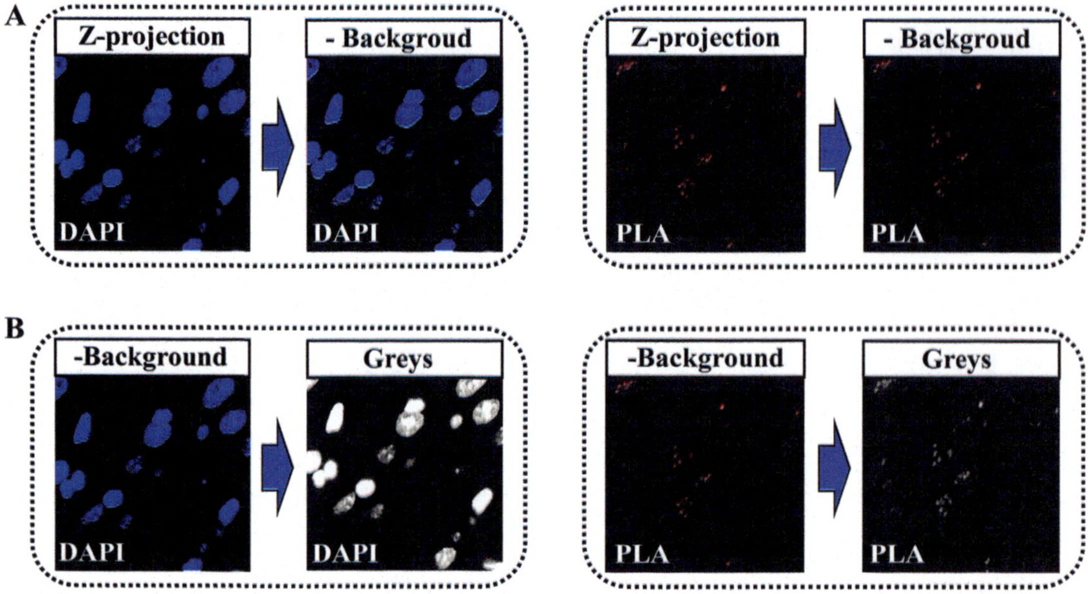

Fig. 3 Image processing. An inset of each maximum projection image (i.e., DAPI-stained nuclei and PLA dots) is shown. (**a**) Background subtraction using 50.0 pixels rolling ball radius approach. (**b**) Gray-scale image

2. For both stacks, generate a maximum intensity projection image at [Image > Stack > Z project... > Max intensity] (Fig. 2).
3. Perform a background subtraction of both resulting z-projections by means of the specific function defined in *ImageJ* at [Process > Subtract background...]. Although nuclei and PLA spot size differently, the default *rolling ball radius* of 50.0 px works well in both sample images (Fig. 3a) (*see* **Note 3**).
4. Showing both z-projections in gray scale (Fig. 3b) may help improve the visual contrast of the images. When necessary, use the mentioned Look Up Table (LUT) option at [Image > Lookup Tables > Grays].

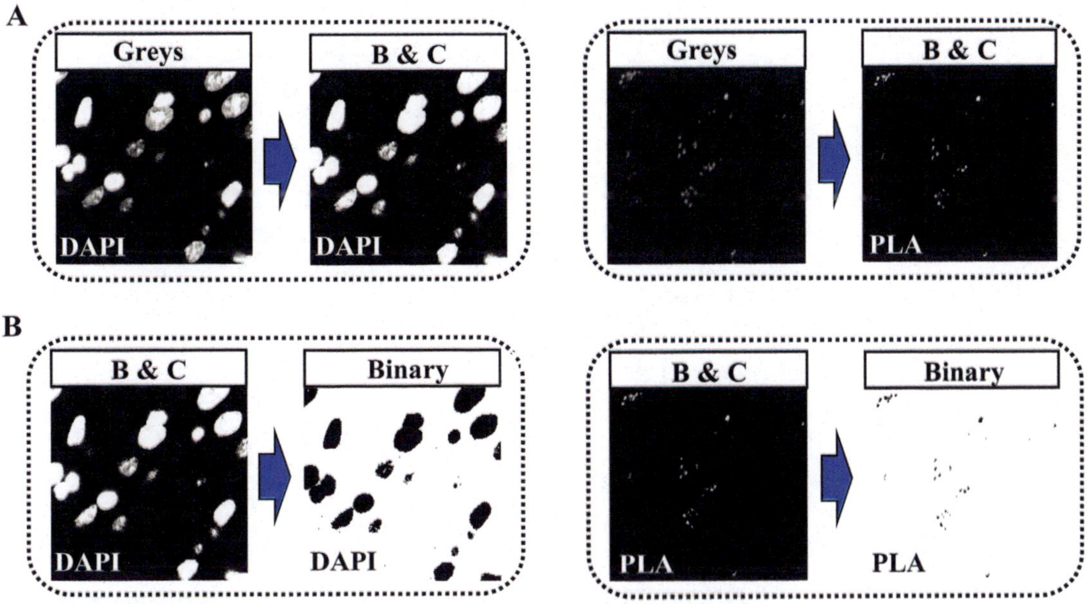

Fig. 4 Image processing. (**a**) Brightness and contrast adjustment. (**b**) Binary image transformation

5. Adjusting the brightness and the contrast of both z-projections can be used to stretch the observable differences between signal and background (Fig. 4a). Use this option at [Image > Adjust > Brightness/Contrast...] when spots are weak and difficult to see, but take special care not to change the cutting values from one image to the next (*see* **Note 4**).
6. If the resulting images need further noise removal (they, for instance, show particles with holes or gaps), a smoothing filter can be applied by using the command at [Process > Smooth] (*see* **Note 5**).
7. Segment both z-projections, and convert them to binary images, i.e., black-and-white image (Fig. 4b) (*see* **Note 6**). To that aim go to [Process > Binary > Make binary].

3.2 Nuclei Analysis

In this section the cells of interest are counted using the binary mask that results from nuclei segmentation.

1. When nuclei overlap or are very close to each other and, therefore, they are difficult to resolve spatially, a watershed transformation can be performed (Fig. 5a) to separate objects that appear contiguous in the binary mask. This operation can be performed at [Process > Binary > Watershed].
2. Count nuclei larger than 30 μm^2 in the resulting binary image by using the *Analyze particles* function available in *ImageJ* at [Analyze > Analyze particles ...] (*see* Fig. 5b and **Note 7**). Set the following parameters: *Size* (μm^2), 30-*Infinity*; *Show*, *Outlines*; and *Display results.*

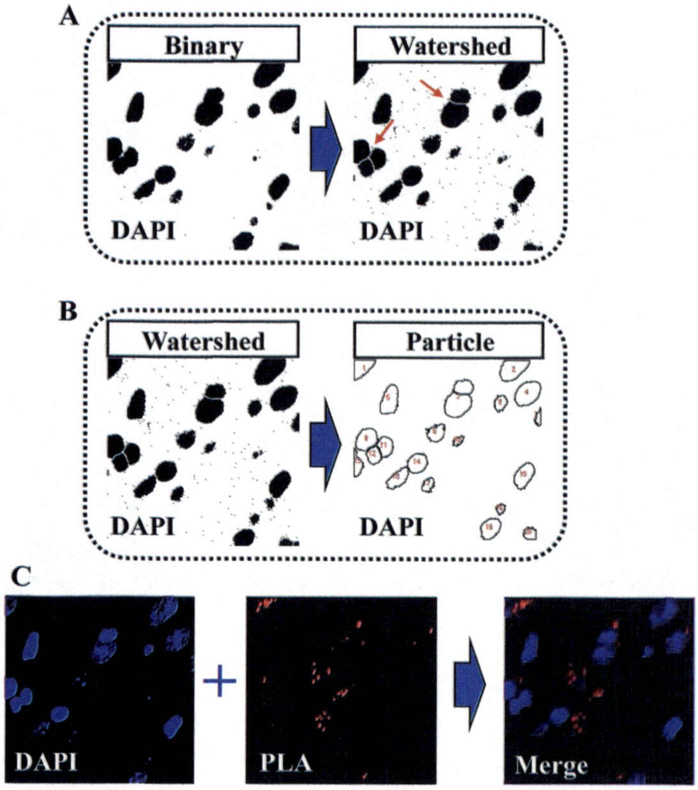

Fig. 5 Particle analysis. (**a**) Inset of the DAPI binary image before (left) and after (right) applying the watershed function. (**b**) Image showing the outlines of the nuclei obtained after the particle analysis. (**c**) Merge of both channels into a composite image

3.3 PLA Signal Analysis

In this section the number of PLA probe particles is counted on its corresponding binary mask. Then, a composite image containing both nuclei PLA spots is created, and the ratio between the PLA probe particles and the number of nuclei calculated.

1. Apply a watershed transformation at [Process > Binary > Watershed] if PLA signal overlaps and thus the spots are difficult to resolve spatially (Fig. 5a).
2. Count PLA probe particles larger than 0.3 μm^2 in the resulting PLA watershed image by using the *Analyze particles* function at [Analyze > Analyze particles . . .] (*see* **Note** 7). Set the following parameters: *Size* (μm^2), 0.3–10; *Show*, *Outlines*; and *Display results*.
3. When needed for visualization purposes, create a composite image containing the DAPI and PLA projected channels by means of the merge channel function defined in *ImageJ* at [Image > Color > Merge channels . . .] (Fig. 5c). Select MAX_PLA image as *C1 (red)* and MAX_DAPI as *C3 (blue)*. Tick the *Create composite* option.

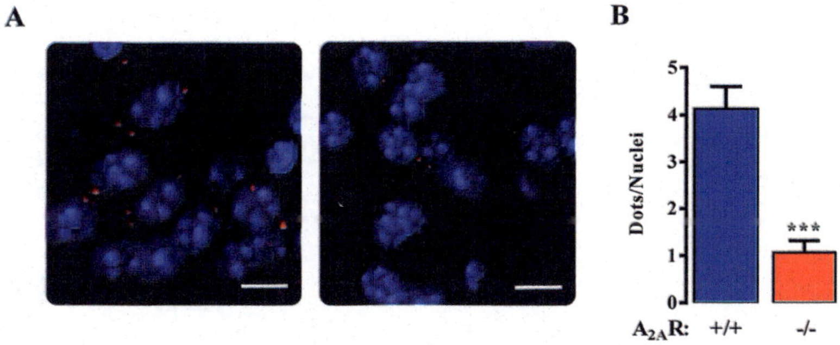

Fig. 6 PLA analysis results. (**a**) Example of PLA images corresponding to the $AT_1R/A_{2A}R$ heteromer detection in $A_{2A}R^{+/+}$ (left panel) and $A_{2A}R^{-/-}$ (right panel) mouse brain. (**b**) Quantification of PLA particles in AT_1R and $A_{2A}R$ proximity confirmed the significant difference of PLA particle density between $A_{2A}R^{+/+}$ and $A_{2A}R^{-/-}$ mice ($***P < 0.05$, student *t* test). Values in the graph correspond to the mean ± SEM (dots/nuclei) of at least five animals for each genotype (adapted from Ref. [11])

4. Calculate the ratio between PLA probe particles and nuclei number (*see* **Note 8**).
5. Graph data containing appropriated negative controls (*see* **Note 9** and Fig. 6a).
6. Perform appropriate statistical analysis of the results (Fig. 6b).

4 Notes

1. The PLA protocol was performed as previously described [9]. Briefly, coronal sections (50 μm) from fixed mouse brains were obtained using a vibratome. Slices were permeabilized with 0.3% Triton X-100 in PBS for 2 h, rinsed with wash solution (0.05% Triton X-100 in PBS), and incubated with blocking solution [10% normal donkey serum (NDS) in wash solution] for 2 h at 22 °C. Subsequently, slices were incubated with the indicated primary antibodies overnight at 4 °C, washed twice before being incubated with the appropriate secondary Duolink in situ PLA probes for 1 h at 37 °C. The following steps were performed following the manufacturer's protocol. In the example provided here, we used a rabbit anti-AT_1R polyclonal antibody and a mouse anti-$A_{2A}R$ monoclonal antibody as primary antibodies to detect $AT_1R/A_{2A}R$ heteromer in mouse brain [6]. The brain slices were observed in a Leica TCS-SP2 confocal microscope with a 63× oil objective lens (NA 1.32).
2. The acquisition of confocal microscopy images was performed as follows: (1) for DAPI-stained nuclei (ex. 364 nm,

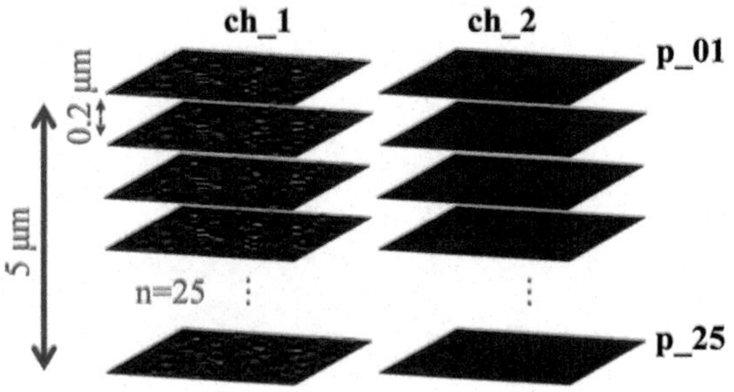

Fig. 7 Schematic representation of the confocal image stacks with $n = 25$ images per stack (from plane p_01 to p_25) acquired for each fluorescence channel (ch_1 and ch_2)

em. 460 nm) and (2) for PLA dots (ex. 561 nm, em. 624 nm). For each experiment a region of interest of 5-µm thick was selected. The pixel size was set to 232.5 nm with a z-interval of 0.2 µm to obtain a z-stack (Fig. 7). Although this pixel size allows for the detection of single PLA dots in these particular experiments, in cases where the size of the particles to identify is smaller, the pixel size should be reduced up to the nyquist criterion. The images were saved as 8-bit tiff images.

3. The radius should be set to at least the size of the largest object that is not part of the background (i.e., nuclei). To find out the radius of the largest objects in your image, you can use the line tool to draw a radius. Then, the length will be reported in the *ImageJ* toolbar. The default rolling ball radius value used by *ImageJ* is 50.0 pixels.
4. Background should appear as dark as possible, and the signal must be as uniform and bright as possible. Keep the corresponding numbers in order to set the same parameters in the rest of the images of the experimental set to analyze. The values used could be different between both channels, but they must be the same for the rest of the corresponding channels of the total of images analyzed. Changing the background and contrast in the images changes the original intensity values. For the counting of objects (i.e., particle analysis), it may be not so critical, but it could affect other parameters and posterior image analysis.
5. If individual particles show gaps that could affect the counting (particle analysis; Subheadings 3.2 and 3.3, **step 2**), then use the smooth filter to blur the active image or selection. This filter replaces each pixel with the average of its 3 × 3 neighborhood.
6. *ImageJ* contains different global thresholding methods that can be checked in a single shot at [Image > Adjust > Auto

Threshold]. For the sample images provided, the *default* automatic thresholding method is used, which works by taking a test threshold and computing the averages of the pixels at or below the threshold and the pixels above. It then computes the average of those two, increments the threshold, and repeats the process iteratively, until the threshold is larger than the composite average.

7. It is well established that within the brain (i.e., striatum), neurons and glial cells may be discriminated by the size of their respective nucleus [10]. Therefore, if the interaction upon study is expected to take place in neurons, as the example provided here, the glial nuclei ($<30\ \mu m^2$) can be excluded from the quantification. Likewise, it has been described that PLA probe particles can be observed as larger than $0.3\ \mu m^2$ particles [9].
8. The number of PLA probe particles is divided by the number of nuclei in each image.
9. Several negative controls should be carefully established. To that aim the same procedure should be performed in (1) the absence of primary antibodies, (2) brain slices from knockout animal for at least one of the targeted receptors, and (3) defined negative brain regions.

Acknowledgments

This work was supported by MINECO/ISCIII (SAF2017-87349-R and PIE14/00034), the Catalan government (2017 SGR 1604), Fundació la Marató de TV3 (Grant 20152031), and FWO (SBO-140028) to FC. We are grateful to Manel Bosch and Benjamin Torrejón from the Scientific and Technological Centers (CCi-TUB) for providing microscopy expertise.

References

1. Pin JP, Neubig R, Bouvier M et al (2007) International Union of Basic and Clinical Pharmacology. LXVII. Recommendations for the recognition and nomenclature of G protein-coupled receptor heteromultimers. Pharmacol Rev 59:5–13. https://doi.org/10.1124/pr.59.1.5
2. Overington JP, Al-Lazikani B, Hopkins AL (2006) How many drug targets are there? Nat Rev Drug Discov 5:993–996. https://doi.org/10.1038/nrd2199
3. Schiöth HB, Fredriksson R (2005) The GRAFS classification system of G-protein coupled receptors in comparative perspective. Gen Comp Endocrinol 142:94–101. https://doi.org/10.1016/j.ygcen.2004.12.018
4. Palczewski K (2006) G protein-coupled receptor rhodopsin. Annu Rev Biochem 75:743–767. https://doi.org/10.1146/annurev.biochem.75.103004.142743
5. Ciruela F, Vilardaga J-P, Fernández-Dueñas V (2010) Lighting up multiprotein complexes: lessons from GPCR oligomerization. Trends Biotechnol 28:407–415
6. Fernández-Dueñas V, Taura JJ, Cottet M et al (2015) Untangling dopamine-adenosine receptor-receptor assembly in experimental parkinsonism in rats. Dis Model Mech

8:57–63. https://doi.org/10.1242/dmm.018143

7. Image J downloading. http://imagej.nih.gov/ij/download.html
8. Image J installation. http://imagej.nih.gov/ij/docs/install
9. Taura J, Fernández-Dueñas V, Ciruela F (2015) Visualizing G protein-coupled receptor-receptor interactions in brain using proximity ligation in situ assay. Curr Protoc Cell Biol 67:17.17.1–17.17.16. https://doi.org/10.1002/0471143030.cb1717s67
10. Matamales M, Bertran-Gonzalez J, Salomon L et al (2009) Striatal medium-sized spiny neurons: identification by nuclear staining and study of neuronal subpopulations in BAC transgenic mice. PLoS One 4:e4770
11. de Oliveira PA, Dalton JAR, López-Cano M et al (2017) Angiotensin II type 1/adenosine A2A receptor oligomers: a novel target for tardive dyskinesia. Sci Rep 7:1857. https://doi.org/10.1038/s41598-017-02037-z

Chapter 4

Introduction to ImageJ Macro Language in a Particle Counting Analysis: Automation Matters

Manel Bosch

Abstract

Image quantitation is as truthful as the reproducibility of the methods involved in experimentation, image acquisition, and analysis. Automation in image analysis is highly recommended as it grants the reproducibility of the resulting data, minimizes the bias and mistakes produced by manual intervention, and optimizes the overall time invested. This chapter focuses on the main concepts of *ImageJ* macro programming with special emphasis on the image analysis pipeline automation.

Key words ImageJ macro language, Image analysis automation, Macro, Fiji

1 Introduction

Image analysis has become a routine in the last decade in the life sciences field. Clearly, the simple observation of images is not enough [1], and some quantitative data needs to be obtained. Moreover, new microscopic techniques such as high-content screening or light sheet microscopy generate large datasets that strongly demand for automated protocols that render unbiased results in a feasible way.

The analysis of image datasets can be done in several ways: from a completely manual step-by-step procedure up to a sophisticated process such as machine learning, where computers are trained to perform all the analysis. Selecting one methodology vs. another depends basically on the number of images to be analyzed and the scientists' computer skills. In this sense, the lack of expertise in computer programming goes in favor of the manual procedure. However, automation has competitive advantages: First, it makes results reproducible and comparable. Second, once a protocol is

---

Electronic supplementary material: The online version of this chapter (https://doi.org/10.1007/978-1-4939-9686-5_4) contains supplementary material, which is available to authorized users.

Elena Rebollo and Manel Bosch (eds.), *Computer Optimized Microscopy: Methods and Protocols*, Methods in Molecular Biology, vol. 2040, https://doi.org/10.1007/978-1-4939-9686-5_4, © Springer Science+Business Media, LLC, part of Springer Nature 2019

set, it clearly reduces the time spent on the analysis. Third, the automated protocol together with its main instructions can be linked to the material and methods of any publication and even added some extra annotations that make it more understandable, which is particularly useful when the procedure needs to be repeated after a long time or by another person.

ImageJ [2] is one of the most widely used computer programs for image analysis. It can handle multidimensional sets of images (hyperstacks) and has lots of tools for their processing and analysis. Moreover, there are many other available tools such as macros and plug-ins that are designed to add new functionalities to the software. These tools can be downloaded from *ImageJ* [3] or from third-party web sites. Besides, *ImageJ* has a forum in the Internet [4] and a mailing list [5] with a large community of developers and advanced users that try to answer and solve any issues users find when performing image analysis.

In addition, *ImageJ* facilitates the image analysis automation by providing a language of its own: *ImageJ* macro language (ijm). Although ijm is not a programming language properly speaking, it allows controlling most aspects of the software. Importantly, no programming experience is required to start automating pipelines with *ImageJ*: the software has a macro recorder, a tool that will record all the steps performed during the analysis. Once finished, the list of commands – functions representing each of the executed steps – shown in the recorder window can be converted into an ijm file. This file may now be run on the same dataset, thus reproducing the same results without all the clicks and parameter inputs that were needed the first time. Running the macro on a different set of images, however, may require further code tuning, including modifications and additions of some lines of code, i.e., programming.

This chapter focuses on the automation of an image analysis protocol once the main steps are recorded using the macro recorder tool. To that aim, the basic concepts in ijm programming are first introduced here and then applied over the Methods section to help in converting an initial macro into a more functional script. The provided protocol is exemplified on an image that has been generated in silico, in order to simplify the image processing pipeline.

1.1 Basic Concepts in Programming

A computer program and a macro are nothing else than a collection of commands or tasks that are sequentially executed reproducing with precision the writing arrangement. The time during which a program is executed is known as "run time." Programming languages generally use identical tools to work with commands, e.g., variables, to store key values; conditional structures, to execute some code when a condition is met; iteration structures, to repeat actions over a collection of data; etc. In addition, programming languages require the use of symbols and punctuation marks. Some of these may be used as operators ("<", "&", "=", etc.), whereas others are delimiters ("{}", "[]", ";", etc.). Although they are basic in any script, they represent one of the major sources of error when writing code. The ijm language is not an exception in this sense, since it requires such programming structures and punctuation marks. It is key, for instance, to write a semicolon ";" at the end of each command, so that it is unequivocally interpreted as an executable function. Besides symbology, the following aforementioned structures are key:

1. *Variables*: They are storage sites of information that will be used in run time. The information contained can be of any data type: numbers, strings (sequences of text characters), Booleans (data type with only two possible values: true or false, i.e., 1 or 0), and arrays (collection of elements). ijm variables are typed less, that is, they do not need to specify what type of information they store, and, moreover, the data type can be changed in run time.

   A variable consists of an identifier, i.e., the name given to the variable, plus a value, i.e., the information stored, both bound by the symbol "=." The combination of variable name and stored value is only valid during run time. The value, however, can be changed during the execution of the code.

   In Fig. 1, the term "title" in line 1 is the identifier, and "Random pattern.tif" is the string stored in "title," which in this case corresponds to the name of the sample image (*see* Subheading 2). Variables are called by their identifier; however, what is really used is the value they contain. In this sense, it is highly recommended to use variable names that clearly reflect the information they contain; in this way, every time a variable is called from the code, it is clear what its function will be (*see* **Note 1**).

   Lines 2 and 3 (Fig. 1) are equivalent, both selecting a window named "C1-Random pattern.tif" (*see* **Note 2**). The difference is that in line 2 the *selectWindow* function (*see* **Note 3** for more details on functions) selects the window by its full name, whereas in line 3 string concatenation is used: the string (C1-) is added to the string contained within the variable "title." Such string manipulations are useful when integrating an *ImageJ* function definition with our own variables. *See* **Note 4** for further information on string manipulation.

```
1 title = "Random pattern.tif";
2 selectWindow("C1-Random
  pattern.tif");
3 selectWindow("C1-" + title);
```

Fig. 1 Block of code illustrating how to define a variable, use it inside a function, and concatenate its content to a string

```
1 if(condition1){
2   //block of code
3 }else if(condition2){
4   //block of code
5 }else{
6   //block of code
7 }
```

Fig. 2 Conditional statements available in ijm

2. *Conditional statements*: These are control structures that allow executing a sequence of commands only when a condition is met. These conditionals are defined with the structure illustrated in Fig. 2.

   If "condition1" (Fig. 2, line 1) is true, meaning it is fulfilled, the code specified between braces "{}" will be executed (*see* **Note 5**). In case "condition1" is false but "condition2" is true, the block of code following the *else if* statement will be executed. When all the previous conditions are false, the code after the *else* statement will be executed. There can be as many *else if* conditions as necessary or none in case there is only one condition to check. The *else* statement may also be dispensable. Conditional statements are very useful in programming. In ijm they allow, for instance, to check the installed version of *ImageJ*, the bit depth of an image, or the existence of a folder before saving images in it. They can also be used to check the results obtained in run time or the values entered by the user during the execution of the macro.

3. *Iteration statements*: These are another type of control structures that allow to repeat the execution of a block of code in case a condition is met. There are three types of loops in ijm, which differ among them in the position where the condition is checked (Fig. 3).

   The *for* loop (lines 1 to 3, Fig. 3) requires an initial numerical value; when this value matches the specified "condition," the code between braces "{}" will be executed. Once this code has been executed, the initial value is added to the specified "increment" so that the defined "condition" is now checked on this new value. The iteration within the loop will end when the incremented value does not meet the defined condition. The *while* loop (lines 4 to 6, Fig. 3) allows to repeat a block of code only if the "condition" is met. It will end once this "condition" is false. The *do...while* loop (lines 7 to 9, Fig. 3) works like the previous one, although in this case the code will be executed

```
for(initial_value; condition; increment){
  //block of code
}
while(condition){
  //block of code
}
do{
  //block of code
}while(condition);
```

Fig. 3 Iteration statements available in ijm

before the "condition" is checked and will keep on being executed until the "condition" becomes false. Iteration statements are crucial in programming. In ijm they are key, for instance, to work on several images in a folder, on all the slices of an image stack or on a list of ROIs.

4. *Code annotations*: They are non-executable lines in the code that can be added for explanatory purposes, e.g., to indicate why a specific value is used in the analysis or why some crucial decisions are taken in the protocol. Annotations are preceded by two slashes "//". If several lines are needed, they can go between "/*" and "*/". Annotations may be crucial for other users to execute the code as they may clarify important steps. Additionally, they may be used before writing any line of code to set the main steps in the protocol (*see* Subheading 3).

2 Materials

1. Image dataset. The example image used and provided in this chapter is named "Random pattern.tif" and has been generated in silico. It contains a random distribution of small blue dots and some bigger red circles. Some of these circles have been randomly filled in green (Fig. 4). The main aim of the analysis is to compare the number of blue dots inside circles with and without green signal. This could, for example, resemble an experiment where the internalization of nanoparticles has to be compared in transfected (GFP positive) vs. non-transfected (GFP negative) cells.
2. Software. One of the main distributions of *ImageJ* has been used: *Fiji* [6], version 1.51d. *See* **Note 6** for some advantages of *Fiji* over *ImageJ*.
3. Macro recorder. The macro recorder can be called from [Plugins > Macros > Record...], and it is highly recommended to

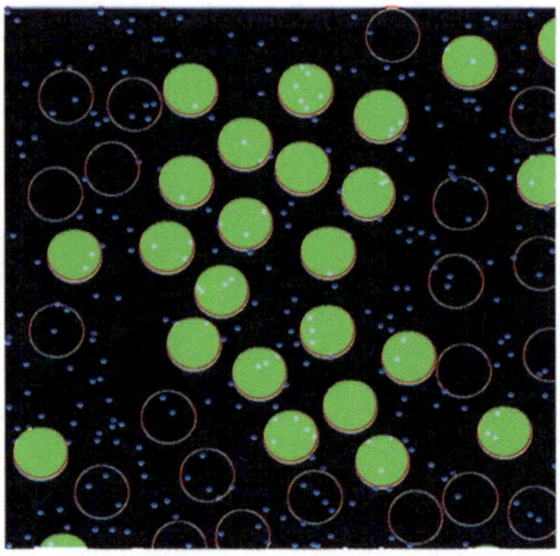

Fig. 4 Example image used in the analysis

keep it open during the analysis to see the ijm instruction of each step.

4. Functions. All the commands, functions of the ijm language, are listed at the built-in macro functions webpage [7]. Additional functions may be required when using third-party plugins (*see* **Note** 7).
5. Macros. The macros used to generate (RandomPatternGenerator.ijm) and analyze (RandomPatternAnalysis.ijm) the image are attached to this chapter and can also be found on GitHub [8]. "RandomPatternGenerator.ijm" is based on the Roberto Sotto-Maior Fortes de Oliveira's RandomSamplePerimeterMethod.txt [9] and Gabriel Landini's RandomSampleAndMeasure.txt [10] macros. The code of this macro is not described in this chapter as it is out of its scope.

3 Methods

3.1 Generate the First Macro Using the Macro Recorder

The analysis of the image dataset described above could be performed as follows:

1. Launch *Fiji*, open the macro recorder at [Plugins > Macros > Record...], and open the image "RandomPattern.tif" at [File > Open...].
2. Split channels at [Image > Color > Split channels], to work with them independently. The resulting three images are binary, i.e., images with only two possible values: 0 and 1 (*see* **Note 8**).
3. Select the channel one, and obtain the regions of interest (ROIs) that represent each red circle. Discard those circles at the edge of the image, as they may be incomplete; this can be done at [Analyze > Analyze Particles...]; tick the boxes *Add to Manager* and *Exclude on edges*.

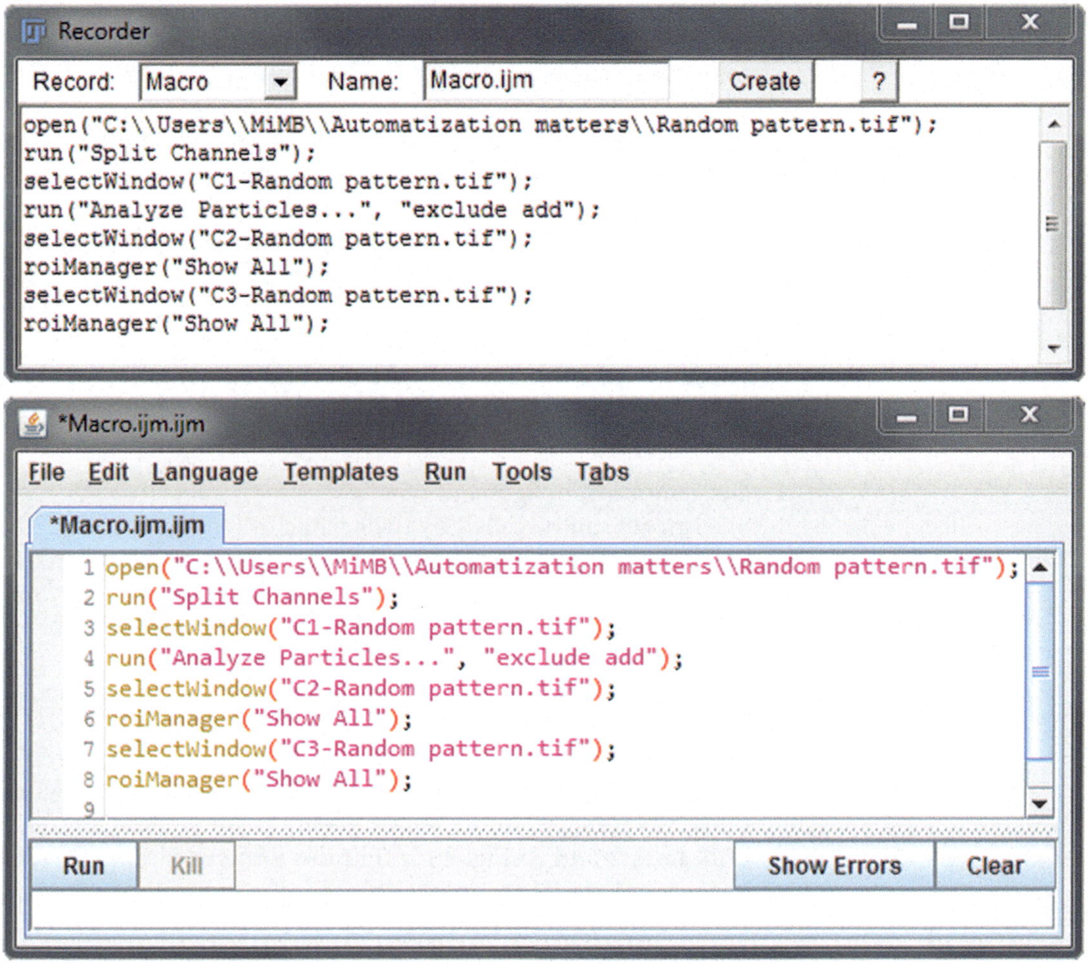

Fig. 5 Macro Recorder (top) and Script Editor (bottom) windows

4. Select channels two and three, and tick the box *Show All* in the *ROI Manager* for each of them to visualize the ROIs obtained in previous step.
5. The macro recorder shows all the commands used (Fig. 5, top). By pressing *Create* at the macro recorder window, all the lines of code will be transferred to the script editor (Fig. 5, bottom), where the macro can be executed over the same image by pressing *Run*. *See* **Note 9** for color coding of the script editor.
6. The final analysis however, requires the selection of each ROI in the *ROI Manager* to count the number of blue dots inside it and check whether it has green signal or not. Doing it manually is time-consuming and may be a source of mistakes if there are many images to analyze. In the following section there is a description of the main ijm instructions that can be added to the initial macro described above to perform the analysis automatically.

3.2 Widening the Usage of a Macro After Recording

Once all the image processing steps are in the script editor, the macro is ready to be executed. However, each new run will be executed over the same images. In the example the macro will always try to select images named "Random pattern.tif" (i.e., C1-Random pattern.tif). It is therefore necessary to modify the initial macro script to ensure it can be applied to any image. Several aspects need to be controlled to that aim, e.g., the images' names or the number/order of the ROIs where measurements will be taken. It is also important to know how to store the intermediate results and how to finally specify the path to save them. All these concepts are introduced in the following steps, which are described as a collection of annotations in the pipeline shown in Fig. 6. The corresponding commands to perform each step should be intercalated between these lines.

1. Get control of images and windows.

   Images can be called by their name or by their ID. The ID is a unique number *ImageJ* gives to any new image displayed. It is highly recommended to collect its name or ID or both when a new image is opened or generated during the processing. To do so the commands shown in Fig. 7 can be used.

   In lines 1 and 2 (Fig. 7), the name and the ID of the active image are stored within the variables "title" and "id," respectively. The image can now be called by its name or by its ID (lines 3 and 4, respectively). The difference between the functions *selectImage()* and *selectWindow()* (seen in the Introduction, lines 2 and 3, Fig. 1) is that the former only works with images and the latter may also work for nonimage windows such as the *Log* window, the *Results* table, the *ROI Manager*, etc.

2. Define array variables to store results.

```
//Get data from image
//Split channels
//Obtain objects from red
 channel
//Define variables to store
 intermediate results
//Count dots and check green
 staining in each ROI
 //Count dots
 //Check green staining
//Create a Results table
//Save results
```

Fig. 6 Main steps required in the current analysis pipeline set as code annotations

```
1 title = getTitle();
2 id = getImageID();
3 selectImage(title);
4 selectImage(id);
```

Fig. 7 ijm instructions to store the name and the ID of an open image (lines 1 and 2, respectively) in two variables and then use these to call the image

```
1 nROIs = roiManager("count");
2 arrayDots = newArray(nROIs);
3 arrayGreen = newArray(nROIs);
```

Fig. 8 Block of code to create the two arrays required for the current analysis, sizing the number of cells previously counted

As said, the analysis of the sample image requires to count the number of blue dots inside red circles and, moreover, to check whether these red circles are filled in green. The counts and checks must be performed for each of the ROIs listed in the *ROI Manager*; it is therefore required to store the results as they are sequentially obtained. Arrays are a type of variable that can store multiple values of different types: numbers, strings, etc. and even a mix of them. They are initialized using the function *newArray(n)* where "n" is the length of the array (Fig. 8).

The analysis at hand requires two arrays, one to store the number of dots found within each cell (red circle) and another to store whether each cell is green positive or negative. One problem is that the number of cells in the image is unknown in advance. This number can be extracted from the number of ROIs obtained after analyzing the particles in the red channel (*see* **Note 10**). The function *roiManager("count")* in line 1 (Fig. 8) will return that number. Lines 2 and 3 in Fig. 8 initialize the two previously described arrays. By using the commands in this way, the length of the arrays will change each time the macro is executed, depending on the number of cells in the particular image analyzed. Either filling in or reading an array requires access to each of its positions one by one, for which a loop is used. Each position is indicated by the array identifier, together with the index of that particular position, enclosed by brackets: arrayDots[i] (line 5, Fig. 9). This index is stored in a variable ("i" in this example) that will be iteratively updated during the execution of the loop. In the following two steps, the code to fill and read both arrays is provided. More array functions can be found in the built-in macro functions [7].

```
1 for(i=0; i<nROIs; i++){
2  selectWindow("C3-" + title);
3  roiManager("Select", i);
4  run("Find Maxima...", "noise=1 output=Count");
5  arrayDots[i] = getResult("Count", 0);
6  run("Clear Results");
7  selectWindow("C2-" + title);
8  roiManager("Select", i);
9  run("Find Maxima...", "noise=1 output=Count");
10 arrayGreen[i] = getResult("Count", 0);
11 run("Clear Results");
12}
```

Fig. 9 Iteration *for* loop to go through all the ROIs in the ROI Manager to count the number of dots (blue and green) per ROI and image. The result obtained after each count is saved in the corresponding array

3. Iterate over a list of ROIs and count.

   The loops are the programming structures devised to go through the list of ROIs. In line 1 (Fig. 9) a *for* loop is initialized with the variable "i" equal to 0 (*see* **Note 11**). The condition to check will be the number of ROIs (variable "nROIs" defined in the previous step), and for each loop iteration, the increment will be 1 (*see* **Note 12**).

   Once the loop is initialized, the channel 3 image (line 2, Fig. 9) and the ROI with the index "i" (line 3, Fig. 9) are both selected. In line 4 (Fig. 9) there is a function that looks for peaks of maximum intensity. This function is found in [Process > Find Maxima...], and it is defined with a noise tolerance of 1 (*see* **Note 13**). It is important to set the output of this function to *Count* in order to obtain the number of peaks (i.e., dots). The code for this function can be obtained from the *Macro Recorder* after manual execution. The result will appear in a *Results* table under a column named "Count." To access any result from this table, there is a function called *getResult (column, row)* (line 5, Fig. 9). This function requires that the "column" name and the "row" position are specified, so that the returned value can be assigned to the corresponding position "i" inside the array. The row number is set to 0 to make sure that the first result from the table is not lost (*see* **Note 11**); to this aim, the Results table must be cleared after each measurement (line 6, Fig. 9).

   The same procedure can be applied on channel 2 image (line 7, Fig. 9) to check whether a ROI is green positive or not.

Again, each ROI will be selected iteratively (line 8, Fig. 9) and the number of maxima counted at each round, using the *Find Maxima* function (line 9, Fig. 9). As the green signal occupies the whole ROI, the result will be 1 or 0 depending on whether the ROI is green positive or not, thus rendering a single maximum. Each result is assigned to a position in the corresponding array (line 10, Fig. 9).

In summary, this block of code repeats the same procedure twice, one on the red image (channel 3, lines 2–6, Fig. 9) and another on the green image (channel 2, lines 7–11, Fig. 9). Later in **step 7** of this section, it will be shown how this code can be reconverted into user-defined functions in order to improve the overall readability.

4. Create a defined Results table.

   Upon the *for* loop execution, both "arrayDots" and "arrayGreen" become filled with the results. The values in the arrays can now be shown by creating a specific table (Fig. 10), or by modifying the existing Results table (*see* **Note 14**). In both cases it is required to go through the array positions reading and printing the stored values.

   In Fig. 10 the table name is set (line 1), using string concatenation. Then, this name is added to the function that creates the table (line 2), together with the latter's width and height pixel specifications. The *print()* function (line 3 and 9) allows to print headings and values separated by a tabulation, specified by "\t." The values are printed using a *for* loop (lines 4–10). Although "arrayGreen" values could be printed directly, another variable – green – is created inside the loop and

```
1 tableName = "[" + title + "_Table]";
2 run("Table...", "name=" + tableName + " width=400 height=400");
3 print(tableName, "\\Headings: Cell number \t Number of Dots \t Green Staining");
4 for(i=0; i<lengthOf(arrayDots); i++){
5   green = "+";
6   if(arrayGreen[i]==0){
7     green = "-";
8   }
9   print(tableName, i+1 + "\t" + arrayDots[i] + "\t" + green);
10 }
```

Fig. 10 Block of ijm code to create and fill a Results table assigned the character "+"; whenever a value in the array equals 0 (checked in a conditional statement, lines 6–8), this new variable will be reassigned to "−." This is done just to add an example of conditional statements in the macro, for the sake of completeness.

5. Save data.

   The results, images, and tables can be saved manually once the macro stops running. However, this can also be programmed from the macro (Fig. 11) as ijm language allows for controlling computer paths to files and folders.

   The path to a folder is obtained in line 1 (Fig. 11). The function *getDirectory(string)* will pop up the computer folder structure for a folder to be selected. A title for the pop-up window can be set in the string between the parentheses of the *getDirectory(string)* function. The directory path returned by this function will be stored in the variable "output." It is highly recommended to save results in a folder different from that containing the original images (*see* **Note 15**).

   In line 2 (Fig. 11) a *ROI Manager* function is used to deselect any ROI. This will grant that all ROIs, not only the ones selected, are saved later on at once. The functions in lines 3 and 5 (Fig. 11) save both the ROIs set and the Results table produced in the analysis at hand. The first function is specific for the *ROI Manager*, whereas the second can be used to save any results such as tables, images, text windows, etc. Both functions require that the computer directory path where the files should be saved is set (*see* **Note 16**) and the format for those files specified (*see* **Note 17**).

6. Save and install the macro.

   Macros can be saved as ∗.ijm files to be later on executed in *Fiji* by drag and drop to the main bar. Alternatively, they can be installed in *Fiji* to be run from the [Plugins] menu. To do so the code needs to be wrapped in a *macro* block and then saved. Wrapping the code inside a macro block is done as shown in Fig. 12.

```
1 output = getDirectory("Select an output directory");
2 roiManager("Deselect");
3 roiManager("Save", output + name + " RoiSet.zip");
4 selectWindow(name + "_Table");
5 saveAs("Results", output + name + ".csv");
```

Fig. 11 ijm instructions to select a folder and save the *ROI Manager* and the Results table inside it

```
1 macro "RandomPatternAnalysis"{
2   //block of code
3 }
```

Fig. 12 Command that wraps all code lines in a macro, for this to be saved in the software

```
1 function count(roi){
2   roiManager("Select", roi);
3   run("Find Maxima...",
    "noise=1 output=Count");
4   result = getResult("Count",
    0);
5   run("Clear Results");
6   return result;
7 }
```

Fig. 13 Block of code wrapping the instructions to count and retrieve the number of particles inside a specified ROI

In line 1 (Fig. 12) the command *macro* specifies that all the content between the braces "{}" is a macro named "RandomPatternAnalysis." The code block should contain all the instructions detailed until now (see the final macro in [8]). To install a macro, first save it and then go to [Plugins > Macro > Install...] and select the macro to install. The macro will now appear in the [Plugins > Macro] menu just for the current *Fiji* session. To keep it permanently in the Plugins menu, the macro should be copied inside the *StartupMacros.fiji.ijm* file located inside the *Fiji* macros folder that contains all the macros installed at start-up.

7. Reuse of code with user-defined functions.

   All functions seen until now are built-in macro functions, i.e., functions defined by the ijm language. This language, however, also offers the possibility to create user-defined functions to be implemented inside macros. This may help reusing some blocks of code and making the macro more readable and understandable.

   Functions need to be wrapped inside a *function* block defined with its name and the arguments the function may need (*see* **Note 3**), as shown in Fig. 13.

   In the figure a function named "count" has been defined, which needs one argument to work (line 1). Such argument refers to the variable "roi" that will be used inside the function (line 2) as the corresponding ROI index in the *ROI Manager*. After counting (line 3), the result from the *Results* table is stored into the variable "result" (line 4). This is done to clear

```
for(i=0; i<nROIs; i++){
  selectWindow("C3-" + title);
  arrayDots[i] = count(i);
  selectWindow("C2-" + title);
  arrayGreen[i] = count(i);
}
```

Fig. 14 Iteration *for* loop to go through all the ROIs in the *ROI Manager* and call the user-defined function "count" to be applied in both channel 2 and 3 images. The result retrieved by the function "count" is saved in the corresponding array

the table (line 5) while being still able to return that particular "result" (line 6).

When the function is called, the real index of the ROIs needs to be passed as the argument. This will be done inside the *for* loop that goes through all the ROIs (Fig. 14).

The block of code shown in Fig. 14 does the same than the one seen in **step 3** (Fig. 9), but now the result retrieved by the "count" function is assigned directly to the corresponding position of the arrays.

User-defined functions need to be located inside the macro to ensure they work when called. On the other hand, if those functions we create may be useful in different macros, we can add them to a library of functions. This library is a txt file inside the macro's directory of *ImageJ*. If it does not exist, it can be created and then added as many functions as required. The functions included in the *Library.txt* file can be called directly from inside the macros without having them specifically written within the code.

3.3 Further Improvements

The macro developed in Subheading 3.2 performs the analysis of the sample image used in this chapter. It splits the three channels, obtains the ROIs from the red channel, and finally counts the number of blue dots in each ROI checking at the same time if each ROI has green signal inside. There are other functionalities not required in this macro that however are worth mentioning, particularly those meant to apply a macro over a collection of images and those needed to add user interaction capabilities. In the following sections, these functionalities are covered.

1. Running the macro over a collection of images.

   Applying a macro to a collection of images can be done in *Fiji* in several ways. One way consists in using the batch processing tool at [Process > Batch > Macro...]. Input and output folders must be specified, and the macro loaded or copied in the batch processing window. A similar option is found at [Process > Multiple Image Processor]. Here the macro code can only be loaded from a file, and the different functions

```
1 input = getDirectory("Select a source directory");
2 list = getFileList(input);
3 for (i=0; i< lengthOf(list); i++){
4   if (endsWith(list[i], ".tif")){
5     open(list[i]);
6     //processing commands
7   }
8 }
```

Fig. 15 ijm instructions to specify a folder, obtain the name of its files, and check if they are images. If the name ends with ".tif," the image will be opened and processed

contained in it cannot use image names to work. A more flexible alternative is to customize the macro by controlling the input directory in order to process its files.

In Fig. 15, the *getDirectory()* function (line 1) gets the path to a directory, so that the files contained in it can now be addressed using the *getFileList()* function (line 2). This function requires that the input directory is given as an argument and returns an array – here stored in the variable "list" – where the names of the files are stored as strings. Then, a loop can be used to work on the listed files. In line 3 (Fig. 15) a *for* loop is initialized which limits the number of iterations to the length of the "list" array. In case the input folder contains other nonimage files, a conditional statement (line 4–7) should be added to ensure that only the images are processed. The code in line 4 checks whether each file retrieved from the array ends with ". tif," meaning it is an image. If true, the file will be opened (line 5) and processed (line 6 represents all the commands needed for the analysis); otherwise it will be discarded, and the next file evaluated.

2. Adding user interaction to create semiautomatic macros.

   Complete automation is sometimes not possible, particularly when some user action or decision is required at some point of the process. Furthermore, there are cases where the input values must be modified for some functions to work. In these cases, it is highly recommended that changing parameters/values inside the code is avoided; instead, functions that allow the user to introduce them in run time should be added. Such interaction can be handled in two ways: (1) by setting a function that halts the macro and asks the user for some action or (2) by writing a block of code that asks in run time for some parameters to be used inside the code.

```
1 waitForUser("Do something");
```

Fig. 16 ijm function that halts the macro to ask the user for some interaction with the analysis

The *waitForUser* function pauses a macro asking for some action to perform (Fig. 16). It will pop up a dialog window with an *OK* button to be pressed once the action is done. The macro will stop if the dialog is closed without pressing the *OK* button.

Additionally, dialog boxes are data input dialog windows that can be used to ask for some values required in run time. Their functions are mainly divided in two categories: (1) those that add empty fields, menus, or choices and (2) those that get the values entered and the options selected. More information about dialog boxes can be found in [7]; an example is also found in the macro used to create the sample image [8].

4 Notes

1. By convention, when using more than one term to define a variable, these are written together, without spaces, all words after the first beginning in uppercase (e.g., nameOfTheImage).
2. To separate the different image channels, there is a function called *Split channels* at [Image > Color > Split Channels]. When applied on composite images (where each channel is treated as a different layer), *Fiji* renames the separated channels with a "C" plus the channel position. A hyphen follows, linked to the original image name: "C1-Random pattern.tif." When the function is applied on RGB images, the output single channels take the original name followed by the corresponding channel color inside parentheses: "Random pattern.tif (blue)."
3. Functions are defined by a name followed by parentheses "()." While some functions work without any additional information, others need arguments. These are parameters that functions will use inside their code and must be specified inside the parentheses. The *selectWindow()* function, for example, requires a string as an argument, which has to be the name of the window to select. When more than one argument is required, they are separated by comas inside the parentheses.
4. Strings can be manipulated to different aims, e.g., to concatenate them with variable identifiers or to extract information out of them, such as the first position where a character appears, the length of a string, or the presence of a specific pattern. In addition, strings can be divided in substrings which may be helpful to remove, for instance, the file extension contained in the image name (Fig. 17).

```
1 title = "Random pattern.tif";
2 name = substring(title, 0,
  lengthOf(title)-4);
```

Fig. 17 ijm function to create a substring

The *substring(string, first, last)* function shown in Fig. 17 shortens the string contained in "title" by cutting the last four characters. The last character of any string is obtained using the function *lengthOf(string)*. The four characters subtracted from the end of the original string correspond to the ".tif" extension. The last position that remains in the resulting substring would be that before the position actually set as "last." As defined in line 1 (Fig. 17), "title" contains "Random pattern.tif"; therefore the variable "name" would store the string: "Random pattern." We can obtain this same result using the function *File.nameWithoutExtension*, but it only works on the last file opened.

5. By convention the lines of the code encircled by curly brackets are indented with respect to the control structure or function that includes them. This improves the readability of the code.

6. Among other advantages, *Fiji* offers a script editor that color codes the different commands, values, and annotations, making it very easy to follow and debug the script. Additionally, the *Fiji* script editor also allows writing in other programming languages, such as Java, Python, etc., which are required when more sophisticated procedures need to be performed.

7. Upon running, a plug-in code appears in the *Macro Recorder* window; however, some third-party plug-ins offer additional functionalities through their macro extensions. An example of this is the Bio-Formats plug-in [11] that allows for reading datasets from different microscope brands (i.e., different file formats) and has a list of macro extensions available at [Plugins > Bio-Formats > Bio-Formats Macro Extensions]. They are helpful, for instance, when dealing with files that contain more than one image series, each of which can be opened and processed independently from a macro.

8. The binary images in *ImageJ* represent the value 1 by 255, which is the maximum value in the 8-bit scale.

9. The color coding used in the script editor of *Fiji* is the following: Green depicts annotations; magenta represents strings; brown indicates built-in macro functions; blue is used for numbers, Booleans, control structures (loops and conditional statements), and commands such as *macro*, *function*, and *return*; and black depicts variable identifiers and user-defined functions.

10. Alternatively, empty arrays can be created using (*newArray(0)*) and then its length modified by using the function *Array. resample(array, length)*, where "array" is the name of the array to enlarge and "length" is the new length it will take.
11. The positions in the list of ROIs start by 0; e.g., if there are 10 positions, they will range from 0 to 9. The same applies to the arrays, the list of results in the *Results* table, and the characters in a string. On the other hand, the slices in a stack of images start by 1.
12. The increment of variable "i" can be set as i = i + 1 which in programming language is shortened to i++. They are equivalent expressions.
13. The tolerance in the *Find Maxima* function is the minimum difference in gray values between a peak of maximum intensity and its surrounding pixels; it is used to decide which peaks are a true maxima. Since in *ImageJ* binary images this difference is higher than 1, this value can be used as the tolerance.
14. The block of code shown in Fig. 18 allows modifying the Results table in case it is already open (checked in line 1).

    First, the results in the table are cleared (line 2). Then, a "for" loop is initiated to go through all the positions in the arrays (line 3). Inside the loop, *setResult("column_name", row, value)* prints the corresponding value in each row of the table. In case the column name does not exist, it is created. The "i" variable will be used as the row number and as the index to retrieve the array values. Next, the "i" index is used to indicate the cell number which is increased by 1 to avoid having 0 as an identifier of a cell (line 4). Upon loop exit, the view of the row numbers is removed (line 8) as the cell number will be used as

```
1 if(isOpen("Results")){
2   run("Clear Results");
3   for(i=0; i<lengthOf(arrayDots); i++){
4     setResult("Cell number", i, i+1);
5     setResult("Number of Dots", i, arrayDots[i]);
6     setResult("Green Positive", i, arrayGreen[i]);
7   }
8   setOption("ShowRowNumbers", false);
9   updateResults;
10 }
```

Fig. 18 Block of code to modify the *ImageJ Results* table

```
1 output = input+"Results"+File.separator;
2 File.makeDirectory(output);
```

Fig. 19 ijm instructions to create a folder named "Results" inside another

```
saveAs("Tiff", output+name+".tif");
```

Fig. 20 ijm function used to save an image in "tif" format

the row indicator. Finally the *Results* table is updated to show all the values.

15. The Results folder can be created inside the macro in case it does not exist (Fig. 19).

    First, a new folder named "Results" is created inside another one stored in the variable "input" (line 1). *File.separator* is the function that adds a new division in the directory path of the system. The new path is stored in the variable "output." Finally, the function *File.makeDirectory* (line 2) creates the folder in the specified path.

16. The name of a file in a computer system consists of its full path inside the computer. In this sense, the path to a directory concatenated with the name of the file used in the macro (output + name) specifies the full path to save a file using ijm. In the macro, the default name that *ImageJ* uses for the list of ROIs, i.e., RoiSet, is concatenated to the "output" and "name" variables.

17. The list of ROIs is saved by default as a zip file, and the results table can be saved as a comma separated value (csv), an excel spreadsheet (xls), or a text (txt) file. Images can also be saved in different formats such as tiff, jpeg, etc. In all cases the *saveAs* function should be used (Fig. 20).

Acknowledgments

I acknowledge Elena Rebollo for her helpful discussion on the manuscript.

References

1. Russ JC (2004) Seeing the scientific image. Proc RMS 39(2):1–15
2. Abràmofff MD, Magalhães PJ, Ram SJ (2005) Image processing with *ImageJ* Part II. Biophoton Int 11:36–43. https://doi.org/10.1117/1.3589100
3. *ImageJ*. http://*ImageJ*.net/index.html. Accessed 9 Mar 2018
4. *ImageJ* Forum. http://forum.*ImageJ*.net/. Accessed 20 Mar 2018

5. *ImageJ* | Mailing List Archive. http://*ImageJ*.1557.x6.nabble.com/. Accessed 20 Mar 2018
6. Schindelin J, Arganda-Carreras I, Frise E et al (2012) *Fiji*: an open-source platform for biological-image analysis. Nat Methods 9:676–682. https://doi.org/10.1038/nmeth.2019
7. Built-in macro functions. http://*ImageJ*.net/developer/macro/functions.html. Accessed 9 Mar 2018
8. Manel Bosch ijm-Macros. https://github.com/manelbosch76/ijm-Macros. Accessed 21 Mar 2018
9. RandomSamplePerimeterMethod macro. https://*ImageJ*.nih.gov/ij/macros/RandomSamplePerimeterMethod.txt. Accessed 19 Mar 2018
10. RandomSampleAndMeasure macro. https://*ImageJ*.nih.gov/ij/macros/RandomSampleAndMeasure.txt. Accessed 19 Mar 2018
11. Linkert M, Rueden CT, Allan C et al (2010) Metadata matters: access to image data in the real world. J Cell Biol 189:777–782. https://doi.org/10.1083/jcb.201004104

Chapter 5

Automated Macro Approach to Quantify Synapse Density in 2D Confocal Images from Fixed Immunolabeled Neural Tissue Sections

Elena Rebollo, Jaume Boix-Fabrés, and Maria L. Arbones

Abstract

This chapter describes an *ImageJ/Fiji* automated macro approach to estimate synapse densities in 2D fluorescence confocal microscopy images. The main step-by-step imaging workflow is explained, including example macro language scripts that perform all steps automatically for multiple images. Such tool provides a straightforward method for exploratory synapse screenings where hundreds to thousands of images need to be analyzed in order to render significant statistical information. The method can be adapted to any particular set of images where fixed brain slices have been immunolabeled against validated presynaptic and postsynaptic markers.

Key words Synapse density, ImageJ macro language, Puncta segmentation, Nuclei segmentation, Chromatic shift correction

1 Introduction

Synapses are the points of communication between neurons whereby electrical or chemical signals are transmitted from one neuron to another. Anatomically, they are composed of a presynaptic and a postsynaptic terminal, respectively, located in different neurons and separated by a gap called synaptic cleft, whose width ranges between 15 and 200 nm. The synaptic vesicles undergo calcium-dependent fusion with the presynaptic membrane, thus releasing their contents to the synaptic cleft, where they interact with receptors located at the postsynaptic membrane. These receptors are held in place by a vast protein scaffold, which contains almost 1500 proteins (reviewed by Harris and Weinberg [1]).

Electronic supplementary material: The online version of this chapter (https://doi.org/10.1007/978-1-4939-9686-5_5) contains supplementary material, which is available to authorized users.

Elena Rebollo and Manel Bosch (eds.), *Computer Optimized Microscopy: Methods and Protocols*, Methods in Molecular Biology, vol. 2040, https://doi.org/10.1007/978-1-4939-9686-5_5,

There is an increasing body of evidence suggesting that altered brain network activity underlays many of the most common neurological diseases [2, 3]. To a large degree, these advances have been made possible thanks to the development of new image analysis methodologies that facilitate the quantification of synapse morphology, number, and distribution in neural cells and tissues visualized under the light microscope [4–7]. Some of the main available quantification techniques focus on different aspects of brain connectivity, such as the distribution of individual synaptic proteins onto a cell of interest, the assessment of cell-to-cell contacts, and the relationship between different synaptic proteins (reviewed in Ref. [8]). The first two are usually confined to a small subset of cells, generally stained by a cell fill marker and a single synapse marker. The last, however, tackles the spatial proximity between a presynaptic and a postsynaptic marker, as a way of identifying *bona fide* synaptic sites.

Fluorescence confocal microscopy remains the workhorse technology for fixed tissue imaging up to 100 μm and, for the foreseeable future, will remain one of the dominant tools for studying synapses and their associated proteins in the central nervous system. On the one hand, many antibodies against different presynaptic and postsynaptic components are available, which have been extensively validated to locate synapsis [9]. On the other hand, modern confocal microscopes are nowadays readily available in most laboratories, and many different customized routines can be easily established to perform multidimensional data acquisitions. At the light microscope, immunolabeled postsynaptic terminals generally appear as well-defined puncta, whereas presynaptic vesicles deliver a more irregularly shaped pattern (*see* Fig. 1). Addressing their spatial association under the confocal microscope often relies, due to the diffraction-limited resolution, on inferring methodologies such as colocalization [8, 10]. However, correlation coefficients between presynaptic and postsynaptic markers tend to be low due to different reasons: (1) presynaptic markers mostly label all synapse vesicles, (2) immature postsynaptic puncta may not yet form synapses [11], and (3) many postsynaptic markers only label a subset of postsynaptic sites. A more direct approach consists in estimating the degree of apposition between the two markers by defining regions of interest (ROIs) around the postsynaptic puncta; the intensity of the presynaptic marker can then be measured within these ROIs in order to select those over an intensity threshold that qualify as synapses. On this line, masking puncta in 3D has probed a highly sensitive technique [5].

This chapter describes an open-source automated procedure, developed as a *Fiji* [12] macro language script, to estimate synapse densities in 2D fluorescence confocal microscopy images from fixed brain slices immunostained against validated presynaptic and postsynaptic markers. Although such approach does not render accurate

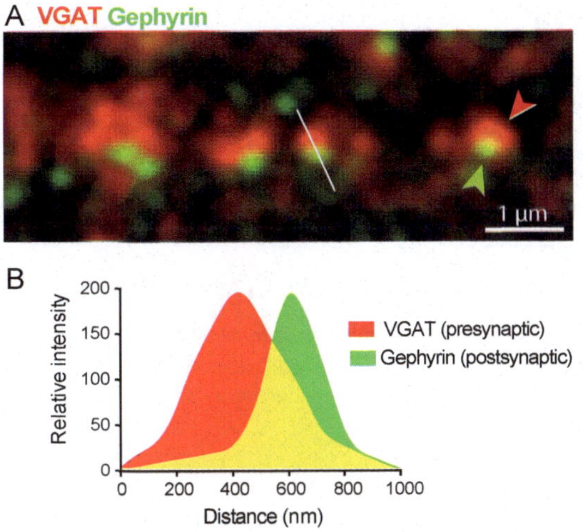

Fig. 1 Synapses identification in confocal microscopy. (**a**) Confocal microscopy image showing inhibitory synapses as stained with antibodies against the postsynaptic scaffold protein Gephyrin (green arrowhead) and the presynaptic vesicle component VGAT (red arrowhead). The white line indicates the ROI used for the intensity plot shown in panel **b**. Bar is 1 μm. (**b**) Graph showing the intensity plots of the presynaptic (red) and postsynaptic (green) signals along the line drawn in panel **a**. The distance between the two peaks lies beyond the resolution limit of the confocal microscope. The overlay region is shown in yellow

information on synapse distribution or morphology, its straightforwardness makes it suitable to address altered connectivity in exploratory and high-content studies where, for instance, different genetic backgrounds or drug treatments are compared. It is known that subtle changes in the structure of neuronal circuits may pass unperceived and nevertheless have seriously detrimental effects [13]. Detecting such slight differences by estimating synapse densities requires systematic random sampling of brain regions involving a high number of images in order to obtain robust comparisons that yield enough statistical information. An automated image analysis procedure is therefore required to extract synapse densities from hundreds to thousands of images.

The image processing workflow (Fig. 2) includes two main segmentation steps. On the first one, the nuclear staining is used to retrieve the nuclear boundaries (Fig. 2b). Since random sampling of tissue sections will most likely generate images having different number of nuclei, this step is necessary to correct the actual area value that will be used to normalize the number of synapses per image (Fig. 2c). The second main segmentation step is focused on postsynaptic spotlike signal. Spot detection is a fundamental procedure for biologists in many imaging applications. One of the best point detection methods is the *Laplacian of*

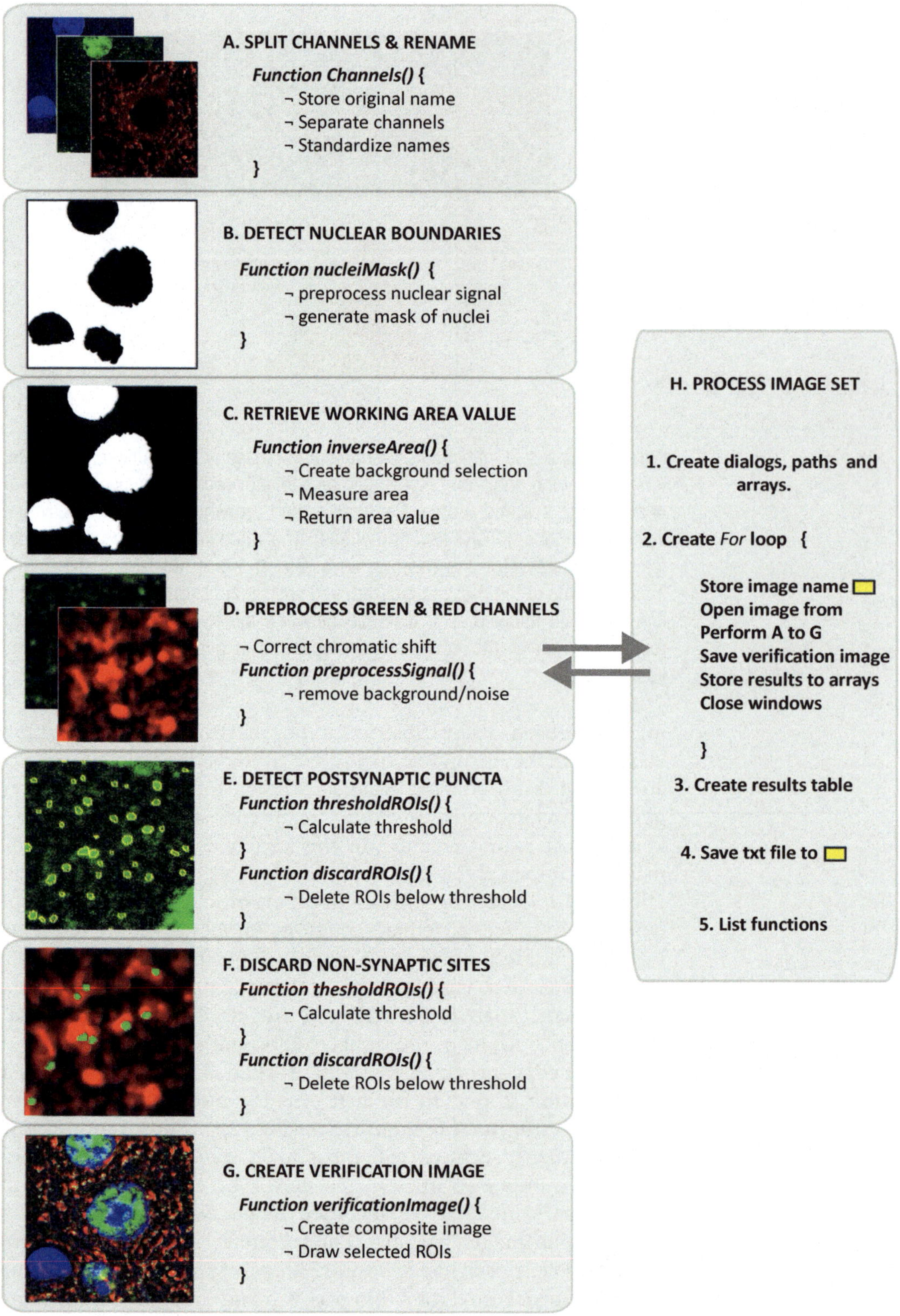

Fig. 2 Image processing workflow. Schematic showing the main steps that compose the imaging pipeline. (**a**–**g**) Main steps necessary to count synapses in a single image. (**h**) **Steps 1–5** represent the automation steps required to apply the pipeline to a set of images

Gaussian (*LoG*) algorithm [14], which is used here to enhance the signal of the postsynaptic puncta. Once detected, further operations between the so-obtained binary masks are used to eliminate unspecific nuclear signal particles (Fig. 2e). After segmentation, synapse detection is performed by a double discrimination test. First, low-quality puncta are discarded based on the intensity density of the particles detected. This step helps eliminate out-of-focus puncta as well as weak off-target particles. Second, the remaining puncta are tested for the presence of the presynaptic label. These two steps are performed using global thresholds, so that the same criteria for synapsis discrimination are applied to all the images and conditions compared.

This chapter is developed using as example inhibitory GABAergic synapses [3, 11, 15], identified by the postsynaptic scaffold protein Gephyrin and the presynaptic vesicular GABA transporter VGAT. The protocol rationale can nevertheless be extended to prospective studies of excitatory glutamatergic synapses [1, 16]. Sample images are given for both synapse types, together with their corresponding macros.

2 Materials

2.1 Images

1. 2D confocal images of mice brain slices immunolabeled against validated presynaptic and postsynaptic components and containing DNA staining. Four sample images are provided with the chapter to test the protocol (Supp. Material and GitHub public repository [17]). The first two ("synapses_inh_01.lsm" and "synapses_inh_02.lsm") show GABAergic synapses as shown by immunostaining using antibodies (Abs) against the postsynaptic scaffold protein Gephyrin and the presynaptic vesicle component VGAT. The other two ("synapses_exc_01.lsm" and "synapses-exc_02.lsm") contain glutamatergic synapses by immunostaining using primary Abs against the postsynaptic scaffolding protein Hommer and the presynaptic vesicular glutamate transporter VGLUT. Alexa Fluor-488 and Alexa Fluor-568 were, respectively, used as secondary Abs. The DNA was stained using DAPI. *See* **Notes 1** and **2** for sample preparation and acquisition set up tips.
2. 2D confocal image of sub-diffraction fluorescent microspheres labeled with fluorophores that absorb/emit in ranges similar to those of the secondary Abs used in the biological samples and acquired in the same conditions (*see* **Note 3**). A sample image ("Beads.tif") is provided as supplemental material that corresponds to the same example experiment.

2.2 Software and Macros

1. *ImageJ* is an open-source image processing and analysis platform [18] originally developed at the National Institutes of Health (Bethesda, Maryland, USA). In this chapter we use the *Fiji Life-line 22 December 2015* distribution [12]. This version can be downloaded at [19] and requires *Java 6*. The description of the *ImageJ* built-in macro functions used can be found at [20].
2. Macros (*see* **Note 4**) developed as indicated in this methods protocol. Two macros are provided as supplemental material: one to calculate the chromatic shift ("Chromatic_Shift_Calculator.ijm") and another that estimates the density of synapses in a set of *n* images ("Synapse_Counter.ijm"). Both these macros are available at [17].

3 Methods

This method first describes how to calculate the *xy* chromatic shift in a control image, to be used in the correction of the biological images if necessary. Second, the main step-by-step manual workflows aimed to estimate synapse density are provided, together with the rationale to build a customized macro that can be applied to any particular set of images. Finally, a section is dedicated on how to perform the analysis using the main macro provided in this chapter.

3.1 Calculating Chromatic Shift

In the lateral dimension, chromatic aberration provokes that different wavelengths are imaged at slightly shifted lateral (*xy*) positions. This section explains the manual steps necessary to obtain, from a 2D reference image of fluorescent beads, the *x* and *y* offset values that will later on be used to correct the biological images using translation (see Fig. 3). Alternatively, the provided macro

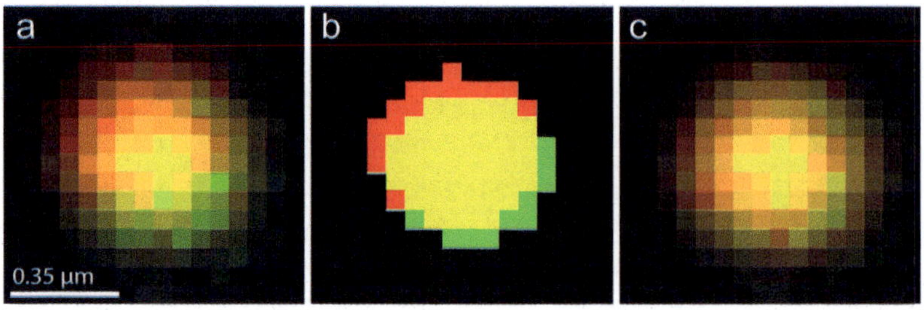

Fig. 3 Chromatic shift correction. (**a**) Confocal microscopy image showing one 100 nm microsphere fluorescing in red and green. (**b**) Overlay of the two channels after signal segmentation; notice how the green signal is diagonally shifted downward and toward the right with respect to the red. (**c**) The red channel has been translated in *x* and *y*, using bilinear interpolation, thus correcting the mismatch. The bar indicates 350 nm

("Chromatic_Shift_Calculator.ijm") can be used to automatically calculate the *x* and *y* offset values on the control image.

1. Open sample image "Beads.tif" in *Fiji* by [File > Open...] or by drag and drop of the file on the *Fiji* bar (*see* **Note 5**). To follow the manual pipeline, go to **step 2**; otherwise jump to **step 15** to use the provided macro.
2. Remove image calibration at [Image < Properties. . .]; set *pixel* as unit of length, and use *1* for both pixel width and height.
3. Separate channels at [Images > Color > Split Channels].
4. Rename each independent channel by [Image > Rename. . .]; we will here name the images "red" and "green", according to their respective channel colors (*see* **Note 6**).
5. Select one channel (e.g., "red"), and apply a *LoG* filter (*see* **Note 7**) at [Plugins > Feature Extraction > FeatureJ > FeatureJ Laplacian]; choose a smoothing scale radius of 2.
6. Convert the image to 8-bit at [Image > Type >8-bit].
7. Threshold the image at [Image > Adjust > Threshold. . .]; choose an appropriate method (the *Default* method works fine for the provided images; otherwise *see* **Note 8**); unclick the option *Dark background* and hit *Apply.*
8. On the binary mask generated, apply the *Watershed* algorithm at [Process > Binary Watershed] to further separate contiguous objects (*see* **Note 9**).
9. Select *centroid* as parameter to measure at [Analyze > Set Measurements. . .].
10. Detect objects at [Analyze > Analyze Particles. . .], using the options *Exclude on edges* and *Add to manager* (*see* **Note 10**).
11. By clicking first *Deselect* and then *Measure* at the *ROI Manager* menu, the results table will open containing the centroid position of all the detected beads. This table can be saved as a text file to be opened later on a spreadsheet for further analysis.
12. Repeat **steps 5–11** on the other channel (e.g., "green").
13. Once the centroid positions of all the beads have been obtained in both channels, the individual *x* and *y* shifts can be computed by subtracting the values from one channel with respect to the other. Then, the *x* and *y* average shifts will be used to translate one of the channels, thus correcting the chromatic shift (*see* **Note 11**).
14. To test the obtained *x* and *y* shift values, select the original channel "red" and translate it at [Image > Transform > Translate. . .]; introduce the *x* and *y* offset values in pixels, and choose bilinear interpolation (*see* **Note 11**). The translated "red" channel and the original "green" channel can now be merged

at [Image > Color > Merge Channels. . .]. The result should be similar to that shown in Fig. 3.

15. To perform the previous steps automatically, open the macro "Chromatic_Shift_Calculator.ijm" by dragging it to the *Fiji* bar. With the image opened and selected, hit the macro editor *Run* button. The macro will retrieve the average *x* and *y* shift values directly in pixels. These values will be used to correct the biological images in the next section.

3.2 Create Macro to Quantify Synapse Densities in a Set of n Images

The following subsections explain the main image processing steps (Fig. 2) in the form of manual step-by-step imaging protocols; as a general rule, the macro recorder can be kept opened during manual execution, so that preliminary scripts are created (revise **Note 4**). Further developed scripts are provided in each subsection, which, sequentially executed, will shape the final macro. Too general or repetitive pipelines have been edited as user-defined functions (*see* **Note 12**). Such scripts can be executed by copy/paste in the *Fiji* script editor.

3.2.1 Channel Preparation

At this initial step, the image needs to be manually opened, and each fluorescent channel must be prepared as an independent image with a short descriptive name that can be easily encoded in the macro script. The refined macro script that automates the steps below is provided in Fig. 4:

1. Open the sample image "Synapses_inh_01.tif" in *Fiji* by [File > Open...] or by drag and drop of the file on the *Fiji* bar. The image includes calibration.
2. Separate channels at [Images > Color > Split Channels].

```
//SEPARATE CHANNELS & STANDARDIZE NAMES
channels();
function channels(){
        //Store original file name as a variable
        rawName = getTitle();
        //Separate channels
        run("Split Channels");
        //Rename channels:
        selectWindow("C3-"+rawName);
        rename("blue");
        selectWindow("C2-"+rawName);
        rename("green");
        selectWindow("C1-"+rawName);
        rename("red");
}
```

Fig. 4 Prepare channels. The script executes a user-defined function called *channels()*; the function is described between curly brackets: first, the variable *rawName* stores the original name of the file; the remaining commands split the channels and rename them. At this point the code still requires the image to be opened manually. To execute this code, copy it in the *Fiji* script editor and hit *Run*

3. Rename each independent channel by [Image > Rename...]; we will use the names "red," "green," and "blue" according to the channel's colors (revise **Note 6**).

3.2.2 Nuclei Boundaries Segmentation

Nuclei segmentation, based on DNA staining, is widely used to retrieve nuclear boundaries in images from fixed tissues or cells. A preprocessing step is necessary in order to homogenize the background and smooth out the shape of the objects to be defined (*see* **Note 13**). Automatic intensity thresholding is then applied to detect the nuclear objects, which are thereafter converted into a binary image also called mask. Irregular DNA staining, due to variations in experimental conditions or cell cycle stage, may hinder the delimitation of the nuclear boundaries (*see* **Note 14**). In cases where a second nuclear signal exists, such as unspecific nuclear Gephyrin in this example, a segmentation strategy can be used that combines the preliminary masks obtained from both channels, so that a more complete nuclear mask is generated (*see* Fig. 5); small nonnuclear particles can be further filtered by size to create the ultimate mask strictly containing the segmented nuclei. The codes to perform this step are provided in Fig. 6.

1. Duplicate channel "blue" at [Image > Duplicate...] in order to leave the original image untouched; rename it at the *Duplicate* menu window so as to identify its purpose, e.g., "blueNucleiMask."
2. Apply a *Gaussian* filter at [Process > Filters > Gaussian Blur...], using a radius of 2.
3. Threshold the image at [Image > Adjust > Threshold...]; the best result for this example image is obtained using the *Li* method [21]. Set the background option so as to produce a binary image where objects are black and have an intensity value of 255. Hit *Apply.*
4. *Fill holes* at [Process > Binary > Fill Holes...]. This step is used to complete the nuclear mask by removing groups of background level pixels within the selected objects.
5. Perform **steps 1–4** in channel "green"; we have renamed this image as "greenNucleiMask" (according to the script pipeline in Fig. 6); use a radius of 5 for the *Gaussian* filter.
6. Combine the created binary images "blueNucleiMask" and "greenNucleiMask" at [Process > Image Calculator...]; choose both images as *input 1* and *2*, respectively, in the pop-up menu, and select *OR* as operator (*see* **Note 15**). By unclicking the option *Create new window*, the result will be generated on top of the first input image; keep this one, and close the second input image "greenNucleiMask."

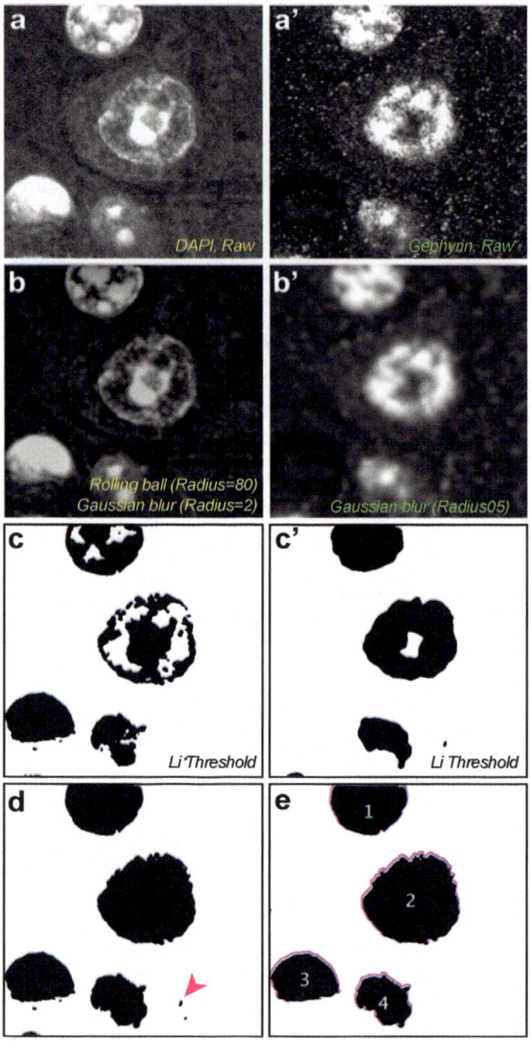

Fig. 5 Nuclei segmentation using two nuclear signals. (**a**–**a**′) Raw "blue" (DAPI) and "green" (Gephyrin) channels, respectively. (**b**–**b**′) Preprocessed "blue" and "green" channels, after background subtraction and filtering. (**c**–**c**′) Binary masks obtained from **b** and **b**′. (**d**) Preliminary mask formed by the sum of **c** and **c**′. (e) Final mask containing only the nuclei, including those touching the image edges; small particles (arrowhead) have been filtered by size

7. On the so modified image "blueNucleiMask," detect the nuclei using the command [Analyze > Analyze Particles…]; select a minimum size for nuclei discrimination (a value of 4 μm^2 is enough for the sample images) in order to discard small particles; untick the option *Exclude on edges* to keep the nuclei that touch an image edge; tick the option *Include holes* to avoid detecting regions within regions; the option *Add to manager* is not needed as the only purpose here is to generate a mask containing just the nuclei; alternatively, select *Mask* as show

```
//DETECT NUCLEAR BOUNDARIES
//Preprocess blue & green channels and convert to mask
nucleiMask("blue", 2);
nucleiMask("green", 2);
//Generate complete mask of nuclei
imageCalculator("OR", "blueNucleiMask", "greenNucleiMask");
run("Analyze Particles...", "size=4.00-Infinity show=Masks clear include");
rename("maskOfNuclei");
//Close extra windows
selectWindow("blueNucleiMask ");
run("Close");
selectWindow("greenNucleiMask ");
run("Close");

function nucleiMask(image, gaussianRadius) {
        selectWindow(image);
        run("Duplicate...", "title="+name+"NucleiMask");
        run("Gaussian Blur...", "sigma=gaussianRadius");
        setAutoThreshold("Li dark");
        run("Convert to Mask");
        run("Fill Holes");
}
```

Fig. 6 Detect nuclear boundaries. The function *nucleiMask(image, gaussianRadius)* is defined at the end of the script encircled in curly brackets; it duplicates an image, identified by its name through the first argument; then, it applies a *Gaussian* filter using the radius provided as a second argument; finally, it thresholds the image using the *Li* method and fills the binary mask holes to complete the objects. From beginning to end, the script code first applies this function to channels "blue" and "green" using the specified arguments; then, the two masks generated are combined and further filtered to create the ultimate mask containing only the nuclei. To execute this code, copy it in the *Fiji* script editor and hit *Run*

option and hit *OK*: a new binary image will appear containing only the desired objects; rename this image as, e.g., "maskOfNuclei" (*see* Fig. 5e), and close the previous mask.

3.2.3 Working Area Calculation

The number of synapses must be converted into a biologically meaningful value, such as synapse density, that can be compared between different conditions. Since the number of nuclei per image changes in a random tissue sampling, the actual working area must be calculated per image, excluding the nuclear regions. This subsection describes how to retrieve the actual working area value in μm^2. Figure 7 provides a script that calculates the area value in μm^2 and stores it to be used at a later step.

1. At [Image > Adjust > Threshold...], threshold the *maskOfNuclei* image created in the previous section; select the *Default* method in the pop-up menu, and tick *Dark background* so that the objects are now selected; do not hit *Apply* yet.
2. Select *area* as measurement's parameter at [Analyze > Set Measurements...].
3. At [Analyze > Analyze Particles...], tick the options *Display results* and *Clear results*, and hit *OK*; the results window will pop up containing the measurement of the inverse nuclei area in μm^2.

```
//RETRIEVE & STORE WORKING AREA VALUE
area = inverseArea(image);
function inverseArea(image) {
        selectWindow(image);
        setAutoThreshold("Default dark");
        run("Set Measurements...", "area redirect=None decimal=2");
        run("Analyze Particles...", "display clear");
        area = getResult("Area", 0);
        return area;
}
```

Fig. 7 Calculate working area value in μm^2. The function *inverseArea(image)* is defined, its code between curly brackets. This function takes a binary image, specified by the name in the argument, thresholds the background, and calculates its area; then, it returns the area value (by means of the return command) to be stored as a variable called *area*, from which the function is executed on the first line. To execute this code, copy it in the *Fiji* script editor and hit *Run*

3.2.4 Synaptic Signal Preprocessing

Before synapse detection, presynaptic ("red" channel) and postsynaptic ("green" channel) signals must be restored. This includes (1) chromatic shift correction, to undo the mismatch between channels, and (2) background correction and filtering, depending on image quality. The script to perform this step is provided in Fig. 8.

1. Select channel "red," and apply translation at [Image > Transform > Translate...]; use the *x* and *y* offsets calculated in Subheading 3.1, and select bilinear interpolations (revise **Note 11**).
2. For background correction, use the *Rolling ball* algorithm at [Process > Subtract Background...]; a radius of 15 is fine for both "red" and "green" images.
3. Further filtering using [Process > Filters > Median] may be beneficial for noisy signals, as it is the case for the "red" channel; a radius of 2 is enough.

3.2.5 Postsynaptic Puncta Detection

Puncta segmentation is based on the dot-enhancing capabilities of the *LoG* algorithm (Fig. 9). The idea behind this strategy is to detect as many dots as possible, so that they can be filtered afterward based on a uniform quality criterion. Binary conversion is directly used to create a mask of the segmented puncta. Also, operations between binary images are explained to eliminate undesired nuclear spots. Further filtering using *size*, *circularity*, and positioning criteria is necessary to eliminate artifactual or incomplete particles. The codes to perform this step are provided in Fig. 10.

1. Select channel "green," and apply the *LoG* filter at [Plugins > Feature Extraction > FeatureJ > FeatureJ Laplacian]; choose a smoothing scale radius of 2 (revise **Note 7**).

```
//PREPROCESS PRE & POSTSYNAPTIC SIGNALS
// Correct chromatic shift
selectWindow("red");
run("Translate...", "x=0.35 y= 0.81 interpolation=Bilinear");
//Preprocess signals
preprocessSignal("green", 15, 0);
preprocessSignal("red", 15, 2);

function preprocessSignal(image, rollingRadius, medianRadius) {
        selectWindow(image);
        run("Subtract Background...", "rolling="+rollingRadius sliding);
        run("Median...", "radius="+medianRadius);
}
```

Fig. 8 Preprocess pre- and postsynaptic signals. First, translation is applied to one of the channels to correct for chromatic mismatch. Then, a user-defined function called *preprocessSignal(image, rollingRadius, medianRadius)* is used to remove noise in both images; the image is selected by name using the first argument, and the parameters for the *Rolling ball* algorithm and the *Median* filter are selected using the other two arguments. To execute this code, copy it in the *Fiji* script editor and hit *Run*

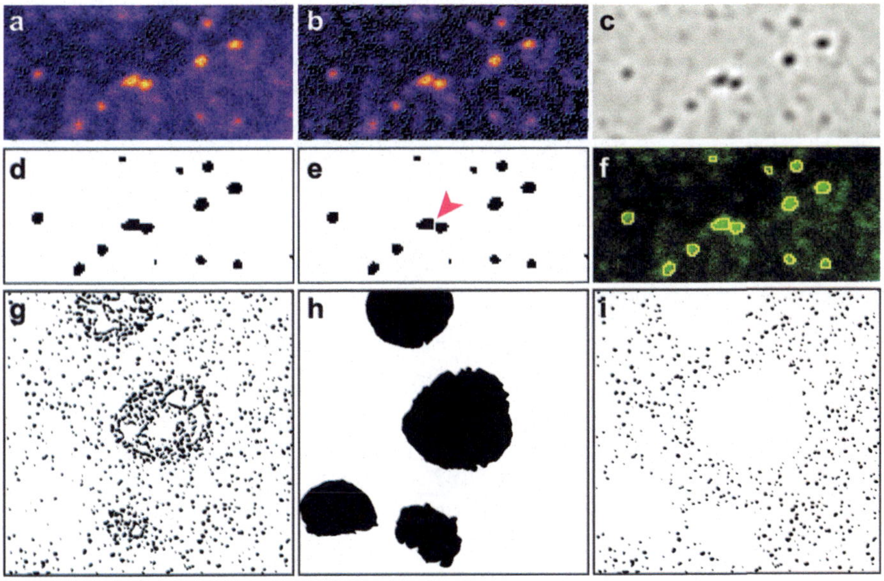

Fig. 9 Puncta segmentation. The upper panel contains an example fragment of the channel image containing postsynaptic puncta (**a**); the *fire* LUT has been applied for better puncta visualization. The background is first removed by a *Rolling ball* algorithm (**b**); then, the *LoG* transformation is applied (**c**); the resulting image is binary converted; (**d**) and the *Watershed* algorithm is further used to separate contiguous objects (**e**, arrowhead); finally, the segmented puncta are detected via the *Analyze particles* menu. The yellow outlines represent the ROI selections, drawn onto the original image. The lower panel shows the mask containing the postsynaptic puncta (**g**) and the mask containing the nuclei (**h**); both masks are combined to deliver the final mask containing only the cytoplasmic puncta (**i**)

```
//DETECT POSTSYNAPTIC PUNCTA
detectPuncta( "green", "maskOfNuclei" , 2);

function detectPuncta(image1, image2, logRadius) {
        selectWindow(image1);
        run("FeatureJ Laplacian", "compute smoothing=" +logRadius);
        setOption("BlackBackground", false);
        run("Convert to Mask");
        run("Watershed");
        //Remove nuclear spots
        imageCalculator("Subtract", image1+" Laplacian", image2);
        roiManager("reset");
        run("Analyze Particles...", "size=0.00-2 circularity=0.7-1.00 exclude clear add");
        //Close masks
        selectWindow(image1+" Laplacian");
        run("Close");
        selectWindow(image2);
        run("Close");
}
```

Fig. 10 Detect postsynaptic puncta. The function *detectPuncta(image1, image2, logRadius)* is described between curly brackets; the two first arguments call the input images by name, being the third the radius to be specified for the *LoG* transformation. Within the function, the *LoG* algorithm is applied to enhance spotlike signals. After binary conversion, the *Watershed* algorithm is used to further separate contiguous dots. The image "maskOfNuclei," previously created, is subtracted to remove nuclear spots. Automatic detection is utilized to filter the puncta and collect their shapes to the *ROI Manager*. The previous masks are then closed, and the detected foci kept in the *ROI Manager* to be filtered in the next step. To execute this code, copy it in the *Fiji* script editor and hit *Run*

2. Convert the resulting image, "green Laplacian," into a binary image (*see* **Note 16**) at [Process > Binary > Convert to Mask].
3. Apply the *Watershed* algorithm at [Process > Binary > Watershed] in order to separate touching spots (revise **Note 9**).
4. Use the nuclear mask generated in Subheading 3.2.2 ("maskOfNuclei") to remove nuclear spots. At [Process > Image Calculator...], select the new mask "green Laplacian" as *input 1* image, the previous "maskOfNuclei" as second input image, and *Subtract* as operator. All spots located in the nuclear region will become white. The "maskOfNuclei" image can now be closed.
5. Detect particles at [Analyze > Analyze Particles...]. At the *Analyze particles* menu, select 2 μm^2 as upper *size* limit and 0.7 as lower *circularity* limit (revise **Note 10**). Tick the options *Add to manager* and *Exclude on edges* and hit *Apply*. The "green Laplacian" image can now be closed.

3.2.6 Synapse Counting

In this subsection, the selected postsynaptic puncta that do not qualify as potential synaptic sites are discarded using a double discrimination step. First, the ROIs having low-quality postsynaptic signal are eliminated; then, among the remaining ROIs, those showing no overlap with presynaptic signal are eliminated. Both discrimination steps are based on the *Integrated density* (*IntDen*) ROI content (*see* **Note 17**). The manual steps below only explain how to extract *IntDen* parameters from a list of ROIs when collated to a given image, in order to inspect for putative discrimination thresholds. However, the actual ROI discrimination necessary at this step cannot be done manually, since it would take too long. A script is provided in Fig. 11 containing two functions, one for threshold selection (*see* **Note 18** and Fig. 12) and another for ROI discrimination; both are first used to discard ROIs where the postsynaptic signal *IntDen* lies below the estimated cutoff value; a second round deletes those ROIs where the presynaptic signal *IntDen* lies below the second estimated cutoff value. The script can be run from the script editor to inspect for putative thresholds on the biological images.

1. Set *IntDen* as parameter of choice at [Analyze > Set Measurements. . .].
2. After selecting the image (e.g., "green"), hit *Deselect* and then *Measure* at the *ROI Manager* window menu. The results window will pop up containing the integrated density measurements of all ROIs. The obtained values may be inspected in any other software in order to analyze the particle distribution and decide for a cutting threshold.

3.2.7 Verification Image

An important issue in image analysis is the visual validation of the protocol. By drawing onto the image the ROIs that remain after synapses discrimination, the outcome of the protocol can be directly observed. This validation step is especially helpful when many images need to be processed, so that any possible outlier result can be tracked back. The steps below explain the manual pipeline to create a verification image; the refined script can be found in Fig. 13:

1. Create a composite image at [Image > Color > Merge Channels. . .] choosing the three independent channels "red," "green," and "blue" in their corresponding dialogs.
2. Convert the image into an RGB type at [Image > Type > RGB Color] so that the selected ROIs can be drawn in the desired color.
3. Set the foreground color to any desired by double-clicking the *Color picker* tool at the tool bar.
4. At the *ROI Manager* window, click *Deselect*, display the *more* menu, and choose the option *draw*. All selections will be automatically stamped in the composite image.

```
//CALCULATE THRESHOLD FOR PUNCTA DISCRIMINATION
thresholdGreen= thresholdROIs("green", 1);
//DISCARD LOW QUALITY PUNCTA & COUNT SELECTED PUNCTA
noPuncta = discardROIs("green", thresholdGreen);
//CALCULATE THRESHOLD FOR SYNAPSE DISCRIMINATION
thresholdRed= thresholdROIs("red", 1);
// DISCARD NON SYNAPTIC PUNCTA & COUNT SYNAPSES
noSynapses = discardROIs("red", thresholdRed);

function thesholdROIs(image, cutoff) {
        selectWindow(image);
        run("Set Measurements...", " integrated redirect=None decimal=2");
        roiManager("deselect");
        roiManager("measure");
        //Create IntDen array & obtain median
        noRois = roiManager("count");
        intDensities = newArray(noRois);
        for (j=0; j<noRois; j++){
                intDensities [j] = getResult("IntDen", j);
                }
        Array.sort(intDensities);
        lengthArray = lengthOf(intDensities);
        if (lengthArray%2==0) {lengthArray = lengthArray+1}
        t = round(lengthArray*(0.6 - 0.1*cutoff));
        threshold = intDensities[t] ;
        return threshold;
}

function discardROIs(image, threshold) {
        selectWindow(image);
        run("Select All");
        run("Set Measurements...", "mean integrated redirect=None decimal=2");
        //Discard ROIs below threshold value
        noRois = roiManager("count");
        for (j=0; j<noRois; j++){
                roiManager("select", j);
                roiManager("measure");
                value = getResult("IntDen", 0);
                if (value < threshold) {
                roiManager("delete");
                j--;
                noRois--;
                }
        run("Clear Results");
        }
        return noRois;
}
```

Fig. 11 Discard non-synaptic puncta and retrieve synapse number. The function *thresholdROIs(image, cutoff)* extracts the median value of the *IntDen* ROI distribution on a given image and returns the threshold value corrected by a cutoff value, provided as argument; such cutoff is specified from 1 to 5, which sets the threshold low at 50–90% of the distribution, respectively, at intermediate increments of 10%; the function *discardROIs(image, threshold)* discards those ROIs whose *IntDen* value on a given image (first argument) lies below a certain threshold (second argument); the function returns the number of ROIs that remain in the *ROI Manager* after its execution. The whole code performs the following tasks: (1) it calculates the median of the postsynaptic *IntDen* ROI distribution (*thresholdGreen*); (2) it uses the obtained value as threshold to discard low-quality puncta and counts the number of remaining puncta (*noPuncta*); (3) it then calculates the median of the presynaptic *IntDen remaining ROI* distribution (*thresholdRed*); and (4) finally, it uses this second threshold to discard those ROIs that do not qualify as synaptic sites and retrieves the final number of synapses (*noSynapses*). To execute this code, copy it in the Fiji script editor and hit *Run*

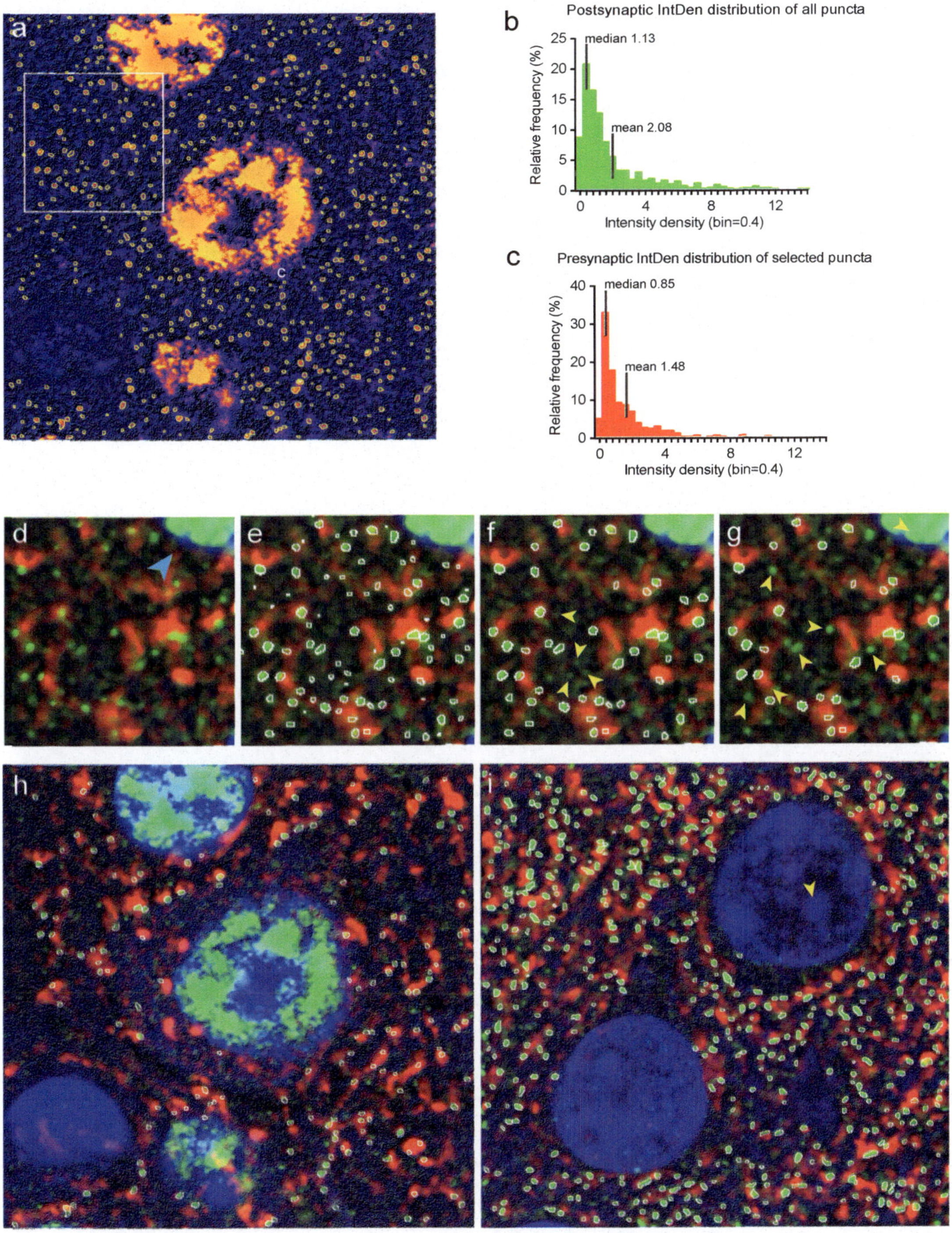

Fig. 12 Synapses threshold calibration and counting. (**a**) Postsynaptic (Gephyrin) channel showing all puncta (yellow circles) detected after segmentation; the *fire* LUT has been applied to the image in order to better show low-intensity signals. (**b**) The upper graph shows the postsynaptic signal *IntDen* distribution histogram of all puncta ROIs detected in panel **a**; (**c**) the lower graph shows the presynaptic signal *IntDen* distribution histogram of the ROIs that remain after puncta discrimination; the main descriptive statistics, mean and

3.2.8 Automate the Protocol for n Images

By sequentially executing the codes provided from Subheadings 3.2.1 to 3.2.7, the main script pipeline is created that allows to process one image. Additional programming features are necessary to be able to automatically process a set of *n* images and retrieve the final results table, where the image names, the areas, the number of selected puncta, and the final number of synapsis will be stored. This subsection provides the minimum requirements for such macro automation. The scripts are provided in Fig. 14, where the different code blocks correspond to the following required actions:

1. Create dialog boxes for the user to choose the folder where the images are stored and the folder where the results will be saved.
2. Create an array containing the list of all images; this array will allow obtaining the total number of images later on.
3. Create a dialog for the user to select the threshold modulation cutoff value.

```
//CREATE VERIFICATION IMAGE
verificationImage("red", "green", "blue");
function verificationImage(image1, image2, image3) {
        //Generate composite image containing the three channels
        run("Merge Channels...", "c1="+image1+" c2="+image2+ " c3="+image3+" create");
        run("RGB Color");
        //Draw selections into verification image
        setForegroundColor(255, 255, 255);
        roiManager("deselect");
        roiManager("draw");
        //Close windows
        selectWindow("Composite");
        close();
}
```

Fig. 13 Create verification image. The function *verificationImage(image1, image2, image3)* creates a composite image by merging three channels, provided as arguments; then it draws, onto the RGB-converted image, all the ROIs contained in the *ROI Manager*, using white color. To execute this code, copy it in the Fiji script editor and hit *Run*

Fig. 12 (continued) median, are shown in each graph. The middle panel contains an inset of the same image (depicted by the white square in a) taken as example to show synapse discrimination; (**d**) is the original three channels image inset; (**e**) shows all the ROIs detected during puncta segmentation (white circles); (**f**) depicts the ROIs that remain after puncta discrimination, using as threshold the median value of the green *IntDen* distribution; the yellow arrowheads depict signal spots that have been discarded during this discrimination step; (**g**) contains the final ROIs that have been selected as synapses after the second round of discrimination, using as threshold the median value of the red *IntDen* ROI distribution; the arrowheads depict spots that do not overlap with red signal. The lower panel contains the output verification images from the supplementary sample images "Synapses_inh_01" (**h**) and "Synapses_exc_02" (**i**), corresponding to inhibitory and excitatory synapse counts, respectively, processed as explained in Subheading 3.3

```
//CALL ORIGEN & DESTINATION FOLDERS
imagesFolder = getDirectory("Choose directory containing images");
resultsFolder = getDirectory("Choose directory to save results");

//CREATE ARRAY LISTING ALL IMAGES IN THE FOLDER
list = getFileList(imagesFolder);

//CREATE DIALOG FOR THRESHOLD MODULATION
Dialog.create("Indicate cut-off value for puncta assingment:");
Dialog.addMessage("1=50% 2=60% 3 =70% 4=80% 5=90%");
Dialog.addSlider("Green cut-off: ", 1, 5, 1);
Dialog.addSlider("Red cut-off: ", 1, 5, 1);
Dialog.show;
cutoffG = Dialog.getNumber();
cutoffR = Dialog.getNumber();

//CREATE ARRAYS TO STORE RESULTS
names = newArray(list.length);
areas = newArray(list.length);
puncta = newArray(list.length);
synapses = newArray(list.length);

//CREATE LOOP TO ANALYSE ALL IMAGES
for(i=0; i<list.length; i++){
        showProgress(i+1, list.length);
        //Open image and get name
        open(imagesFolder+list[i]);
        name = File.nameWithoutExtension;
        // Add main code here  (3.2.1. to 3.2.7)
        // Save verification image
        saveAs("TIFF", resultsFolder+name+"_processed.tif");
        // Close windows
        close("\\Others");
        roiManager("Reset);
        // Fill arrays
        names[i] = name;
        areas[i] = area;
        puncta[i] = noPuncta;
        synapses[i] = noSynapses;
}

//CREATE RESULTS TABLE
run("Table...", "name=[Results] width=400 height=300 menu");
print("[Results]", "\\Headings:"+"Image\t Area\t No Puncta\t No Synapses");
for(i=0; i<list.length; i++){
        print("[Results]",""+names[i]+"\t"+areas[i]+ "\t"+puncta[i] "\t"+synapses[i]);
}

//SAVE RESULTS TEXT FILE IN RESULTS FOLDER
selectWindow("Results");
saveAs("Text", resultsFolder+"Results.txt");
run("Close");

//LIST ALL FUCNTIONS BELOW
```

Fig. 14 Minimum codes to analyze *n* images and retrieve results. The codes below will allow automating the main image processing pipeline for a set of *n* images; the user-defined functions are not included in the figure and should be listed at the end of the code

4. Create four arrays to store the results (image name, area, number of selected puncta, and final number of synapses) obtained per image analyzed.
5. Create a processing loop to apply the image processing workflow to all the images in the folder. This loop must contain (1) a command to open each image; (2) a variable to store its name,

to be later on used to identify the verification image; (3) the main body code developed from Subheadings 3.2.1 to 3.2.7; (4) a command to save the verification image to the results folder; (5) the commands to fill the results' arrays; and (6) the command to close all images/windows that remain open before the next loop round starts.

6. Create the table that will contain the results from all the images.
7. Fill the results table with the values stored in the results' arrays.
8. All the functions can be listed at the end of the code.

3.3 Quantify Synapse Density Using the Provided Macro "Synapse_Counter.ijm"

This section provides the instructions to use the full macro, "Synapse_Counter.ijm," created as explained in Subheading 3.2. For general purposes, we have incorporated two additional features: (1) an option to segment the nuclei either based on one ("blue") or two ("blue+green") channels (*see* **Note 19**) and (2) a dialog box for the user to introduce the *x* and *y* parameters that will grant chromatic shift correction. The following steps explain how to use the macro, using as example the supplementary images provided with the chapter. For particular image sets, the specific parameters required for image preprocessing can be easily adjusted by changing the corresponding values in the variables that call the corresponding function arguments (*see* Figs. 6, 8, and 10). The order of the channels can be also easily corrected in the corresponding function (*see* Fig. 4):

1. Create a folder and copy all the raw images into it; the images should be 3-channel single images in a format that *Fiji* can directly open (*see* **Note 5**). Also, create a folder for the results to be stored.
2. Open the "Synapse_Counter.ijm" supplementary macro by drag and drop on the *Fiji* bar, and hit *Run*.
3. In the pop-up browser, select the folder containing the images, and hit *Open*. Then, select the folder where the results will be stored and hit *Open*.
4. A dialog box will pop up; introduce here the *x* and *y* shift correction parameters, estimated as explained in Subheading 3.1. For the provided sample images, these are $x = 0.35$ and $y = 0.81$.
5. Next, a new dialog will ask *Is nuclear staining also present at green channel?* To process the sample images "synapses_inh_01.lsm" and "synapses_inh_02.lsm," choose *Yes*; otherwise choose *No* if only the "blue" channel contains nuclear staining, which should be the option for images "synapses_exc_01.lsm" and "synapses-exc_02.lsm."
6. In the next pop-up window, two sliders appear, where the cutoff values for particle discrimination of both puncta

("green") and vesicular transporter ("red") should be modulated. A range from 1 to 5 can be selected, which means that 50–90% of the particles will be discriminated by the threshold, at 10% increments. For sample images 1 and 2, leave both values to 1; use both 5 for sample images 3 and 4. By hitting *OK*, the macro will be executed.

7. Once the macro has finished analyzing all the images, the results folder will contain all the verification images where the synapse selection is drawn (*see* Fig. 11), plus a text file containing the measured parameters, area, number of puncta, and number of synapses, all identified by image name. The synapse density can be estimated dividing the number of synapses by the corresponding working area.

4 Notes

1. For immunofluorescence vibratome sections were used, obtained from adult mice that were transcardially perfused with 4% paraformaldehyde. Abs against Gephyrin, Homer, VGAT, and VGluT1 required antigen retrieval treatment; sections were treated by boiling for 10 min in sodium citrate buffer (2 mM citric acid monohydrate, 8 mM tri-sodium citrate dihydrate, pH 6.0) before blocking and incubation with primary Abs. Signal amplification was required for Abs against Gephyrin and Homer; after washing primary Abs, sections were incubated with the corresponding biotinylated secondary Abs, washed, and incubated with Alexa-488-conjugated streptavidin.
2. The images were acquired using a confocal LSM Zeiss780 microscope, equipped with an apochromatic 63× oil (NA = 1.4) objective (*see* **Note 2**). The three channels were sequentially acquired using a pinhole aperture of 0.8 AU. Pixel size was adjusted to 50 × 50 nm. Images (sampling units) of 25.95 × 25.95 micron (512 × 512 pixels) were randomly taken at the primary somatosensory cortex, within layer 4. Any multichannel confocal microscope is appropriate as long as it contains the laser lines necessary to excite the different dyes and/or fluorophores used. Line sequential acquisition of the different channels is recommended to be able to speed up the image capture while avoiding cross-talk emission contamination; nevertheless, when possible, simultaneous acquisition should be the fastest choice. Pinhole aperture can be minimized in order to reduce the thickness of the sections, as long as the detection efficiency is not severely affected; this should be easier in systems having high quantum efficiency and low-noise detectors, such as those based in the gallium

arsenide phosphide technology. Pixel size should be at least adjusted according to the Nyquist-Shannon criteria, as a function of the objective numerical aperture (*see* Ref. [22]). A slight oversampling is nevertheless recommended in order to improve the intensity profile of individual spotlike signals.

3. Chromatic aberrations are wavelength-dependent artifacts that occur because the refractive index of every optical glass formulation varies with wavelength. In order to minimize the chromatic aberration, (1) apochromatic objectives are preferred; (2) coverslip thickness should be the one recommended for the particular lens used, unless the latter has the proper thickness correction collar; and (3) the refraction index of the mounting media should match that of the objective immersion medium. The samples used in this chapter were mounted using a mix of 97% 2,2′-Thiodiethanol (TDE, Sigma-Aldrich) and 3% H_2O in order to match the refraction index (IR = 1.52) of the objective immersion oil [23]. In any case, since a slight chromatic *xy* shift may be unavoidable, the use of beads is imperative in order to perform the final shift correction before collating the channels. Microspheres should be acquired under the exact same detection windows and sampling conditions as the biological samples to be corrected. Any sub-diffraction beads (100–200 nm) having labels similar to those used in the biological samples should be all right. Finally, to avoid changes in the orientation of the chromatic aberration, the scanning area should be centered and never rotated, so that the shift of the different images is identical.

4. *ImageJ* macro language (IJM) is a scripting language built into *ImageJ*, which allows writing simple programs called ***macros*** that automate series of image processing actions. By opening the macro recorder at [Plugins > Macros > Record...], the manually performed actions are sequentially recorded into an IJM preliminary script. Hitting the button ***create*** will open the recorded instructions into the script editor where additional editing can be made in order to make the code usable. For more information about macros, see [24].

5. *ImageJ/Fiji* can open many different file formats along with their important metadata, via [File > Import > Bio-formats] or directly at the [File > Open] menu. The list of *ImageJ*-supported file formats can be checked at [25]. Original formats will by default contain image calibration; it is important to take this into account when measurements are going to be performed, so that the resulting values can be interpreted in either pixels, micrometers, or any other unit. Image calibration can be checked at [Image > Properties...]. In this chapter, for the chromatic shift calculation, we prefer to remove image

calibration first, as the final units to be used in image translation will be in pixels.

6. Images can be identified in the macro scripts by their names. In general, renaming images will help shorten the names and identify the current image processing step, thus simplifying the script. The names used in the provided protocol are arbitrary. If other names are chosen, the corresponding code should be changed accordingly.
7. The *LoG* filter is used to enhance spotlike signals. It highlights regions of rapid intensity change, which are the pixels where the *Laplacian* function changes sign, also called zero crossing points. By smoothing the image first with a *Gaussian* filter, the number of zero crossing points will change according to the inverse of the *Gaussian* radius, being the optimal smoothing scale radius related to the radius of the spots to be enhanced. *ImageJ* contains a *LoG* algorithm located within the *FeatureJ* plugin [26]. The operator retrieves a new 32-bit image where the enhanced dots appear dark on a light background. By default, the term "Laplacian" will be added to the image name in the resulting image window.
8. Thresholding is used to extract objects in an image based on their intensity, by setting one or two (upper and lower) cutoff value(s) that separate specific pixel intensities from each other. Most algorithms used for automatic thresholding are developed for a specific purpose or extraction problem. Thus, performance depends on the image content and quality and the intended use of the pattern extracted. *ImageJ* provides 16 different methods to automatically compute global thresholds, all of which can be tested at once using the *Autothreshold* plugin at [Image > Adjust > Autothreshold].
9. The *Watershed* algorithm implemented in *ImageJ/Fiji* is a region-based segmentation approach that separates different objects that touch. It first calculates the peaks of local maxima of all objects, based on their Euclidean distance map. It then dilates each peak as far as possible, either until the edge of the particle is reached or the edge of another growing peak is found. For more information about *Watershed* algorithm implementations, *see* Ref. [27].
10. The *Analyze particles* plugin counts and measures objects in binary or thresholded images, based on an algorithm that finds the edges. *Size* and *circularity* ranges are generally used to select the objects of interest that will form the binary mask. Size refers to area, in squared micron unless specified, and circularity refers to the calculation $4\pi \times [Area]/[Perimeter]^2$. The option *Exclude on edges* is used the discard incomplete objects that lie on the edge of the image. The option *Add to*

manager will load all selected regions into the *ROI Manager* tool, where they can be further analyzed.

11. The *Translate* function at [Image > Transform > Translate...] moves the image in *x* and *y* by a set number of pixels. The input values must be pixels. If the *x* and *y* shifts were calculated on a calibrated image, they should be divided by the pixel size, which can be obtained at [Image > Properties...]. In the present pipeline, the shifts are directly calibrated in pixels. Since the resulting pixel shifts are not integers, a resampling method can be applied during translation to refine the correction; this can be done by selecting an interpolation method, either *bilinear* or *bicubic*, at the translate dialog box.
12. A user-defined function is a block of code that performs a general or repetitive task always in the same way. It can be passed values (defined as arguments), and it can return a value by means of the *return* statement. Functions can be installed or simply written at the end of the macro. They are called from the script by *name (arguments separated by comma)*. For more information about functions, visit [28].
13. Uneven image backgrounds can be successfully restored using the rolling ball algorithm implemented in *ImageJ*, which derives from the *Rolling ball* algorithm described in Stanley Sternberg's article [29], modified to use a paraboloid of rotation instead of a ball. The *Rolling ball* radius is the radius of curvature of the paraboloid and should be at least as large as the radius of the largest object in the image that is not part of the background. The *ImageJ* implementation includes some additional code to avoid subtracting object corners (this choice is activated by selecting the *Sliding paraboloid* option at the *Subtract background* dialog box). More information can be found at [30]. Applying filters such as *Gaussian* or *Median* filters may also help in denoising and therefore facilitate object segmentation. Care need to be taken as to choose the filter that better preserves the object properties and details to be analyzed. For more tips on image denoising using filters, see [31].
14. Most likely each particular set of images will need a customized approach for nuclei segmentation. Depending on the quality of the labeling, global thresholding methods may not work well enough. Adaptive thresholding may be a good alternative; this method changes the threshold dynamically over the image by computing local parameters confined to smaller regions, which are more likely to have homogeneous illumination. The *Auto Local Threshold* plugin implemented in *ImageJ* and *Fiji* contains up to nine different local thresholding methods. The *Try all* option allows for exploring how the different algorithms perform on a particular image. For more information, *see* Ref.

[32]. A second possibility (the one used in the provided pipeline) is to complement the nuclear mask using a second image where the nucleus is also stained and distinguishable. Note that, for each particular image data set, the nuclear segmentation strategy will have to be adapted and modified in the corresponding script.

15. The *Image Calculator* command performs arithmetic and logical operations between two images selected from pop-up menus, by applying one of 12 possible operators. Among these, the logical *OR* operator generates an image where either pixel contained in the source images is included. By selecting the option *Create a new window*, the source images are maintained; otherwise the result is overwritten on the first input image.
16. Direct binary conversion of the *Laplacian* filtered 32-bit image is chosen here. The result is identical to that produced by 8-bit conversion plus thresholding, using the *Default* method. More severe thresholding is not used at this step since the idea is to detect as many spots as possible.
17. Using mean intensity as cutoff value helps disregard weak objects in general but may also remove biological meaningful structures. For instance, a tiny but very bright particle may contain less protein content than another particle that is bigger but less bright. Integrated density is calculated as area of the ROI multiplied by the mean intensity of the ROI; using this parameter as cutoff value helps remove those detected foci where protein content is lower. The same argumentation is used to discriminate those puncta whose degree of apposition to the presynaptic vesicles lies below a threshold, based on the intensity density of the presynaptic signal.
18. After dot enhancing and segmentation, many detected ROIs will most likely contain low signal intensity values; a mixture of low protein content, out-of-focus signal, or unspecific background most likely account for these low intensities. As a result, the distribution of *Intensity density* values of the whole ROI population is skewed to the right (Fig. 11b, c). The parameter that better reflects the central tendency is skewed distributions in the median, which is used here as a first guess discrimination threshold. Such statistical approximation provides good results for large-scale comparative studies. However, this strategy may yield biased results when the 50% discarded population contains qualitatively different particles. The possibility to adjust the percentage of "good quality" ROIs has been implemented in the provided threshold selection function; a cutoff value from 1 to 5 modulates the percentage of selected particles, at intervals of 10%, from 50 to

90%. This facilitates the threshold determination to be used in the final comparative study.

19. Nuclei segmentation in the option using only the "blue" channel is based on the local thresholding strategy explained in **Note 14**. Any other strategy can nevertheless be easily implemented in the provided code to adapt it to any particular set of images.

Acknowledgments

We are grateful to Juan Arranz for immunofluorescence experiments and Manel Bosch for useful advice on macro programming. MLA provided samples and biological input. JBF contributed the chromatic shift calculation macro. ER performed the imaging, designed and programmed the macro analysis pipeline, and wrote the paper. Images were acquired at the Molecular Imaging Platform IBMB (CSIC) with the support from the Spanish Ministry of Economy and Competitiveness and the European Regional Development Fund CSIC13-4E-2065.

References

1. Harris KM, Weinberg RJ (2012) Ultrastructure of synapses in the mammalian brain. Cold Spring Harb Perspect Biol 4(5): a005587. https://doi.org/10.1101/cshperspect.a005587
2. Rakic P, Bourgeois JP, Goldman-Rakic PS (1994) Synaptic development of the cerebral cortex: implications for learning, memory, and mental illness. Prog Brain Res 102:227–243. https://doi.org/10.1016/S0079-6123(08)60543-9
3. Henstridge CM, Pickett E, Spires-Jones TL (2016) Synaptic pathology: a shared mechanism in neurological disease. Ageing Res Rev 28:72–84. https://doi.org/10.1016/j.arr.2016.04.005
4. Mata G, Heras J, Morales M, Romero A, Rubio J (2016) SynapCountJ: a tool for analyzing synaptic densities in neurons. Proceedings of the 9th International Joint Conference on Biomedical Engineering Systems and Technologies (BIOSTEC 2016) 2: BIOIMAGING. p 25–31
5. Fish K, Sweet R, Deo A, Lewis D (2008) An automated segmentation methodology for quantifying immunoreactive puncta number and fluorescence intensity in tissue sections. Brain Res 1240:62–72
6. Danielson E, Lee SH (2014) SynPAnal: software for rapid quantification of the density and intensity of protein puncta from fluorescence microscopy images of neurons. PLoS One 9 (12):e115298. https://doi.org/10.1371/journal.pone.0115298
7. Mokin M, Keifer J (2006) Quantitative analysis of immunofluorescent punctate staining of synaptically localized proteins using confocal microscopy and stereology. J Neurosci Methods 157:218–224
8. Hoon M, Sinha R, Okawa H (2017) Using fluorescent markers to estimate synaptic connectivity in situ. Methods Mol Biol 1538:293–320. https://doi.org/10.1007/978-1-4939-6688-2_20
9. Weiler NC, Collman F, Vogelstein JT, Burns R, Smith SJ (2014) Synaptic molecular imaging in spared and deprived columns of mouse barrel cortex with array tomography. Sci Data 1:140046. https://doi.org/10.1038/sdata.2014.46
10. Cordelières F, Bolte S (2014) Experimenters' guide to colocalization studies: finding a way through indicators and quantifiers, in practice. Methods Cell Biol 123:395–408
11. Dobie FA, Craig AM (2011) Inhibitory synapse dynamics: coordinated presynaptic and postsynaptic mobility and the major contribution of

recycled vesicles to new synapse formation. J Neurosci 31(29):10481–10493. https://doi.org/10.1523/JNEUROSCI.6023-10.2011

12. Schindelin J, Arganda-Carreras I, Frise E, Kaynig V, Longair M, Pietzsch T, Preibisch S, Rueder C, Saalfeld S, Schmid B, Tinevez J, White D, Hartenschtein V, Eliceiri K, Tomancak P, Cardona A (2012) Fiji: an open-source platform for biological-image analysis. Nat Methods 9:676–682
13. Dickstein D, Kabaso D, Rocher A, Luebke J, Wearne S, Hof P (2007) Changes in the structural complexity of the aged brain. Aging Cell 6 (3):275–284
14. Smal I, Loog M, Niessen W, Meijering E (2010) Quantitative comparison of spot detection methods in fluorescence microscopy. IEEE Trans Med Imaging 29(2):282–301. https://doi.org/10.1109/TMI.2009.2025127
15. Sassoe-Pognetto M, Panzanelli P, Sieghart W, Fritschy JM (2000) Colocalization of multiple GABA(A) receptor subtypes with gephyrin at postsynaptic sites. J Comp Neurol 420 (4):481–498
16. Arranz J, Balducci E, Arato K, Sanchez-Elexpuru G, Najas S, Parras A, Rebollo E, Pijuan I, Erb I, Verde G, Sahun I, Barallobre M, Lucas J, Sanchez M, de la Luna S, Arbones M (2019) Impaired development of neocortical circuits contributes to the neurological alterations in DYRK1A haploinsufficiency syndrome. Neurobiology 127:210–222
17. Github website MI, Synapse Counter. https://github.com/MolecularImagingPlatformIBMB/Synapse_Counter.git
18. Schneider CA, Rasband WS, KW E (2012) NIH Image to ImageJ: 25 years of image analysis. Nat Methods 9(7):671–675
19. Fiji download website. https://imagej.net/Fiji/Downloads
20. ImageJ macro functions website. https://imagej.nih.gov/ij/developer/macro/functions.html
21. Li CH, Tam PKS (1998) An iterative algorithm for minimum cross entropy thresholding. Pattern Recogn Lett 19(8):771–776
22. Pawley J (2000) The 39 steps: a cautionary tale of quantitative 3-D fluorescence microscopy. BioTechniques 28(5):884–886. 888
23. Staudt T, Lang MC, Medda R, Engelhardt J, Hell SW (2007) 2,2′-thiodiethanol: a new water soluble mounting medium for high resolution optical microscopy. Microsc Res Tech 70 (1):1–9. https://doi.org/10.1002/jemt.20396
24. ImageJ macro programming. https://imagej.nih.gov/ij/docs/guide/146-14.html
25. formats Is. https://docs.openmicroscopy.org/bio-formats/5.7.3/supported-formats.html
26. FeatureJ. http://imagescience.org/meijering/software/featurej/
27. Roerdink J, Meijster A (2001) The watershed transform: definitions, algorithms and parallelization strategies. Fundamenta Informaticae 41:187–228
28. functions Iu-d. https://imagej.nih.gov/ij/developer/macro/macros.html#functions
29. Sternberg S (1983) Biomedical image processing. Computer 16(1):22–34
30. ImageJ's subtract background. https://imagej.nih.gov/ij/developer/api/ij/plugin/filter/BackgroundSubtracter.html
31. Singh I, Neeru N (2014) Performance comparison of various image denoising filters under spatial domain. Int J Comp Appl 96(19):21–30
32. Auto local threshold. https://imagej.net/Auto_Local_Threshold

Chapter 6

Automated Quantitative Analysis of Mitochondrial Morphology

Anna Bosch and Maria Calvo

Abstract

Mitochondria are dynamic organelles that in most cells behave as a dynamic network and can change their biogenesis and structure depending on the cell needs or as a response to different conditions. Analyzing the architecture of mitochondria is determinant to describe their state and function. In this chapter, image processing techniques are applied in a workflow manner to segment the mitochondrial network and extract the most relevant parameters that enable an accurate morphology analysis. This workflow is programmed with *ImageJ* macro language and can be applied to automatically analyze multiple cells from multiple images or tiles. When combined with multiwell plates and automated microscopy, this method may allow to perform high content image analysis of hundreds of cells under different conditions.

Key words Mitochondria, Mitochondrial network, Image analysis, High content screening, Segmentation, Form factor, Aspect ratio

1 Introduction

Mitochondria are dynamic organelles that in most cells behave as a dynamic network and can change their biogenesis and structure depending on the cell needs or as a response to different conditions [1]. Both network morphology and state at a particular moment result from the balance between fission and fusion processes, mitophagy-regulated homeostasis, tethering, and swelling and shrinking [1, 2]. Mitochondria have multiple cellular functions, and most of them are highly related to their dynamics, shape, and turnover [3]. Moreover, the dysfunction or alteration of mitochondria dynamics has been progressively related to a high number of diseases [4, 5]. Therefore, the analysis of mitochondria architecture is determinant to describe their state and function.

Electronic supplementary material: The online version of this chapter (https://doi.org/10.1007/978-1-4939-9686-5_6) contains supplementary material, which is available to authorized users.

Elena Rebollo and Manel Bosch (eds.), *Computer Optimized Microscopy: Methods and Protocols*, Methods in Molecular Biology, vol. 2040, https://doi.org/10.1007/978-1-4939-9686-5_6,

In order to analyze the mitochondrial network morphology, different labellings can be used. Mitochondria morphology can be visualized in live cells using vital fluorescent cationic dyes such as TMRM or rhodamine 123 [2]. As these dyes are membrane potential dependent, they provide a well-defined pattern of healthy polarized mitochondria; however, labelling of the full network cannot be guaranteed in cells where mitochondria polarization may have been altered. For this reason, in order to analyze exclusively the morphology of all the mitochondria, independently on their membrane potential, any labelling should preferably be performed by immunofluorescence against resident proteins of the external mitochondrial membrane, which furthermore have not been altered in the specific experimental model.

The present protocol deals with the 2D mitochondrial morphology analysis in monolayer adherent cell cultures. For the sake of simplicity and acquisition optimization, laser scanning confocal microscopy (LSCM) images were taken from a middle plane of the cell, where most of the mitochondrial mass is concentrated, under the assumption that this plane will be representative of the general mitochondria distribution. With the purpose of comparing the measurements performed in different cells and conditions, the cells must be delimited and their area calculated in the selected plane. In order to achieve such segmentation, a membrane or cytoplasmic cell tracer should be used. Then, image processing techniques are applied in a workflow fashion to segment the mitochondrial network and extract the most relevant parameters needed to analyze its morphology, mainly (1) number of mitochondria per cell, (2) mitochondrial size, (3) mitochondrial mass, (4) mitochondrial aspect ratio, and (5) mitochondrial form factor. Selection of relevant parameters that better describe mitochondria morphology are based on the work of Koopman et al. [2, 5].

This workflow, programmed with *ImageJ* macro language [6], can be applied to analyze multiple cells from multiple images or tiles in a folder. Therefore, when combined with multiwell plates and automated microscopy, it may allow to perform high content image analysis of hundreds of cells under different conditions.

This chapter explains a strategy aimed to segment mitochondria and cells and details how the subsequent image processing steps are programmed in *ImageJ* macro language to ultimately automate the morphological analysis of the mitochondrial network in both a single image or a set of *n* images contained within a folder.

2 Materials

2.1 Sample Preparation

1. Rabbit anti-TOM20, translocase of the outer membrane-20, 200 μg/mL.
2. Donkey anti-rabbit Alexa Fluor 488, 2 mg/mL.

3. High Content Screening Cell Mask Red Stain, 10 mg/mL.
4. Phosphate-buffered saline (PBS) pH 7.4.
5. Bovine serum albumin.
6. Triton X100.
7. Paraformaldehyde 16% in H_2O (highly pure).
8. Mounting medium ProLong Gold (*glycerol-based* hard mount medium).
9. 8-well glass slides (1 cm^2/well).
10. Glass coverslips (24× 60 mm, thickness 0.13–0.17 mm).

2.2 Image Acquisition

Laser scanning confocal microscope (TCS-SP5, Leica Microsystems, Mannheim, Germany) equipped with:

1. Motorized stage.
2. High numerical aperture (NA) and magnification objective lens (Plan Apo 63× 1.4NA) (*see* **Note 1**).
3. Argon laser.
4. Diode-pumped solid-state laser 561 nm.
5. Acoustic optical tunable filter.
6. Acoustic optical beam splitter.
7. LAS AF Leica proprietary software.
8. Matrix Screening Leica proprietary application for image tile acquisition.

2.3 Image Processing and Analysis

1. Sample images ("MitochondrialNetwork-C = 1" and "MitochondrialNetwork-C = 2") can be downloaded from the Springer website (*see* **Note2**).
2. Image processing, analysis, and macro scripting has been performed in *Fiji*, a distribution of *ImageJ* [6]. *ImageJ* 1.51 s or a newer version is required.
3. The macro "MitochondrialMorphologyAnalysis_Folder.ijm" can be downloaded from [7].

3 Methods

3.1 Fluorescence Labelling of Fibroblast Cells

1. Plate the cells on a 8-well glass slide so that each well contains 8000 cells approximately in 300 μl of culture medium and cultured for 24 h in an incubator at 37 °C and 5% CO_2.
2. Wash cells quickly in PBS.
3. Incubate with paraformaldehyde 4% in PBS for 15 min at room temperature (rt).
4. Wash cells in PBS three times for 5 min at rt.

5. Block and permeabilize cells with 1% BSA and 0.1% Triton X100, respectively, for 20 min at rt.
6. Incubate cells with primary antibody TOM20 at a dilution of 1:200 for 1 h at rt.
7. Wash cells in PBS three times for 5 min.
8. Incubate cells with secondary antibody donkey anti-rabbit Alexa Fluor 488 at a dilution of 1:500 and HCS Cell Mask Red Stain at a dilution of 1:2000 for 45 min at rt.
9. Wash cells in PBS three times for 5 min and a final wash in H_2O.
10. Remove the gasket of wells, add 10 μl of mounting medium in each well and mount with a 0.17 mm thickness glass coverslip.

3.2 Fluorescence Imaging

1. Scan each well by using a motorized laser scanning confocal microscope. The LAS AF and the Matrix Screening application from Leica were used for tile scanning (*see* **Note 3**).
2. Sequential image acquisition of the Alexa Fluor 488 and the HCS Cell Mask Red labellings is performed exciting with the 488 and 561 nm laser lines and detecting their emission at 500–555 nm and 570–630 nm, respectively. One optical section with the emission pinhole sizing 1 Airy unit was scanned at the middle plane of the nucleus. Images were acquired at 400 Hz, with a pixel size of 58 nm and a line scanning average of 4 (*see* **Note 4**).

3.3 Image Processing and Mitochondrial Network Analysis of a Single Two-Channel Image

In order to analyze automatically the morphology of mitochondria in cells, an *ImageJ* macro has been developed: "MitochondrialMorphologyAnalysis_Folder.ijm." This macro has been structured in functions in order to simplify the code and make it more readable and to allow reusing them in other macros. To use the macro, drag and drop it to the *ImageJ* bar and press *Run* in the scripting window.

The principal steps in the workflow (Fig. 1) will be described through the functions of the macro. First, images from a folder are opened and renamed (Subheading 3.3.1). Second, individual cells are segmented (Subheading 3.3.2). Third, the mitochondrial network from all cells is segmented (Subheading 3.3.3). Finally, the aforementioned morphological parameters of the mitochondrial network are analyzed in each cell (Subheading 3.3.4).

3.3.1 Open Images from a Folder and Standardize Names

1. Select the folder with images to analyze. When using the provided sample images, the macro will open the images "MitochondrialNetwork C-1," corresponding to mitochondria labelling in green, and "MitochondrialNetwork C-2," corresponding to Cell Mask labelling in red (*see* **Note 5** and Fig. 2).

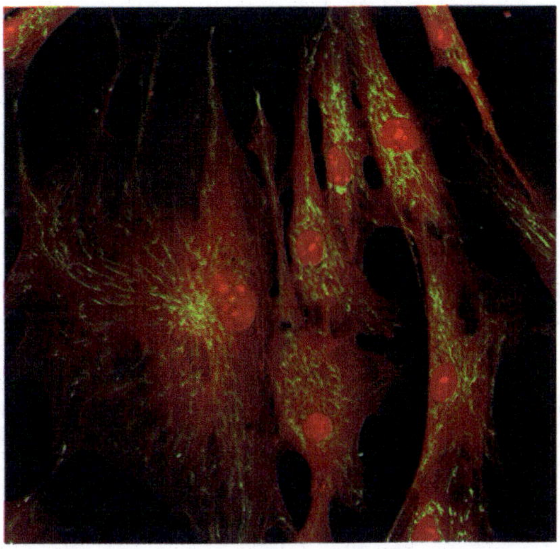

Fig. 1 Overview of the image processing and analysis workflow. Raw data are images of two channels, mitochondria labelling in green and whole cell labelling in red. The presented approach consists of three distinct parts: segmentation of individual cells, segmentation of mitochondria network from all cells, and analysis of morphological parameters of mitochondrial network from each cell

2. The macro will automatically select and rename the channels (Fig. 3). Channel 1 (C = 1) and Channel 2 (C = 2) will be renamed "MitoOriginal" and "CellMaskOriginal," respectively.

3.3.2 Cell Segmentation

In this first part of the workflow, the boundaries of the cells will be defined; this will allow each cell to be individualized as a region of interest (ROI) and listed in the *ROI Manager* later on (Fig. 4). For this purpose, the HCS Cell Mask Red Stain channel (renamed as "CellMaskOriginal"), which contains the general cell staining, will be used. However, before intensity thresholding, different steps must be done in order to correct intensities, track cell extensions, and separate cells at the contact regions (Fig. 5).

1. Depending on the staining conditions (*see* **Note 5**), HCS Cell Mask can be highly concentrated in the nuclei; a nonlinear reassignment must be done, and the data filtered to correct for uneven staining intensity inside each cell. In the present example, a *Gamma* nonlinear mathematical operation is used at [Process > Math > Gamma...], using a value of 0.5 (*see* **Note 6**). Subsequent *Mean* filtering at [Process > Filters > Mean...], using high kernel (e.g., 7), renders a uniform labelling.
2. Fibroblasts are very large and flat cells with extended processes. Segmenting such complicated shapes may benefit from the addition of several channels containing different cell labellings; the consequent signal improvement will facilitate both the delimitation of the cell extensions and the separation between

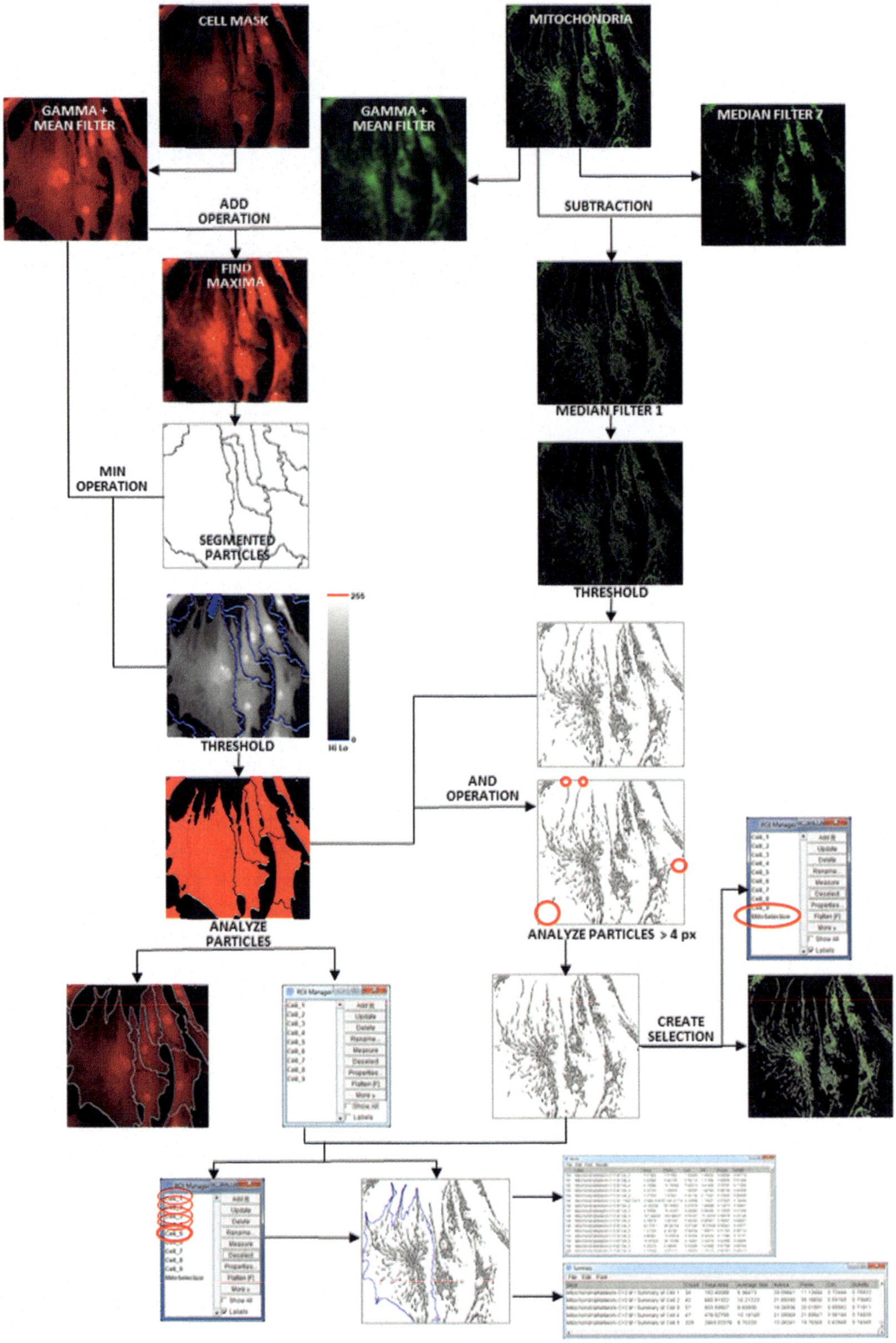

Fig. 2 Original input images: mitochondria network channel in green and cell mask channel in red. See methods for details. Bar represents 10 μm. Image credit: Dra. Glòria Garrabou-IDIBAPS

```
//Select Images Folder
dir = getDirectory("Choose images folder");
list=getFileList(dir);
//Create a Results Folder inside the Images Folder
dirRes=dir+"Results"+File.separator;
File.makeDirectory(dirRes);
//to delete previous ROIs
roiManager("reset");
//Start
for(i=0;i<list.length;i=i+2){
      open(dir+list[i]);
      open(dir+list[i+1]);//it opens two channels
      //automatic selection of channels
      for (j=1; j<=nImages; j++) {
                  selectImage(j);
                  t=getTitle();
                  if (matches(t,".*C=2.*")==1){
                  cellMask=getImageID();
            }
                  if (matches(t,".*C=1.*")==1){
                  mito=getImageID();
            }
      }
      //Rename original images
      selectImage(cellMask);
      rename("CellMaskOriginal");
      selectImage(mito);
      rename("MitoOriginal");
```

Fig. 3 Code corresponding to the opening of images and the standardization of names. In green, commented out code

cells. Before the addition, the mitochondria labelling channel ("MitoOriginal") has to be also processed; the *Gamma* non-linear reassignment at [Process > Math > Gamma...], using a value of 0.4, renders a good result on this particular image. Further *Mean* filtering at [Process > Filters > Mean...] using a large kernel (e.g., 20) makes the signal more uniform. The resulting image is renamed ("MitoMask") and added to the previously processed HCS Cell Mask Image ("CellMask") at [Process > Image Calculator...], choosing the *Add* command and the images "MitoMask" and "CellMask" as input images 1 and 2, respectively. The image resulting from this addition is named "TotalMask" (after total labelling).

3. The *Find Maxima* tool is then used to divide contacts between cells at [Process > Find Maxima...], using *a Tolerance* of 60 (*see* **Note 7**). This tool will output a binary image containing the segmented particles from the image "TotalMask." It creates a binary image, "TotalMask Segmented," containing the lines that correspond to the limits between cells. Limits will have 0 value, and particle regions will have 255 value.

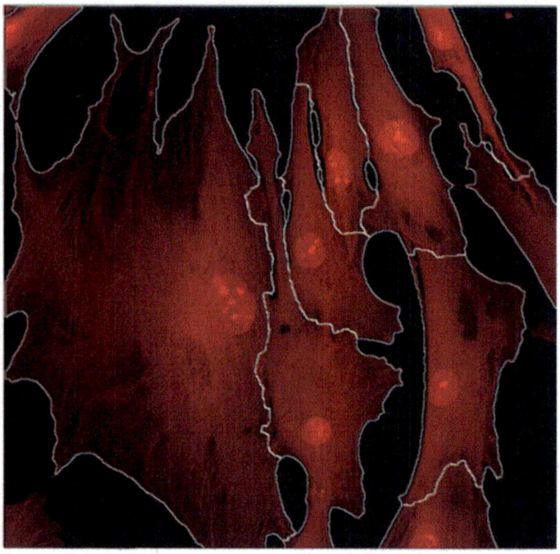

Fig. 4 Segmentation and individualization of cells. Bar represents 10 μm

```
function cellSegmentation(){
      selectWindow("CellMaskOriginal");
      // Correction of uneven staining in HCS Cell Mask Image
      run("Gamma...", "value=0.40");
      run("Duplicate...", "title=CellMask");
      run("Mean...", "radius=7");
      // Addition of channels
      selectWindow("MitoOriginal");
      run("Duplicate...", "title=MitoMask");
      run("Gamma...", "value=0.40");
      run("Mean...", "radius=20");
      imageCalculator("Add", "CellMask","MitoMask");
      rename("TotalMask");
      // Find Maxima- Segmentation of contact regions between cells
      run("Find Maxima...", "noise=60 output=[Segmented Particles]");
      // Minimum Calculation
      imageCalculator("Min create", "TotalMask","TotalMask Segmented");
      selectWindow("Result of TotalMask");
      // Intensity Threshold- Segmentation of cells
      setAutoThreshold("Huang dark");
      setOption("BlackBackground", false);
      run("Convert to Mask");
      // Analyse Particles- Individualization of Regions of Interest
      run("Analyze Particles...", "size=10000-Infinity pixel show=Masks add");
      count=roiManager("Count");
        for(j=0;j<count;j++){
            roiManager("Select", j);
            roiManager("Rename", "Cell_"+j+1);
      }

}
```

Fig. 5 Code corresponding to the cell segmentation steps. Comments in the code (green text preceded by //) indicate the different steps of the macro

4. Once the contact regions between cells have been defined ("TotalMask Segmented"), these delimiting territories need to be transferred to the "TotalMask" image. This is achieved by performing a minimum operation between these two images at [Process > Image Calculator…]. By choosing the *Min* operator at the pop-up menu, the calculator will create a new image ("Result of TotalMask") containing the result of the minimum operation pixel by pixel (*see* **Note 8**).
5. After the cell territories have been defined, cells can be precisely segmented by intensity level thresholding at [Image > Adjust > Threshold…], using the *Huang* thresholding algorithm [8] (*see* **Note 9**).
6. The *Analyze Particles* tool is then used to individualize the ROIs. At [*Analyze > Analyze Particles…*], a certain particle size range (i.e., 10,000- Infinity pixels) can be set in order to discard small particles; *the option Add to Manager* will send the selected regions to the *ROI Manager* tool (*see* **Note 10**). This step is necessary for later cell by cell analysis. The macro will count the ROIs that it finds and rename them as "Cell_" followed by a number.

3.3.3 Mitochondria Segmentation

The third part of the workflow is aimed to segment the mitochondrial network (Fig. 6). Before intensity thresholding, different steps must be done in order to subtract the local background and smooth the labelled structures. The mitochondria labelling channel ("MitoOriginal") will be used (Fig. 7) and processed according to the steps described below. The corresponding code to this step is provided in Fig. 7.

1. Local background will be subtracted from the mitochondria channel "MitoOriginal." When labelling mitochondria, the resulting background is not uniform throughout the cell, being higher in those regions where mitochondria density is greater. To correct for this inhomogeneous background, an image of the general illumination is obtained by *Median* filtering the "MitoOriginal" image at [Process > Filters > Median…], using a high kernel radius, e.g., 7; the resulting image is then renamed as "Mito Flat Field" (due to the analogy with flat-field illumination correction) and finally subtracted from the original one at [Process > Image Calculator…], choosing the *Subtract* operator and the images "MitoOriginal" and "Mito Flat Field" as input images 1 and 2, respectively. By doing so, the high background is subtracted from high-density mitochondria regions.
2. In order to remove the salt and pepper noise generated from subtraction and at the same time smooth mitochondrial signal, a median filter is applied to the image resulting from **step 1** at [Process > Filters > Median...], using a radius of 1.

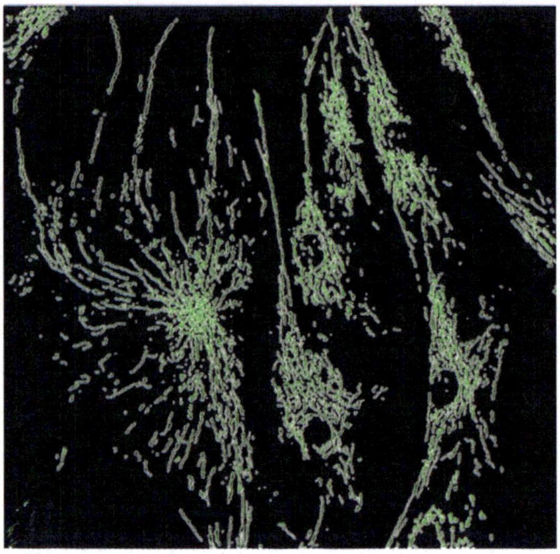

Fig. 6 Segmentation of mitochondrial network. Bar represents 10 μm

```
function mitoSegmentation(){
      //Local background subtraction
      run("Duplicate...", "title=[Mito Flat Field] duplicate range=[]");
      run("Median...", "radius=7 stack");
      imageCalculator("Subtract create stack", "MitoOriginal","Mito Flat
Field");
      //Median Filtering and Intensity Thresholding
      run("Median...", "radius=1 stack");
      setAutoThreshold("Triangle dark");
      setOption("BlackBackground", false);
      run("Convert to Mask");
      //Delete signal out of the segmented cells:
      imageCalculator("AND", "Result of MitoOriginal","Mask of Result of
TotalMask");
      mitoMask=getImageID();
      //Analyse Particles- Size Filter
      run("Analyze Particles...", "size=4-Infinity pixel show=Masks");
      rename("Mito Mask");
      //Close unnecessary images
      selectImage(mitoMask);
      close();
      selectWindow("Mito Flat Field");
      close();
      selectWindow("Mask of Result of TotalMask");
      close();
      //Visualization of the mitochondria network segmentation over the original
image
      selectWindow("Mito Mask");
      run("Create Selection");
      roiManager("Add");
      //Rename mitochondria ROI
      count=roiManager("Count");
      roiManager("Select", count-1);
      roiManager("Rename", "MitoSelection");
}
```

Fig. 7 Code corresponding to the mitochondria segmentation. In green, commented out code

3. Finally, mitochondrial signal is intensity thresholded at [Image > Adjust > Threshold...], using the *Triangle* algorithm [9], and converted to a binary image (*see* **Note 9**).
4. The *Analyze Particles* size filter (i.e., 4-infinity) is then used in order to exclude small particles from the final quantification. Visualization of the mitochondria network segmentation over the original image is used in order to check if the segmentation has been done correctly; a selection of the mitochondrial network from the "MitoMask" image will be generated at [Edit > Selection > Create Selection...], added to the *ROI Manager* and overlapped on the "MitoOriginal" image.

3.3.4 Analysis of Mitochondria Cell by Cell

The next goal is to analyze mitochondria network morphology. The most relevant descriptors of the shape and distribution of mitochondria [2] are number of mitochondria per cell, mitochondrial area, mitochondrial mass, mitochondrial aspect ratio, and mitochondrial form factor.

1. The macro will measure, from the mitochondrial network of each cell, the following parameters [10]:
   (a) Area: Area of selection (in this case in calibrated units, square microns).
   (b) Perimeter: The length of the outside boundary of the selection.
   (c) Circularity (Circ): *4π × area/perimeter^2*. A value of 1.0 indicates a perfect circle. As the value approaches 0.0, it indicates an increasingly elongated shape (Fig. 8).
   (d) Aspect ratio (AR): *major_axis/minor_axis*. Enable "Fit Ellipse" in Analyze>Set Measurements to have the major and minor axis displayed.
   (e) Round (roundness): *4 × area/(π × major_axis^2)*, or the inverse of the aspect ratio.
   (f) Solidity: *area/convex area*.
   (g) Mitochondrial mass is calculated as follows: total area of mitochondria/total area of cell.
   (h) Form factor (FF) is calculated as *1/Circularity*, *perimeter^2/4π × area*, and it varies from 1- Infinity. A FF value of 1 corresponds to a circle. For a given surface, the circle is the 2D shape with the lowest perimeter.

   The parameters from (a) to (f) subsections are selected in [Analyze > Set Measurements...]. Circularity, AR, Round, and Solidity are included in the *Shape descriptors* option (*see* **Note 11**).
2. Macro will count the cells ROIs listed in the *ROI Manager* and will quantify the set measurements from each cell on "Mito Mask" image by a looping statement till the end of the list (Fig. 9).

Shape	Area	Per.	Circ.	FF	AR
	30.24	19.48	1.00	1.00	1.00
	30.25	22.00	0.79	1.27	1.00
	30.25	49.80	0.15	6.52	1.09
	30.24	52.80	0.14	7.34	21.00
	30.25	86.57	0.05	19.72	5.67

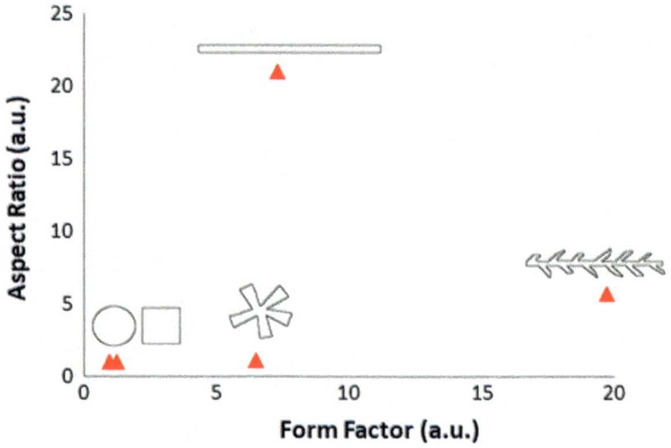

Fig. 8 Reference particles of different geometric shapes used to test morphological parameters. (**a**) All particles have the same area (except a slight difference) and have been drawn under the same scale. Perimeter (Per.), circularity (Circ.), form factor (FF, 1/Circ.), and aspect ratio (AR) have been calculated for each geometric shape. (**b**) Correlation of AR and FF separate shapes depending on their degree of branching and length. A high degree of branching and length will involve high FF values and AR values, whereas round structures will involve FF and AR values close to 1. FF and AR are dimensionless parameters

```
function analysis(){
      run("Set Measurements...", "area perimeter shape display
      redirect=None decimal=5"); //Set measurements. Measurements cell
      by cell
      selectWindow("Mito Mask");
      rename(t);//to obtain the original name into the results table
      count=roiManager("Count");
      for(k=0;k<count-1;k++){
            roiManager("Select", k);
            roiManager("Measure");
            row=nResults();
            setResult("Label", row-1, t+" / Total Cell "+k+1);
            run("Analyze Particles...", "size=4-Infinity pixel
show=Nothing display summarize");
      }
      //Organize Results & Summary tables (Rename label in summary
table)
      IJ.renameResults("Results","ResultsWindow");
      IJ.renameResults("Summary","Results");
      rowSummary=nResults();
      count=count-1;//ROIs except mitochondria ROI
      ini=rowSummary-count;
      iter=count;
      cell=1;
      do{
            setResult("Slice", ini, t+" / Summary of Cell "+cell);
            ini++;
            cell++;
            iter--;
      }while(iter>0);
      IJ.renameResults("Results","Summary");
      IJ.renameResults("ResultsWindow","Results");
}
```

Fig. 9 Code corresponding to the analysis of mitochondria cell by cell. In green commented out code

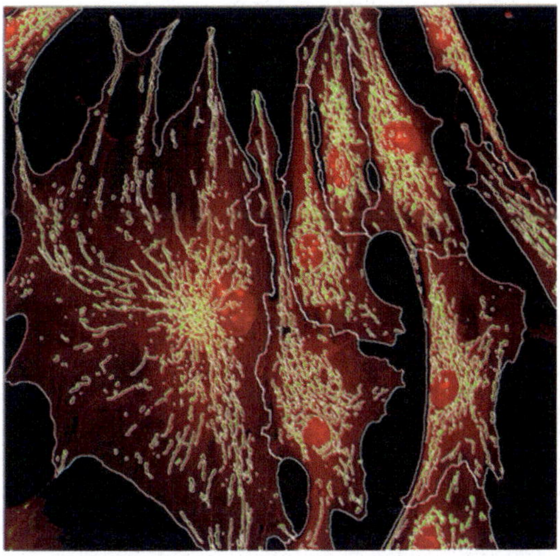

Fig. 10 Overlay of segmented cells and mitochondrial network. Bar represents 10 μm

3.3.5 Results Visualization and Saving

1. Once the analysis is completed, the macro will show the two original channels merged together with the ROIs (Fig. 10).
2. Then the macro will save automatically all the results in the "Results" folder created at the beginning.
3. Finally, the macro will ask to close all windows so that the user can check the results before they are closed (Fig. 11).
4. Data are represented as correlation of form factor and aspect ratio (*see* **Note 12** and Fig. 12).

4 Notes

1. To resolve mitochondria morphology, it is necessary to have the best imaging resolution; therefore high numerical aperture is required.
2. Images "MitochondrialNetwork-C = 1.tif" and "MitochondrialNetwork-C = 2.tif" are two split channels of the same original image corresponding to mitochondria labelling in green and cytoplasm labelling in red.
3. Matrix Screening application performs a topographic map of the slide in order to optimally automatize in focus image acquisition of all the wells.
4. Slow scanning and line averaging produce high signal to noise images that will improve segmentation of mitochondria.
5. Depending on the permeabilization or labelling concentrations, the HCS Cell Mask labelling is concentrated in the

```
function visualization(){
	//Visualize Original Merged Image with all ROIs
	selectWindow("CellMaskOriginal");
	run("Merge Channels...", "c1=CellMaskOriginal c2=MitoOriginal keep
ignore");
	roiManager("Show All without labels");
	waitForUser("Check Results", "Click OK to continue");
}
function saveResults(dir){
	selectWindow("Results");
	saveAs("measurements", dir+"Results.txt");
	selectWindow("Summary");
	saveAs("measurements", dir+"Summary.txt");
}
function closeImagesAndWindows(){
	run("Close All");
	if(isOpen("Results")){
		selectWindow("Results");
		run("Close");
	}
	if(isOpen("ROI Manager")){
		selectWindow("ROI Manager");
		run("Close");
	}
	if(isOpen("Threshold")){
		selectWindow("Threshold");
		run("Close");
	}
	if(isOpen("Summary.txt")){
		selectWindow("Summary.txt");
		run("Close");
	}
	if(isOpen("B&C")){
		selectWindow("B&C");
		run("Close");
	}
	if(isOpen("Log")){
		selectWindow("Log");
		run("Close");
	}
}
```

Fig. 11 Code corresponding to the functions to visualize the overlapped original image with all ROIs, save the results, and close all images and windows

nuclei, and sometimes it produces bright spots in them. We have observed better results with saponin permeabilization and reducing concentration of labelling.

6. The following formula is used by the *Gamma* correction on 8 bit images: $f(p) = (p/255)^{\gamma}$, where p is the intensity value of a pixel and the gamma (γ) value is chosen by the user. Any other gamma value different from 1 will change the intensity values, mainly the medium ones. A gamma value lower than 1 will increase medium intensity values, while higher ones will not be affected significantly. In this case a gamma value of 0.5 was applied to increase medium values corresponding to cytoplasm and therefore to reduce differences with pronounced nucleus staining [6].

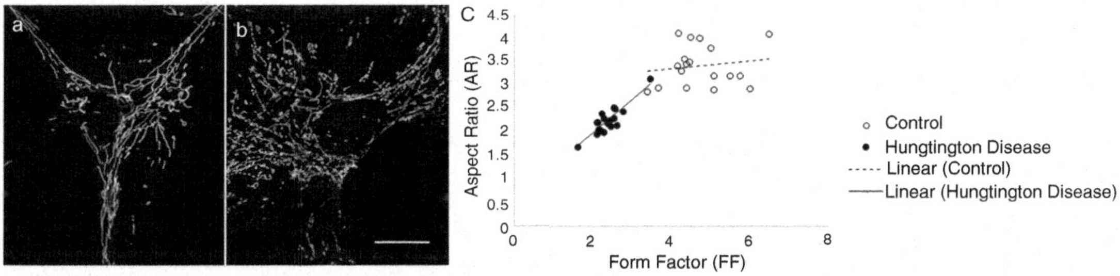

Fig. 12 Quantitative analysis of mitochondrial network in control and Huntington disease (HD) human fibroblasts. Representative confocal images showing mitochondrial morphology in cells from a healthy control subject (**a**) and a HD patient (**b**). Mitochondria have been stained by immunofluorescence with anti TOM20 as explained in methods. In control cell image, mitochondria are long and connected, whereas in the HD cells, mitochondria appear more discontinued and rounded. Bar represents 10 μm. (**c**) Coordinates in the graph represent FF and AR mean values from each cell (18 cells per genotype). Linear regression of represented data is shown (dotted line in control and continuous line in HD). Some features observed in the images (length, circularity, branching) can be quantified by FF and AR parameters. FF and AR values are higher in control cells rather than in disease ones. FF and AR values of control and HD images shown are 4.18/4.10 and 2.24/2.04, respectively. Image credit: Dra. Silvia Ginés and Laura López Molina- UB

7. The *Find Maxima* command determines the local maxima in an image. Local maxima are pixels which intensity value stands out from neighbor pixels by more than the noise tolerance (60 in this example). This tool can create watershed segmented particles. In this case it assumes that each maximum belongs to a particle and segments the image by a watershed algorithm applied to the values of the image.
8. The minimum operation between two images compares them pixel by pixel and finds the minimum pixel value between them which is written to the corresponding location in the output image. This operation allows to encode limits between cells with 0 value and keeps the original values for the rest of the image.
9. Thresholds can be set manually however for analysis reproducibility and automatization; it is recommended to choose an automatic threshold method at [Image > Adjust > Auto Threshold] to segment images using the same algorithm. Pixels with intensities between the two cutoff values will be set at 255 value, and pixels outside these limits will be set at 0, thus converting the image into a binary one.
10. The *Analize Particles* tool needs a binary or thresholded image. It scans the image individualizing regions of connected pixels that meet the filtering criteria set in the *Analize Particles* configuration window.
11. In order to test how FF and AR are affected by morphology changes of the objects, synthetic particles of different geometric shapes but with the same area have been used as can be seen

in Fig. 8. FF increases as branching and length increase, whereas AR reports primarily about length (Fig. 8a). The graph (Fig. 8b) shows that correlation of AR and FF is able to discriminate between the different morphologies of the particles based on their degree of branching and length.

12. Parameters correlation to discriminate different mitochondrial morphologies. As previously described [2, 5], form factor and aspect ratio have revealed as good discriminators of different mitochondrial morphologies [5]. In Huntington disease, where striatal cells exhibit aberrant mitochondria with a high degree of fission [11], these parameters have been useful for characterization and quantification of this phenotype. Mitochondrial network morphology from human fibroblasts from a healthy control subject and a Huntington disease patient has been analyzed and compared by quantification of form factor (Inverse of *Circularity*) and aspect ratio (Fig. 12). Form factor and aspect ratio from each cell are represented as *x*, *y* coordinates, respectively. The two phenotypes can be clearly differentiated and quantified by correlating both descriptors (Fig. 12). In control cells, mitochondria are long and connected (higher AR and FF), whereas in Huntington disease, cells are more discontinued and rounded (lower AR and FF).

Acknowledgments

Authors would like to thank Elisenda Coll from the Advanced Optical Microscopy Facility from CCiTUB from the University of Barcelona for her help in reviewing the manuscript and preparation of samples; Dra. Silvia Ginés and Laura Lopez Molina from Biomedical Science Dept., Institut de Neurociències from the University of Barcelona and IDIBAPS (in collaboration with Movement Disorder Unit from Hospital Santa Creu i Sant Pau, Barcelona); and Dra. Glòria Garrabou and Dr. Francesc Cardellach from Muscle research and mitochondrial function Unit from Cellex- IDIBAPS, CIBERER, Internal Medicine Dept. Hospital Clínic Barcelona School of Medicine-Universitat de Barcelona for providing samples and images.

References

1. Lackner LL (2014) Shaping the dynamic mitochondrial network. BMC Biol 12:35. https://doi.org/10.1186/1741-7007-12-35
2. Iannetti EF, Smeitink JA, Beyrath J, Willems PH, Koopman WJ (2016) Multiplexed high-content analysis of mitochondrial morphofunction using live-cell microscopy. Nat Protoc 11(9):1693–1710. https://doi.org/10.1038/nprot.2016.094. Epub 2016 Aug 18
3. Van der Bliek AM, Shen Q, Kawajiri S (2013) Mechanisms of mitochondrial fission and

fusion. Cold Spring Harb Perspect Biol 5(6): a011072. https://doi.org/10.1101/cshperspect.a011072

4. Giedt RJ, Fumene Feruglio P, Pathania D, Yang KS, Kilcoyne A, Vinegoni C, Mitchison TJ, Weissleder R (2016) Computational imaging reveals mitochondrial morphology as a biomarker of cancer phenotype and drug response. Sci Rep 6:32985. https://doi.org/10.1038/srep32985
5. Koopman WJ, Visch HJ, Smeitink JA, Willems PH (2006) Simultaneous quantitative measurement and automated analysis of mitochondrial morphology, mass, potential, and motility in living human skin fibroblasts. Cytometry A 69(1):1–12
6. Rasband WS (1997–2018) ImageJ, U. S. National Institutes of Health, Bethesda, Maryland, USA, https://imagej.nih.gov/ij/
7. https://github.com/BoschCalvo2018/MitochondrialMorphologyAnalysis_Folder.ijm.git
8. Huang LK, Wang MJJ (1995) Image thresholding by minimizing the measures of fuzziness. Pattern Recogn 28:41–51
9. Zack GW, Rogers WE, Latt SA (1977) Automatic measurement of sister chromatid exchange frequency. J Hitochem Cytochem 25:741–753
10. ImageJ User Guide. https://imagej.nih.gov/ij/docs/guide/index.html
11. Cherubini M, Puigdellívol M, Alberch J, Ginés S (2015) Cdk5-mediated mitochondrial fission: a key player in dopaminergic toxicity in Huntington's disease. Biochim Biophys Acta 1852(10 Pt A):2145–2160

Chapter 7

Structure and Fluorescence Intensity Measurements in Biofilms

Bertrand Cinquin and Filipa Lopes

Abstract

Confocal laser scanning microscopy (CLSM) is one of the most relevant technologies for studying biofilms in situ. Several tools have been developed to investigate and quantify the architecture of biofilms. However, an approach to accurately quantify the intensity of a fluorescent signal over biofilm depth is still lacking. Here we present a tool developed in the *ImageJ* open-source software that can be used to extract both structure and fluorescence intensity from CLSM data: BIAM (Biofilm Intensity and Architecture Measurement). This is of utmost significance when studying the fundamental mechanisms of biofilm development, differentiation, and in situ gene expression or when aiming to understand the effect of external molecules on biofilm phenotypes.

Key words Biofilms, Fluorescence intensity, Structure, Confocal microscopy, Bioimage analysis

1 Introduction

Biofilms are ubiquitous in nature, representing the prevailing mode of microbial life. They are widely used for bioremediation and industrial process purposes and are associated to several serious medical and industrial problems.

A biofilm is an assemblage of microbial cells associated with a surface and enclosed in an extracellular matrix (ECM), being the latter composed of exopolysaccharides, proteins, nucleic acids, and lipids. The biofilm architecture, namely, the spatial arrangement of cells, colonies, particulates, and extracellular polymers, is highly heterogeneous both in time and space and greatly depends on environmental parameters such as nutrient availability, species diversity, and gene expression. Within this context, complex 3D architectures are reported [1, 2] whereby microcolonies are

Electronic supplementary material: The online version of this chapter (https://doi.org/10.1007/978-1-4939-9686-5_7) contains supplementary material, which is available to authorized users.

Elena Rebollo and Manel Bosch (eds.), *Computer Optimized Microscopy: Methods and Protocols*, Methods in Molecular Biology, vol. 2040, https://doi.org/10.1007/978-1-4939-9686-5_7,

separated from one another by interstitial voids, that is, water channels where diffusion and convection of nutrients and metabolites occur.

Biofilms have been intensely studied over the past decades, and confocal laser scanning microscopy (CLSM) has been a standard tool for characterizing them in situ [3]. Indeed, images acquired by CLSM have been used to evaluate biofilm architecture and to quantify structural parameters such as biomass, area, and thickness [4]. As a result, several imaging software packages have been developed by the scientific community to extract quantitative biofilm features from the CLSM images. Among these, the *COMSTAT ImageJ* plug-in [5–8], the *daime* computer program [9], and the *PHLIP Matlab* toolbox [10] represent a set of reference tools. These tools are efficient and reliable to characterize biofilms in terms of structural parameters such as biovolume, surface coverage, surface area, thickness distribution, roughness coefficient, diffusion distance, fractal dimension, or porosity. However, the relationship between biofilm structure and the fundamental cellular/molecular mechanisms involved in biofilm development remains a challenge, as already pointed out by Beyenal and Lewandowski in 2004 [11]. To a certain extent, this is due to the fact that most studies have been limited to either a qualitative or a relative description of genetic reporters inside biofilm CLSM [12–16]. However, to our knowledge, a general method aimed to quantify and relate both the fluorescence intensity of a dye/genetic reporter and the structural features of the labeled population is still lacking.

To bridge this gap, we describe in this work a new tool called *BIAM* (Biofilm Intensity and Architecture Measurement), developed with the open-source *ImageJ* analysis software. *BIAM* is able to extract both the structural information and the fluorescence intensity of fluorophores and/or genetic reporters in a biofilm. The measurements given by this tool include intensities, areas, and perimeters of the segmented biofilm from which we can further obtain parameters such as biovolume and mean fluorescence intensity as a function of biofilm depth. Moreover, the studied biofilm can be segmented and analyzed as a unique/whole entity or as a pool of microcolonies. Additionally, the *BIAM* script can be used by the scientific community in a semiautomated way.

2 Materials

2.1 Datasets

1. Experimental biofilm image. It should be a single channel z-stack of tiff files encoded in 8 bits, acquired using a confocal microscope. A sample dataset "DataForAnalysis.tif" is provided as supplemental material to test the protocol.

2. Control image. In order to retrieve relative intensity information, a biofilm grown in a similar way and homogenously stained with a standard fluorophore needs to be acquired. A sample control dataset "DataForCorrection.tif" is also provided.
3. Correction text file. It is obtained from the control image, as explained in **Note 1** and of course in the original work by Baudin and coworkers [17]. The sample file "Corrected_Coefficient.txt" used in the real example is provided. As a remark, the coefficients contained in this file will be different from those you will measure using the control image, because they have been generated by averaging the measurements from several datasets.

2.2 Software and Tools

1. *BIAM* macro. The script "Biofilm_Analysis.ijm" can be downloaded from GitHub [18]. The macro runs under *ImageJ* (version 1.51 k minimum is required) and *Java7* (it works under the last version of Java). No additional *ImageJ* plug-ins are necessary to run the macro.
2. Additional code. The little macro "CleaningImageJ.ijm" is provided here to help cleaning up ImageJ windows that are not images (*see* **Note 2**). In case the macro stopped during the process, it is useful to get a "fast cleaning" of ImageJ.
3. Computer specifications. As the dataset is usually between 200 and 400 Mb, we advise to run the macro on a computer with minimum 4 Gb of RAM to ensure faster processing. When ImageJ works with opened images, the total memory taken at the end of the process is about three times the dataset size (*see* **Note 2** for additional information concerning the data management under ImageJ).

3 Methods

We will try to be as explicit as possible without describing in too much detail each line of the code. Excerpts of the actual code will be presented for the most challenging parts. An overview of the process is presented in Fig. 1.

3.1 User Interface and File Opening

To ease the use of the macro, we have created a simple user interface that will guide the analysis of the biofilm (*see* **Note 3**). The following steps are necessary to start the analysis:

1. Launch the macro by dragging it to the *Fiji* bar and pressing *Run* in the macro editor window; a browsing window will open where the user is asked to select the data folder that contains the actual dataset and where the different parts of the analysis will be saved.

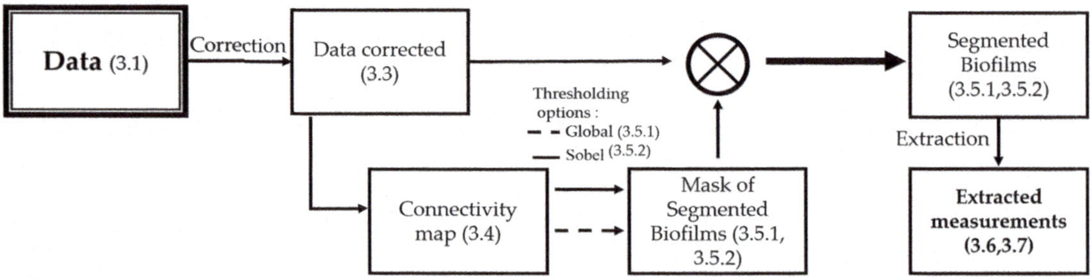

Fig. 1 General *BIAM* workflow. The numbers in parenthesis refer to the specific paragraph where the details are provided

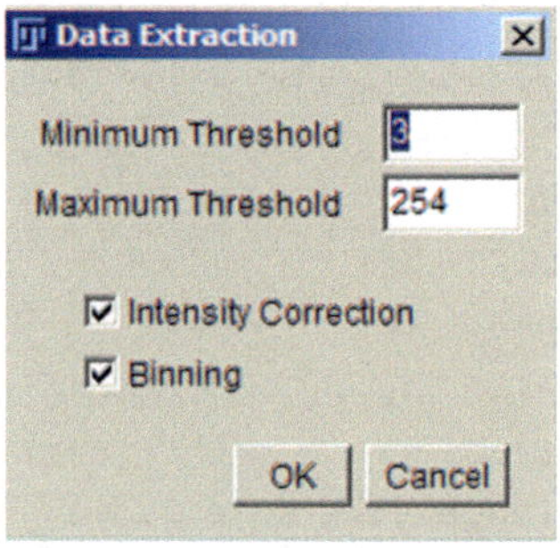

Fig. 2 First user interface box showing the *Minimum* and *Maximum Threshold* values set by default and the *Intensity Correction* and *Binning* options selected

2. The first user interface box displays two values (Fig. 2), minimum and maximum thresholds. By default, *Minimum Threshold* is set to 3 as the current data have been acquired using a noiseless detector. This value should be modified to fit the noise level of the current dataset. The *Maximum Threshold* is set to 254 that is the maximum allowed for an 8-bit image once the saturated pixels are removed (they have a value of 255). If you are working with images with different encoding, this value will need to be changed.
3. Check the *Intensity Correction* option; this will allow for the correction of the dataset by the previously produced correction file, so that the subsequent analysis is performed on the so corrected dataset. Otherwise, the analysis will be realized on the raw data.
4. Check the *Binning* option; this will resize the raw dataset to 1024 × 1024 pixels. The resized dataset will be saved under the name "Namedata_1024.tif," and the original dataset will be closed. If unchecked, the analysis will continue on the current dataset.

3.2 Dataset Size Normalization

It is difficult to guarantee that the lower and upper bounds chosen for the acquisition volume will be the same from one biofilm to another. This step is meant to solve this issue by considering as first slice the plane where the average intensity is maximum. The code block performing this operation is written in lines 48–70 of the supplementary macro. This code performs the following steps:

1. It measures the average intensity of each plane using a *for* loop from 1 to *nSlices* (number of slices in the file) and feeding an array, called, e.g., "IntensityArr."
2. Then, it looks for the slice index that has the maximum level of intensity in the array using the function *Array.findMaxima*.
3. Finally, it creates a substack between this slice and the final slice of the dataset. In this way, all the datasets analyzed will have the same intensity profile and will be corrected homogeneously. In the provided code, the new image is called "Substack."

3.3 Intensity Correction of the Dataset

If the *Intensity Correction* option has been checked, the macro will correct the dataset using the correction file obtained as explained in **Note 1**. The macro will automatically ask for the location of the .txt file containing the correction coefficients. When so, load the correction file. To try the pipeline, the provided correction file "Corrected_Coefficient.txt" can be used to correct the sample file "DataForAnalysis.tif." The script encoding this pipeline contains two main blocks:

1. Script to load the coefficients and the correction file (supp. script lines 72–89). It is important to take into account that the correction coefficients are arranged as a single column. Also, the *parsefloat* command (line 82) will guarantee that the exact value is retrieved, as we will divide an integer by a real number. For this reason, it is necessary to transform the actual 8-bit dataset into 32 bit.
2. Script to perform the correction applying the coefficients (supp. script lines 90–110). A *for* loop will allow to select the slice to be divided by the corresponding coefficient. Since the experimental dataset may be bigger or smaller than the provided correction file, the script includes a test that insures an optimum correction: if the dataset is bigger than the correction file, the last slices will be not corrected; if the dataset is smaller than the corrected file, the dataset will be fully corrected, and the macro will cease to read the correction file after the correction of the last slice.

3.4 Biofilm Connectivity

In this section a "connectivity map" is created that will enhance the contrast of those pixels that belong to the biofilm. This will improve the segmentation step that comes afterward.

Table 1
Pixel value meaning in the connectivity dataset

Pixel value in the connectivity dataset	Given pixel
−2	No connection
−1	The pixel above is part of a microcolony
1	Not connected with the pixel above
2	The pixel above is connected

The belonging of a pixel to the biofilm is identified by the pixel's connection to other pixels that are part of the biofilm as well. Different values, between "−1" and "2," will be given to the pixels depending on their connectivity (Table 1): "1," for pixels that belong to the biofilm and have no pixel above that belongs to the biofilm; −1, for pixels that do not belong to the biofilm and have a pixel above that belongs to the biofilm; "2," for pixels that are below pixels that belong to the biofilm; and finally "0," for pixels that are not part of the biofilm and have pixels above that are neither part of the biofilm. We are currently not using −1 value pixels. For more information regarding the connectivity map, *see* **Note 4**. A segmentation of the basal plane will be realized. After this step, the only information that can be retrieved is the basal area (area at the substratum) for each microcolony, but this is the most important step in this process to extract three-dimensional information (Subheading 3.6).

1. Create a three-dimensional mask (supp. script lines 115–123) of the corrected dataset discarding the pixels with value inferior to 3 and superior to 254 (given our low noise detector, this value needs to be tuned according to your own dataset).
2. Duplicate the mask and shift it by 1 slice (creating a substack) and add 1 "blank" slice at the end. We want the mask of the dataset and the mask shifted by 1 to have the same number of slices to perform mathematical operations (supp. script lines 125–127).
3. Add one mask to the other, remove the value 1, and multiply the result by 2 (supp. script lines 129–132).
4. Subtract one mask to the other (supp. script line 134).
5. Add the result of **step 3** to the result of **step 4** (supp. script line 137).
6. Project the result of 5 by summing each slice to the other (supp. script line 144).
7. Save the result as Connected Map (supp. script line 146).

3.5 Segmentation of the Biofilm

It is often difficult to assess with absolute confidence which thresholding method should be used for biofilm segmentation. The global methods will most likely work on "young" biofilms, which may present a relatively homogeneous signal. However, this is no longer the case for "older" biofilms, where merges of microcolonies may occur. In our case, after having analyzed several biofilms, we realized that when dealing with microcolonies of mixed sizes, a *Sobel* filtering step previous to the segmentation usually gives a better accuracy. The provided macro offers the user three different segmentation strategies. Two of them will be described in the following subsections. The third one simply calls the existing plug-in *Robust Automatic Threshold Selection* (RATS) developed by Ben Tupper [19] and based on the work of Wilkinson [20]. This method is more described in **Note 5**. Nevertheless, we advise the user to perform the different methods and decide in accordance with the microcolony definition used. The codes to call the different methods from the macro are depicted in Fig. 3. A simple addition to this code is provided in **Note 6** in order to color the microcolonies (Fig. 4). This step is optional and esthetic but can help the user tune the analysis.

3.5.1 Manual Global Thresholding

The first strategy will apply a global thresholding between the lower value set up at 0 and the upper value chosen by the user, to segment the biofilm. The steps below describe the operations performed by the macro (*see* Fig. 5) to achieve this goal. Within the actual macro pipeline, only the selection of the thresholding method is made by the user.

1. The macro first selects the connectivity map (result of Subheading 3.4).
2. A dialog box is then programmed to pop up, where the user can select the thresholding method of choice.

```
//Segmentation Option
Dialog.create("Segmentation Selection");
items =newArray("Manual", "RATS", "Sobel");
Dialog.addRadioButtonGroup("Segmentation Options", items, 1, 3, "One")
Dialog.show;
Option = Dialog.getRadioButton;
print("Segmentation: "+Option);
```

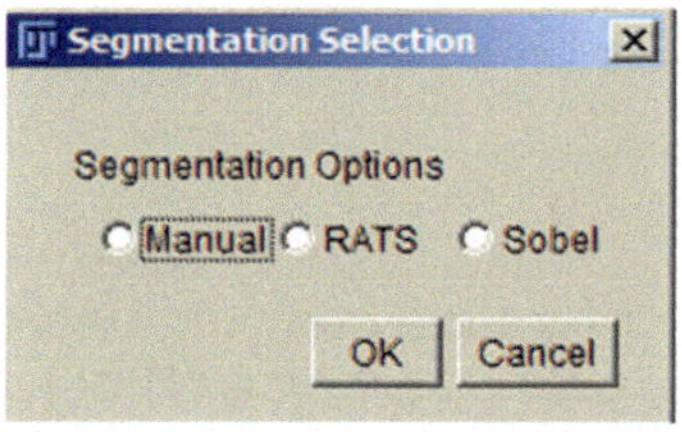

Fig. 3 Excerpt of script encoding the segmentation options (above) and the interface for the user to choose among them (below)

```
248 //color the colonies
249     for(i=0; i<=ROIcount-1;i++){
250         selectWindow("Weighted");
251         roiManager("Select",i);
252         getRawStatistics(nPixels, mean, min, max, std, histogram);
253         maxC = max;
254         minC = min;
255         diffC = maxC- minC;
256         selectWindow("Weighted");
257         run("Subtract...", "value="+maxC);
258         run("Add...", "value="+16*(i+1));
259     }
260     run("16 colors");
261     SUM_Connect_path2 = Worked_Data_Path+"SUM_Connect_Colored_Sobel.tif";
262     saveAs("Tiff", SUM_Connect_path2);
```

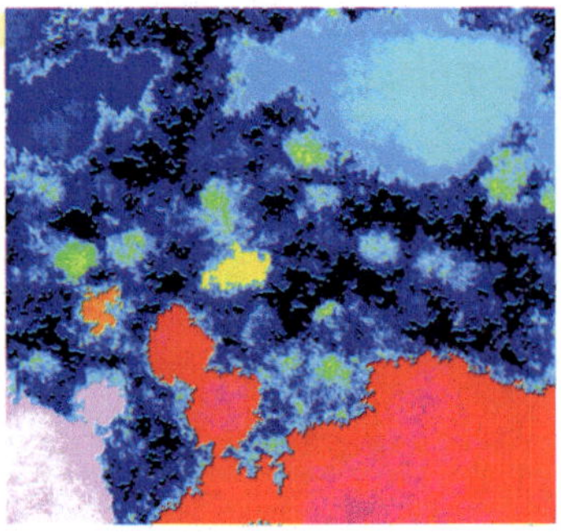

Fig. 4 Excerpt of script for color-coding up to 16 different segmented microcolonies, after Subheadings 3.5.1 or 3.5.2. For each microcolony, the maximum and minimum intensities ("maxC" and "minC") are extracted, and their difference "diffC" computed; this value is subtracted from the image "Weighted" (line 257), and the resulting value is added 16 times the colony index (since the LUT 16 colors is used in a 16-bit image, there is 16 times 16 gray levels). Finally, the image is saved under the name "SUM_Connect_Colored" plus the thresholding method used. The snapshot on the right panel shows an example of colored microcolonies

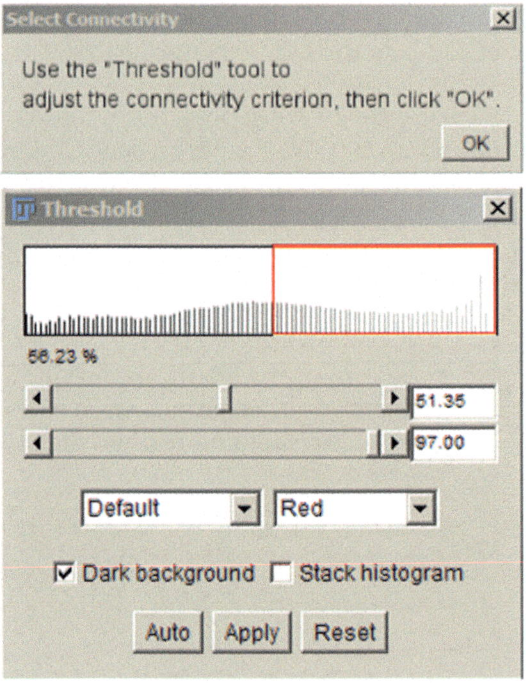

Fig. 5 Message displayed in the "Manual" option (above) and threshold adjustment window (below)

3. Then the *Analyze Particles* function at [Analyze > Analyze Particles. . .] is called from the code, to be applied using the following parameters: *Size* (100 μm^2 to Infinity); *Display results*; *Include holes*; and *Add to Manager*. The Size filtering option should be programmed in accordance to the criterion

$$Vert = \begin{bmatrix} 2 & 1 & 0 & -1 & -2 \\ 3 & 2 & 0 & -2 & -3 \\ 4 & 3 & 0 & -3 & -4 \\ 3 & 2 & 0 & -2 & -3 \\ 2 & 1 & 0 & -1 & -2 \end{bmatrix} Vertical\ Kernel \otimes Connected\ Map$$

$$\text{Hori} = \begin{bmatrix} 2 & 3 & 4 & 3 & 2 \\ 1 & 2 & 3 & 2 & 1 \\ 0 & 0 & 0 & 0 & 0 \\ -1 & -2 & -3 & -2 & -1 \\ -2 & -3 & -4 & -3 & -2 \end{bmatrix} Horizontal\ Kernel \otimes Connected\ Map$$

$$Weighted = (Vert^2 + Hori^2) * (Connected\ Map)^2$$

Fig. 6 Horizontal and vertical Sobel filters used in this protocol. They are written the mathematical operations made to obtain the "Weighted" image, a normalized image that can be thresholded by a global approach

used for microcolony definition. The others will help organize the regions of interest (ROIs) in order to extract results afterward.

4. Finally, the macro saves the ROI list for further examination or comparison studies between different analyses.

3.5.2 Edge Detection Strategy

In this strategy the provided code uses a combination of *Sobel* filters, especially developed for edge detection problems. Two kernels are used to gain sensitivity on vertical and horizontal edges (*see* **Note** 7 and Fig. 6 for further details on the used kernels).

1. First, it selects the connectivity map image (result of Subheading 3.4) and duplicates it.
2. The image is then convolved by a vertical kernel using the function [Process > Filters > Convolve...], having the *normalize* option checked.
3. **Steps 1** and **2** are repeated for the horizontal kernel convolution.
4. The results of **steps 2** and **3** are squared (each image being multiplied by itself), selecting the option *32 bit*, as the outcome may be bigger than the 8-bit depth.
5. The two squared images are then summed up, the result representing the gradient and therefore a better view of the microcolonies.
6. Next, the result of 5 is multiplied twice by the connectivity map image. This results in a connectivity map weighted by the gradient. This step increases the contrast of the microcolony edges, which allows to use an automated global thresholding approach to segment them.
7. The *MinError AutoThreshold* method (*Dark background* option checked) is used and applied, thus resulting in a binary image.

8. The binary operations *Close* and *Fill Holes* at [Process > Binary] are utilized to get plain whole microcolonies.
9. Next, the *Analyze Particles* function is called from the code using the following parameters: *Size* (100 µm to Infinity); *Display results*; *Include holes*; and *Add to Manager* (*see* Subheading 3.5.1, **step 3**).
10. Finally, the code saves the ROIs list for further examination or comparison studies between different analyses. Please read **Note 6** on "Coloring the colonies."

3.6 Extracted Measurements

At this step, the basal areas of the microcolonies have been segmented and stored into the *ROI Manager*. They will now be used to extract the three-dimensional information for each one of the microcolonies. Cropping the original dataset using the obtained "basal area" ROIs will insure that the upper layers of each microcolony will not contain bacteria from any other microcolony. In case different microcolonies are merged, the extraction will be trickier and more difficult to assess without further biological hypotheses. To gain accuracy into this critical step, we will process the data into two major steps: (1) First, several three-dimensional masks will be created for each microcolony that will contain different information; and (2) a filtering to further clean the microcolonies will be applied (the topology of the microcolony can be complex with multiple "peaks" or bacteria floating nearby on top of the studied microcolony).

1. The microcolonies will be cropped out of the biofilm using the ROIs obtained during the thresholding steps. An enlargement factor is introduced here to take slightly more area than the actual ROI. This will not change the quality of the segmentation for well-defined microcolonies, as its surroundings will most likely be clean; however, the situation will definitely improve for microcolonies that are more complex (i.e., neighboring another one). Since a bacterium found at the border between two microcolonies cannot be attributed to one or the other, we decided to rather count such bacteria twice, once for each microcolony. In the provided code, the enlarge parameter has been set at 1 µm for it is usually a good value. A value above 2 µm will usually enlarge the ROI too much so that it will take in account other parts of the closest microcolonies. But in case of necessity, the user can easily change it in the code. This part of the code will generate, for each given microcolony "i," several three-dimensional stacks (Fig. 7): the first one, named "objects_ROI" + x (line 275), contains the original value (either raw or corrected depending on whether the correction factors have been applied); the second one, named "Masks_-ROI" + x (line 278), is a simple binary mask; the third one,

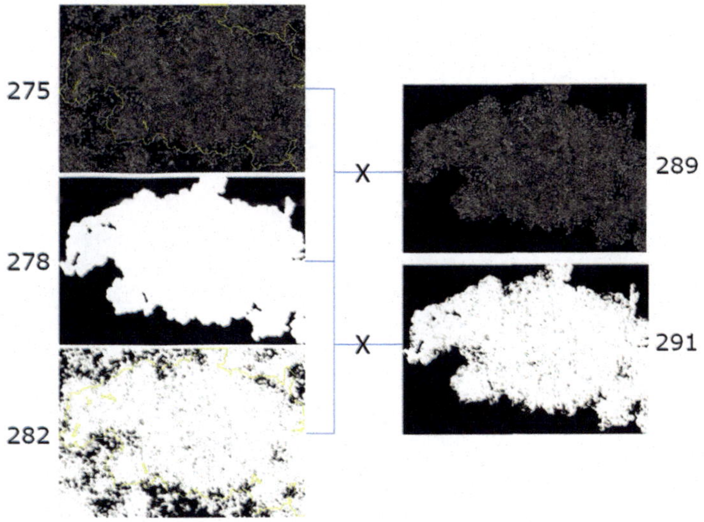

Fig. 7 On the left there are the images named "objects_ROI," "Mask_ROI," and "Connect_ROI." We use the "Mask_ROI" to generate two cleaned 3D maps named "MASK_ROI" and "Substack" (right), the first one containing the connectivity information and the second one the intensity of fluorescence. The numbers to the sides of the images refer to the macro's line of code where the images are renamed

named "Connect_ROI" + x (line 282), is a cropped version of the connectivity map. From these three 3D cropped stacks, the macro will perform the following operations: First, it will multiply "objects_ROI" + x by "Masks_ROI" + x and then "Connect_ROI" + x by "Masks_ROI" + x (supp. script lines 288 and 290). As a result, two 3D maps will be created, one containing the real intensity values and the second containing the connectivity information.

2. Next, for filtering the microcolonies, a size criterion is used in order to exclude small areas that are not part of the biofilm (e.g., floating bacteria, debris, etc.). It is critical to program this step as interactive, so that the user can actually introduce the proper size criterion (*see* Fig. 8).

3. In order to measure non-filtered microcolonies, the area for each microcolony at each layer of the biofilm is retrieved, the biofilm being seen as a plain object with no holes (Fig. 9).

4. Finally, to measure the filtered microcolonies, only the area occupied by the bacteria belonging to the studied microcolony is considered, as well as the average intensity of the fluorescence emitted by them. This means that the voids and the extracellular matrix that are not detected are not measured. This is done following five steps. First, a refined mask is created from "Substack" generated in **step 1** (line 291). We are insuring that the

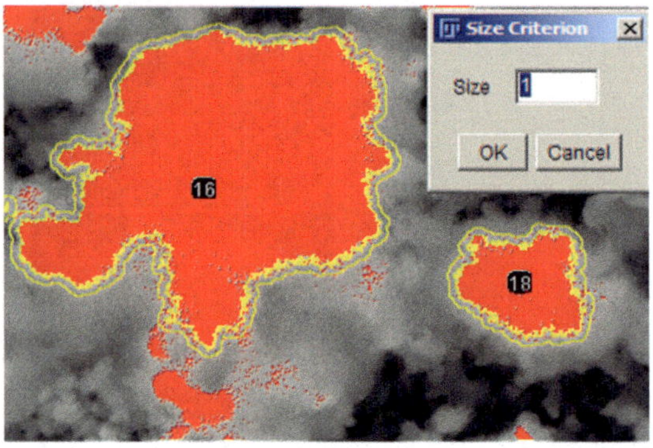

Fig. 8 Size criterion user interface used to sort out isolated bacteria. It is used during the particle analysis (line 317). As several ROIs can exist that belong to the same microcolony at a given depth (or *z*), the macro code lines 321 to 339 have been added in order to combine those regions and obtain for each depth (or each *z*) a single region of interest

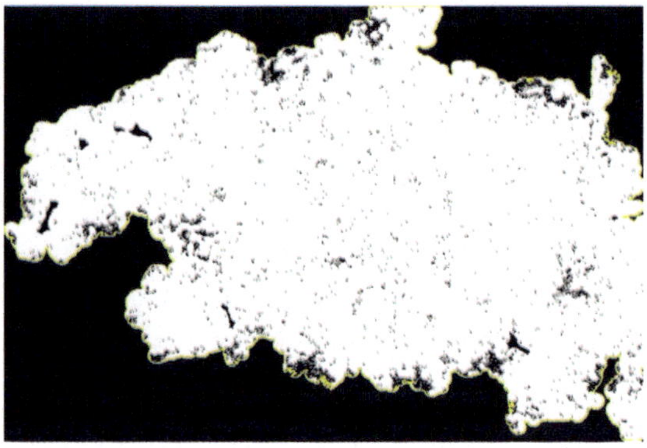

Fig. 9 Mask of a segmented microcolony where to measure area and perimeter

number of slices of this novel stack is consistent with the one of "Substack" as we use the region of interests to create the stack (supp. script lines 374–378). Second, we generate the filtered stack we called "SubstackCleaned" + x by multiplying the mask with the "Substack" (line 382–385). To extract the intensity from this filtered microcolony, we generate new ROIs. We have to ensure that for each slice, there is intensity to retrieve. If the intensity contained in the new ROI is above 0, a selection is created, and it will be used to extract information. If the intensity is 0, then no selection can be created, and the macro will exit of the loop. To do this, we set up a variable called "exitloop" (line 390) to force the exit of the loop in case the average intensity is 0. It usually means that the top of the microcolony is reached. Finally, once the ROIs for each microcolony have been generated and validated (Fig. 10), a simple reading can be done and saved into a table results (supp. script lines 411–422).

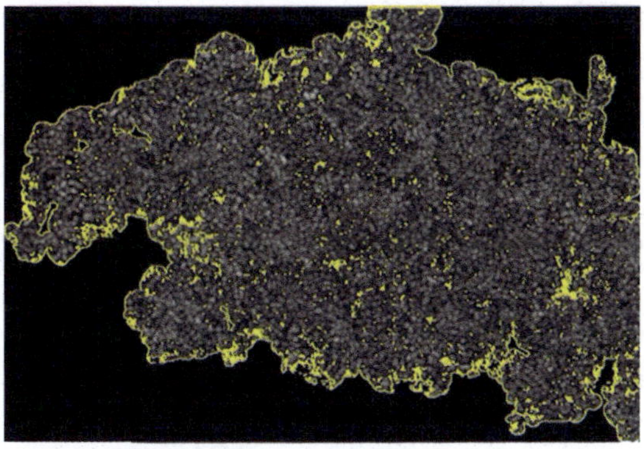

Fig. 10 Biofilm image where to measure area, perimeter, and intensity. The final segmentation (yellow outline) excludes the voids, as compared to Fig. 8

3.7 Final Results, Display, and Saving

At the end of the analysis, we will have many opened windows. In addition to the corrected stack image file of the biofilm (the connectivity map and the colored segmented map), we have for each microcolony the three-dimensional mask, the intensity image, the detailed three-dimensional mask, and the two result tables: filtered and non-filtered. The images are saved into a subfolder, "Saved_Results." Then, all useless windows are closed, and the filtered and non-filtered tables are concatenated into a single one so that only one table per segmented microcolony (instead of two) is delivered and saved as a .csv file. There are as many result files as found microcolonies. The ratio of the filtered and non-filtered column will give access to the measure of porosity for each slice of each microcolony. Averaging the ratio for each slice for a given microcolony will give the porosity for this microcolony. Finally, the tables are closed as well, leaving the macro ready for another round without further manual cleaning.

4 Notes

1. About the correction dataset. Since the primary goal here is to retrieve the fluorescence intensity of the experimental biofilm, careful attention should be paid to the image acquisition process itself. First, it should be pointed out that both the experimental and the control biofilm: the one stained with the standard fluorophore [17] should be grown in a similar manner and acquired with the exact same settings (laser power, detection window, depth step, pinhole, etc.). Second, in order to generate the correction file, an average intensity measurement of each slice of the stack should be done. The series of values is then normalized by the maximum value. These new values will be used as correction coefficients as explained in the correction section (Subheading 3.3).

2. Cleaning non-necessary results. We could have called this little note "Memory management versus Time management." As *ImageJ* generally works on active windows and opened images, duplicating the first dataset literally means doubling the memory space needed. Closing all non-necessary images is an important step. Of course, we could save and reopen the images when needed, but this usually requires more time. The whole process for a dataset can take between 4 and 10 min, mainly depending on the amount of RAM memory available at the computer; nevertheless, we preferred the option of saving time by not saving too much intermediate results and only the ROI lists that can be reopened later to get a better feedback on the analysis. We turn on and off the batch mode using *SetBatchmode(true)* and *SetBatchmode(false)* for parts of the code where operations done on the images do not create more images. This trick tremendously speeds the process. In addition to that, you will find in the repository a second file called "CleaningUpImageJ.ijm" that closes any *ImageJ* window that is not an image, i.e., *Log* window, *Results* tables, and *ROI Manager.*
3. User interface. Since the code presented here may further develop in the future, we have found it easier to guide the data mining along the process, asking the user to input certain parameter values rather than having a single panel with all the parameters included as a black box. Moreover, we have let *ImageJ* work on "opened" images that allows to see "what is going on" and "when it is going on." This feeling of performing the analysis "live" may be very useful for microbiologist who may "use the code" in a future without having an extensive understanding of the programming insights, just by experimenting with the different code blocks.
4. The connectivity map and the −1 pixels. This connectivity stack is obtained by doing a few operations with two stacks: the original dataset and the same stack shifted by one slice (*see* Subheading 3.4). The slice z will be added to the slice $z + 1$. This sum will be multiplied by 2. The subtraction will be done as well; the slice z will be subtracted by the slice $z + 1$. Finally, the connectivity for each slice is given by the sum of these two stacks. Each pixel of this particular map can take the value −2, −1, 1, or 2 (*see* Table 1). For an easier analysis, we chose to set the negative-valued pixels to the value "0." Such a connectivity three-dimensional map is highly informative. Weighting each pixel according to how they are connected highlights the "landscape" of the biofilm. The sum projection of the map helps the process of segmentation.
5. *RATS. Robust Automatic Threshold Selection* shows some suitability for images with inconsistent background. It is based on

Sobel filter edge detection with an addition segmentation using quadtree architecture. The *RATS* plug-in is thoroughly described in [19]. A few parameters can be tuned. In *BIAM*, *RATS* is called with parameter noise = 1 lambda = 3 min = 205 that have presented good results. As the data have been acquired with a hybrid detector with virtually no noise, we set up noise =1 as the minimum thresholding. We kept lambda the scaling factor at the default value 3 as it corresponds to the size of bacteria. We set up the minimum leaflet size, that is, the minimum size of structure we want to segment at 200 pixels as it corresponds to a small structure that can be considered as a microcolony. These parameters would need to be tuned depending on some other considerations: level of intensity of the background, resolution of the image, and size of the smallest structure that needs to be segmented.

6. Coloring the microcolonies. We wrote this part of code to help the visualization of the segmented microcolonies as the outlines given by the *ROI Manager* don't offer enough contrast. This piece is simply esthetics, but we believe that it is of great help when the user wants to tune the different segmentation parameters and chose the best thresholding method.
7. Sobel-based thresholding strategy. Convolving each pixel of an image by the kernel below will attribute it a new value depending on the surrounding ones.

$$\begin{bmatrix} 2 & 3 & 4 & 3 & 2 \\ 1 & 2 & 3 & 2 & 1 \\ 0 & 0 & 0 & 0 & 0 \\ -1 & -2 & -3 & -2 & -1 \\ -2 & -3 & -4 & -3 & -2 \end{bmatrix} \text{Horizontal Kernel or H}$$

When dealing with horizontal edges, the matrix will force a strong contrast between those pixels within the edge (which will become value 0) and those pixels aside (which will have values above 0). However, when dealing with vertical edges, the same matrix will not highlight any edge, as the operation of convolution will be null.

As we desire to detect all edges, one kernel will not be sufficient, so we use also the vertical kernel below.

$$\begin{bmatrix} 2 & 1 & 0 & -1 & -2 \\ 3 & 2 & 0 & -2 & -3 \\ 4 & 3 & 0 & -3 & -4 \\ 3 & 2 & 0 & -2 & -3 \\ 2 & 1 & 0 & -1 & -2 \end{bmatrix} \text{Vertical Kernel or V}$$

A normalization operation will follow the application of both kernels. This step will enhance the edge contrast depending on the actual structures and not their intensity and will be performed by the operation below.

$$(V^2 + H^2) \times (\text{Connected Map})^2$$

The result is directly used to achieve the segmentation of the microcolonies, using the ***minimum error*** thresholding automatic method [21].

Acknowledgments

The authors would like to thank NEUBIAS, the Network of EUropean BIoimage AnalystS, for the great discussions and support around bioimage processing and analysis. This work was financially supported by the NanoGenoBiofilm Project from the Alembert Institute (IDA) of Ecole Normale Supérieure Paris-Saclay and a doctoral grant from CentraleSupélec.

References

1. De Beer D, Stoodley P (2006) Microbial biofilms. In: Dworkin M, Falkow S, Rosenberg E et al (eds) The prokaryotes, 3rd edn. Springer New York, New York, NY, pp 904–937
2. Bridier A, Tischenko E, Dubois-Brissonnet F et al (2011) Deciphering biofilm structure and reactivity by multiscale time-resolved fluorescence analysis. In: Linke D, Goldman A (eds) Bacterial adhesion: chemistry, biology and physics. Springer Netherlands, Dordrecht, pp 333–349
3. Tolker-Nielsen T, Sternberg C (2014) Methods for studying biofilm formation: flow cells and confocal laser scanning microscopy. In: Alain F, Ramos J-L (eds) Pseudomonas methods and protocols. Springer New York, New York, NY, pp 615–629
4. Yang X, Beyenal H, Harkin G, Lewandowski Z (2000) Quantifying biofilm structure using image analysis. J Microbiol Methods 39:109–119
5. Heydorn A, Nielsen AT, Hentzer M et al (2000) Quantification of biofilm structures by the novel computer program COMSTAT. Microbiology 146:2395–2407
6. Vorregaard M, Lyngby K (2008) Comstat2 -a modern 3D image analysis environment for biofilms. Dissertation, Technical University of Danemark
7. Sternberg C, Christensen BB, Johansen T et al (1999) Distribution of bacterial growth activity in flow-chamber biofilms. Appl Environ Microbiol 65:4108–4117
8. http://www.comstat.dk/
9. Daims H, Lücker S, Wagner M (2006) daime, a novel image analysis program for microbial ecology and biofilm research. Environ Microbiol 8(2):200–213
10. Mueller LN, De Brouwer JFC, Almeida JS, Stal LJ, Xavier JB (2006) Analysis of a marine phototrophic biofilm by confocal laser scanning microscopy using the new image quantification software PHLIP. BMC Ecol 6:1–15
11. Beyenal H, Donovan C, Lewandowski Z, Harkin G (2004) Three-dimensional biofilm structure quantification. J Microbiol Methods 59 (3):395–413
12. Christensen BB, Haagensen JAJ, Heydorn A, Molin S (2002) Metabolic commensalism and competition in a two-species microbial consortium. Appl Environ Microbiol 68:2495–2502
13. Franks AE, Glaven RH, Lovley DR (2012) Real-time spatial gene expression analysis within current-producing biofilms. ChemSusChem 5:1092–1098
14. Møller S, Sternberg C, Andersen JB et al (1998) In situ gene expression in mixed-culture biofilms: evidence of metabolic interactions between community members. Appl Environ Microbiol 64:721–732
15. Serra DO, Richter AM, Klauck G et al (2013) Microanatomy at cellular resolution and spatial order of physiological differentiation in a bacterial biofilm. MBio 4:e00103

16. Teal TK, Lies DP, Wold BJ, Newman DK (2006) Spatiometabolic stratification of Shewanella oneidensis biofilms. Appl Environ Microbiol 72(11):7324–7330
17. Baudin M, Cinquin B, Sclavi B (2017) Understanding the fundamental mechanisms of biofilms development and dispersal: BIAM (Biofilm Intensity and Architecture Measurement), a new tool for studying biofilms as a function of their architecture and fluorescence intensity. J Microbiol Methods 140:47–57
18. https://github.com/Arktthul/Biofilm_Analysis
19. Wilkinson MHF (1998) Optimizing edge detectors for robust automatic threshold selection: coping with edge curvature and noise. Graph Models Image Process 60:385–401
20. https://imagej.net/RATS:_Robust_Automatic_Threshold_Selection
21. Kittler J, Illingworth J (1986) Minimum error thresholding. Pattern Recogn 19(1):41–47

Chapter 8

3D + Time Imaging and Image Reconstruction of Pectoral Fin During Zebrafish Embryogenesis

Hanh Nguyen, Jaume Boix-Fabrés, Nadine Peyriéras, and Elena Kardash

Abstract

Morphogenesis is the fundamental developmental process during which the embryo body is formed. Proper shaping of different body parts depends on cellular divisions and rearrangements in the growing embryo. Understanding three-dimensional shaping of organs is one of the basic questions in developmental biology. Here, we consider the early stages of pectoral fin development in zebrafish, which serves as a model for limb development in vertebrates, to study emerging shapes during embryogenesis. Most studies on pectoral fin are concerned with late stages of fin development when the structure is morphologically distinct. However, little is known about the early stages of pectoral fin formation because of the experimental difficulties in establishing proper imaging conditions during these stages to allow long-term live observation. In this protocol, we address the challenges of pectoral fin imaging during the early stages of zebrafish embryogenesis and provide a strategy for three-dimensional shape analysis of the fin. The procedure outlined here is aimed at studying pectoral fin during the first 24 h of its formation corresponding to the time period between 24 and 48 h of zebrafish development. The same principles could also be applied when studying three-dimensional shape establishment of other embryonic structures. We first discuss the imaging procedure and then propose strategies of extracting quantitative information regarding fin shape and dimensions.

Key words Zebrafish, Pectoral fin, 3D shape analysis, *Fiji/ImageJ*

1 Introduction

Proper shaping of the body and its organs during embryonic development is crucial for the survival of a new organism. To understand how different shapes are established during morphogenesis, live observation of the growing embryos through a microscope is the best approach. During live imaging, three-dimensional datasets over time are acquired, and the quantitative information is extracted by applying image processing algorithms. Typically,

Electronic supplementary material: The online version of this chapter (https://doi.org/10.1007/978-1-4939-9686-5_8) contains supplementary material, which is available to authorized users.

Elena Rebollo and Manel Bosch (eds.), *Computer Optimized Microscopy: Methods and Protocols*, Methods in Molecular Biology, vol. 2040, https://doi.org/10.1007/978-1-4939-9686-5_8, © Springer Science+Business Media, LLC, part of Springer Nature 2019

embryonic tissues are built through a combination of cell division and cell migration processes. Dividing cells provide the bulk material to build embryonic body structures, while migrating cells colonize the growing tissues adding different cell types during organogenesis. Together, these two processes refine the final shape.

An excellent model to study three-dimensional shape establishment in the developing embryo is that of a pectoral fin in zebrafish, which serves as a vertebrate model for limb development [1]. Zebrafish has several advantages over other vertebrate animal models: external embryonic development, translucent embryos, robustness of the developmental process, and it being equipped with most exquisite genetic tools. Pectoral fin structure is relatively simple in early development consisting mainly of two different cell types: the mesodermal domain forming the inner portion of the fin and a layer of ectodermal cells covering the mesoderm. Because pectoral fin is an external organ, it is easily accessible for microscopy imaging and manipulations.

The main molecular mechanisms controlling fin and limb development are conserved among fish, birds, and mammals [2]. Fin bud initiation occurs at about 18 h post fertilization (hpf) at the region of the lateral plate mesoderm, where transcriptional factors, such as Tbx5a [3], regulate cell proliferation within the fin mesenchyme. Downstream to Tbx5a are important morphogen molecules such as those belonging to the family of fibroblast growth factors including Fgf4, Fgf8, Fgf10, Fgf16, and Fgf24 [4–6]; bone morphogenetic factors (BMPs) [7]; and Wnt signaling molecules [8–10]. These, together with other morphogens and signaling molecules, coordinate cell proliferation and migration into the fin area to ensure proper patterning and shaping of the fin bud [4–6, 11, 12].

The main challenge of studying the early stages of pectoral fin formation is to find suitable imaging conditions to monitor its growth over a long-term period that extends up to 24 h. That requires a highly optimized imaging routine ensuring the embryo remains stable and develops properly for the duration of data acquisition. Specifically, the challenges associated with the study of pectoral fin formation are (1) the lack of early-stage markers for identifying fin cells; (2) the lack of mounting strategies to secure embryos for a long-term live imaging; (3) handling large datasets during data processing; and (4) the lack of strategies for automated 3D shape analysis.

Pectoral fin bud becomes morphologically distinct after 2 days' post fertilization; however, identification of the fin tissue at the onset of its formation (18 hpf) remains challenging because of the scarcity of the early-stage markers. Most studies circumvent this problem by utilizing immunostaining and in situ hybridization of fixed samples to observe pectoral fin at early stages. However, these strategies are not suitable for live imaging. To solve this problem, we use either the transgenic line *Tg(mPrx1:EGFP)* to observe fin

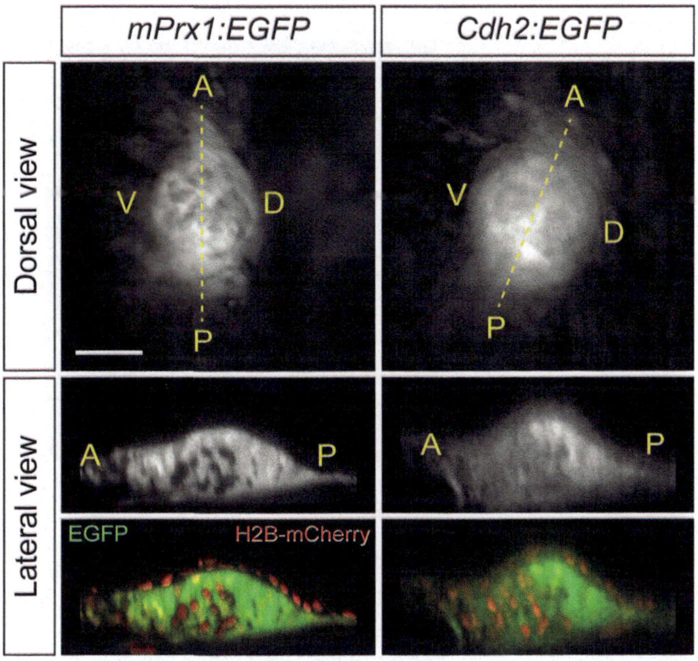

Fig. 1 Pectoral fin labeling in two different transgenic lines. The left column shows the fin bud in a double transgenic embryo *Tg(mPrx1:EGFP)* and *Tg(Xla.Eef1a1:H2B-mCherry)*, and the right column shows the fin in a double transgenic embryo *Tg(Cdh2:EGFP)* and *Tg(Xla.Eef1a1:H2B-mCherry)*. H2B-mCherry labels the nuclei. The dorsal view shows a z-projection along the proximal-distal axix of the fin bud. The lateral view was obtained by applying the "reslice" command along the AP axis (yellow dashed line) of the fin bud. Stage: 36 hpf. Scale bar: 40 μm

mesodermal cells starting from 18 hpf [13] or the transgenic line *Tg*(*Cdh2:EGFP*), in which both the mesodermal and the ectodermal tissues of the fin are labeled (Fig. 1, personal communication). In the Tg(*mPrx1:EGFP*) embryos, only a subpopulation of the mesodermal cells in the fin bud is labeled. The EGFP-negative cells in the fin tissue correspond to the somitic mesodermal cells that migrate into the fin area from the somites 4–7 [11, 12, 14]. *Tg* (*mPrx1:EGFP*) line is useful for studying fin tissue patterning because it allows to distinguish between different cell types within the fin tissue. In the *Tg*(*Cdh2:EGFP*) embryos, both the mesodermal and the ectodermal tissues of the fin are labeled, and the overall signal appears more uniform as opposed to that in the *Tg*(*mPrx1:EGFP*) line (Fig. 1) making this line more suitable for characterizing fin geometry. In this protocol, we analyze fin shape using data obtained with *Tg*(*Cdh2:EGFP*) embryos.

One of the challenges when studying pectoral fin growth is immobilize the embryo ensuring it remains stable for an extended time without causing artifacts such as body deformations, especially when the embryos are older than 18 hpf. Normally, embryos are immobilized in the low melting agarose (LMA) with

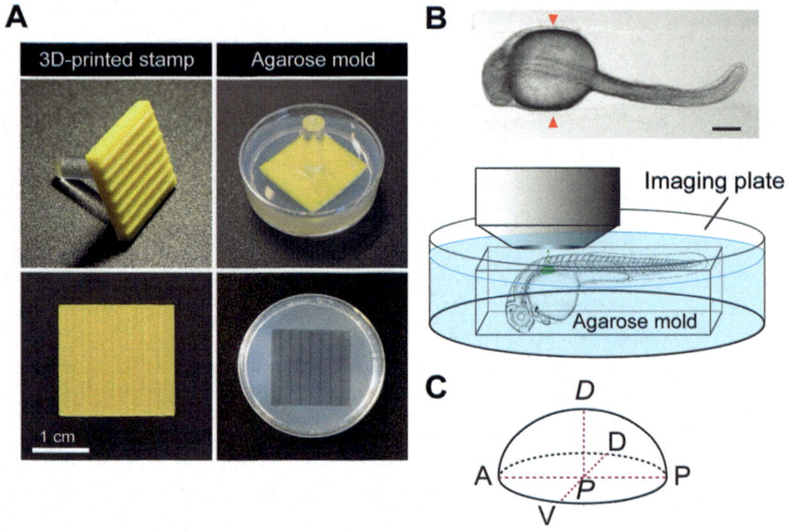

Fig. 2 The mounting setup for imaging pectoral fin growth with the upright microscope. (**a**) The 3D-printed stamp and the imaging mold made with 1.5% agarose. The dimension of the mold base is 2 cm × 2 cm × 2 mm, which fits in a 3.5 cm diameter petri dish. Each groove has the width of 0.64 mm and the height of 0.6 mm. (**b**) The top image shows a bright-field image of the 24-h-old embryo positioned dorsally inside the groove in the agarose mold so that the yolk is inserted between two agarose walls (red arrows). The scheme below demonstrates the orientation of the embryo during live imaging at the confocal microscope relative to the objective. Scale bar: 200 μm. (**c**) The axes of the fin are defined. AP anterior-posterior, DV dorsal-ventral, *PD* (in italics) proximal-distal

concentrations ranging between 0.4 and 1%; however, in our hands, 0.4% LMA was sufficient to compress the fin resulting in deformations after 30 hpf (our unpublished observations). Therefore, ideally, the embryo should be secured during imaging with no compression on its fins and the entire body. This can be achieved using an agarose mold produced with the help of a custom-made 3D-printed stamp (Fig. 2a). The embryo is placed into the mold with its yolk inserted into and held by the groove in the agarose (Fig. 2b). This type of mounting immobilizes the embryo while it allows each part of its body to develop without physical constrains. Another important aspect for acquiring high-quality digital data well-suited for further processing is the proper orientation of the fin. Three principal axes of the fin bud are anterior-posterior (AP), dorsoventral (DV), and proximal-distal (*PD*) (Fig. 2c) [15, 16]. During 3D shape analysis, the three principal axes of the acquired image object are defined with its *z* axis corresponding to the laser path (vertical in most microscopes) and the *xy* axes defining the scanning plane in a confocal microscope. When the embryo is mounted dorsally as shown in Fig. 2b, the fins can be located at both sides of the notochord, on top of the yolk, in the region corresponding to the area between the second and the third

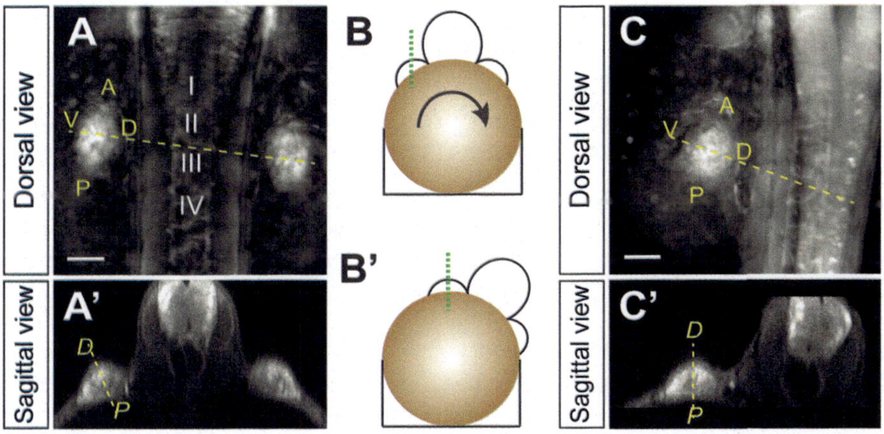

Fig. 3 An optimized mounting of the zebrafish embryo for long-term pectoral fin observation. (**a**) Representative confocal images of the pectoral fin at 32 hpf in the *Tg(Cdh2:EGFP)* embryo. A dorsal projection was done on the confocal *z*-stack of the fin along its *PD* axis by [Image > Stacks > Z project > Sum slices]. Two fins are visible at both sides of the notochord. Anterior of the embryo is to the top. Scale bar: 40 μm. The Roman numbers label the somites: I, II, III, and IV. (**a′**). The sagittal view was obtained by applying the "reslice" command along the yellow dashed line marking the DV axis in (**a**). (**b**) The scheme shows embryo orientation in the agarose, the sagittal view. The *PD* axis of the fin is tilted relative to the laser path (in green dashed line). The arrow indicates the direction of the tilt. (**b′**) The scheme shows the same embryo as in (**b**) after it was tilted to the right. In this orientation, the *PD* axis of the fin is parallel to the laser path, which is shown as the green dashed line. (**c**) A dorsal projection of the confocal *z*-stack of the fin along its *PD iaxis* in the *Tg(Cdh2:EGFP)* embryo that was tilted to the right orienting *the PD* axis of the fin parallel to the laser path. (**c′**) A sagittal view of the embryo in (**c**). Stage: 32 hpf. Scale bar: 40 μm

somites (Fig. 3a). Data acquired in such manner would result in fins' *PD* axes tilted relatively to the laser path, which could lead to an inaccurate estimation of the fin geometry. While in certain experimental setups, it could be beneficial observing both fins simultaneously, we recommend tilting the embryo to orient one of the fins with its *PD* axis parallel to the laser path (Fig. 3b, c′).

Finally, processing large datasets is particularly challenging because they tend to exceed the capacity of most commercially available computers and require expensive and specialized programs. In light of these obstacles, *Fiji* software remains a favorite of the community because it is (1) freely accessible for everyone, (2) user friendly, (3) frequently updated, (4) supported by a large active community of users and developers offering advice, and (5) capable of handling large datasets and reading different raw data formats. There is no universal solution in the field of 3D imaging for analyzing shapes of the growing 3D embryonic structures. In this protocol, we proposed a simple systematic workflow for extracting quantitative information from the raw data. During 3D shape analysis, the typical parameters obtained are: dimensions along the three principal axes; volume; shape descriptors such as roundness; and geometry if an object can be fitted into simple geometrical shapes such as a sphere, ellipsoid, or cylinder.

Below, we discuss the time-lapse live imaging routine of pectoral fin formation in zebrafish at the early stages and the image processing strategies to obtain quantitative information of the fin shape. The present protocol is adapted to zebrafish embryos and pectoral fin; however, the principles listed here can be applied for other model organisms or biological systems.

2 Materials

2.1 Zebrafish Work

1. Zebrafish embryos. To label the cytoplasm in pectoral fin cells, we use either of the two transgenic lines: *Tg(mPrx1:EGFP)* to label the mesodermal domains or *Tg(Cdh2:EGFP)* to label both the mesoderm and the ectoderm of the fin tissue. To label nuclei to follow cellular dynamics, we use the transgenic line *Tg(Xla.Eef1a1:H2B-mCherry)* or any other ubiquitous promoter to drive the global expression of H2B-mCherry. Alternatively, transgenic embryos expressing EGFP in the fin tissue are injected with sense mRNA encoding H2B-mCherry at one-cell stage. Nuclei labeling is optional (*see* **Notes 1** and **2**).
2. Embryo medium. Zebrafish embryos are maintained in the embryo medium as described in [17, 18].
3. Custom-made 3D-printed stamps to make agarose molds for imaging in 35 mm petri dishes (Fig. 2a).
4. 1.5% agarose in embryo medium to make the imaging molds.
5. In case of injecting mRNA, drugs, or morpholinos (optional): microinjection needles are made by pulling glass capillaries 1.0 mm OD.
6. Tricaine to anesthetize embryos (3-aminobenzoic acid ethyl ester).
7. A stereomicroscope equipped with a fluorescence excitation source for embryo selection, dechorionation, and mounting, i.e., Leica MZ10F.

2.2 Molecular Biology

1. mMessage machine for mRNA synthesis with SP6 or T3 polymerases depending on the minimal promoter used.
2. Phenol red (optional).

2.3 Time-Lapse Imaging Setup Using Upright Confocal Fluorescence Microscope

1. Lasers: 488 nm for EGFP excitation and 561 nm for mCherry excitation (*see* **Note 3**).
2. The confocal microscope needs to be equipped with at least two detectors (PMT or GaAsP) to detect EGFP and mCherry emission signals simultaneously.
3. Objectives: Plan-Apochromat water-dipping lens, e.g., 20x, 25x, or 40x (1.0 NA). 20x or 25x objective should be used for imaging the entire fin and the surrounding tissue, and the

40x objective should be used for observing pectoral fin and cellular structures at high magnification.

4. Temperature control setup that allows to keep temperature range between 26 °C and 28 °C during data acquisition.

2.4 Software, Image Analysis Tools, and Sample Images

1. The current algorithm is based on *Fiji* version lifeline 2015 [19, 20] and requires the following plugins (normally should be installed with the current version): *Correct 3D drift* [21, 22] and *Bleach Correction* [23].
2. A sample file "Cdh2_EGFP_Fin.tif" for practicing the protocol presented here is available at the Springer website.
3. Macro for automated processing "Fin_Segmentation_Check.ijm" is available at GitHub [24].

3 Methods

3.1 Preparation of Transgenic Embryos for Overnight Live Imaging

1. Day 1: Set up mating crosses between transgenic fish lines *Tg* (*mPrx1:EGFP*) or *Tg*(*Cdh2:EGFP*) and *Tg*(*Xla.Eef1a1:H2B-mCherry*). Alternatively, set up the crosses using transgenic fish lines either *Tg*(*mPrx1:EGFP*) or *Tg*(*Cdh2:EGFP*) for subsequent microjection of mRNA encoding H2B-mCherry into one-cell stage embryos.
2. Day 2: Collect embryos from the crosses between transgenic lines *Tg*(*mPrx1:EGFP*) and *Tg*(*Xla.Eef1a1:H2B-mCherry*) or inject the *Tg*(*mPrx1:EGFP*) embryos with mRNA encoding for H2B-mCherry to label the nuclei. Alternatively, use the transgenic line *Tg*(*Cdh2:EGFP*) instead of *Tg*(*mPrx1:EGFP*). A detailed protocol for zebrafish embryo microinjection is described elsewhere [25].
3. Raise the embryos to the desired stage: In this protocol, we start the imaging at 28 hpf, when pectoral fin field is clearly visible and appears as two spots located on top of the yolk on both sides of the notochord.
4. Day 3: Using a dissecting stereoscope equipped with the fluorescence excitation source, select positive embryos expressing both EGFP signal in the pectoral fin area, which is visible on both sides of the notochord corresponding to the second and third somites, and H2B-mCherry in every cell. If using double transgenic embryos, the optimal screening time should start at around 20–24 hpf. If the embryos are injected with mRNA for H2B-mCherry, the optimal screening time to select for the homogenous nuclear signal is between 3 and 4 hpf.

3.2 Preparation of Agarose Molds for Overnight Live Imaging

This can be done 1 day before or on the day of imaging (*see* **Note 4**).

1. Prepare 1.5% agarose solution by boiling the agarose powder dissolved in embryo medium and cool it down to 50–55 °C.
2. Add tricaine into the agarose solution to the final concentration of 0.04%.
3. Pour this mixture into the petri dishes, insert the custom-made stamps (Fig. 2a), and wait until the agarose is solidified.

3.3 Embryo Mounting and Image Acquisition Setup with a Confocal/Two-Photon Microscope

1. Using a stereoscope with the fluorescence excitation source, push the embryos into the grooves in the agarose mold as shown (Fig. 2b). Use the EGFP signal in the fin mesenchyme to orient the embryo by tilting it slightly to the right or to the left so that the fin bud to be imaged is facing the objective in such a way that the laser path is parallel to the proximal-distal axis of the fin (*see* Fig. 3a–c′ and **Note 5**).
2. Add sufficient volume of tricaine dissolved in the embryo medium to the final concentration of 0.04% to ensure complete anesthesia and immobilization of the embryos during live imaging.
3. Move the sample from the stereomicroscope to the confocal microscope.
4. If a temperature control setup is available, set the temperature to 26 °C (*see* **Note 6**).
5. For the *z*-stack acquisition, we recommend using 512 × 512 resolution (corresponding to 409.6 × 409.6 μm when zooming 1×), 2 μm section width by adjusting airy unit (AU) to 1.5 in Zeiss LSM 780, and *z*-spacing of 1 μm. The scanning speed should by adjusted depending on the signal quality and the desired time interval (*see* **Note 7**). When defining the range along *z*-axis, include extra distance above and below the fin to accommodate drifts due to morphological changes in the embryo and fin growth along proximal-distal axis (*see* **Note 8**).
6. Zooming 1.5x or 1.2x with 20x or 25x objectives, respectively, would allow to image one fin at the sufficient resolution to extract quantitative information about its shape. For the data presented here, the *xy* resolution is 0.8 μm/px (*see* **Note 9**).
7. Choose a suitable time interval. Depending on the question asked and the *z*-stack thickness, the time interval can be between 2 min and 1 h (*see* **Note 10**).
8. Define laser power to allow for a good-quality signal without saturation of the detectors or massive bleaching during acquisition (*see* **Note 11**).
9. Acquire data.
10. Save the data in the native format of a microscope retaining the metadata, which contains all the quantitative parameters used during data acquisition including *xyz* resolution, time interval, zooming, laser power, and channel information (*see* **Note 12**).

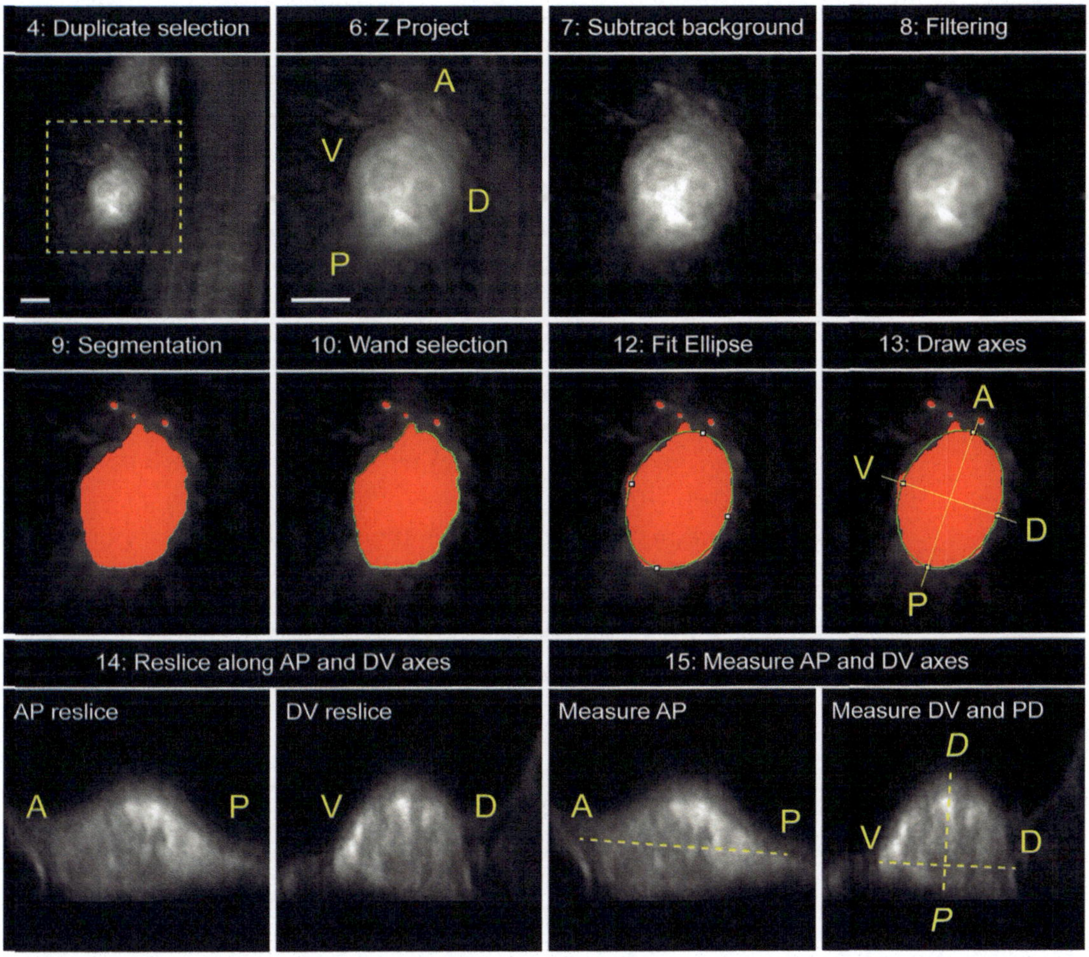

Fig. 4 A step-by-step image processing and the 3D-shape analysis of the pectoral fin using *Fiji*. A selection of steps listed in Subheading 3.4 analyzing the fin shape in the *Tg(Cdh2:EGFP)* embryo at 36 hpf. **Step 4**: Select the region of interest (yellow dotted square) and duplicate it. **Step 6**: Make a z-projection, AP and DV axes are indicated with yellow letters. **Step 7**: Subtract background. **Step 8**: Filtering. **Step 9**: Segmentation, the segmented area is highlighted in red. **Step 10**: Wand selection, the selected red segmented area is enclosed and outlined in green boundary. **Step 12**: Fit Ellipse, the free-form green border transforms into elliptical shape. **Step 13**: Draw axes, AP and DV axes are drawn through the ellipse's vertices using the line tool and labeled in yellow. **Step 14**: Reslice along AP and DV axes. AP and DV are indicated with yellow letters. **Step 15**: Measure AP, DV, and PD axes, as indicated with yellow dotted lines. Scale bar: 40 μm

3.4 Image Processing and 3D Shape Analysis of Pectoral Fin

Steps 1–8 can be conducted on the entire hyperstack, while **steps 9–16** should be applied to each time frame individually to extract fin dimensions at the desired time step. Selected steps from the protocol below are shown in Fig. 4.

1. Launch *Fiji*.
2. Import the image files into the *Fiji* at [File > Import > Bio-Formats...] or [Plugins > Bio-Formats > Bio-Formats Importer] to open as a virtual stack to conserve memory

[26]. In the *View stack with*, select *Hyperstack* (should be selected by default). Select *Use virtual stack*. In the hyperstack, there should be one or two emission channels depending on the number of the fluorophores used, a *z*-dimension and a time dimension. If there are two channels in the original file, do not split them prior to the registration step.

3. Registration. During long-term imaging lasting over several hours, the embryo undergoes significant morphological changes, which cause *xyz* drifts. To correct for these drifts, use the registration plugin at [Plugins > Registration > Correct 3D Drift]. If two emission channels are present, select the channel, which should be used for the registration. Select these options *Edge enhance images for possibly improved drift detection?* and *Use virtual stack for saving the results to disk to save RAM?* (*see* **Note 13**).
4. Duplicate selection. In this step, only the relevant portion of the original image containing the fin is selected to reduce data size (*see* **Notes 14** and **15**). After registration, it is common to have extra frames along *z*-axis, which do not contain actual data and can be removed. To select the portion of a hyperstack, go to [Image > Duplicate...]. Select *Duplicate hyperstack*; choose the channel to duplicate, the *z*-range, and the range between time frames. If there is more than one channel, duplicate each channel separately.
5. Bleach correction (optional). During long acquisition sessions, the signal intensity can be drastically reduced due to photodestruction of the fluorophore. The bleach correction algorithms at [Image > Adjust > Bleach correction] correct this reduction by equalizing signal intensity over time (*see* **Note 16**). For the *Correction Method*, select *Simple Ratio* [27].
6. Z-projection. Make a projection of the stack in the EGFP channel at [Image > Stacks > Z Project...], and select the suitable projection type for the images: *Sum Slices*, *Standard Deviation*, or *Max Intensity* (*see* **Note 17**). Here, we used the *Sum Slices* option. Select *All time frames*. The following **steps 7–9** are performed on *z*-projected stack.
7. Subtracting background (optional). Depending on the promoter used, there might be a non-specific background signal. Background subtraction improves the signal-to-noise ratio in the image of the fin for the *Tg(Cdh2:EGFP)* line. Go to [Process > Subtract Background...]. For *Rolling ball radius*, choose 100 px. Press *OK* (*see* **Note 18**).
8. Filtering. Filtering is often used during image processing routines to reduce the background noise and improve overall image quality for images with low signal intensity. It also optimizes the image for the subsequent thresholding step. Here,

we use a median filter. Go to [Process > Filters > Median . . .]. For *Radius*, use 2 px.

9. Segmentation. This step is necessary to define the fin area to be used for shape analysis. During image segmentation or masking, the minimum and the maximum intensity values are defined, and only the portion of the image between these values is selected. To segment the image, use [Image > Adjust > Threshold. . .]. Select *Default*, *Red*, and *Dark Background*. If necessary, adjust the threshold manually (*see* **Note 19**). The following **steps 10–16** are performed for each time frame.
10. Use *Wand tool* from the main menu to select the segmented area as a region of interest (ROI). With the *Wand tool* selected, place the mouse anywhere on the fin area.
11. Define the parameters to be measured at [Analyze > Set Measurements. . .]. Select *Shape descriptors* and *Fit ellipse*. To obtain the parameters specified for the selected shape, use [Analyze > Measure] or press the key *m* on the keyboard. These parameters can be used as readout of shape when comparing between different developmental stages and mutant conditions (*see* **Note 20**).
12. Fit ellipse using [Edit > Specify > Fit Ellipse]. To keep the ellipse overlay permanently on the ROI, use the "Flatten" command [Image > Overlay > Flatten], or press the combination of "Ctrl+Shift+F" on the keyboard. This will generate another image with the elliptical shape outlined and its vertices visible.
13. Using the image with elliptical shape outlined generated during **step 12**, draw the AP and DV axes with the line tool from the main menu using the vertices of the ellipse. Store each axis line by adding it to the *ROI Manager* at [Edit > Selection > Add to Manager] or by pressing *t* on the keyboard.
14. Duplicate the *z*-stack corresponding to the time frame of the *z*-projected image used in **steps 10–13**. Copy AP axis line on it by selecting it from ROI Manager. Obtain the lateral view corresponding to the optical section along the AP axis of the fin using [Image > Stacks > Reslice(/). . .] or by pressing "/" on the keyboard; choose *Flip vertically* (*see* **Note 21**). Repeat for the DV axis line to obtain the sagittal view.
15. Measuring the axes dimensions. The lateral and the sagittal dissections of the fin are used to obtain fin's dimensions along its three axes. To do that, draw a straight line along the base of the fin using either lateral or sagittal resliced images and select [Analyze > Measure] or press *m* to obtain the values of the AP and DV axes lengths. Draw a straight line from the bottom to the top to measure the *PD* axis length using either lateral or sagittal resliced images.

16. 3D reconstruction. Use 3D Viewer plugin at [Plugins > 3D Viewer] to create a 3D representation of a *z*-stack. Choose *Volume* or *Surface* as an option and one of the pseudocolors. Yellow and white usually are the best (*see* **Note 22**). To take a snapshot from a desired orientation, go to [View > Take snapshot]. To generate a movie from 3D reconstructed image, go to [View > Record 360 deg. rotation] or use free-hand recording at [View > start freehand recording > stop freehand recording] (*see* **Note 23**).
17. Repeat **steps 10–16** for every time frame of interest (Fig. 5). For cases where many time points need to be analyzed, we provide the macro "Fin_Segmentation_Check.ijm," which runs **steps 10–15** automatically on a preprocessed time-lapse dataset, as explained in Subheading 3.5.

3.5 Macro for Automated Image Segmentation

"Fin_Segmentation_Check.ijm" macro provided in this chapter serves for automated analysis of shape over time. Here, we use this macro for analyzing pectoral fin shape; however it can be used to analyze similar geometrical shapes as well. The macro automatically acquires the dimensions of AP and DV axes and generates lateral and sagittal views (or orthoslices along these axes) for every time step in a 3D + time image. Processing steps

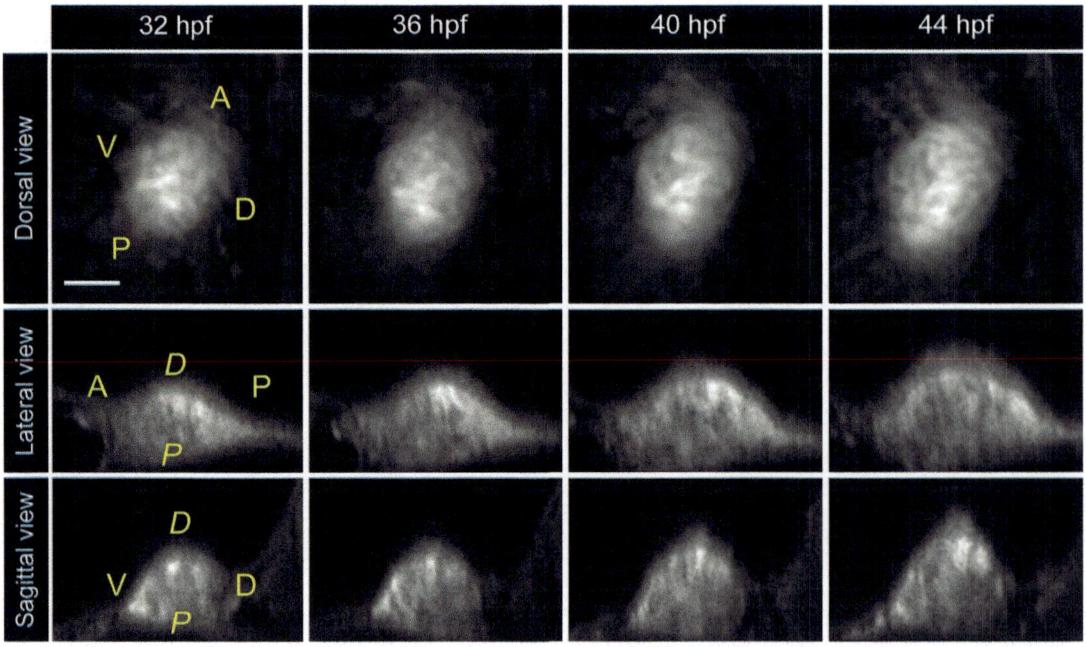

Fig. 5 Evolution of the pectoral fin shape along its three axes. The evolution of the growth along three axes of the fin in a transgenic embryo *Tg(Cdh2:EGFP)*. The *z*-stacks from a time-lapse movie of the growing fin were analyzed at four different time points following the steps listed in a protocol. AP, DV, and PD axes are indicated with yellow letters. *PD* is in italics. Scale bar: 40 μm

described in Subheading 3.4 are recorded to create an initial code using the macro recorder in *Fiji* at [Plugins > Macros > Record...]. Then, programming functions such as variables, loops, and arrays to store the data are added to optimize the automation. Follow the steps below to operate the macro.

1. Open the file "Cdh2_EGFP_Fin.tif" by dragging it into the *Fiji* software.
2. Open the macro by dragging the file "Fin_Segmentation_Check.ijm" into *Fiji* software and press *Run* within the macro dialog box.
3. A successive dialog boxes will pop up asking to specify the destination folder for storing the results. Choose the folder to store the results. Another dialog box will pop up asking whether lateral and sagittal views should be flipped vertically; choose one option and press *OK* (*see* **Note 21**).
4. Press *OK* when prompted "Process all 9 images."
5. The macro will generate the files corresponding to the AP and DV orthoslices over time called "AP.tif" and "DV.tif," respectively; the segmentation file called "Segmentation_Check.tif"; the file containing the selection of the AP and DV axes for every time step called "Zproj_series.tif"; and the Excel file called "FERET.xls" containing the AP and DV values for each time frame.

4 Notes

1. Labeling the tissue. When studying three-dimensional shapes, it is best to use the fluorescent protein of choice targeted to the cytoplasm. That would allow a uniform labeling of the structure, which is required for the segmentation step during image processing. Using additional markers for nuclei and cell membrane could be useful for analyzing cellular dynamics and cell shape outlines. When studying three-dimensional morphogenesis of embryonic shapes over time, in addition to local cell proliferation which supports growth, cell migration may further contribute by adding more material into the growing tissue, which is the case for pectoral fin. In that case, cell tracking is essential to complete the growth analysis; therefore, cell nuclei labeling should be included.
2. Choosing a fin-specific promoter. The choice of the promoter depends on the question posed in a study such as individual cell behavior, tissue patterning, or organ 3D shape. In Fig. 1, we compared two different promoters that drive EGFP expression in the fin tissue. In the *Tg(mPrx1:EFGP)* embryos, a portion of fin mesodermal cells that originate through local divisions in

the fin area is labeled. In the *Tg(Cdh2:EFGP)* embryos, we observed a more uniform EGFP signal in both mesodermal and ectodermal tissues allowing for a more accurate estimation of pectoral fin dimensions and shape.

3. Using red fluorophores for nuclei labeling. Depending on the red fluorophore used for nuclei labeling, the excitation peak may vary between 554 nm for DsRed Express 2 and 598 nm for mRaspberry [28].
4. Toxicity and anesthesia. We recommend using freshly made agarose molds for every session to minimize bacterial and fungal contamination. If the embryo exhibits abnormal development as compared to the control embryos after the imaging session, decreasing the concentration of tricaine in the agarose and in the medium is an option. Depending on the embryonic stage, the tricaine concentration may vary between 0.016 and 0.04%.
5. Embryo orientation/defining the axes. Fin bud exhibits a symmetry resembling a semi-ellipsoid. Since the fin's proximal-distal axis is slightly tilted relative to the laser path when the embryo is mounted dorsally, we suggest tilting the embryo after mounting to ensure the laser path is aligned with the fin's proximal-distal axis (Fig. 3a–c′). Such orientation ensures optimal imaging conditions conforming to the fins' geometry at early stages of development.
6. Temperature control. It is preferable to ensure constant temperature of 26–28 °C in the imaging plate during the imaging session by using a temperature control setup (e.g., OkoLab). If not possible, it is best to keep the room temperature above 23–24 °C during the imaging session.
7. Compromise between different parameters during data acquisition. Depending on the microscope setup used (confocal or two photon), the AU may vary. From our experiences, using the proposed combination of parameters (spatial resolution, time interval, and *z*-spacing) for long imaging session (up to 20 h) is a good starting point to achieve a compromise between signal intensity, image quality, and time resolution to obtain good quality images for data analysis.
8. 3D drift due to morphological changes. From 24 to 48 hpf, we observe the initial upward shift of the fin area along the *z*-dimension, which is followed by the downward shift along the *z*-axis as the embryo continues to grow and undergo morphological shape changes in the body. Additional shift in *xy* plane are observed as well. To account for shifts in the *z*-

dimension and ensure that the region of interest would remain in the frame at the end of the movie, we recommend adding an additional range of at least 100 μm above and below the fin structure along *z* axis. The total *z*-range should be between 250 and 300 μm, of which the estimated fin thickness should not exceed 100 μm along PD axis by 48 hpf. Becuase of the possible *xy* drifts during the imaging, we do not recommend using high zooming factors (not higher than 2x for 20x objective).

9. 3D shape analysis requires lower resolution as compared to nuclear tracking. To quantify 3D shape changes over time, it is not essential to obtain high pixel resolution in the image as long as the cytoplasmic labeling is sufficiently uniform for segmentation. On the other hand, in the growing pectoral fin, we observe a significant increase in cellular density and compaction over time; therefore, successful tracking of all nuclei requires much higher resolution in the image (e.g., 0.5 μm/px) to be able to distinguish individual cells at the later stages. The field of view with the fin should include no more than somites 1–4 in order to achieve sufficient resolution for visualizing the fin in the context of its environment and to characterize nuclei dynamics.
10. Time interval between the frames would depend on the specific question. Fin growth is a relatively slow process, and the time interval can be up to 1 h in certain cases. Such long-time interval would allow simultaneous observation of several embryos using a motorized stage. When monitoring cellular dynamics, however, 2–4 min intervals between consecutive time steps are preferred for cell tracking algorithms to follow each nuclei successfully.
11. Bleaching. To limit photobleaching, we recommend reducing laser power when possible. Other options to reduce bleaching include: increasing time interval between frames; using wider pinhole and therefore thicker optical section requiring less illumination to obtain a good signal; and scanning at a faster speed; the latter will reduce image quality but would allow for a longer imaging time.
12. Formats to save raw data. For Zeiss microscopes, we recommend saving the data in the File_Name.lsm format. For Leica microscopes, we recommend saving the data in the File_Name.lif format. Both formats are readable by the *Fiji* software.
13. The Correct 3D drift script in *Fiji* uses positional information to register time points to each other and subsequently adds

extra *xyz* spacing to keep the region of interest at relatively the same position over time [22]. Depending on the specific subcellular structure labeled, the results might differ. Therefore, if performed separately on individual channels, they would be corrected differently, and the channels would be mismatched. When performing registration on a hyperstack consisting of two channels, one channel will be chosen as the guide so that the remaining channel would be registered in the same fashion/configuration.

14. Memory issues can arise because of the insufficient memory and storage available for image processing in the computer. When studying morphogenetic processes over longer time, it is common to acquire large sets of data (tens and hundreds of gigabytes). In addition, adding extra *xyz* space to compensate for 3D drifts would further increase the file size. For pectoral fin, the size of time-lapse image files would range typically between 20 and 80 Gb (covering between 15 and 24 h of the developmental process). The size of these files exceeds the memory and processor capacity of most commercially available personal computers. To reduce data size, we recommend selecting only the relevant region of interest cutting off the empty *xyz* space. To identify the relevant ROI containing the object to be analyzed, first we make a *z*-projection over time. This can be done on the virtual stack and therefore would not consume a massive amount of memory. [Image > Stacks > Z Project...], choose "Sum Slices" or "Average Intensity." Select [Edit > Selection > Specify...] to generate a ROI with defined width and height including the area to be analyzed. Check that the selection includes the object throughout the entire time series.

15. Using the same ROI on the virtual stack, duplicate the ROI containing the object [Image > Duplicate ...]. When duplicating, select only the range of *z*-slices containing the object. This procedure will help reduce the size of datasets. Avoid using [Image > Crop] command on virtual stacks, as it will not keep the *t* and *z*-dimensions.

16. Bleach correction. We recommend trying different algorithms of bleach correction available with *Fiji*: *simple ratio*, *exponential fitting*, and *histogram matching* [26]. Bleach correction is very useful for visualization and improved nuclear detection; however, the data cannot be used for intensity-based quantitative analysis because this algorithm alters the intensity values in the images.

17. Z-projection types. Depending on the purpose of the visualization, choose the most appropriate *z*-projection types: we recommend *Max Intensity*, *Sum Slices*, or *Standard Deviation*

depending on the specific situation. For example, *Sum Slices* and *Standard Deviation* would be the preferred options for uniform signal such as cytoplasmic labeling. *Max Intensity* would be useful to highlight specific structures, which have higher intensity levels. We advise to try different options and choose the most suitable one for the specific sample [29].

18. Subtract background algorithm employs a *Rolling ball* algorithm [30]. The rolling ball radius (in pixels) should be larger than the largest object in the image that is not part of the background. Preliminary manual measurement of the largest object size helps to decide which size would be optimal for the rolling ball.

19. Image segmentation. In case of variations in intensity levels, the default algorithm may ignore portions of the image that contain important information/the structure of interest. In that case, we suggest adjusting the threshold manually by defining the range between the min and max values using the sliders in the *Threshold* window. Alternatively, there is an option to try all thresholding algorithms simultaneously provided by *Fiji*, which only works on 8-bit or 16-bit images. To do that select [Image > Type] and choose *8-bit* or *16-bit*. Then select [Image > Adjust > Auto Threshold]. Keep the default settings, which are *Method: Try all* and *White objects on black background* (if the image is grayscale). This produces a montage with results from all the methods that allows to experiment with different algorithms on a particular image. We recommend using this approach on a single projected image instead of a *z*-stack to avoid running out of memory [31].

20. Measuring different parameters. In this protocol, in order to minimize human errors and bias when deciding on the position of the axes, we used *Shape descriptors* and *Fit ellipse* as most informative features of the shape. Other possible parameters may include *Area*, *Perimeter*, *Feret's diameter* for irregular shapes, and other options.

21. Depending on the microscope configuration during *z*-stack acquisition, *Flip vertically* should be selected during reslicing.

22. Differences between *Volume* and *Surface* options in 3D Viewer plugin. When signal intensity is low or in case of a homogenous signal, choose *Surface* for a better contrast in a 3D reconstructed image (*see* the comparison in Fig. 6).

23. Improving the quality of 3D reconstruction. We recommend running the smooth filter on the *z*-stack at [Process > Smooth].

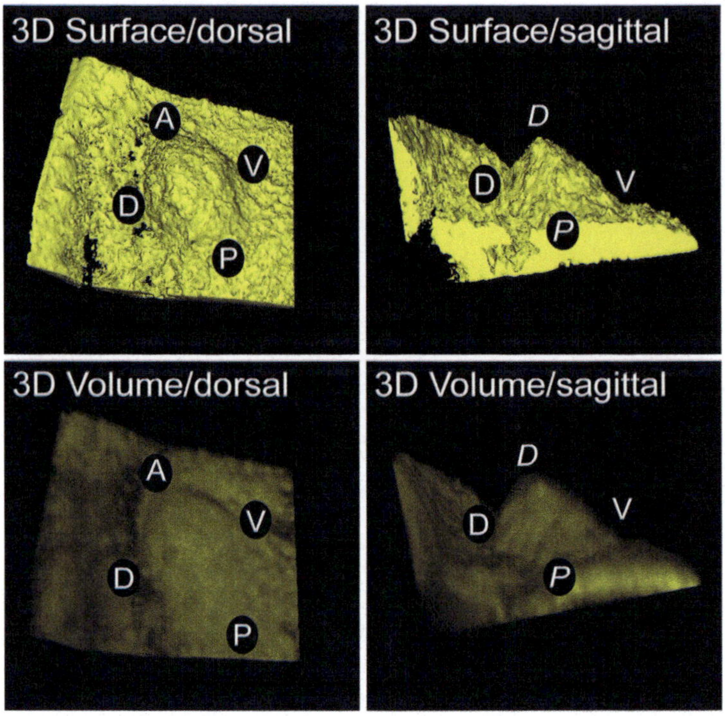

Fig. 6 3D reconstruction. The comparison between two different styles during volume reconstruction with 3D Viewer plugin for the fin at 42 hpf in the transgenic embryo *Tg(Cdh2:EGFP)*. The *z*-stack was filtered using "Smooth" filter prior 3D Viewer plugin application

Acknowledgements

Elena Kardash is grateful to Dr. Nadine Peyriéras and her laboratory for providing supporting environment for working on the present chapter. E.K. is supported by ANR-10-INBS-04 through the National Infrastructure France-BioImaging supported by the French National Research Agency, and H.N is supported by 2017-ITN-721537 as part of the ITN ImageInLife Marie Skłodowska-Curie Actions. Jaume Boix-Fabrés is supported by a PTA contract from the Spanish Ministry of Economy and Competitiveness at the Molecular Imaging Platform IBMB-PCB.

References

1. Mercader N (2007) Early steps of paired fin development in zebrafish compared with tetrapod limb development. Develop Growth Differ 49:421–437
2. Zeller R, López-Ríos J, Zuniga A (2009) Vertebrate limb bud development: moving towards integrative analysis of organogenesis. Nat Rev Genet 10:845–858
3. Ahn D-G, Kourakis MJ, Rohde LA et al (2002) T-box gene tbx5 is essential for formation of the pectoral limb bud. Nature 417:754–758
4. Fischer S, Draper BW, Neumann CJ (2003) The zebrafish fgf24 mutant identifies an additional level of Fgf signaling involved in vertebrate forelimb initiation. Development 130:3515–3524

5. Norton WHJ, Ledin J, Grandel H et al (2005) HSPG synthesis by zebrafish Ext2 and Extl3 is required for Fgf10 signalling during limb development. Development 132:4963–4973
6. Nomura R, Kamei E, Hotta Y et al (2006) Fgf16 is essential for pectoral fin bud formation in zebrafish. Biochem Biophys Res Commun 347:340–346
7. Christen B, Rodrigues AMC, Monasterio MB et al (2012) Transient downregulation of Bmp signalling induces extra limbs in vertebrates. Development 139(14):2557–2565
8. Nagayoshi S, Hayashi E, Abe G et al (2007) Insertional mutagenesis by the Tol2 transposon-mediated enhancer trap approach generated mutations in two developmental genes: tcf7 and synembryn-like. Development 135(1):159–169
9. Ng JK, Kawakami Y, Büscher D et al (2002) The limb identity gene Tbx5 promotes limb initiation by interacting with Wnt2b and Fgf10. Development 129(22):5161–5170
10. Ober EA, Verkade H, Field HA et al (2006) Mesodermal Wnt2b signalling positively regulates liver specification. Nature 442 (7103):688–691
11. Wyngaarden LA, Vogeli KM, Ciruna BG et al (2010) Oriented cell motility and division underlie early limb bud morphogenesis. Development 137:2551–2558
12. Mao Q, Stinnett HK, Ho RK (2015) Asymmetric cell convergence-driven zebrafish fin bud initiation and pre-pattern requires Tbx5a control of a mesenchymal Fgf signal. Development 142:4329–4339
13. Hernández-Vega A, Minguillón C (2011) The Prx1 limb enhancers: targeted gene expression in developing zebrafish pectoral fins. Dev Dyn 240:1977–1988
14. Masselink W, Cole NJ, Fenyes F et al (2016) A somitic contribution to the apical ectodermal ridge is essential for fin formation. Nature 535 (7613):542–546
15. Zeller R, Duboule D (1997) Dorso-ventral limb polarity and origin of the ridge: on the fringe of independence? Bioessays 19:541–546
16. Recher G, Jouralet J, Brombin A et al (2013) Zebrafish midbrain slow-amplifying progenitors exhibit high levels of transcripts for nucleotide and ribosome biogenesis. Development 140:4860–4869
17. Kimmel CB, Ballard WW, Kimmel SR et al (1995) Stages of embryonic development of the zebrafish. Dev Dyn 203(3):253–310
18. Westerfield M (2000) The zebrafish book: a guide for the laboratory use of zebrafish (Danio rerio). http://zfin.org/zf_info/zfbook/cont.html
19. Schneider CA, Rasband WS, Eliceiri KW (2012) NIH image to ImageJ: 25 years of image analysis. Nat Meth 9(7):671–675
20. Fiji. https://imagej.net/Fiji/Downloads
21. Parslow A, Cardona A, Bryson-Richardson RJ (2014) Sample drift correction following 4D confocal time-lapse imaging. J Vis Exp. https://doi.org/10.3791/51086.
22. Correct 3D drift. https://github.com/fiji/Correct_3D_Drift/releases/tag/Correct_3D_Drift-1.0.1
23. Miura K (2004) CorrectBleach. https://zenodo.org/record/30769 - .W4xTHX59hJw
24. Fin Segmentation automated macro. https://github.com/MolecularImagingPlatformIBMB/ZebrafishFin.git
25. Rosen JN, Sweeney MF, Mably JD (2009) Microinjection of zebrafish embryos to analyze gene function. J Vis Exp 25:e1115
26. Virtual Stack. https://imagej.nih.gov/ij/docs/guide/146-8.html - sub:Virtual-Stacks
27. Bleach Correction. https://imagej.net/Bleach_Correction
28. Piatkevich KD, Verkhusha VV (2011) Guide to red fluorescent proteins and biosensors for flow cytometry. Methods Cell Biol. Elsevier 102:431–461
29. Z Project. https://imagej.net/Z-functions
30. Rolling ball. http://imagejdocu.tudor.lu/doku.php?id=gui:process:subtract_background
31. Threshold. http://imagej.net/Auto_Threshold - Available_methods

Chapter 9

Automated Macro Approach to Remove Vitelline Membrane Autofluorescence in *Drosophila* Embryo 4D Movies

Jaume Boix-Fabrés, Katerina Karkali, Enrique Martín-Blanco, and Elena Rebollo

Abstract

This chapter provides an *ImageJ/Fiji* automated macro approach to remove the vitelline membrane autofluorescence in live *Drosophila* embryo movies acquired in a 4D (3D plus time) fashion. The procedure consists in a segmentation pipeline that can cope with different relative intensities of the vitelline membrane autofluorescence, followed by a developed algorithm that adjusts the extracted outline selection to the shape deformations that naturally occur during *Drosophila* embryo development. Finally, the fitted selection is used to clear the external glowing halo that, otherwise, would obscure the visualization of the internal embryo labeling upon projection or 3D rendering.

Key words Autofluorescence, Vitelline membrane, *Drosophila* embryo, Membrane segmentation, 3D projection, ImageJ macro, Live imaging, Fitting algorithm

1 Introduction

Live imaging of *Drosophila* embryos has greatly contributed to the understanding of fundamental questions that concern morphogenesis, differentiation, and growth control [1, 2]. Continuous advances in imaging probes, fluorescence microscopy techniques, and image analysis tools have progressively enabled to peer more deeply into the inhomogeneous and constantly moving three-dimensional embryo architecture, at increasing temporal and spatial scales. Among the multifold difficulties associated with embryo preparation and imaging [3], autofluorescence stands as a naturally occurring optical barrier which, to make matters worse, mostly arises at the excitation wavelengths (violet to yellow) routinely used to image common fluorescent protein (FP) variants.

Electronic supplementary material: The online version of this chapter (https://doi.org/10.1007/978-1-4939-9686-5_9) contains supplementary material, which is available to authorized users.

Elena Rebollo and Manel Bosch (eds.), *Computer Optimized Microscopy: Methods and Protocols*, Methods in Molecular Biology, vol. 2040, https://doi.org/10.1007/978-1-4939-9686-5_9, © Springer Science+Business Media, LLC, part of Springer Nature 2019

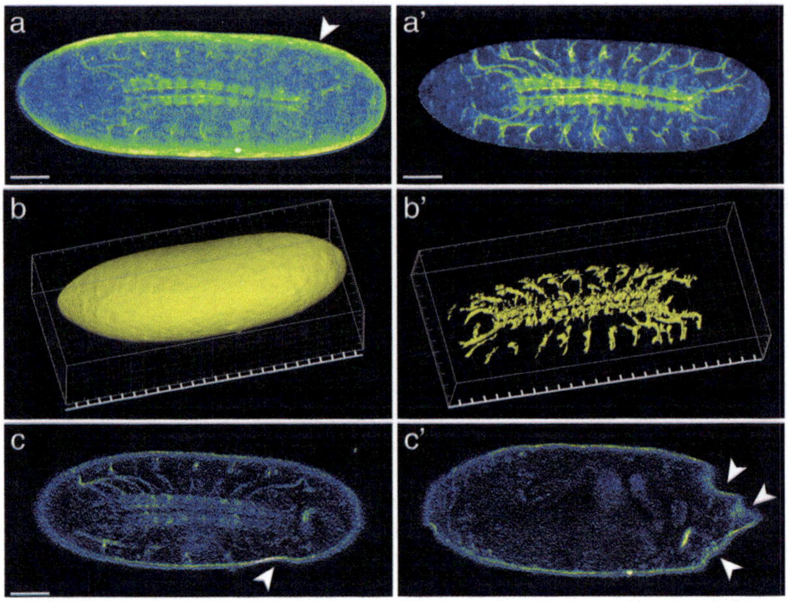

Fig. 1 Different representations of a 3D stack from a *Drosophila* embryo expressing an endogenous GFP-tagged variant of the protein Fasciclin 2 (Fas2-GFP) which localizes to motor axons and peripheral glia. The upper panel shows the 3D maximum projection before (**a**) and after (**a′**) the vitelline membrane autofluorescence has been removed using the protocol provided in this chapter; notice how the projected glowing halo that obscures the signal all around the specimen in (**a**) (arrowhead) has disappeared in (**a′**); *green to fire* LUT is used to enhance intensity grading; the bar is 50 μm. The intermediate panel shows a 3D rendering of the same embryo 3D stack, both before (**b**) and after (**b′**) outline autofluorescence removal; notice how the internal structures become visible in (**b′**); the ruler marks depict 50 μm. The lower panel contains two single images at different time points throughout embryo development; notice the lobe-like invaginations (arrowheads) that typically occur due to embryo twitches (**c**) and peristaltic movements (**c′**)

Peculiarly, the vitelline membrane shows up as a glowing halo at the embryo surrounding, which interferes with proper 3D rendering and obscures the visualization of the internal labeled structures (see Fig. 1a, b). Furthermore, light scattering at the interface between different mediums increases with the refraction index mismatch [4]; accordingly, the use of breathable mounting mediums such as mineral or halocarbon oil [5, 6], characterized by its high refraction index as compared to the watery embryo tissues, does enhance the undesired glowing at the embryo outline.

Vitelline membrane removal [7], either mechanical or chemical, is technically challenging, reduces embryo viability, and disrupts its 3D architecture. On the other hand, the optical sectioning capabilities of confocal laser scanning (CLSM), multiphoton (MPM), or light sheet fluorescence microscopy (LSFM) greatly reduce background autofluorescence; however, they do not fully eliminate the surrounding embryo halo, and, often, the 3D analysis of the acquired specimens is conditioned to selecting areas far from the embryo edge or reducing the 3D stack dimensionality. In addition, FPs in the red spectral region, which have much less autofluorescence contribution [8], are not always at hand for the topic of interest and, when available, do not cover a spectrum wide enough to accomplish multichannel imaging on their own. Last but not least, spectral imaging and linear unmixing can successfully remove autofluorescence by separating its spectra from that of any overlapping FP; however, limitations to this approach often appear in live imaging, where the signals of interest are usually weak and photodamage by intense irradiation must be avoided [9].

In case all the aforementioned constrains come about, removing the embryo surrounding autofluorescence a posteriori becomes a fair alternative to optimize signal projection and 3D visualization. Image segmentation techniques allow for this goal but face several difficulties on this particular matter (see Fig. 1). First, the ratio between signal and autofluorescence may change, even over a single acquisition experiment, depending on several circumstances such as (1) brightness and photostability of the FPs used, (2) imaging depth, or (3) expression level modulation of the tagged protein over time. Second, signal proximity to the embryo edge may result in overlap of the latter with the glowing halo, thus hampering proper segmentation. Finally, the twitches and peristaltic waves (Fig. 1c–c′) that typically occur during early and late *Drosophila* embryo development, respectively [10], deform the edge shape, thus precluding the use of algorithms developed to extract fixed 3D volumes [11].

Aimed to circumvent all the above limitations, this chapter provides an *ImageJ/Fiji* [12, 13] macro approach tailored to successfully remove the vitelline membrane autofluorescence in live *Drosophila* embryos imaged in a 4D (3D over time) fashion. The method first applies a preprocessing pipeline that reduces noise and equalizes the signal intensity variations, thus precluding a good segmentation of the vitelline membrane autofluorescence no matter how it compares with the embryo label in use. Then, global thresholding and segmentation are performed, plus some additional particle filtering to clear up debris and other embryo-derived specks floating in the mounting medium. The resulting binary mask is further treated to extract the embryo external outline which, after some adjustments, is used as template to remove the autofluorescent halo that surrounds the embryo without affecting any of the internal labeling. A fitting algorithm has been designed that adjusts the outline selection to the particular lobe-like shape deformations that occur toward the end of the embryonic development. Both the general segmentation pipeline and the fitting algorithm have been integrated in a macro called "DmNutcracker.ijm" that performs the processing automatically and allows to adjust several parameters to be able to process embryos labeled with a wide range of FPs. This chapter provides the step by step explanation and rationale of the method, together with the macro and sample images to run it.

2 Materials

1. Images. CLSM and MPM 4D movies of live *Drosophila* whole-mounted embryos expressing a protein of interest tagged with any FP that excites within the blue-to-yellow spectral range (*see* **Note 1**). Any embryonic developmental stage is suitable for this protocol, even late stages. This chapter provides sample images, both single snapshots and 4D stacks, from three *Drosophila* strains: (1) the exon trap *fas2*$^{GFP\text{-}CB03613}$, where motor axons and peripheral glia are GFP labeled at the central nervous system (CNS) [14]; (2) *w; UAS-GFP; pucE69I–Gal4/TM6B*, containing a GFP reporter of *puc* activity, localized to some neuron subsets of the central and peripheral nervous system [15]; and (3) *w;;UAS-[Moe::GFP]/TM3 Ser* that overexpresses a GFP-tagged version of Moesin that labels the cortical cytoskeleton [16], crossed by *w;;repo-Gal4/ TM3,Sb* to drive expression to the CNS [17]. Embryos were mounted in a hanging drop of halocarbon oil and imaged using a LSM Zeiss780 confocal system equipped with a 25× (NA = 0.8) apochromatic multi-immersion objective (*see* **Note 2**). The movies are provided as 4D hyperstacks in tiff format (*see* **Note 3**).
2. Software. *ImageJ*, an open-source image-processing and analysis platform [12], originally developed at the National Institutes of Health (Bethesda, Maryland, USA). This protocol has been developed in the *Fiji Life-line 22th December 2015 ImageJ's* distribution [13]. This version can be downloaded at [18] and requires *Java 6*. The description of the *ImageJ* built-in macro functions used can be found at [19].
3. Tools. Macros (*see* **Note 4**) are developed as indicated in the current protocol. The fully automated macro "DmNutcracker.ijm" that processes 4D stacks is provided in this chapter and is also available at [20] for future version updates. Two intermediate macros "Sect.3.1.ijm" and "Sect.3.2.ijm" are also provided that will facilitate understanding the methods explained in the respective sections.
4. Computer. In principle the protocol will work in any operative system (MacOs, Windows, Lynux) as long as the right *Java* and *Fiji* versions are installed. The minimum requirements will depend on the size of the file to be processed, being the main one having a RAM memory at least 2.5× the file size.

3 Methods

The first section describes the sequence of image-processing steps that will be used to remove, in a single 2D image, the vitelline membrane autofluorescence that outlines the embryo. An upgrade to this pipeline, using the fitting algorithm developed in this chapter, is explained in Section 2. These two protocols will serve to both process individual images and test the parameters to be applied on whole 4D stacks that will be handled as explained in the last section.

3.1 Removing Vitelline Membrane Autofluorescence in a Single Image

The main image-processing steps required to remove the glowing embryo outline are illustrated in Fig. 2. First, a preprocessing pipeline (**steps 2–5**, Fig. 2b–e) has been carefully designed to be able to equalize the vitelline membrane autofluorescence intensity with respect to the internal fluorescence of the labeled structures. This pipeline is key since, often, the autofluorescence is weaker than the fluorescence signal, thus hampering proper vitelline membrane segmentation. This preprocessing pipeline will make possible to work on different sets of images, having very different autofluorescence relative levels, just by adjusting a few parameters. After preprocessing, global thresholding and segmentation is applied (**steps 6–8**, Fig. 2f–h), including some particle filtering that will eliminate floating specks that may distort the definition of the embryo shape. Last, the region of interest (ROI) containing the embryo outline is extracted, shrunk up to fit the internal halo edge, and used to clear the glowing halo on the original image (**steps 9–12**, Fig. 2i–l). All these steps can be manually performed after opening the *Fiji*'s macro recorder at [Plugins > Macro > Record...], in order to create a basic script that performs the task automatically. Such basic code, slightly changed to improve usability, is provided in this chapter as "Sect.3.1.ijm." To use this script, go to **step 12**; otherwise follow the steps below.

1. Open image "DmFasII_early.tif" in *Fiji* by [File > Open...] or by drag and drop on the *Fiji* bar (the raw image is shown in Fig. 2a).
2. Apply *Subtract background* at [Process > Subtract Background] (*see* **Note 5** and Fig. 2b). Use this option for images having a strong vitelline membrane signal, such as the current image or the image "DmPuckered_early.tif," where approximately a radius of 20 needs to be used. However, avoid this step if the vitelline membrane to be removed is very weak, as occurs in "DmMoesin_early.tif."
3. Enhance the image contrast using [Process > Enhance Local Contrast (CLAHE] (*see* **Note 6** and Fig. 2c). This option uses an adaptive equalization that will help equalize the levels of the outside autofluorescence with respect to the internal labeled

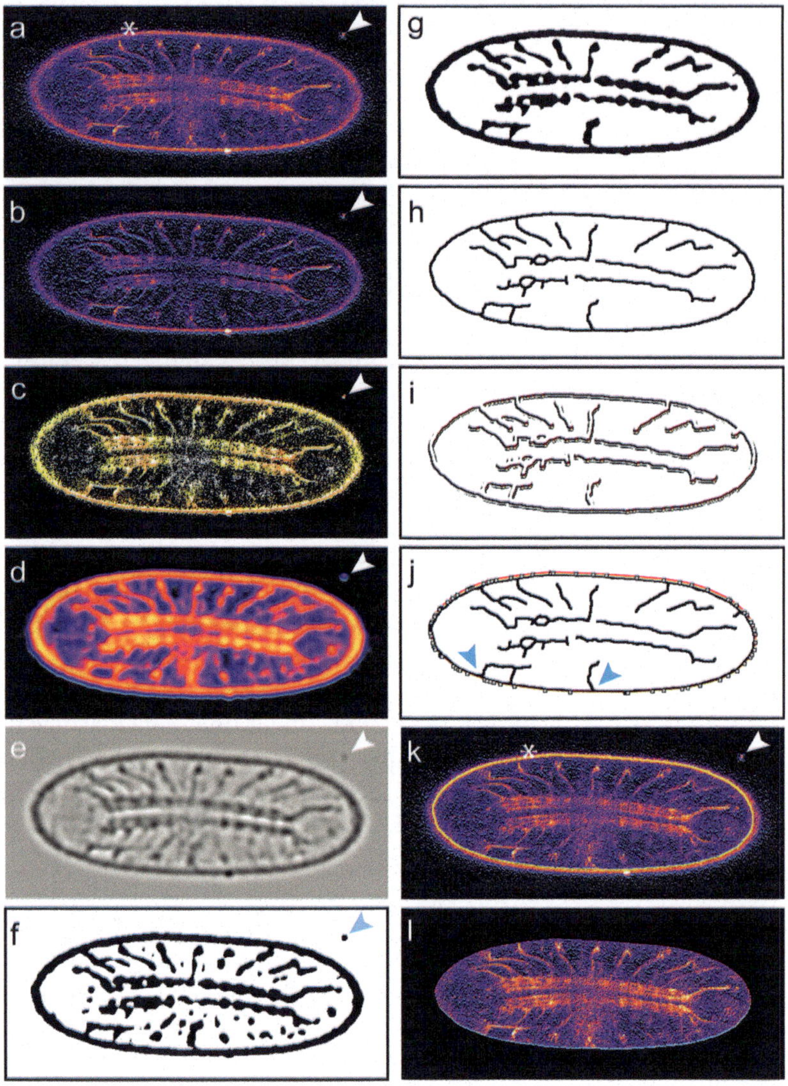

Fig. 2 Snapshots representing the processing steps required to remove vitelline membrane autofluorescence (Subheading 3.1). (**a**) Raw single image from a *Drosophila* embryo expressing Fas2-GFP; the *fire* LUT has been applied to visually enhance the weak signals; notice the glowing membrane (asterisk) and the small autofluorescent particle in the medium (arrowhead). (**b**) *Subtract background* reduces overall noise. (**c**) *CLAHE* enhances contrast and equalizes the signal. (**d**) *Gaussian blur* normalizes the signal profile. (**e**) *LoG* filter enhances the edges. (**f**) *Otsu* thresholding and segmentation render the binary mask. (**g**) Particle filtering removes outlier specks. (**h**) *Skeletonize* and *Dilate* perfect the outline. (**i**) *Points from mask* extracts the skeleton coordinates (red dots). (**j**) *Convex hull* extracts the external selection (red line with dots); notice how the internal skeleton lines that touch the external edge are excluded from the selection (arrowheads). (**k**) *Enlarge* (negative value) shrinks the selection to adjust it to the internal edge of the glowing halo (yellow line). (**l**) *Clear outside* removes the membrane autofluorescence on the original image

structures. Therefore, it should be adjusted to the particular image in use. A *block size* of 30 should be all right for the current image and for "DmPuckered_early.tif," where vitelline membrane is relatively strong and noisy. Smaller block sizes (8–10) should be used for weaker autofluorescent outlines like that in "DmMoesin_early.tif."

4. Apply a *Gaussian blur* filter at [Process > Filters > Gaussian Blur...] to remove the noise created in the previous step (*see* **Note 7** and Fig. 2d). For all the sample images provided use a radius of 3.
5. Enhance edges by applying a *Laplacian of Gaussian* (*LoG*) filter at [Feature extraction > FeatureJ > FeatureJ Laplacian], using a smoothing radius of 3 (*see* **Note 8** and Fig. 2e).
6. Convert to 8-bit at [Image > Type >8-bit] and invert at [Edit > Invert] (*see* **Note 9**).
7. Apply the global thresholding *Otsu* method [21] at [Image > Adjust > Threshold...] (*see* **Note 10**); tick the option *Black background*, and click *Apply*, so that a binary mask is obtained (Fig. 2f).
8. Remove undesired specks from the binary mask (*see* **Note 11** and Fig. 2g). First detect all roundish objects at [Analyze > Analyze Particles...]; select a circularity range from 0.8 to 1, and tick the option *Add to Manager*. The selected particles are added to the *ROI Manager*. Run again the particle detector, this time discriminating by size (0–100); again, add the new particles to the *ROI Manager*. To remove the detected objects, fill them in the mask using the color of the image background. To that aim double click on the *Color picker* tool and select white (255, 255, 255) as foreground color. Then, hit *Deselect* at the *ROI Manager* window, unfold the *More* menu, and select *Fill*. You can now close the *ROI manager window*.
9. To extract the embryo outline first skeletonize the binary mask at [Process > Binary > Skeletonize] and dilate it by one pixel at [Process > Binary > Dilate] (*see* Fig. 2h); then, convert the lines into points at [Edit > Selection > Points from mask], and fit the resulting point selection into its convex hull at [Edit > Selection > Convex Hull] (*see* **Note 12** and Fig. 2i, j). Press *t* to add the new selection to the *ROI Manager*. The result can be visualized by showing the selection on top of the original image or the skeleton mask. The outline ROI should appear on top of the vitelline membrane autofluorescent halo. In the remaining steps, this ROI will be further adjusted to the inner edge of the glowing halo in order to remove the latter by clearing all the signal outside the ROI.
10. Adjust the ROI to the internal edge of the autofluorescent halo (Fig. 2k). Select the ROI at the *ROI Manager* window, go to

[Edit > Selections > Enlarge...], tick *Pixel Units*, and introduce the number of pixels (in negative value) required to shrink the shape until it fits the innermost position of the glowing halo. This step can be repeated as many times as needed by choosing the corresponding negative or positive value in pixels.

11. Once the selection is adjusted, choose black (255, 255, 255) as background color in the *Color Picker* tool and go to [Edit > Clear Outside]. The vitelline membrane autofluorescence will be removed from the original image (Fig. 2l).

12. To perform the previous steps, automatically open the macro "Sect.3.1.ijm" by drag and drop on the *Fiji* bar. Then open the image and hit *Run* in the script editor window. The macro delivers the processed image and the mask, called "skeleton." By changing directly on the code the *Rolling ball radius*, the *CLAHE block size*, the *Enlarge* negative value and the parameters to remove outlier particles, this macro is ideal to try settings on sample single images selected from a 4D stack, before processing the latter automatically using the final macro "DmNutcracker.ijm."

3.2 Automated Shape Fitting

Up to here, the protocol will work for the majority of the images in a *Drosophila* embryo 4D time stack but will most likely fail in those late timepoints where the twitches or peristaltic movements create lobe-like invaginations on the vitelline membrane (*see* Fig. 1c–c′). In this section we introduce a new fitting algorithm (*see* Fig. 3) that uses the convex hull selection created as in the previous section and further adjusts it to the embryo shape by using as seed the skeleton mask also produced as in Subheading 3.1. The fitting algorithm converts the convex hull into an ellipse (Fig. 3d), which contains 72 default coordinates that will be used as reference points (x_{ref}, y_{ref}) to search for the fine-tuned coordinates, located at the skeleton outline (*see* Fig. 3f). For each ellipse coordinate (x_{ref}, y_{ref}), the algorithm creates an imaginary line that crosses the skeleton perpendicularly (Fig. 3e). The crossing points will become the ultimate fitted coordinates (x_{fit}, y_{fit}) shown in Fig. 3f. Such coordinates will be used to build a new polygon ROI which, after being reduced to just leave the vitelline membrane around it (as performed in Subheading 3.1) will serve to clear the image autofluorescence outside the ROI (Fig. 3g, h). The fitting algorithm code (Fig. 4) executes the steps below using the rationale summarized in Fig. 5. The provided macro "Sect.3.2.ijm," based on this script, can be used to perform this section automatically (go to **step 10**).

1. Once the convex hull selection is generated (Subheading 3.1), the fitting algorithm starts by converting it into an ellipse using the built-in function *Fit Ellipse* (*see* Figs. 3d and 5). The resulting ROI contains 72 default coordinates that will be used as reference points (x_{ref}, y_{ref}) to search for the fine-tuned coordinates, located at the skeleton outline.

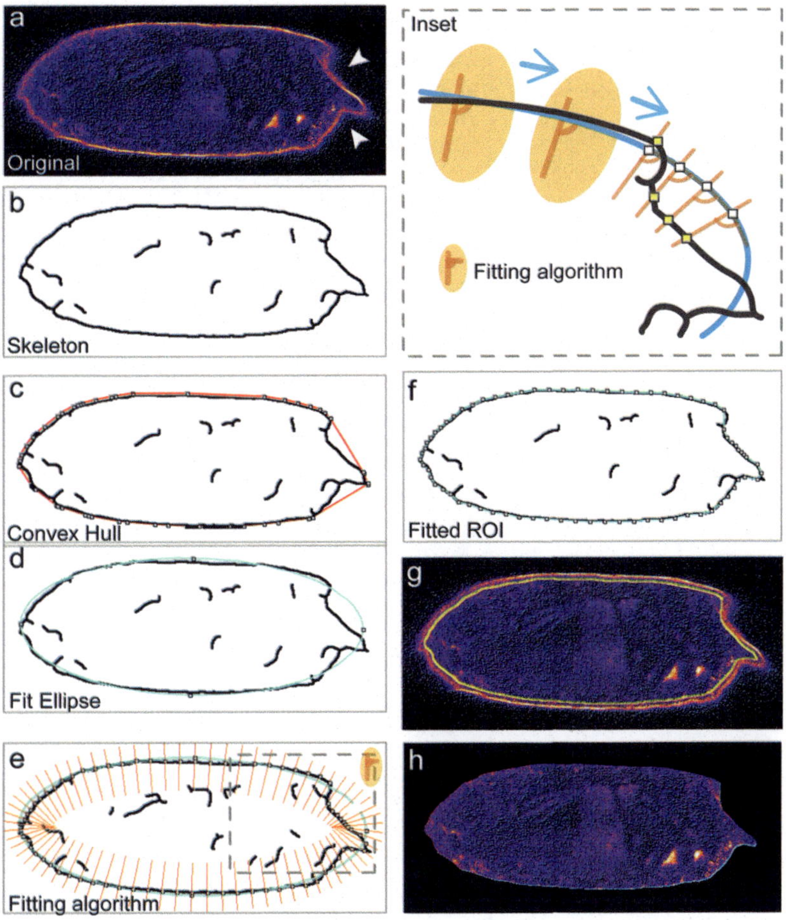

Fig. 3 Fitting algorithm. (**a**) Raw single image from a late-stage *Drosophila* embryo expressing Fas2-GFP; the *fire* LUT has been applied to visually enhance the weak signals; notice the shape invaginations produced by peristaltic movements (arrowheads). (**b**, **c**) "Skeletone" mask (black lines) and "Convex hull" selection (red line) that result from Subheading 3.1. (**d**) *Fit ellipse* converts the previous selection into an ellipse (cyan line). (**e**) Representation of the perpendicular lines (orange) devised by the fitting algorithm at the coordinates of the ellipse (cyan line); the point where each line coincides with the skeleton line becomes the new coordinate (white dots); the dashed line denotes the inset where the fitting algorithm working procedure is represented (upper right panel); the orange icon symbolizes the fitting function that goes over the ellipse coordinates (white squares on blue line) and adjusts them to the skeleton line (yellow squares on black line). (**f**) The resulting selection (cyan line) tightly fits the external embryo edge. (**g**) The ROI is reduced (yellow line) until it locates at the inner edge of the vitelline membrane autofluorescence. (**h**) The image is cleared outside the ROI, thus removing the external glowing halo

```
//Select skeleton mask and fit to an ellipse
run("Fit Ellipse");
roiManager("add");
/*Get Image Pixel Size and calculate in pixels the lenght (60 µm) of the
perpendicular line used by the searchMax function*/
getPixelSize(unit, pixelW,pixelH);
pixelsLine=round(60/pixelW);
// Get ellipse coordinates and store in array
Roi.getCoordinates(X,Y);
Xref=Array.concat(X,X[0]);
Yref=Array.concat(Y,Y[0]);
// Create arrays to store fitting  and searching coordinates
Xfit=newArray();
Yfit=newArray();
Xline=newArray(pixelsLine);
Yline=newArray(pixelsLine);
//Find fitting coordinates and store them in their corresponding arrays
for(k=0;k<(Xref.length-1);k++){
        dx=Xref[k+1]-Xref[k];
        dy=Yref[k+1]-Yref[k];
        a=atan2(-dy,dx);
        z=a-(PI/2);
        b=searchmax(Xline,Yline,Xref[k],Yref[k],z,pixelsLine);
        if(b<pixelsLine){
                Xfit=Array.concat(Xfit,Xline[b-1]);
                Yfit=Array.concat(Yfit,Yline[b-1]);
        }else{
                Xfit=Array.concat(Xfit,Xref[k]);
                Yfit=Array.concat(Yfit,Yref[k]);
        }
}
//Convert "fit" coordinates into polygon and add it to the ROI Manager
makeSelection("polygon",Xfit,Yfit);
roiManager("reset");
roiManager("add");
// "searchMax" user-defined function
function searchMax(Xline,Yline,Xref,Yref,z,pixelsLine){
  b=0;
Int=0;
do{
        Xline[b]=round(Xref+(b-pixelsLine/3))*cos(z));
        Yline[b]=round(Yref+(b-lineLengt/3))*sin(-z));
        selectWindow("skeleton");
        Int=getPixel(Xline[b],Yline[b]);
        b++;
}while(Int<255&& b<pixelsLine)
returnb,Xline,Yline
}
```

Fig. 4 Fitting algorithm's basic script. The code, applied on the skeleton mask produced in Subheading 3.1, delivers a single ROI selection that tightly fits the embryo shape. The main *for* loop runs through all the ellipse coordinates calculating angles and calling the function *searchMax* that builds the imaginary

2. The script then extracts the image pixel size to calculate, in pixels, the length of the perpendicular line that will be used to search for the fine-tuned coordinates (*see* **Note 13**). This value is stored in a variable called "pixelsLine."
3. Next, the x and y ellipse's reference coordinates are stored in two respective arrays called "Xref" and "Yref" (*see* **Note 14**).
4. Four additional arrays are created, "Xfit" and "Yfit," to be filled later on with the final coordinates (x_{fit}, y_{fit}) obtained after fitting, and "Xline" and "Yline" (*see* **Note 15**) that will store the coordinates (x_{line}, y_{line}) of the perpendicular line, calculated as explained in the steps below.
5. The main script body is then built as a *for* loop that will go over all the ellipse's coordinates. For each of them, the algorithm calculates the parameters necessary to build a line perpendicular to the ellipse. First of all, the algorithm obtains the angle (z) of such imaginary line with respect to the x axis, by using the *arctangent* function (*see* Fig. 5a and **Note 16**).
6. Next, the algorithm uses the resulting z angle as reference to build the coordinates (x_{line}, y_{line}) of an imaginary perpendicular line passing through (x_{ref}, y_{ref}). To that aim, the algorithm calls a user-defined function, called *searchMax*. This function has been built as a *while* loop that calculates, at each round, a new coordinate (x_{line}, y_{line}) of the imaginary line, using the formula summarized in Fig. 5b and explained in **Note 17**.
7. For each coordinate that *searchMax* creates along the imaginary line, the function also verifies whether the skeleton is located at that particular position, by checking the pixel intensity value of the corresponding point in the skeletonized mask image produced in Subheading 3.1; if the value 255 is found, the function will stop looping and return the current (x_{line}, y_{line}) coordinate; if no skeleton line is found, it will continue looping until the end of the line is reached, and then the function will return this last line coordinate, which will be eliminated in the next step.
8. The coordinates returned by the *searchMax* function will be sequentially stored in the corresponding "Xfit" and "Yfit" arrays; a conditional statement has been implemented to

Fig. 4 (continued) line along with the right coordinates are found. These, are filled to the final arrays "Xfit" and "Yfit." The user-defined function *searchMax*, included at the end of the script, requires five arguments: the line coordinates, the reference coordinates, and the z angle. This function runs as a *while* loop that stops when the coordinate that coincides with the skeleton is found or when the loop reaches the end of the imaginary line. The code is depicted in the colors used by the *Fiji* script editor

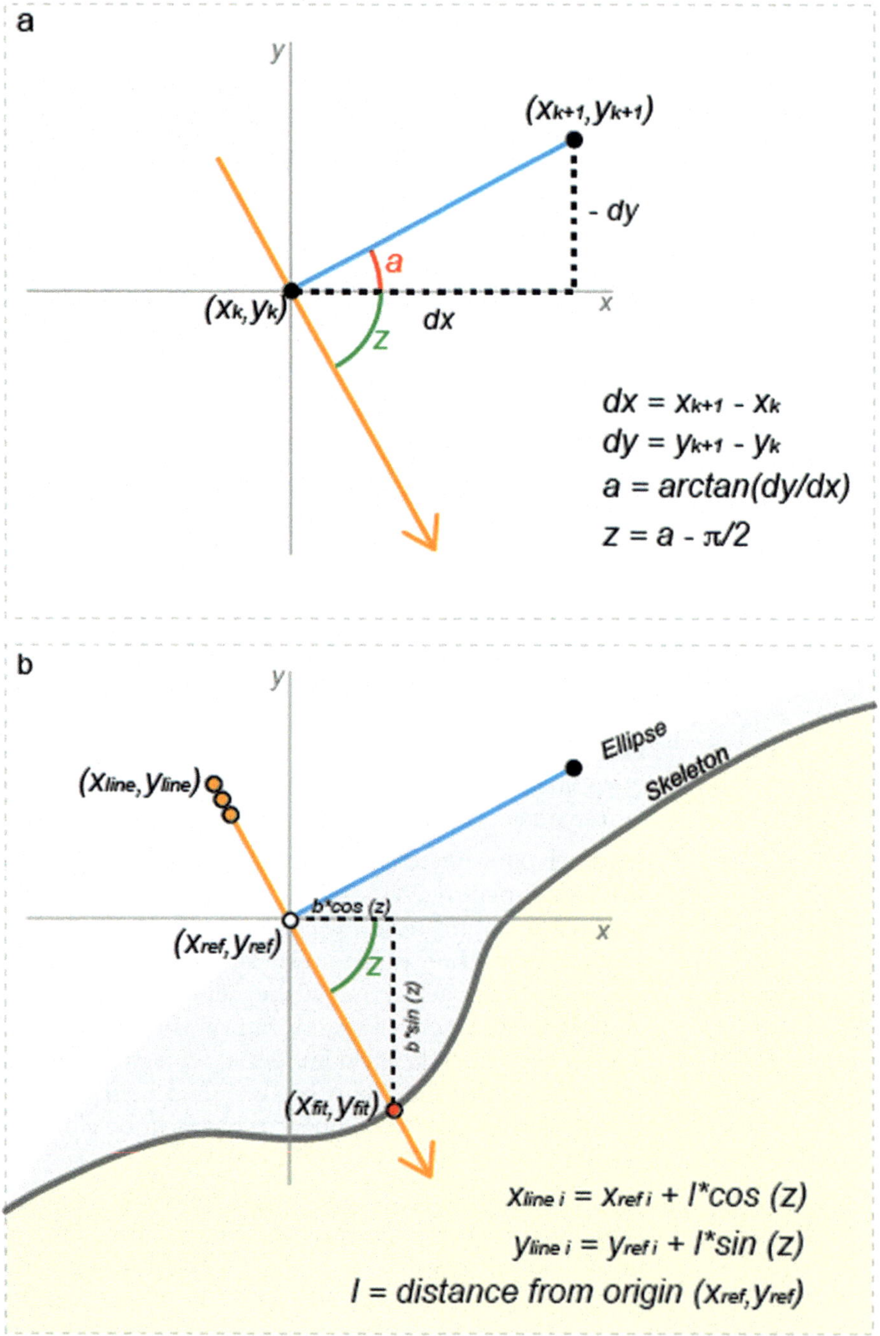

Fig. 5 Calculations used by the "fitting algorithm." (**a**) Representative scheme and formulas used for the *z* angle calculation; (x_k, y_k) and (x_{k+1}, y_{k+1}) depict two consecutive points in the ellipse; *dy* and *dx* are, respectively, the numerator and denominator of the tangent calculation; the angle *a* is calculated by the inverse tangent (*arctang*); z results from subtracting $\pi/2$ (that is, 90° in radians) to *a*. (**b**) Representative scheme and formulas used to calculate the coordinates (x_{line}, y_{line}) of an imaginary line that passes through the reference coordinate (x_{ref}, y_{ref}) at an angle *z* relative to the *x* axis; within this line, (x_{fit}, y_{fit}) is the coordinate that coincides with the position of the skeleton line of the binary mask

avoid storing wrong coordinates, which may happen, for instance, when the skeleton line contains a gap, and no 255 value is found by the *searchMax* function. Such conditional statement stores in the "Xfit" and "Yfit" arrays the fitted coordinates returned by the *searchMax* function, except when they correspond to the last position of the line. In this case, a gap was found, and the reference coordinate (x_{ref}, y_{ref}) is filled instead. At the end of the process, all the coordinates stored in the "Xfit" and "Yfit" arrays will be part of a ROI that tightly fits the embryo outline.

9. In the last step, the algorithm creates a polygon line using the (x_{fit}, y_{fit}) coordinates by using the *polygon Fiji* function and adds it to the *ROI Manager* window (*see* Figs. 3f and 4). From this step on, the fine-tuned selection can be used to remove the embryo's autofluorescence glowing halo on the original image as performed in Subheading 3.1, **steps 10** and **11** (*see* Fig. 3g, h).
10. To use this protocol in single images, open both macros "Sect.3.1.ijm" and "Sect.3.2.ijm" simultaneously by drag and drop on the *Fiji* bar. Then open the image, and run one macro after the other. The second one will use the skeleton mask produced by the first, fit the convex hull to it, and clear the membrane autofluorescence on the original image.

3.3 Processing a 4D Movie Using "DmNutcracker.ijm"

The full macro "DmNutcracker.ijm" will process a 4D movie using the pipelines explained in Subheadings 3.1 and 3.2 but applied automatically to all the images in the hyperstack. The main parameters needed to adjust particular 4D images can be introduced in dialog boxes created to that aim. It is recommended to know these values in advance, by checking a single image according to Subheadings 3.1 and 3.2, before processing the final stack, since it may take a while. To use this macro, follow the steps below.

1. Open a hyperstack (*see* **Note 3**), for instance, "DMFasII.tif," and open the macro "DmNutcracker.ijm," both by drag and drop on the *Fiji* bar.
2. Upon hitting *Run* at the script editor window, a dialog box pops up containing the parameters/options below. After completing, hit *OK*.
   (a) *Subtract Background Radius* refers to the radius of the *Rolling ball* algorithm (*see* **Note 5**), mainly used for noisy and strong vitelline membrane signals. Use 20 for the sample movies "DmFasII.tif" and "DmPuckered.tif"; use a high value (e.g., 1000) for movies where no *Subtract background* should be applied, such as "DmMoesin.tif."
   (b) *CLAHE Block Size* refers to the size of the local region used for histogram equalization (*see* **Note 6**). Use 30 for the sample movies "DmFasII.tif" and "DmPuckered.tif";

use a smaller value (e.g., 8) for movies where the vitelline membrane signal is weak, as occurs in "DmMoesin.tif."

(c) *Sigma Gaussian blur* is the radius for the smoothing kernel (*see* **Note 7**). Use 3 for all the provided sample movies.

(d) *LoG* radius refers to the radius of the *Gaussian* function that precedes the *laplacian* filter (*see* **Note 8**). Use 3 for all the provided sample movies.

(e) *Fitting line length* refers to the length of the perpendicular line created by the fitting algorithm at each coordinate of the ellipse, which should to be set to find the skeleton line (*see* **Note 13**). The default value (60 μm) should work for most *Drosophila* embryos.

(f) *ROI fitting* (*px*) refers to the value (in pixels) that the final ROI will have to be reduced until it is located at the inner edge of the vitelline membrane before clearing the signal outside. This value will correspond to approximately half the halo thickness, pixelwise. The default value (−6) should work for all the provided movies.

(g) *Remove floating particles?* is an option that should only be ticked if the medium around the embryo contains floating autofluorescent specks. Otherwise it should be avoided since it lengthens the execution time.

(h) *Embryo close to edge?* is an option that should be ticked in case the embryo is very close or touches the edges of the image; this option increases the images canvas size to avoid artifacts during the segmentation procedure.

3. If the option *Remove floating particles?* was activated, a new dialog appears to introduce the lower circularity and upper size limits for particle discrimination. The default values 0.8 and 100 will work for the current images. Other values may be required in other image sets, which should be defined by checking the size and shape of the floating particles.

4. Next, the message "Select plane to delete extra slices" appears that will allow to select, at the image bar, the Z plane from which the Z slices should be removed (*see* **Note 18**). Once selected, hit *OK* on the message window.

5. From here on, the macro runs automatically keeping most processing steps in batch mode (*see* **Note 19**). Only two images are shown: the original 4D hyperstack, transformed into a 3D stack, and a temporal image ("temp") where all processing steps are being performed. The macro progress can be verified in the stack's progress bar. Total execution will actually take some time, depending on the workstation capacity.

6. The macro finally delivers a processed hyperstack that needs to be saved for further projection or 3D rendering.

4 Notes

1. Protein fusion to green fluorescent protein (GFP) and its variants, either using transgenesis or at the gene's endogenous locations [22], is widely used to study protein subcellular localization and dynamics in live *Drosophila* embryos. Other methods include microinjection of cell permeable dyes or fluorescent reagents. For general information regarding *Drosophila* embryo-labeling techniques, *see* Refs. [3, 23]. All live embryo recordings using fluorescent markers that require excitation in wavelengths near the blue spectral range will most likely suffer from vitelline membrane autofluorescence contamination and may therefore benefit from this imaging protocol.
2. The 4D acquisitions included as sample images contained a minimum of 40 sections at 2 μm *Z*-step, spanning most the embryo thickness, acquired every 5–10 min interval and spanning several developmental stages. As mounting/immersion medium, a mix 3:1 of halocarbon oil S700 and halocarbon oil S27 was used, which has a refraction index (RI) of 1.41. On the objective lens (a 25× multi-immersion objective), the correction ring was set to a position close to that of glycerol. Because of embryo thickness (more than 100 μm), conventional wide-field microscopy produces poor images that suffer from out-of-focus light. CLSM is one of the best methods of choice for high-resolution/high-contrast acquisition of live *Drosophila* embryos. Care needs to be taken as to choose the right objectives that provide good chromatic aberration correction and let match the refraction indexes of the immersion (lens) and mounting (specimen) media used [3]. In addition, low excitation laser intensity compensated by higher pinhole aperture helps prevent phototoxicity and photobleaching during long-term acquisitions, at the cost of reducing spatial resolution. Fast scanning confocal systems, such as spinning disk microscopes, allow for higher temporal resolution acquisitions with minimum photodamage. MPM yields deeper optical sectioning and increased viability and may be the method of choice for long-term recordings of internal embryo structures at low-to-medium temporal resolutions [3, 6]. LSFM has recently emerged as a very competitive technology for imaging *Drosophila* embryos [24], offering a good compromise between resolution and survival.
3. The present imaging protocol has been specifically tailored for single channel 4D (3D plus time) CLSM and MPM movies although, in principle, it can handle any hyperstack, that is, a stack comprising the four dimensions x (width), y (height), z (slices), and t (time frames). *ImageJ/Fiji* can right away

open as hyperstacks many different file formats along with their important metadata via [File > Import > Bio-formats] or directly at the [File > Open] menu.

4. *ImageJ* macro language (IJM) is a scripting language built-in into *ImageJ*, which allows writing simple programs called *macros* that automate series of image-processing actions. By opening the macro recorder at [Plugins > Macros > Record...}, all actions performed manually are sequentially recorded into a IJM preliminary script. Hitting the button *create* will open the recorded instructions into the script editor were additional editing can be made in order to make the code usable. For more information about macros, *see* Ref. [25].

5. Uneven image backgrounds can be successfully restored using the rolling ball algorithm implemented in *ImageJ* at [Process > Subtract Background...]. The rolling ball radius should be at least as the radius of the largest object in the image that is not part of the background. For this protocol, the radius should be adjusted according to the thickness of the vitelline membrane halo, which is the object of interest that needs to be segmented. However, in images where the vitelline membrane autofluorescence is very weak, this step should be avoided, since it will worsen segmentation. A measurement of the halo thickness may help define the radius, which can be done by painting a line through it and extracting the intensity profile at [Analyze > Plot Profile].

6. The plugin *Enhance Local Contrast* (*CLAHE*) at [Process > Enhance Local Contrast (CLAHE)] uses an adaptive histogram equalization method called *Contrast Limited Adaptive Histogram Equalization* [26]. The resulting image will have more contrast, but also the intensity will become balanced on the different regions. In the current method, this plugin allows to bring up the intensity level of the autofluorescent outline to a point where its segmentation is possible. The size of the local region around a pixel for which the histogram is equalized is called *block size*; this parameter will help adapt the method to different embryo images where autofluorescent signals may be relatively weak or strong as compared to the signal of the FP imaged. It will also have to be adapted depending on the particular image pixel size.

7. The *Gaussian blur* filter is used here to reduce the extra noise created during contrast enhancing. The radius chosen for the filter kernel should be adjusted to each particular image, depending on its pixel size, to a value that ideally allows to have a continuous edge but does not over attenuate the membrane signal.

8. The *LoG* edge detector filter highlights image regions based on rapid intensity change [27]. It detects pixels where the *Laplacian* function changes sign, also called zero crossing points. A smoothing filter is included in the *LoG* function, prior to the *Laplacian* in order to reduce the range of scales over which intensity changes take place. In this pipeline, we run the *LoG* filter after the previous *Gaussian* blur, which is done in purpose because it helps define the edge better. Therefore, we choose the same radius definition for both, although variations may be tolerated in images having smaller pixel sizes. The *LoG* filter delivers a 32-bit image that will need to be converted to 8-bits in order to proceed with segmentation.
9. The inversion of the image values is optional as long as the right foreground vs. background values are selected in the thresholding step.
10. The adaptive filtering options used during the preprocessing pipeline rise the local intensity levels of the vitelline membrane autofluorescence, which now becomes easily segmentable using the *Global thresholding* methods included in *Fiji*. Among these, the *Otsu*'s method is a clustering algorithm that maximizes the separability of the resulting gray-level classes and can therefore be applied to very general problems that require unsupervised decision [21], such as this one. When thresholding at [Image > Adjust > Threshold...], take special care with the option *Dark background*. This option will highlight in black the selected pixels, while giving the white background a value of zero. This is normally the way for fluorescent images, where fluorescence appears as light on dark background. However, in this pipeline, thresholding is applied on a *Laplacian* converted image, where the signal appears dark on a light background. This is why the image values are previously inverted. Alternatively, the option *Dark Background* could be deactivated.
11. The binary mask produced may contain undesired particles surrounding the embryo, like autofluorescent dirt or debris floating in the mounting medium. Such particles need to be removed in order to prevent artifacts when extracting the embryo shape. Most of these particles will have become roundish after the *LoG* transformation, in contrast to the elongated embryo shape, and can therefore be easily selected by *Circularity* using the *Analyze Particles* dialog box. The calculation for circularity is 4π (area/perimeter^2), where a value of 1.0 indicates a perfect circle. By selecting a range of 0.8–1 in the dialog box, such particles can be selected and added to the *ROI Manager* for its elimination. Some more undesired structures may also appear during binarization, especially in regions where the vitelline membrane intensity was high, and the *LoG*

filtering created a small double edge at the embryo outline. A second round of selection, using size as criterion, is enough to select these particles and add them to the *ROI Manager* pool that will be eliminated. The value used as size criterion should be adjusted so that no relevant piece of the binary mask is removed.

12. The *Skeletonize* function at [Process > Binary > Skeletonize] reduces the objects in a binary mask to single-pixel lines, thus allowing in the current pipeline to accurately define the embryo limits. However, such outline may often appear fused to some internal lines coming from the segmented fluorescent signal (*see* Fig. 2h) and, more importantly, may eventually have small gaps, depending on how successful the segmentation has been. A strategy to isolate the outline from the internal lines and, at the same time, obtain a complete outline, is to first convert the skeletonized mask into a selection of points at [Edit > Selection > Points from Mask]. Then, the overall point selection is converted into a convex selection at [Edit > Selection > Convex Hull] and added to the *ROI Manager*. The resulting convex object selection is now separated from any internal structure and, moreover, is close. This will allow to preserve the internal embryo signal when using the outline selection to delete the undesired autofluorescence in the subsequent steps (*see* Fig. 3i–l). The skeleton dilation step, although not really necessary at this point, is introduced here to improve pixel connectivity along the skeleton line, which will be essential for the fitting algorithm performance in the next section.

13. The length of the perpendicular line (*see* Fig. 3e) has been established in 60 μm, being one third outside the ellipse. This way, the searching coordinates inside and outside the ellipse will extend long enough to find, according to the embryo structure and dynamics, any possible skeleton line than deviates from the pure convex shape. By converting this 60 μm searching range into pixels, using the corresponding image pixel size, the script will become adaptable to any images having different pixel sizes.

14. When creating the arrays "Xref" and "Yref" containing the ellipse's reference coordinates, the first array position is duplicated in the last position of the array, using the function *Array.concat*. This will allow the fitting algorithm to go through all the ellipse's coordinates two by two, which is its modus operandi, until the circle is closed.

15. When creating the arrays "Xline" and "Yline," the number of elements composing the array (specified between parenthesis) corresponds to the number of pixels that compose the perpendicular imaginary line, stored in the variable "pixelsLine."

16. The *arctangent* function converts rectangular coordinates (x, y) into polar coordinates (r, a), where r is the distance from the origin (in this case the distance from the reference point (x_k, y_k) to the next ellipse point (x_{k+1}, y_{k+1})) and a is the angle from the x axis (*see* Fig. 5a), expressed in radians. Once obtained a, the angle with respect to the x axis that would form a line perpendicular to the ellipse's points would be calculated by subtracting 90° (which in radians is expressed as $\pi/2$) from a, that is, $z = a - \pi/2$. The *arctangent* function computes angles with respect to the Cartesian coordinates. In an image, however, the Y axis is inverted with respect to the Cartesian system. As a result, the calculation of $dy = y_{k+1} - y_k$ delivers a value of opposite sign, therefore affecting the sign of the *arctangent* function. This ambiguity can be solved by simply multiplying by -1 the first entry value of the *arctangent* function: dy. In programming language, this calculation will be expressed as $a = a\tan2\,(-dy, dx)$.
17. The formula used to calculate the coordinates of a line with respect to a reference point using a reference angle is $(x_i, y_i) = (x_{ref} + l \times \cos(z), y_{ref} + l \times \sin(z))$, where l is the distance to the reference point and z is the given angle (*see* Fig. 5b). In the *searchMax* function (Fig. 4), the number of loops corresponds to the total number of coordinates necessary to build the imaginary line, estimated as the number of pixels in "pixelsLine." A loop counter b is established that advances one coordinate by one, starting at a distance $l = b -$ pixelsLine/3, that is, at a position outside the ellipse separated from the reference point by one third of the total line length. As a result, the script that calculates the coordinates (x_{line}, y_{line}) inside the loop is expressed by $x_{line} = \text{round}(x_{ref} + (b - \text{pixelsLine}/3) \times \cos(z)$ and $y_{line} = \text{round}(y_{ref} + (b - \text{pixelsLine}/3) \times \sin(z)$, where the function *round* is used to deliver integer numbers. The established searching range of 60 μm can be easily changed in the code to adapt the algorithm to different contexts.
18. The segmentation procedure does not work for external Z planes that mainly contain membrane autofluorescent. Since such planes do not contribute any relevant signal to the final result, they should be deleted.
19. During macro execution the images are generally updated after being applied an operation, a filter, etc. This update takes some time to process. Showing the results of any intermediary steps can be avoided by using the commands *setBacthMode*(*true*) and *setBatchMode*(*false*), respectively, at the beginning and the end of a piece of code that does not require any user interaction. Setting the batch mode will speed up the process up to 20 times.

Acknowledgments

We acknowledge Christian Klämbt for kindly providing the *fas2*$^{GFP-CB03613}$ fly strain. KK and EMB prepared and recorded *Drosophila* embryos. JBF programmed the fitting algorithm script. ER designed and programmed the macro pipeline and wrote the article. Imaging was performed at the Molecular Imaging Platform IBMB-PCB in a system financed by CSIC13-4E-2065 (FEDER/MINECO). JBF is financed by a PTA contract from the Spanish Ministry of Economy and Competitiveness.

References

1. Mavrakis M, Pourquie O, Lecuit T (2010) Lighting up developmental mechanisms: how fluorescence imaging heralded a new era. Development 137(3):373–387. https://doi.org/10.1242/dev.031690
2. Pantazis P, Supatto W (2014) Advances in whole-embryo imaging: a quantitative transition is underway. Nat Rev Mol Cell Biol 15 (5):327–339. https://doi.org/10.1038/nrm3786
3. Mavrakis M, Rikhy R, Lilly M, Lippincot-Schwartz J (2008) Fluorescence imaging techniques for studying drosophila embryo development. Curr Protoc Cell Biol Chapter 4:4.18.11–14.18.43
4. Richardson DS, Lichtman JW (2015) Clarifying tissue clearing. Cell 162(2):246–257. https://doi.org/10.1016/j.cell.2015.06.067
5. Parton RM, Valles AM, Dobbie IM, Davis I (2010) Live cell imaging in Drosophila melanogaster. Cold Spring Harb Protoc 2010(4): pdb top75. https://doi.org/10.1101/pdb.top75
6. Rebollo E, Gonzalez C (2010) Time-lapse imaging of embryonic neural stem cell division in Drosophila by two-photon microscopy. Curr Protoc Stem Cell Biol Chapter 1:Unit1H.2. https://doi.org/10.1002/9780470151808.sc01h02s13
7. Wieschaus E, Nüsslein-Volhard C, Roberts D (1998) Drosophila: a practical approach, 2nd edn. University Press.; Looking at embryos, Oxford, pp 179–214
8. Heppert JK, Dickinson DJ, Pani AM, Higgins CD, Steward A, Ahringer J, Kuhn JR, Goldstein B (2016) Comparative assessment of fluorescent proteins for in vivo imaging in an animal model system. Mol Biol Cell 27 (22):3385–3394. https://doi.org/10.1091/mbc.E16-01-0063
9. Zimmermann T (2005) Spectral imaging and linear unmixing in light microscopy. Adv Biochem Eng Biotechnol 95:245–265
10. Pereanu W, Spindler S, Im E, Buu N, Hartenstein V (2007) The emergence of patterned movement during late embryogenesis of Drosophila. Dev Neurobiol 67(12):1669–1685. https://doi.org/10.1002/dneu.20538
11. Doube M, Klosowski MM, Arganda-Carreras I, Cordelieres FP, Dougherty RP, Jackson JS, Schmid B, Hutchinson JR, Shefelbine SJ (2010) BoneJ: Free and extensible bone image analysis in ImageJ. Bone 47 (6):1076–1079. https://doi.org/10.1016/j.bone.2010.08.023
12. Schneider CA, Rasband WS, KW E (2012) NIH Image to ImageJ: 25 years of image analysis. Nat Methods 9(7):671–675
13. Schindelin J, Arganda-Carreras I, Frise E, Kaynig V, Longair M, Pietzsch T, Preibisch S, Rueder C, Saalfeld S, Schmid B, Tinevez J, White D, Hartenschtein V, Eliceiri K, Tomancak P, Cardona A (2012) Fiji: an open-source platform for biological-image analysis. Nat Methods 9:676–682
14. Buszczak M, Paterno S, Lighthouse D, Bachman J, Planck J, Owen S, Skora AD, Nystul TG, Ohlstein B, Allen A, Wilhelm JE, Murphy TD, Levis RW, Matunis E, Srivali N, Hoskins RA, Spradling AC (2007) The carnegie protein trap library: a versatile tool for Drosophila developmental studies. Genetics 175 (3):1505–1531. https://doi.org/10.1534/genetics.106.065961
15. Pastor-Pareja JC, Grawe F, Martin-Blanco E, Garcia-Bellido A (2004) Invasive cell behavior during Drosophila imaginal disc eversion is mediated by the JNK signaling cascade. Dev Cell 7(3):387–399. https://doi.org/10.1016/j.devcel.2004.07.022

16. Chihara T, Kato K, Taniguchi M, Ng J, Hayashi S (2003) Rac promotes epithelial cell rearrangement during tracheal tubulogenesis in Drosophila. Development 130:1419–1428. https://doi.org/10.1242/dev.00361
17. #BL7415 BSC. https://bdsc.indiana.edu/
18. Fiji download website. https://imagej.net/Fiji/Downloads
19. ImageJ macro functions website https://imagej.nih.gov/ij/developer/macro/functions.html
20. Molecular Imaging Platform (IBMB) website http://www.ibmb.csic.es/groups/molecular-imaging-platform
21. Otsu N (1979) A threshold selection method from gray-level histograms. Trans Syst Man Cybernet 9(1):62–66
22. Kanca O, Bellen HJ, Schnorrer F (2017) Gene tagging strategies to assess protein expression, localization, and function in Drosophila. Genetics 207(2):389–412. https://doi.org/10.1534/genetics.117.199968
23. Rebollo E, Karkali K, Mangione F, Martin-Blanco E (2014) Live imaging in Drosophila: the optical and genetic toolkits. Methods 68 (1):48–59. https://doi.org/10.1016/j.ymeth.2014.04.021
24. Khairy K, Lemon WC, Amat F, Keller PJ (2015) Light sheet-based imaging and analysis of early embryogenesis in the fruit fly. Methods Mol Biol 1189:79–97. https://doi.org/10.1007/978-1-4939-1164-6_6
25. Macros. https://imagej.nih.gov/ij/docs/guide/146-14.html
26. Zuiderveld K (1994) Contrast limited adaptive histogram equalization. In: Graphics gems IV. Academic Press Professional, Inc., San Diego, CA, pp 474–485
27. Marr D, Hildreth E (1980) Theory of edge detection. Proc R Soc Lond 207:187–217

Chapter 10

Which Elements to Build Co-localization Workflows? From Metrology to Analysis

Patrice Mascalchi and Fabrice P. Cordelières

Abstract

Co-localization analysis is one of the main interests of users entering a facility with slides in hands and nice analysis perspectives in mind. While being available through most, if not all, analysis software, co-localization tools are mainly perceived as black boxes, fed with images, that will, hopefully, return (the expected) numbers.

In this chapter, we will aim at deconstructing existing generic co-localization workflows, extracting elementary tools that may be reused and recombined to generate new workflows. By differentiating work cases, identifying co-localization reporters and the metrics others have been using, we aim at providing the audience with the elementary bricks and methods to build their really own co-localization workflows. A special emphasis is given on the preparatory phase where the acquisition system is assessed, using basic metrological tests.

Key words Co-localization, Co-expression, Co-occurrence, Correlation, Co-distribution, Elements, Workflow, Image processing, Image analysis

1 Introduction

1.1 Co-localization or Co-localizations: One Word, Many Meanings

From the biologist perspective, co-localization often appears as a word conveying several meanings. Its precise definition is highly linked to the phenomenon the experimenter is trying to characterize (Fig. 1).

When dealing with large-scale samples, such as slices of tissues, the word "co-localization" is generally used in the sense "co-expression." In this case, the aim is to determine whether a same set of cells are positive for two proteins of interest. This experimental situation does not presuppose the two molecular actors to be at the same location. One could expect "co-localization" while, for example, working on a nuclear transcription factor and the product

Electronic supplementary material: The online version of this chapter (https://doi.org/10.1007/978-1-4939-9686-5_10) contains supplementary material, which is available to authorized users.

Elena Rebollo and Manel Bosch (eds.), *Computer Optimized Microscopy: Methods and Protocols*, Methods in Molecular Biology, vol. 2040, https://doi.org/10.1007/978-1-4939-9686-5_10,

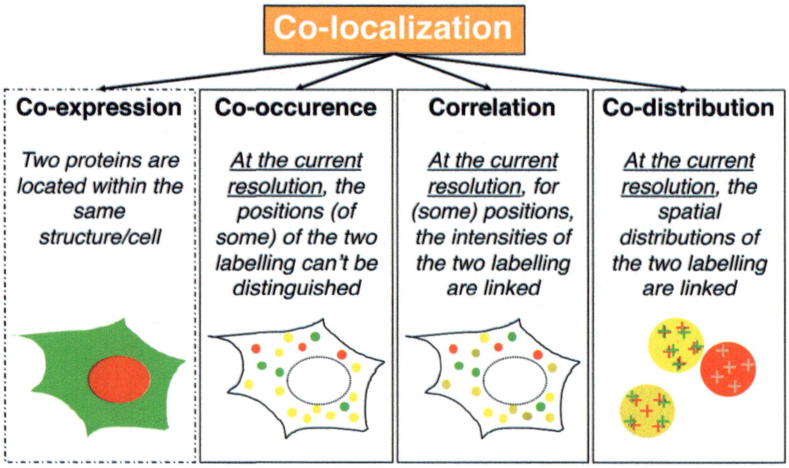

Fig. 1 Co-localization or co-localizations: one word, many meanings

of its targeted gene to be located in the cytoplasm. The same goes for studies where the focus is rather on anatomically distinguishable structures: two transcripts might be carried by two distinct populations of cells, while conclusion to be drawn is "both co-localize within the same cell layer." "Co-localization" is most generally used to describe the interplay of molecular actors at subcellular level. While sitting on the same structures, two situations may occur. On the one hand, the two players may have a common location while not being dependent on one another's stoichiometry. This situation might be referred to as "co-occurrence." It should be emphasized that the terms "common location" may differ in meaning, depending on the resolution used for image acquisition. Using a low-resolution acquisition system will lower the power of discrimination, therefore enforcing "co-occurrence." "Co-occurrence" might be seen as a correlation of location, not paying attention to the relative quantities of both molecular cues. On the other hand, in addition to sharing the same location, the molecules might be interdependent in terms of quantities. This situation is often referred to as "correlation," and appropriate methods such as the use of correlation coefficients are employed to characterize it. The emergence of the so-called "super-resolved" microscopies, and especially the methods based on single molecule detection, has drastically changed the co-localization paradigm. For the really first time in light microscopy, the analysis unit is not the pixel anymore, but it takes the shape of a set of spatial coordinates, accompanied by a precision of localization. Working on individual molecule's localizations, co-occurrence/correlation might only be used after degrading the information in order to get back to a regular image. When dealing with primary detections, tools coming from the field of spatial statistics are employed [1, 2], looking for "co-distribution" of labels, as pure coincidence (relative to the working resolution) of detection is not likely anymore. Obviously, such a partitioning of the co-localization field is highly artificial: we [3–5] and others [6, 7] have proposed different views, based either on the type of phenomenon to describe or on the means to describe it.

1.2 Image Processing Is Like a Brick Game: Know Your Bricks

When looking at different image processing workflows dealing with co-localization, although the methods used are different, the overall steps almost stay the same from one strategy to another. Knowing the variety of biological questions, in our view, it would be too adventurous to try and propose generic co-localization protocols. Instead, building an image processing workflow should always be thought relative to a precise problematic, knowing the domain of application, and therefore limits, of the tools to be used. The inspiration for the shape of the following sections has been driven by the work of Miura and Tosi [8], who use the term "components" to define the individual bricks assembled to compose an image processing/analysis "workflow." Analyzing published co-localization strategies brought us to the conclusion that this type of image analysis makes no exception to their concept. We therefore decided to revisit the field with their concept in mind, in the aim of helping the reader to explore existing components and conceive alternative ways to perform co-localization analysis.

2 Materials

Several methodological papers have already been published, proposing practical protocols [9–11] for the design of co-localization workflows. The current list summarizes the materials required to perform the analysis according to the proposed protocols.

2.1 Samples

1. Mono-labeled samples: to evaluate cross-excitation and cross-detection (*see* Subheading 3.1.1), samples carrying only one chromophore are required.
2. Double-labeled sample: on coverslip or any other supporting layer (*see* **Note 1**).
3. Reference slides: ready to use reference slides, designed at least to check for field illumination homogeneity. Argolight™ [12, 13]: carries many patterns for assessing the imaging system (field homogeneity, resolution, intensity response, stage repositioning, etc). Autofluorescent plastic slides (*see* **Note 2**), e.g., Chroma™ slides. Brakenhoff's slide [14, 15] or FluorCal™ optical calibration slide (Valley Scientific): made of a thin fluorescent layer, this slide allows performing a general characterization of axial response of the imaging system.

2.2 Disposable Materials

1. Regular slides.
2. Coverslips: type 1.5, i.e., 0.17 mm thick, or 1.5H for super-resolution and TIRF microscopies, either polylysine coated or not.
3. Multi-well plates: wells should be large enough to accommodate the coverslips to be coated with fluorescent beads.
4. Tubes: 1.5 mL Eppendorf™ tubes to perform beads' dilutions and larger tubes (15 mL) to perform Poly-L-lysine dilutions (*see* **Note 3**).
5. Tweezers: used to manipulate coverslips.
6. Wipers.
7. Parafilm™.

2.3 Reagents

1. Two types of fluorescent beads can be used:
   (a) Uniformly labeled beads, of two diameters: below the optical resolution, to measure the latter, and well above the resolution of the system, to test for channels' registration, e.g., Molecular Probes' 0.1 μm TetraSpeck™ and 4 μm TetraSpeck™.
   (b) Nonuniformly labeled beads (inner core carrying one fluorescence, outer ring another): although optional, they provide a good alternative to perform co-registration tests, e.g., Molecular Probes' FocalCheck™ Microspheres, 6 μm, fluorescent green/orange/dark-red ring stains.
2. Chromophore in solution: fluorescein (green), rose bengal or acid fuchsin (red), and acid blue 9 (far red) could be used to check illumination's homogeneity (*see* Ref. [16]).
3. Poly-L-lysine solution (0.1% (w/v) in H_2O): used to charge the surface of coverslips; it may help in getting the fluorescent beads attached to the support.
4. Ethanol.
5. Regular distilled water.
6. Mounting medium: should be the same as for the regular, biological samples (*see* **Note 4**).
7. Nail polish: required only in case the mounting solution is not a setting medium.

2.4 Software

1. Image processing and analysis software: *ImageJ/Fiji* [17] or *Icy* [18].
2. *MetroloJ* plugin [19, 20] for illumination co-registration and resolution analysis in 2D/3D.

3. Macro for co-registration analysis in 2D, available from reference link [21] and provided as supplemental material in this chapter.
4. *3D suite* plugin for co-registration analysis in 3D [22].
5. Macro for resolution analysis in 2D, available from reference link [23] and provided as supplemental material in this chapter.
6. Spectral unmixing plugin such as *Spectral Unmixing Plugin* [24], *Spectral Unmixing of Bioluminescence Signals* [25], or *PoissonNMF* [26].
7. Uneven illumination correction tool such as the *BaSiC* tool [27].
8. Image restoration: *PSF Generator* [28, 29] and *Deconvolution-Lab* [30–32].
9. Co-localization analysis packages/plugins: a comprehensive, non-exhaustive list is given in Table 1.

3 Methods

In co-localization analysis, data extraction and choice of a metric come at the end of the process. At the start, the experimenter should make sure the microscope response gives a representation of the sample as close as possible from reality, keeping in mind that even with the most advanced microscope, the "image is not the object," as pointed out by Rene Magritte in his 1929 painting "the treachery of images." In this section detailed step-by-step protocols are provided to help check for system's integrity (Subheading 3.1). Hints on data preprocessing, including images' correction and restoration, are introduced in Subheading 3.2. Finally, a short panel of co-localization metrics and tools to apply them is discussed in Subheading 3.3.

3.1 Checking Data Integrity

Co-localization workflow starts before accessing the microscope, on the bench side. In particular, great care should be taken when preparing the sample. Common issues may arise during this crucial step: nonspecific labeling, ending up in either background (when the signal is diffuse) or mis-localization of the antibodies. In both situations, unfortunately, not much can be done except improving the sample's preparation. What can be dealt with is the characterization (a process known as metrology; *see* Refs. [33–37]) and, to some extent, correction of the optical system-related aberrations that may impair interpretation. Although many issues may arise from using fluorescence microscopy, three main drawbacks should be looked for.

Table 1
Some freely accessible add-on examples to compute co-localization reporters

Reporter type	Package/plugin [software]	Analyzed entity	Coordinates as input	References
Correlation-based indicators	Coloc2 [*ImageJ*/Fiji]	Whole image or ROI	No	[80]
	JACoP [*ImageJ*/Fiji]	Whole image	No	[3, 4, 81]
	Co-localization Studio [Icy][d]	Whole image or ROI	No	[77]
	MeasureCorrelation [CellProfiler]	Whole image or ROI individual objects	No	[59]
Intensities' overlap-based quantifiers	Coloc2 [*ImageJ*/Fiji]	Whole image or ROI	No	[80]
	JACoP [*ImageJ*/Fiji]	Whole image	No	[3, 4, 81]
	Colocalization Studio [Icy][d]	Whole image or ROI	No	[77]
	MeasureCorrelation [CellProfiler]	Whole image or ROI individual objects	No	[59]
Pixels'/voxels' overlap-based quantifiers	DiAna [*ImageJ*/Fiji][c, d]	Individual objects	No	[82]
	JACoP [*ImageJ*/Fiji]	Whole image	No	[3, 4, 81]
	Squassh/MosaicSuite [*ImageJ*/Fiji][a, c, d]	Whole image or ROI individual objects	No	[83]
	GcoPS [Icy][c, d]	Whole image or ROI	No	[84]
	CalculateImageOverlap [CellProfiler]	Whole image or ROI individual objects	No	[59]
Center/objects' overlap-based quantifiers	JACoP [*ImageJ*/Fiji]	Individual objects	No	[3, 4, 81]
	ExpandOrShrinkObjects using "Shrink objects to a point" option [CellProfiler][a–c]	Individual objects	No	[59]
Distance-based quantifiers	DiAna [*ImageJ*/Fiji][c, d]	Individual objects	No	[82]
	JACoP [*ImageJ*/Fiji]	Individual objects	No	[3, 4, 81]
	ThunderSTORM [*ImageJ*/Fiji][a–c]	Individual objects	Yes	[85]
	Colocalizer [Icy]	Individual objects	Yes	[86]
	Colocalization Studio [Icy][d]	Individual objects	Yes	[77]

(continued)

Table 1
(continued)

Reporter type	Package/plugin [software]	Analyzed entity	Coordinates as input	References
	SODA [Icy]	Individual objects	Yes	[87]
	MeasureObjectNeighbor [CellProfiler]	Individual objects	Yes	[59]

This list of software plugins is proposed in freely accessible environments (see square brackets). The latter were chosen in accordance with their automation and scripting ability as well. The design of a full workflow for image-processing and co-localization analysis is then practicable, for one or many input datasets. Besides, some tools have been designed to analyze whole images or regions of interest (e.g., intensities' overlap-based quantifiers). Similar analysis at the level of individual objects would require some scripting or may be included as an option for some tools (e.g., "MeasureCorrelation" in CellProfiler). In that case, special care should be taken about the significance of objects' size (*see* Subheading 3.3.3). Finally, some coordinate-based tools are dedicated to Single-Molecule Localization Microscopy (SMLM) but are adaptable to spotty objects, as long as object coordinates are retrieved as a list from original images

[a]Includes preprocessing

[b]Includes image corrections

[c]Includes detection options

[d]Includes tools for comparison/statistical tests

3.1.1 Cross-Excitation and Cross-Detection

Selection of appropriate fluorescent tags and acquisition parameters is a crucial step in performing co-localization experiment. Care should be brought to the spectral properties of the chromophores and their adequacy to the characteristics of the acquisition system. Two main drawbacks should be avoided. First, the overlap between the probes' excitation spectra should be minimized, avoiding a potential cross-excitation. Second, the emission spectra should be well separated so that unambiguous detection is performed, minimizing the so-called bleedthrough. The choice is not trivial. The experimenter could opt for chromophores that display well-separated spectra. By doing so, one of the two dyes may lie in the close-infrared domain, where lower optical resolutions are expected, therefore impairing the precision of localization. Is the loss of resolution an issue? As always, choices should be made wisely, having in mind the biological question. While working on objects larger than the optical resolution, it may not significantly impair the co-localization diagnostic. When precision of localization is the funding parameter of the co-localization metrics, for instance, in pointillist methods, the question should be addressed. Although image processing methods exist to counteract cross talk and bleedthrough (*see* Subheading 3.2.1), they generally rely on characterizing the overall contaminating contribution and reverting it by linear unmixing. This tool may not take into account local modification of photo-physical properties of the dyes. Hence, image processing-based methods should be employed only once all possibilities offered on the sample preparation side have been exhausted. Evaluating the cross talk and bleedthrough and adapting the detection parameters could be performed as follows:

1. Before starting: get to know your dyes and microscopy setup. Websites exist (*see* **Note 5**) that provide a way to get the probes' spectra displayed together with the elements used for excitation/emission discrimination. Some will generate "spillover tables," which characterize the expected combination of cross talk and bleedthrough. When choosing a couple of dyes, such tool might help minimizing side effects while optimizing the chromophores' choice relative to the microscopy system the experimenter is planning to use.
2. Start imaging the double-labeled sample. Tune acquisition parameters for both channels so that the ideal spatial sampling rate is achieved, the full dynamic range of the detector is used, and optimal discrimination is obtained. With these parameters set, check that both signals are not decreasing over time, indicating that no photo-bleaching occurs.
3. Load the images under *ImageJ/Fiji* [17, 38], and run a co-localization plugin (e.g., *JACoP* or *Coloc 2*—"2D intensity histogram") to generate the cytofluorogram (*see* **Note 6**). A plot is generated where the two intensities of each single pixel from the couple of images are used as coordinates.
4. Inspect the overall shape of the graph: how many populations can you distinguish? (*see* Fig. 2 and refer to **Note 7** to interpret the result).
5. In case of potential cross-detection, image the mono-labeled samples, in both channels, using the same settings as for the double-labeled sample.
6. Repeat **steps 3** and **4**: in the absence of cross-detection, each pixel should have a non-null intensity in one channel and almost zero (modulo noise) in the other channel. As a result, only one cloud of dots should be visible, located on one axis. In case an additional population is detected, imaging parameters should be tuned.
7. If needed, re-tune the acquisition parameters, trying the following refinements: lower the intensity of the light source; on a filter-based microscope, consider using a different combination of filters; and on a microscope equipped for spectral detection, consider decreasing the bandwidth of the detection windows and/or shifting the detection windows to lower wavelengths for the highest-energy emitter and to higher wavelengths for the lowest-energy emitter.
8. Repeat **steps 3** and **4**: in case the cross-detection still occurs, consider the following options, increasing the dilution factor of antibodies (*see* **Note 8**) or choosing a different combination of dyes.

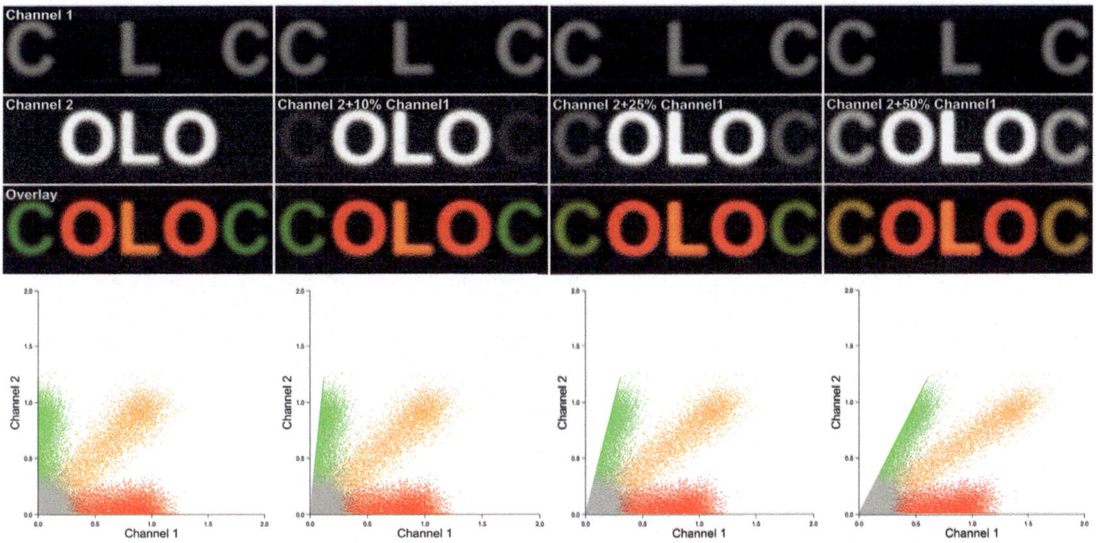

Fig. 2 Simulated cross-detection data: influence on the cytofluorogram. Using *ImageJ*/Fiji, the letters "COLOC" have been drawn and separated into two images: channel 1 containing "CLC" and channel 2 containing "OLO". This generates a total of four classes of pixels: background (outside of the letters), pure green (the two Cs), pure red (the two Os), and full co-localization (the central L). Each channel has been Gaussian blurred to mimic the microscope's convolution, and noise was added (left column). While going to the rightmost column, part of Channel 1's signal was added to Channel 2 (10, 25, and 50%). The bottom row displays the effect of such a contamination on the cytofluorogram where points are colored according to the class they belong to. As more cross-detection is simulated, the "pure green" class progressively rotates toward the Channel 1 axis, revealing a progressive increase of artefactual co-localization

3.1.2 Even Illumination

When dealing with either intensity measurements or image segmentation, even illumination is mandatory. However, depending on the light source coupling to the microscope and on the nature of the device used for illumination, this requirement might not be totally fulfilled. Uneven illumination should therefore be characterized and may, to some extent, be corrected. Once more, plugins such as *MetroloJ* [19, 20] offer a way to quantify and visualize inhomogeneities (*see* Fig. 3). The following protocol is provided as a comprehensive guide to what is being done, behind the scene. Simple reference images may be acquired to correct the actual samples' images and quantify unevenness of illumination. All are built making a simple assumption: the response of the dye being the same all over its surface, any inhomogeneity seen, mainly accounts for uneven illumination.

1. Prepare a reference slide: two types of reference samples might be used. On the one hand, fluorescent plastic slides, being uniformly fluorescent, seem to be an ideal sample for such characterization (*see* **Note 9**). On the other hand, fluorescent preparation made of a diluted solution of dyes could be used as an alternative. Place a drop of dye onto a glass slide, cover with

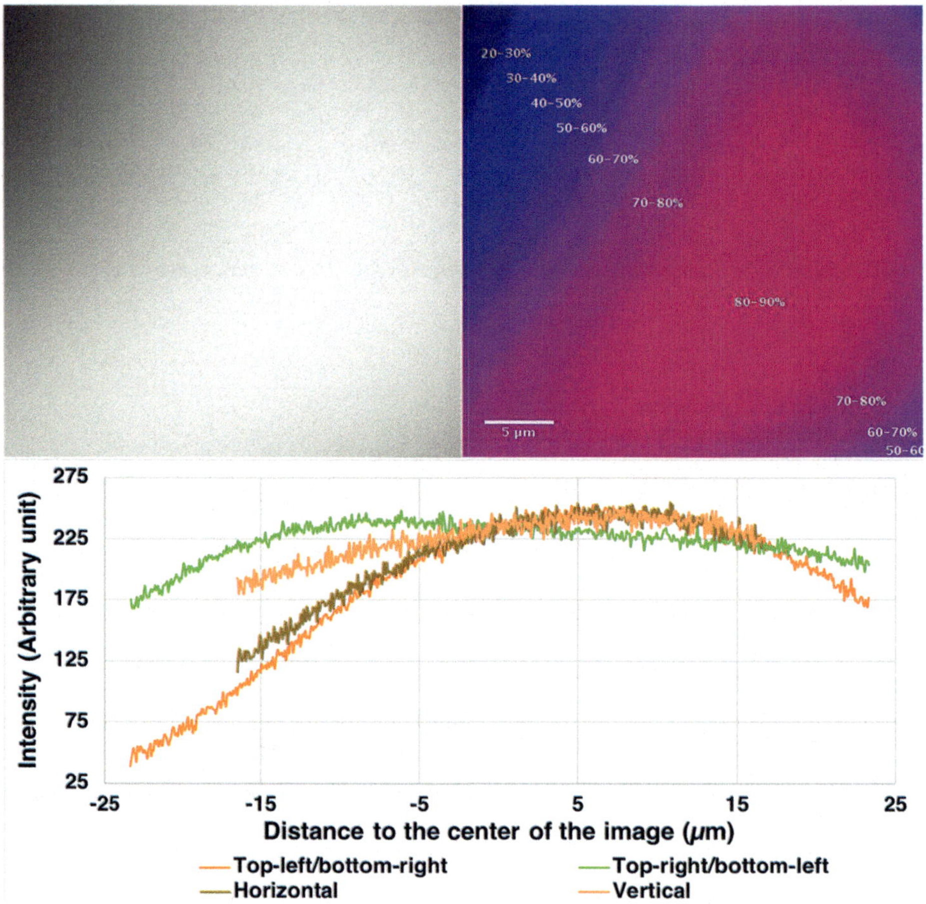

Fig. 3 Even illumination: using a uniformly labeled sample, an image is acquired (top row, left) which reports for the homogeneity of illumination. After image normalization, quantification reveals uneven illumination, ending up in a loss of up to 80% in the upper left corner of the field (top row, right). Plotting intensity profiles allows determining misalignment of the source, as the four characteristic intensity profiles (horizontal, vertical, and the two diagonals) fail to intersect at the center of the image (bottom row). All measurements were made using the [*MetroloJ*] plugin [19, 20]

a coverslip, and seal using nail polish (*see* **Note 10**). To avoid the hassle of getting proper focus, a reference point can be drawn onto the coverslip inner surface with a permanent non-water-soluble marker. Finding its outline should set an easy-to-find reference.

2. Image reference slide: make sure the acquisition parameters are set according to the final settings defined in Subheading 3.1.1.
3. Focusing on the sample can be tricky: try using the reference mark (*see* **step 1**), if applicable.
4. If needed, adapt settings to avoid saturation.
5. Launch acquisition.

6. Visual inspection: using the brightness and contrast toolbox, adapt the display at [Image > Adjust > Brightness/Contrast]. Smaller differences are better seen using appropriate lookup tables (LUTs): from the [Image > Lookup tables] menu, pick a color table that would gradually encode lower to higher intensities using a gradient of colors so that you can directly identify the extrema values and look for intensity variations.
7. Quantification: simple processing can be applied to transform the intensity image into a map displaying local variations. One possible way to go is to normalize the data. This operation consists in scaling the image to get the maximum intensity set to one. Once obtained, normalized intensities from characteristic points can be retrieved. Make sure no ROI is selected: [Edit > Selections > Select None].
8. Retrieve the maximum intensity of the image: use the [Analyze > Set measurements] to set data to retrieve and select *Min* and *maxgray value* as parameters. Use the [Analyze > Measure] menu to perform measurement (*see* also **Note 11**).
9. Before normalizing the image, one must bear in mind its depth. The image, most probably being 16 bits, cannot accommodate decimal values (*see* also **Note 12**). The image type must first be set to 32 bits to avoid data clipping: use the [Image > Type > 32-bit] function to convert the image.
10. Normalization is performed by dividing the image at [Process > Math > Divide...], using the maximum intensity retrieved at **step 8**.
11. As for visual inspection, the use of LUTs can help figuring out intensity variations within the field. Adding an intensity calibration bar is also a way to refer each hue to a value. This operation can be performed using the [Analyze > Tools > Calibration bar...] menu.
12. To go further in the characterization, normalized intensities from specific points of interest could be determined. Those values can either be used to determine the degree of misalignment of the light source or to monitor illumination fluctuations, should several measurements along time have been done. In particular, coordinates of the maximum or center of mass are of interest: both should lie close to the image's center. Intensities within the corners of the images are also of interest (*see* **Note 13**). Although those locations are generally less illuminated than the center, their values should be close to the centers.

In case of high discrepancy between central and peripheral values, appropriate actions should be taken. Checks of the optics and light source alignment, when possible, should be the first way to go. As a second step only, image processing-based corrections may help counteract uneven illumination.

3.1.3 Co-registration

In order to report for true co-localization, the experimenter has to make sure the device used to image the sample is actually reporting the two signals to be at the same location. The origins of a fail are diverse: misalignment of the light sources (when several are used), mis-positioning of the filter cubes, and various optical aberrations (chromatic, spherical, etc.). Prior to starting the analysis, the microscope's response should be characterized. To evaluate proper co-registration, a sample made of multi-labeled objects is used. As multiple dyes are grafted onto a single object, pure co-localization is expected. Having several objects in the field of view allows getting a spatial overview of the microscope's response. Generally speaking, the objectives are well corrected for chromatic shift in their central field, which is a limited territory of one fourth of their total field of view. Recent popularization of the sCMOS cameras, although giving benefit of a higher sensitivity, came at a price: the detection area being larger than for regular (EM-)CCD, camera is able to reach regions that are poorly or not corrected (*see* Fig. 4). Therefore, co-registration check outcomes might be of two kinds: (1) characterize a chromatic shift, and if simple, try to revert it by image processing; (2) define the region of interest where the objective's correction is optimized, and therefore restrict the analysis to this specific region. Checking the co-registration requires a sample where dyes of similar nature as the actual samples are known to be co-registered. Two options are available: using reference slides (e.g., Argolight™ slides used in Fig. 4) or freshly prepared fluorescent bead slides. In this section, we will give instructions on how to prepare, image, and analyze the latter.

1. Prepare reference slide: clean the slides and coverslips using ethanol and wipers.
2. Place the coverslips either in a multi-well plate or on a sheet of parafilm.
3. Dilute Poly-L-lysine to a final concentration of 0.05% (w/v) using distilled water.
4. Use a sufficient amount of the diluted solution to cover each coverslip (typically, 200 μl for a 24 × 24 mm coverslip).
5. Using the tip of the pipet, stretch the drop so that it covers the full glass area.
6. Allow the solution to settle on the coverslip for 15–30 min.

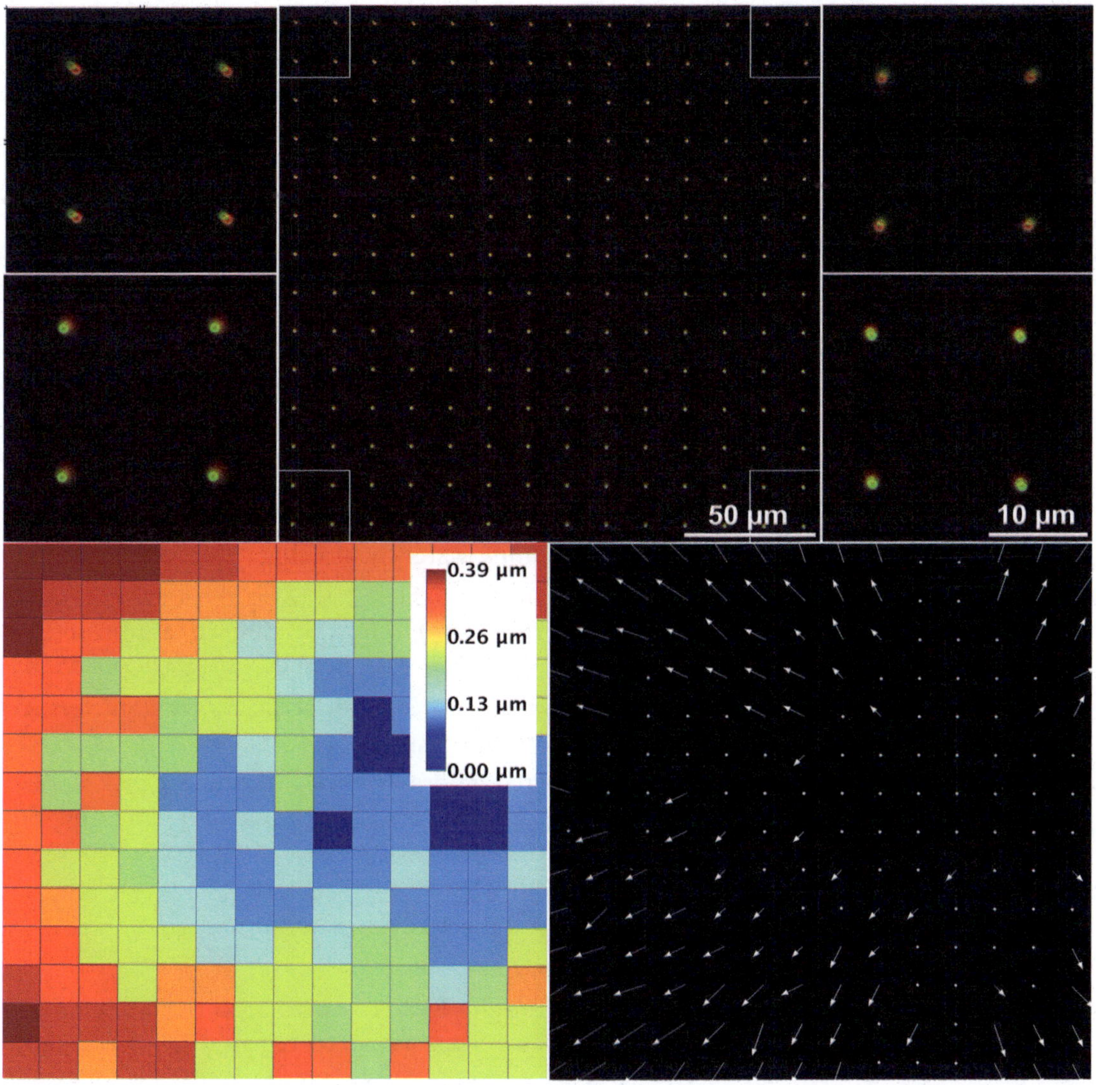

Fig. 4 Co-registration: using multi-labeled reference objects (Argolight™ HM slide), images are taken for the two channels and analyzed for co-registration with the co-registration *ImageJ/Fiji* macro [21]. Top row: full field of view as observed using a 63×/1.4 objective, image is taken using a sCMOS camera without an appropriate correction lens in the c-mount coupler (see the four corners magnification). Several types of visual representation are possible: mis-registration distance can be used to produce either a regional heat map (bottom row, left) or a field of vectors (bottom row, right)

7. Remove the Poly-L-lysine solution: it can be reused up to three times.
8. Rinse three times using a large volume of distilled water.
9. Remove the remaining water by bloating. Approach one side of the coverslip to a wiper: the water should be transferred to the paper.

10. Dilute fluorescence beads in distilled water. As the dimensions of the beads we use are close to the resolution limit, it might be hard to focus on the coverslip when imaging. Therefore, it is advised to mix them with larger diameter beads. Tetraspeck™ 170 nm and 4 μm diameter beads might be used. In order to define the dilutions to be used, tests should be carried out with each batch (*see* **Note 14**).
11. Use a sufficient amount of the beads' suspension to cover each coverslip (typically, 200 μl for a 24 × 24 mm coverslip).
12. Using the tip of the pipet, stretch the drop so that it covers the full glass area.
13. Let the beads sediment for at least 30 min: depending on the beads' diameters, time will influence the concentration of deposited material.
14. Rinse three times using a large volume of distilled water.
15. Remove the remaining water by bloating.
16. Mount the coverslips onto the slides, using the same mounting medium as for the sample. In the case of a non-setting medium, seal the coverslip onto the slide using nail polish.
17. Image reference slide: make sure the acquisition parameters are set according to the final settings defined in Subheading 3.1.1.
18. For freshly prepared fluorescent bead samples, pick a field where beads are sparsely distributed and covering the full area. For manufactured reference slides, refer to the user manual and select the most appropriate pattern/area to image.
19. If needed, adapt settings to avoid saturation.
20. Launch acquisition.
21. Analyze data: although plugins exist to analyze co-alignment, they generally rely on defining the co-registration over a single reference spot. As a consequence, optical disturbances may not be visible, especially when the reference point is set on the center of the field, where optical corrections are at their best. Therefore, we describe here one possible workflow that can be easily automated using *ImageJ/Fiji* (a possible implementation is provided as supplemental material, listing 1 [21]). This example workflow is targeted at 2D analysis. It can however be adapted using *Fiji'* 3D plugins, such as *3D maxima finder* and *3D ROI Manager*, found in Thomas Boudier's 3D suite [22]. Open the images to be analyzed using *ImageJ/Fiji*.
22. Detect the reference elements (might be beads or reference patterns) on the first channel: as they are expected to have dimensions close to the resolution, one could use the [Process > Find Maxima...] detection function. Make sure the

Light background checkbox is unticked, as this work is being carried out on fluorescence images, and tick the preview box.

23. Adapt the *Noise tolerance* parameter to visualize single crosses over each single reference element.
24. Once parameters have been set, select the *Point selection* as an output: a region of interest is overlaid to the image, containing the detected points.
25. Use [Edit > Selection > Enlarge...] to transform each detection cross into a circle. The radius parameter should be set to a value larger than the expected channel's displacement.
26. Activate the *ROI Manager* at [Analyze > Tools > ROI Manager...], and use its *More/Split* option to individualize each detected area.
27. Review each ROI: having defined the surroundings of the reference spots, we will now pair the detections from channel i with the closest detection in channel $i + 1$. Select the image from the first channel.
28. From the *ROI Manager*, activate the *ith* ROI.
29. Re-launch the [Process > Find Maxima...] function (the detection is now restricted to the active ROI), and use the *List output* option to log the coordinates of the current reference spot in channel 1.
30. Repeat **steps 27–29** for the other channels.
31. Finally, for the all sets of coordinates, compute the paired distances.

This workflow, being repetitive, would benefit from some automations (*see* supplemental material, listing 1 [21]). In addition, a macro-encoded workflow would ease the process of generating outputs such as heat maps showing the direction of displacement or its amplitude (namely, the distance between detections in channel i and $i + 1$; *see* Fig. 4). In this protocol, the local maxima detection is used, which leads to a precision of detection estimated to one pixel. Using the same strategy for a rough primary detection, one could use either mass center retrieval or even Gaussian fitting to refine localization when more accurate measurements are required.

3.1.4 Resolution

Although theoretical optical resolution might be calculated, measuring it within the framework of the actual experiment is highly recommended. Parameters such as the overall refractive index encountered by light while traveling the sample, the structures' positions along the optical axis, or the effective numerical aperture of the objective influence the ability to distinguish close objects. It should be reminded the importance of knowing the optical system's performances: co-localization hypothesis, or more precisely rejection of this hypothesis, is made relative to resolution. The following workflow details how to prepare and analyze fields of individual, sub-resolution objects in the aim of measuring local optical resolution:

1. Prepare reference slide: procedure to produce reference slides for resolution measurement is the same as for co-registration (*see* Subheading 3.1.3). Choice of appropriate beads is crucial (size, spectral properties, and intensity: *see* **Note 15**).
2. Image reference slide: use the same recommendations as aforementioned in the co-registration section (*see* Subheading 3.1.3).
3. Analyze data: from the image data, individual beads can be isolated and analyzed. The measured values can then be replaced within the context of the field, using its spatial coordinates to generate map presenting the local variations of resolutions. Multiple beads' detections can be performed using, for instance, local maxima as mentioned above (*see* Subheading 3.1.3). The principle of this analysis is to determine resolution along all axes (*see* **Note 16**). The following procedure references functions aimed to retrieve 2D resolutions measured on one bead. Using automations such as *ImageJ* macro, the concept can be extended to a full field of view containing many beads (*see* the provided supplemental material 2 [23]). Using the line selection tool, point at a single bead and draw a horizontal line. The line's extremities should be well away from the bead.
4. Use the [Analyze > Plot profile] function to display the intensity profile over the line.
5. Under the graph, press the *List* button: a table is displayed, containing the coordinates of pixels along the line (X) and their intensities (Y).
6. Use the [File > Save] menu to save data to the disk.
7. Open *ImageJ*'s curve fitter at [Analyze > Tools > Curve Fitting...].
8. In the bottom window, numbers are displayed: clear its content, and fill it using the *Open* button to load saved data.
9. From the top drop-down list, select *Gaussian*, and then press *Fit*: a new graph is displayed, carrying both original and fitted data.
10. In the log window, fitted parameters are displayed. The full width at half maximum, and therefore the approximated resolution, might be calculated from the *d* parameter (standard deviation): $\text{FWHM} = 2\sqrt{2\ln(2)}d$.
11. Repeat **steps 3–11** for the remaining axis.

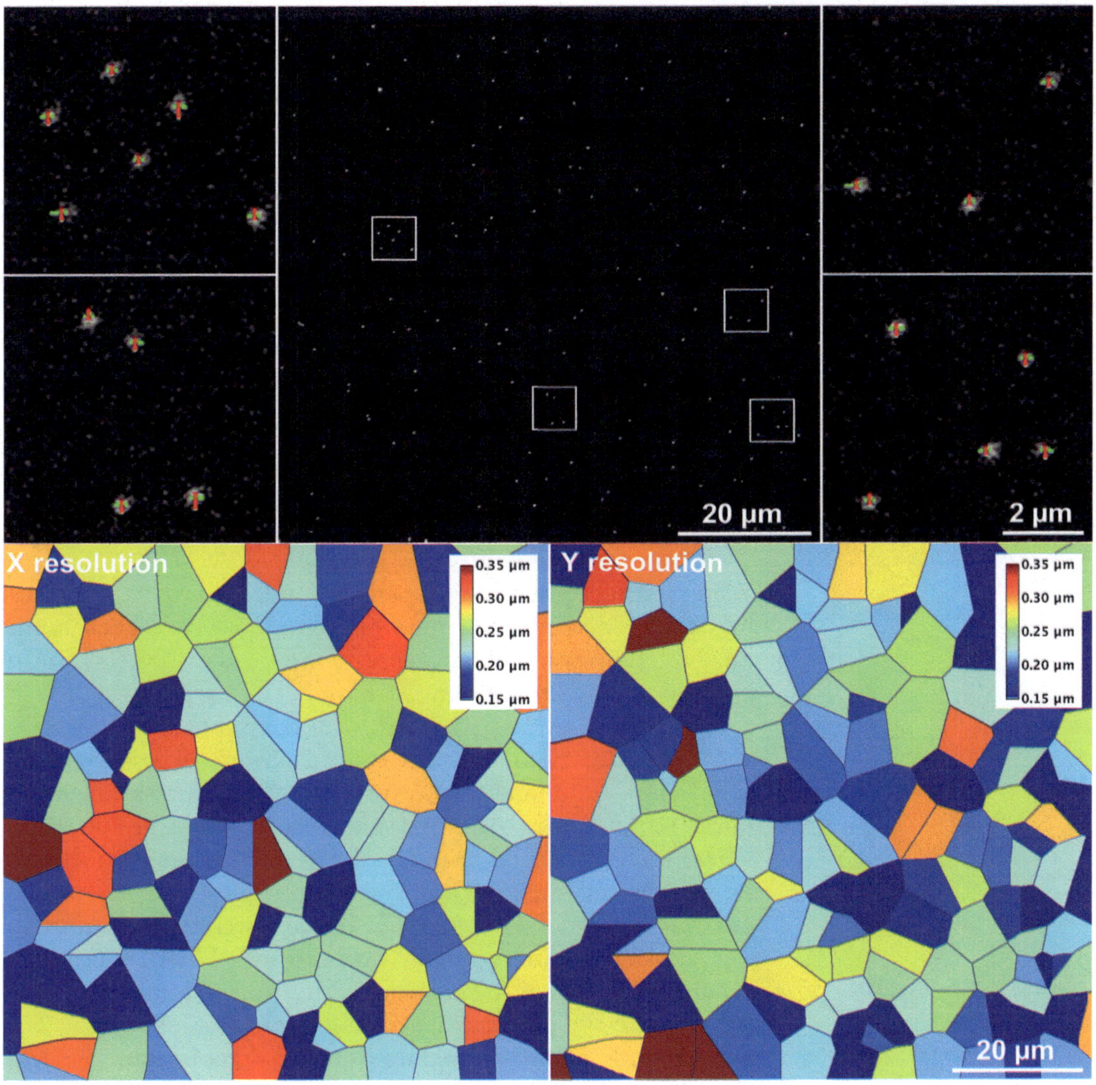

Fig. 5 Resolution: using multi-labeled sub-resolution objects (beads), an image is recorded using a confocal microscope (top row, middle panel). The microscope's optical response is then analyzed along two directions, horizontal and vertical. Top row, four lateral panels: intensity profile is extracted along the two directions and then approximated by a Gaussian profile. For each, full width at half maximum is determined and used as the length of a segment overlaid to the original image (green, horizontal; red, vertical). Additional output can also be generated (lower row). Using each detected bead, a Voronoï diagram is computed. Resolution value is used to fill each tile of the diagram. Lower row displays such a representation for X (lower left) and Y (lower right) resolutions. Slight discrepancy between X and Y resolutions might be explained by both imprecision of measurement combined with polarization effects of the light source as reported by Li et al. [79]

This workflow, being repetitive, would benefit from some automations (*see* the provided supplemental material listing 2 [23]). In addition, outputs could be generated, such as heat maps showing the differences of resolutions all over the fields of view and potential differences depending on the axis (*see* Fig. 5).

3.2 Data Preprocessing

The next step in preparing the data for co-localization aims at correcting for measured aberrations, restoring the information content, and, if object-based methods are to be used, segmenting the images.

3.2.1 Corrections

Relevance of image corrections should always be raised [39, 40]: what is the most appropriate option between correcting the images and re-acquiring them? It all depends on the latitude the experimenter has on tweaking the acquisition system, the availability/lifetime of the sample, and the domain of application of available image manipulation tools. Only if applicable, the experimenter could walk that path: once the characteristics of the optical system have been determined, the experimenter might also restrict the analysis to areas of the sample where aberrations are minimized. Image-processing-based corrections should therefore not be thought as mandatory but rather as a tool, which may or may not be used.

1. When setting up acquisition parameters according to Subheading 3.1.1, both bleedthrough and cross talk have been minimized. However, minimizing might not always mean excluding. If reference samples have been acquired, correction is still possible by first making the assumption that the "leakage" from one channel to the other is linear; second, by characterizing it; and finally, by reverting it. The process, known as linear unmixing, is based on solving a set of linear equations where each corrupted channel (the output image) is obtained by linearly mixing contributions from some/all dyes, in proportions to be defined [41–43]. The first step of linear unmixing aims at defining those proportions. Final images are obtained by solving the equations, thus re-assigning each contribution to the proper image. Several plugins for *ImageJ/Fiji* exist to both define the coefficient and correct images (*Spectral Unmixing Plugin* [24], *Spectral Unmixing of Bioluminescence Signals* [25], *PoissonNMF* [26]), to be installed and used according to the provider's instructions.
2. Uneven illumination can be corrected by using two strategies. The first uses a reference image, as described on Subheading 3.1.2. Since the reference image might not be scaled as the actual channel's image, the correction should be performed by pixel-wise division (**steps 3–7**). Alternatively, when dealing with evenly sized objects, in the absence of a proper reference image, using a low-pass filter may help produce an estimate (**steps 8–14**). This strategy is based on the assumption that uneven illumination is the major contribution in the low-frequency domain. Of course, both strategies should be used with great care, as they will impact the recorded intensities. For instance, misevaluating the objects' sizes and the

cutoff frequency of the low-pass filter may highly impair the image. While we propose two protocols, *ImageJ/Fiji* plugins exist that provide direct correction tools such as the *BaSiC* tool [27].

3. Strategy 1, using a reference image: open both reference and target images.
4. Launch the [Process > Image Calculator...] function.
5. Select the image to correct as "Image 1" and the reference image as "Image 2."
6. Under "Operation" drop-down list, select *divide*.
7. Make sure both *Create a new window* and *32-bit (float) result* options are selected (*see* **Note 12**).
8. Strategy 2, using a low-pass filter to generate an estimate of a reference image: open the image.
9. Duplicate it using the [Image > Duplicate...] menu, giving it an obvious name such as "estimateReference."
10. Assuming the objects of interest are small in size, relative to the extent of uneven illumination, run the [Process > Filter > Gaussian Blur...] function.
11. Enable the preview.
12. Progressively increase the radius value to make objects just disappear from the scene.
13. Apply by pressing Ok.
14. Proceed as for strategy 1 (**steps 4**–7) to correct the actual image using the reference image.
15. Depending on the sources of aberrations, ways to correct the channels' mis-registration should be adapted. Especially, non-linear distortions might not be the easiest to model and correct. When correcting images, attention should be drawn at signal impairment: is the cure generating additional artifacts? Should the images be shot again, after proper realignments have been made on the setup? When channels' mis-registration is linear such as when dichroic mirrors fail to position all at the same place/orientation, a simple translation might be applied, as described below. Pertinence of corrections should always be raised: in case the measured displacement is smaller than the optical resolution, corrections might not be required. First, characterize the co-registration as described in Subheading 3.1.3.
16. Open the images to correct.
17. The reference image will not be corrected, but for the other channels, process as follows: from the table obtained as a result

of procedure Subheading 3.1.3, identify the displacement values along the X and Y axes.

18. Launch *ImageJ/Fiji* [Image > Transform > Translate] menu.
19. Input the displacement values (*see* **Note 17**).
20. Press *Ok*.

3.2.2 Restoration

The process of image acquisition gives us a representation of the object that is modulated by the transfer function, which characterizes the full acquisition chain. In optical microscopy, this response function is called the point spread function (PSF) and defines how a unit point, sized below the optical resolution, will look like on the image. Strategies exist to recover a more precise representation of objects from the image, called restoration methods.

In this aim, deconvolution is one possible approach. It requires some knowledge about the acquisition chain: PSF should be known (*see* Subheading 3.1.4 for PSF measurements) or might alternatively be estimated from the system's characteristics. In most of the deconvolution algorithms, an objects' estimator is built and tuned iteratively, taking the PSF into consideration (for reviews, *see* Refs. [44, 45]). This estimator is then convolved by the PSF and compared to the actual microscope's output. From this comparison, a correction vector is built, which measures the distance between the convolved estimate and the effective output. A new iteration takes place by applying the correction vector to the estimator. Depending on the algorithm, the number of iterations could be user-defined or submitted to examination of the correction vector's amplitude.

When performed properly, co-localization analysis may benefit from two of the deconvolution properties. First, as the signal is reassigned to its origin, contrast is restored. Second, as the PSF on the deconvolved image is narrowed, resolution is improved [46, 47]. Taking as an example the *Richardson and Lucy* algorithm [48, 49], a slowly converging algorithm that displays strength on noisy images, an example procedure is proposed to define the number of iterations to be used. *PSF Generator* [28, 29] and *DeconvolutionLab* [30–32] *ImageJ/Fiji* plugins are selected to perform it.

1. Performing deconvolution: open the image to deconvolve. Keep track of its dimensions; especially, write down its number of slices.
2. Either open the corresponding PSF or generate one using the *PSF Generator* plugin: a manual can be found on the plugin's webpage [29]. An experimental PSF could be obtained by first proceeding as explained in Subheading 3.1.4, isolating individual PSFs and averaging them.

3. Launch the *DeconvolutionLab1* plugin: to test the number of iterations, the version 1 of the plugin allows the recording of intermediate results, whereas the current version 2 only displays the final result as a stack.
4. Set the deconvolution parameters by reviewing each "Module" entry.
5. Select the algorithm, and enter a large number of iterations (250–500 for Richardson-Lucy).
6. Allow the intermediate estimates to be saved by ticking the corresponding box under "Video" module and selecting an output folder.
7. Pay attention to the FFTW module: ticking the box improves the execution speed by using the FFTW library [50, 51] instead of *ImageJ*'s internal *Fast Fourier Transform* class.
8. Launch deconvolution by pressing the *Run* button: a deconvolved stack should appear in the output folder after each iteration has been completed.
9. Assembling a hyperstack for evaluation: once deconvolution has been performed, drag-and-drop the output folder onto *ImageJ/Fiji's* toolbar. In the window that should pop up, press *Ok* to load all the images as a single stack.
10. For each deconvolution iteration, the newly loaded stack is made of all slices. To visualize the effect of the number of iterations on the deconvolution process, transform the stack onto a hyperstack using the [Image > Hyperstacks > Stack to Hyperstack...] function.
11. Select *xyczt(default)* as data order.
12. Leave the number of channels to 1.
13. Use the value recorded at first step for number of slices.
14. Use the number of iterations as number of timepoints.
15. Press *Ok*: a new hyperstack is displayed, allowing navigating through the z axis and along iterations (time slider).
16. Evaluating the hyperstack: navigate through the hyperstack toward the final iteration (last "timepoint," lower slider).
17. Navigate through the Z dimension toward the most contrasted slice (upper slider, labeled "Z").
18. Using the line tool, draw a line across an object of interest. Ideally, it should be picked among the smallest ones, with dimensions close to the optical resolution.
19. Plot the intensity profile along the line using the [Analyze > Plot Profile] menu.
20. From the plot window, click on *More* > *Set range*, and select *Fix Y Range While Live*.

21. Press the *Live* button: it should turn red. Using the two previous settings, the plot will be updated while navigating the stack but will not be scaled depending on intensity values: it will help visually inspecting the contrast and resolution improvements, as a function of the number of iterations (*see* Fig. 6 for an example output).
22. Move back to the first frame (lower slider, corresponds to iteration 1); then, progressively move toward the higher-index frames.
23. While navigating the hyperstack, inspect the plot: the signal should increase, while the peak should look shaper (*see* Fig. 6, middle-left panel).
24. Remember the first frame when no significant improvement is visible both on intensities and peaks' width.
25. From the current frame, try to identify deconvolution's artifacts (some common artifacts are illustrated on Fig. 6, bottom panel; also refer to **Note 18**). In case any artifacts have been identified, try adapting the parameters used to generate the PSF or acquiring a new PSF if measured and/or lowering the number of iterations.
26. The number of iterations to use is obtained when the lowest number of artifacts has been found and the highest contrast/resolution enhancement has been reached.

When using deconvolution, one should be aware that the result is an estimate of the object, built from its measurement (the image), not the actual object. Its purpose is to propose a version of the image that is actually closer to the object than the raw image. To build this estimate, assumptions are made, especially about the PSF. In most algorithms, it is supposed to be non-variant in space. Of course, this is an approximation that reveals to be false when thick samples (>30–40 μm) are imaged. Should optical aberrations be present, they will still appear on the deconvolved image and may as well be enhanced. Therefore, restoration using deconvolution should not be seen as a way to counteract them, rather to enhance both contrast and resolution on properly acquired images, on a well-set system.

With the recent developments of machine learning in the light microscopy field, some new methods of image restoration have become available. Although time is still required to get more insights about their benefits and pitfalls, the field could benefit from methods taking into account more parameters, such as the PSF variability along the optical axis (see for instance [52]).

3.2.3 Preprocessing: What Else?

Image correction and restoration are only part of a larger range of tools that may be used for image preprocessing prior to co-localization analysis. Only straightforward preprocessing

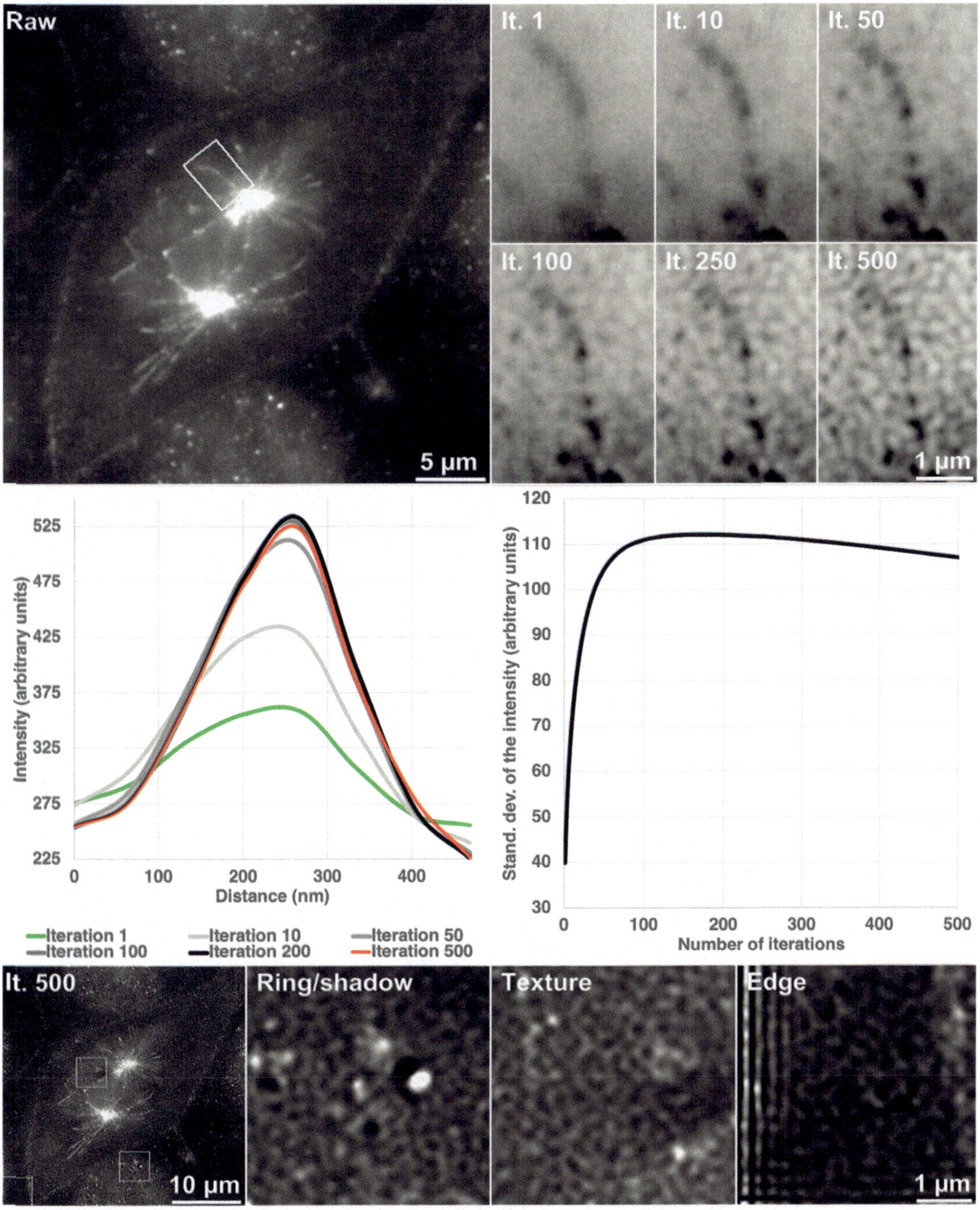

Fig. 6 Evaluating the restoration process: an example workflow. A sub-region containing a structure close to the resolution limit is analyzed (top panel). On the top-right panel, contrast has been adjusted to enable texture comparison: intensity ranges may vary between thumbnails. A line is drawn across the structure and its intensity profile is plotted (middle-left panel). Two parameters should be monitored: increase of the signal-to-noise ratio (SNR) and improvement of the resolution (full width at half maximum). Standard deviation is also a parameter to monitor (middle-right panel). Note how beyond 100/150 iterations the SNR stops improving while a strong texture is progressively appearing. A compromise should be found, aiming at an improved resolution and SNR, limiting the artifacts depicted on the lower panel (mainly dark rings, texture, and Gibbs' phenomenon)

workflows have been presented, and additional ones may require more knowledge to get started. For instance, noise is always an issue when objects are to be delineated. Statistical methods exist that would improve the image's signal-to-noise ratio, taking into consideration the Poisson and Gaussian noises as being part of the imaging process. Although theoretical background might not be the easiest thing to apprehend, some of the tools come as *ImageJ/Fiji* plugins, such as *Pure Denoise* [53, 54]. Depending on the type of co-localization analysis to be performed, delineating the objects might be required. This step, known as segmentation, will often require a full workflow by itself, made of individual steps such as denoising, restoration, filtering, thresholding, etc. Isolating structures based on spatial and intensity-based features will require the combination of multiple approaches that directly depend on the nature of the images, the strategy employed to acquire them, and the type of objects they carry. Although thresholding is generally the first method used to partition the image into object's versus non-object's pixels, setting up a threshold is generally highly subjective and may reveal a bias in the output, unless proper rules are set. Many methods exist that take into consideration the image's histogram, making assumptions about its shape and its supposed distribution of intensities [55]. More advanced strategies exist based on machine learning that benefit from the extraction and recombination of multiple features [56, 57]. Relying on measurements made in regions pre-classified by the user, the computer extracts features and computes which ones are discriminant (and to which extent) to build a classifier that may later on be used (see, for instance, the *ImageJ/Fiji Trainable Weka Segmentation* plugin [58, 59], or the stand-alone, yet *Fiji* and *CellProfiler* [60] connected, *iLastik* software [61]).

3.3 Data Extraction: A Comprehensive Guide to Picking the Proper Reporter/Metrics

3.3.1 Choosing a Reporter/Metric

Co-localization assessment, unless simple overlay is used, is performed by computing a reporting numerical value. Two types of values might be calculated.

On the one hand, the term "indicator" could be used for methods that will return a relative value, which may vary on a predefined scale (*see* Fig. 7). An indicator is not directly measuring an amount of co-localization, rather giving a tendency. In case a unique experimental condition is used, conclusions can be tricky to draw, especially for values lying in the mid-range zone. Additional methods would then be required to artificially produce a reference dataset and draw conclusions (*see* **Note 19**).

On the other hand, the term "quantifier" could be used to qualify direct measurements of co-localization. They rely on first defining a criterion and then building a metric. For instance, a criterion might be the following: pixels are considered as co-localized when "positive" (i.e., above a certain intensity threshold) on both channels. Here the corresponding metrics would be

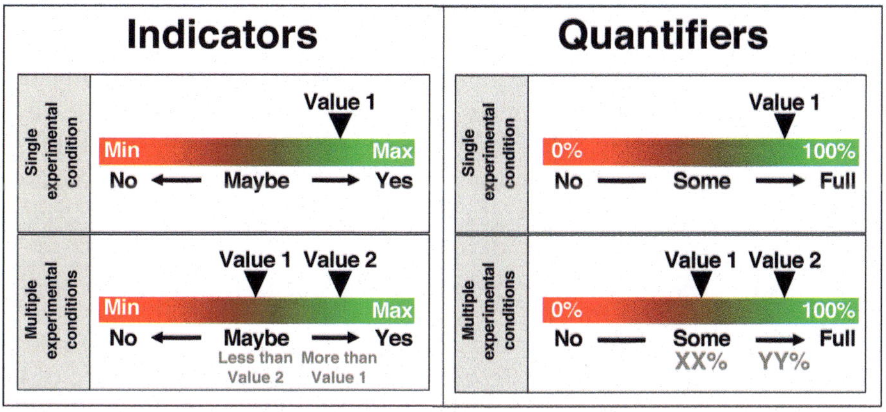

Fig. 7 Choosing a reporter/metric: indicator versus quantifier. Indicators are placed relative to a predefined scale, while quantifiers give a more direct readout of co-localization

defined as the percentage of positive pixels. Overlap is only one particular case of quantifiers. Another criterion might be that co-localized particles from channel A are lying within a defined distance from particles in channel B, the metrics being the percentage of particles for which the rule applies. In the latter case, the metric used is a direct quantification (hence the term "quantifier") while not being a measure of overlap.

No matter the terminology, the most important part of co-localization studies is to pick or design a proper metric. Choice or conception must rely on the type of input data, on the image's content, and on the envisioned outcome. For instance, in case several experimental conditions are to be compared, relative measurements are fine, and indicators are appropriate to use. In case an absolute statement is to be made about a single condition, quantifiers are more indicated, although properly designed indicators may also be used.

3.3.2 Correlation-Based Indicators

As being the most widely implemented within manufacturer's software, the correlation-based methods are the most widely used. They rely on the assumption the quantity of labeling on the two channels is related by a mathematical rule. Generally, a linear relation is supposed to link both quantities. In such case, its strength can be evaluated using the Pearson's correlation coefficient (PCC, Eq. 1) [62–64]. No matter their spatial distribution, intensities of the pixels are used as coordinates to plot a cytofluorogram just as in flow cytometry, where this method is imported from:

$$\mathrm{PCC} = \frac{\sum_i (R_i - R_{\mathrm{aver}}).(G_i - G_{\mathrm{aver}})}{\sqrt{\left[\sum_i (R_i - R_{\mathrm{aver}})^2 . \sum_i (G_i - G_{\mathrm{aver}})^2\right]}} \tag{1}$$

By construction, PCC is expected to sit within the [−1, 1] range, 1 describing a perfectly linear relation between the two sets of intensities (e.g., when the same image is being analyzed twice), whereas −1 means the two sets of intensities are linearly, but inversely, related. The mid-range 0 value reflects a total absence of structured relation between the two sets of intensities. While extreme values are straightforward indicators to place over the generic range, mid-range values might be trickier to interpret. Additional experimental conditions should therefore be used to tune down the PCC's full range to a shorter range encompassing minimum and maximum expected values. Being an indicator, the PCC does not provide a direct quantitative estimate of the degree of co-localization: it only provides a trend. However, PCC is a good tool to describe the evolution of co-localization while comparing several experimental situations where, for instance, co-localization is expected to increase/decrease. In 2011, Dunn et al. [65] pointed out the relation between the PCC and the "coefficient of determination" (R^2), the latter being the square of the former. R^2 quantifies the "percentage of variability in G that can be explained by its linear regression with R." This value can therefore be classified as a quantifier.

Use of PCC raises several drawbacks that can be addressed individually. First, PCC is highly sensitive to noise. To circumvent underestimation of PCC due to noise, Adler et al. [6] proposed using image replicates and evaluate on images' doublets from the same channel PCC's discrepancy from the expected 1 value. From there, a correction factor is built that is used to correct the observed PCC. Second, using PCC, one makes the assumption a linear relationship links both channels' intensities. For instance, if a saturable association exists between elements depicted in the two images, PCC is not appropriate anymore, the relationship being nonlinear and reaching a *plateau*. In case the association is monotonic, an alternative can be found, using the intensities' ranking instead of directly using their values. A new version of PCC is then calculated from the ranks and named Spearman's coefficient (SC) [66], which started to be used in the microscopy field in 2008 [6, 9]. Third, the stoichiometry linking both channels' intensities is expected to be unique. In such a case, evaluating PCC over the full image might lead to an unexpected low PCC, the intensity pairs not following a unique line. In the lucky event the different stoichiometry of association is spatially coherent, restricting the analysis to properly chosen regions of interest can help refining the co-localization diagnostic. Finally, McDonald et al. [67] pointed out significance of PCC could be statistically tested by calculating $t = \mathrm{PCC}\sqrt{(N-2)/(1-\mathrm{PCC}^2)}$ which is expected to be t-distributed with $N - 2$ degrees of freedom. As mentioned by the authors, by construction, any image taken on a microscope fails to meet the requirement for the statistical test to be applicable: the independence of all measures. Due to the PSF that convolves the object onto images, and to the Shannon-Nyquist sampling that results in having at least three measures within the PSF unit, local correlation occurs, hence resulting in multiple measurements of each data point. As previously mentioned, the work-around stands in having more than one sample and more than a single experimental condition to analyze.

3.3.3 Intensities' Overlap-Based Quantifiers

Although two markers of interest can be present on one structural set, their stoichiometry might not be unique, i.e., the same on each structure. Therefore, PCC might not be the most appropriate indicator, and an alternative has to be built. Instead, a well-known set of quantifiers is usually employed: the Manders' coefficients (MC) [64]. Those values result from calculating the portion of signal involved in co-localization. In their first definition [64], the assumption is made that an absence of labeling gives zero intensity on the image. Manders' coefficient M_1 is then calculated as the sum of intensities from channel A that overlap with non-zero pixels in channel B, divided by the total intensity from channel A, and M_2 reporting for the reverse combination (*see* Eqs. 2 and 3, Fig. 8 left panel):

$$M_1 = \frac{\sum_i R_{i,\mathrm{coloc}}}{\sum_i R_i} \text{where } R_i\text{,coloc} = R_i \text{ if } G_i > 0 \tag{2}$$

$$M_2 = \frac{\sum_i G_{i,\mathrm{coloc}}}{\sum_i G_i} \text{where } G_i\text{,coloc} = G_i \text{ if } R_i > 0 \tag{3}$$

With the development of new detectors, and especially improved sensitivity, the zero for nonpositive pixels' rule progressively revealed to be untrue. Therefore, the zero/non-zero criterion has

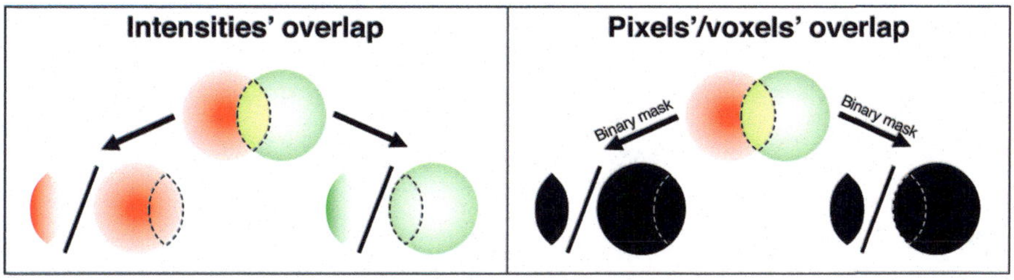

Fig. 8 How to build a co-localization reporter? In this example, two types of quantifiers are built, based on overlap measurements. On the left side, the quantifiers are calculated based on the proportion of overlapping intensities. This is the foundation of MC (*see* Subheading 3.3.3). On the right panel, the quantifier estimates the physical overlap: the same calculations are made but taking into account only the proportion of overlapping localizations (*see* Subheading 3.3.4)

progressively been replaced by 1 based on a threshold value. In some softwares, Manders' coefficients calculated based on a threshold are called *thresholded Mander's* coefficients. Setting such a value, when user-defined, is prone to high subjectivity. To circumvent this drawback, strategies have been developed such as the method proposed by Costes et al. where thresholds are set to minimize correlation within the non-thresholded pixel's population [68].

Once more, the experimenter has to keep in mind the scope, the limits of the quantifiers used, and therefore the possible interpretations to be made. As pointed out by McDonald et al. [67], the message carried by MC should be carefully interpreted. For instance, in case one signal occupies the vast majority of the imaging area, it seems obvious the second signal will share with it a large part of its positive pixels. In such a case, although one MC might be of high value, the question should remain as to whether this co-localization occurs on a determined or random basis. Proper statistical testing should therefore be carried out. Although being described as image-wise quantifiers, the MC might also be determined object by object. While reflecting regional co-localization, one should keep in mind such local measurements are made with less values than in the global case: proper statistical assessments should therefore be implemented.

3.3.4 Pixels'/Voxels' Overlap-Based Quantifiers

In some experimental conditions, the quantities of molecules might not be related to one another. Therefore, although both species of interest may be located on the same structures, no intensity correlation is visible even if a localization correlation occurs. Another strategy should then be built, relying on the MC approach. Pixels' intensities should not be used directly but rather processed in advance to define their state relative to a binary scale: positive or negative. Most of the time, a simple threshold is used but should not be thought as the only possible mean. The aim is to transform the intensity image onto a binary mask using the most adapted segmentation method. Assuming the binary mask to have intensities of 1 for positive pixels and 0 otherwise, modified MC can be used to determine the physical degree of overlap (*see* Eqs. 4 and 5, Fig. 8 right panel):

$$\mathrm{MM}_1 = \frac{\sum_i R_{i,\mathrm{coloc}}}{\sum_i R_i} \text{where } R_i\text{,coloc} = 1 \text{ if } G_i > 0 \qquad (4)$$

$$\mathrm{MM}_2 = \frac{\sum_i G_{i,\mathrm{coloc}}}{\sum_i G_i} \text{where } G_i\text{,coloc} = 1 \text{ if } R_i > 0 \qquad (5)$$

Being derived from the MC approach, those quantifiers suffer from the very same drawbacks. Especially, segmentation is a crucial step, as a permissive approach would end up dilating the objects and extending their area, whereas a too restrictive one would lead to an underestimation of co-localization. Finally, as stated above for intensity-based overlap, those quantifiers can also be computed on a local basis, providing the experimenter with a spatial mapping of co-localization, at the object level.

3.3.5 Centers/Objects' Overlap-Based Quantifiers

MC and MC-derived quantifiers are best fitted when objects of similar shapes are studied. How to proceed when small elements from one channel are to be co-localized with large elements from the second channel? In case of disparity, alternative methods can be employed. Lachmanovich et al. [69] proposed to segment the large objects, while the smaller ones would be replaced by a single set of coordinates, supposed to report for their locations. A co-localization criterion is defined, by considering it to occur when the representative point of objects from one channel falls onto the segmented area/volume from the second channel (*see* Fig. 9). The metric is then built considering the proportion of points fulfilling this condition. Interestingly, this metric could be used out of its original scope. Considering a set of plain objects in the first channel surrounded by continuous, hollow structures in the second channel, pixels/voxels overlap-based method would report no co-localization. Exploring the relationship between centers in the second channel over the objects in the first one would reflect coincidence. This basic theoretical example illustrates how combining the use of two strategies may help unraveling topological links between intricate objects.

3.3.6 Distance-Based Quantifiers

Dealing with small objects whose size is close to the resolution limit is challenging, especially when thinking of the segmentation and the errors this processing may carry. The use of MC and MC-derived quantifiers can be impaired by the low amount of material (pixels/voxels) available to work on, a situation that would lead to a rather binary distribution of values when dealing at the object-to-object level. Extracting only one significant element per object could perform a simplification of the scene. A first data extraction step could be carried out by computing all the object's centers (intensity-based or geometrical). This summarization is only possible when the objects are small enough and their intensities are evenly distributed. The second analysis step consists in computing the Euclidean distances between the sets of centers from the two images. A decision has then to be made: which criterion should be used to classify two centers as co-localizing? One possible criterion is to consider the imaging system's resolution (*see* Ref. [4]): in case the computed distance lies below this limit, one should consider both to be indistinguishable by the current mean of observation (*see* Fig. 9). A metric could then be

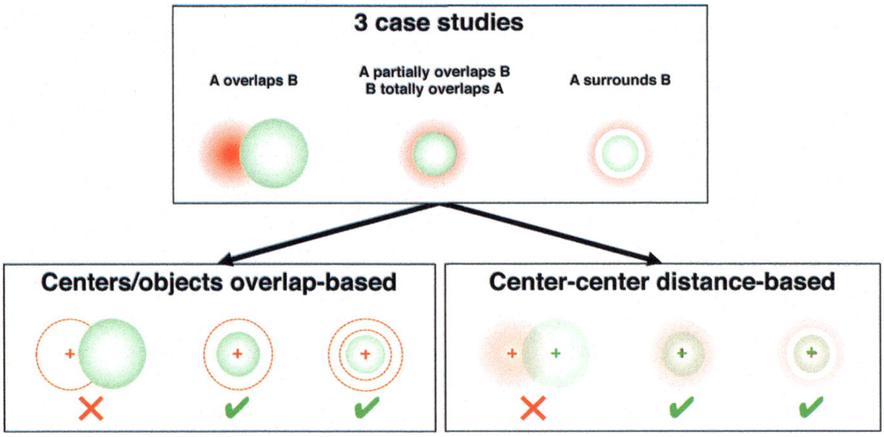

Fig. 9 How to build a co-localization reporter? Three case studies are shown, from left to right: A (red) partially overlaps B (green), A partially overlaps B while B is included within A, and A surrounds B. Two metrics are used: centers-objects overlap (*see* Subheading 3.3.5) or center-center distance (*see* Subheading 3.3.6). Using those metrics, co-localization diagnostic is set even in situations where no physical overlap occurs. They may be useful, for instance, to describe one signal being surrounded by another. In such case, centers/objects are not equivalent when considering A centers versus B object and A objects versus B centers, already providing a hint about their topological arrangement

built by, for instance, determining for each channel the proportion of co-localizing centers. Taking back the previous example of plain structures surrounded by hollow objects (*see* Subheading 3.3.5), a major drawback of this method can be emphasized. Concentric but non-overlapping objects would likely have their centers close to one another. Using distance-based quantifiers, co-localization would be reported, while no overlap is occurring. As previously stated, proper inspection of the data is always required and combination of several co-localization metrics applied to decipher such a complex situation.

3.4 Conclusion: Picking the Right Co-localization Tool(s)

A turnkey software solution to retrieve "indicators" or "quantifiers" might seem attractive but will unfortunately not adapt to all possible experimental conditions and biological questions. Thus, defining precisely your own question or hypothesis is of much importance before looking for any co-localization analysis tools. Simple problematics, not necessarily "black and white" approaches, will help limiting the seeking range for adapted software. Using previous description of metrics in relation with your images, start determining your working constraints: is there any interest in isolating objects of interest in the images? What is your biological object of interest (whole cell, subcellular compartments, etc.)? Do you need to identify two sub-populations exhibiting two different co-localization states? Thus, step by step, you will end with a framework for which precise types of metrics will fit.

In some cases, computation of simple reporters, such as the percentage of "positive" pixels in both channels (*see* Subheading 3.3.1), could be easily set up from raw images with many different software, in a manual manner. However, other metrics described above may require more complex and automated steps, often involving pixel-by-pixel calculations that would never be performed by hand. Several tools/add-ons have been proposed to fulfill this need, based on various software environments. We present here a non-exhaustive list of freely available tools (*see* Table 1) that contain dedicated modules for co-localization analysis and that would offer you the opportunity to extract some of the metrics presented herein. It is important to consider trying several of them and comparing different workflows in a fair way, as some processing options might exist as initial steps for one given tool. Such image-processing workflows can either be found in literature or in online search engines such as the BISE website [70] or designed from scratch to best adapt to your problematics.

As these "bricks" for metrics computation (*see* Subheading 1.2) usually come as intermediate steps in co-localization analysis workflows, i.e., after image processing and before data interpretation, it becomes important to check if inputs and outputs of the chosen tool(s) fit with remaining workflow bricks. Particularly, it could be very helpful to keep working with a common computing platform all along the different steps needed for your analysis, saving a lot of time and efforts with potential conversions. Another criterion of interest to choose a tool is to check if automation can be performed, and its importance depends on the number of images/ replicates to compute. In case repetitive co-localization analyses are planned, inspect how automation can be implemented. In some cases, prior knowledge with scripting language is mandatory, and learning it from the beginning might be heavily time-consuming. In that case, interacting with other scientists who already know the basics about the given software environment represents a good starting point.

Among freely accessible software, a lot of efforts have been carried out to allow for inter-compatibility and thus foster the design of powerful and automated workflows (ex: *ImageJ—Icy* [18], *ImageJ/Fiji—R* [71], *CellProfiler—ImageJ*, etc.). Finally, advanced bioimage analysts might consider using *KNIME*™ platform [72], which represents another genuine and versatile environment to create workflows and include your favorite image processing and analysis tools (e.g., *ImageJ*, *CellProfiler*, *Ilastik*, etc.), and even expand protocols from data acquisition to its management on servers (e.g., *OMERO* [73]).

4 Notes

1. If the addition of fiducial markers is planned, for instance, to quantify and correct for channels' mis-registration, the sample should not be mounted in a sealed environment yet.
2. Autofluorescent plastic slides tend to be hard to use as a reference: first, the absence of a glass/material interface makes it hard to set a precise focus; second, the material tends to bleach under mid-to-high illumination powers.
3. Depending on the material constitutive of the storage tubes, diluted solution of beads should not be kept for a long period as they tend to get adsorbed onto the tubes' walls.
4. Avoid using DAPI-containing mounting media.
5. The experimenter might be guided in choosing appropriate chromophores using online spectra viewers such as https://searchlight.semrock.com/. One should keep in mind the theoretical spectra provided by those tools may slightly vary within the sample's environment.
6. An alternative method exists to explore the cross talk by adopting a representation where the cytofluorogram is literally teared open. In this histogram view, populations close to the extrema are likely identified as cross talk (*see* Ref. [74] for more details).
7. In case the answer is 1, and shaped around a central line, it is likely co-localization occurs, intensities being linearly related between channels. In case the answer is 1, shaped around a nonlinear backbone, it is likely co-localization occurs, intensities being nonlinearly related between channels. In case the answer is more than 1, look closely to the populations close to the axes: linear relationship lying close to one axis or the other is a signature of cross-detection.
8. Cross-detection might result from a too high amount of dye, enabling the contribution from the side parts of the spectrum to be excitable/detectable.
9. Care should be drawn about the potential bleaching encountered with some of them. When dealing with immersion objectives, the experimenter should also bear in mind the need for a proper cover glass to be mounted onto the slide.
10. Concentration should be adapted, depending on the dye's quantum efficiency, the illumination source's power, and the detector's sensitivity. Due to Brownian motion, even if the imaged area suffers from bleaching, dye concentration should equilibrate quickly as bleached molecules move away from the field while emitting ones that enter it.

11. Alternatively, basic histogram's statistics are also available through the [Analyze > Histogram] menu.
12. As division has been selected, the resulting values are likely to be decimal. Storing the result in a 8- or 16-bit image will result in clipping of the data. Those two formats are only meant to store integer data.
13. In specific cases, such as when using first-generation spinning disk microscopes, a loss of up to 60% of the illumination could be detected, thus enhancing the need of characterizing this parameter.
14. Generally speaking, dilutions of 1/5000–1/1000 are appropriate to get well-separated beads while ensuring that at least one large bead is visible in the field to focus on.
15. Beads' sizes should lie below the expected optical resolution so as the pure optical response is collected. Using bigger objects would result in collecting a composite response. The second aspect dictating the beads' choice is its intensity. In this matter, using multi-labeled beads may seem more convenient when trying to characterize resolution over a large spectral range. However, having several dyes coupled to a single bead may mean having less actual dye representative for each channel and hence less intensity. Therefore, it is advisable to rather use mono-labeled beads, either preparing one slide per spectral range or mixing several spectral variants on one sample.
16. This process is done by adjusting the intensities on a Gaussian profile: resolution is obtained by computing its full width at half maximum. Although not being the perfect model of the PSF (point spread function), this method allows a fast characterization of the local optical resolution.
17. Interpolation options are available that may be required, especially in cases where the displacements are not integer values. Using interpolation will however modify the numerical values and therefore impact intensity-based co-localization investigations.
18. For each deconvolved structure, "seeds" should be visible on the original image: compare both to make sure the process is not generating artificial structures, especially on noisy images. Try reviewing the results of lower numbers of iterations: in case the artificial structures remain, try a different algorithm, or take new acquisitions. Shadow-like areas may appear on deconvolved images: bright structures surrounded by black regions. This may result from using an inappropriate PSF, being signal reassignment therefore erroneous. One way to visualize this phenomenon is to use the [Image > Adjust > Brightness/Contrast...] tool, saturate the brightest pixels, and look for such black borders around structures.

19. Reference datasets could be obtained through at least four means. First, one of the two raw images could be rotated by 90° (*ImageJ/Fiji* function: [Image > Transform > Rotate 90 Degrees Left] or [Image > Transform > Rotate 90 Degrees Right]): this will keep the original data and avoid interpolation (*see* **Note 17**) while providing a reference where objects' positions have been modified. However, local correlation is still present. Second, instead of rotation, translation could be used (*ImageJ/Fiji* function: [Image > Transform > Translate]). This principle has been used by Van Steensel et al. [75] to explore PCC significance using the cross-correlation function. The same limitations as for rotation apply. Third, randomization can be applied. One of the two images is cut out into blocks, being each block's size fitted on the PSF's FWHM and attributed to a new, random location. Repeating this operation a large number of times, each time computing PCC, always retrieves a distribution of PCC values that reports for random overlap [68]. Finally, exploration of co-localization reporters could also be made relative to analytical models (*see* Refs. [76, 77] for a more detailed explanation).

Acknowledgment

The Bordeaux Imaging Center is a service unit of the CNRS-INSERM and Bordeaux University, member of the national infrastructure France BioImaging supported by the French National Research Agency (ANR-10-INBS-04). FPC is a member of NEUBIAS (Network for European Bioimage Analysts), COST Action CA15124. Figures have been assembled using *ImageJ*'s FigureJ plugin [78].

References

1. Malkusch S, Endesfelder U, Mondry J et al (2012) Coordinate-based colocalization analysis of single-molecule localization microscopy data. Histochem Cell Biol 137:1–10
2. Malkusch S, Heilemann M (2016) Extracting quantitative information from single-molecule super-resolution imaging data with LAMA – LocAlization Microscopy Analyzer. Sci Rep 6:34486
3. Bolte S, Cordelières FP (2006) A guided tour into subcellular colocalization analysis in light microscopy. J Microsc 224:213–232
4. Cordelières FP, Bolte S (2008) JACoP v2.0: improving the user experience with co-localization studies. In: ImageJ User & Developer Conference. pp 174–181
5. Cordelières FP, Bolte S (2014) Experimenters' guide to colocalization studies: finding a way through indicators and quantifiers, in practice. Methods Cell Biol 123:395–408
6. Adler J, Pagakis SN, Parmryd I (2008) Replicate-based noise corrected correlation for accurate measurements of colocalization. J Microsc 230:121–133
7. Adler J, Parmryd I (2010) Quantifying colocalization by correlation: the Pearson correlation coefficient is superior to the Mander's overlap coefficient. Cytometry A 77:733–742
8. Miura K, Tosi S (2017) Epilogue: a framework for bioimage analysis. In: Standard and super-resolution bioimaging data analysis. John

Wiley and Sons Ltd, Hoboken, NJ, pp 269–284
9. French AP, Mills S, Swarup R et al (2008) Colocalization of fluorescent markers in confocal microscope images of plant cells. Nat Protoc 3:619–628
10. Zinchuk V, Zinchuk O (2008) Quantitative colocalization analysis of confocal fluorescence microscopy images. Curr Protoc Cell Biol Chapter 4:Unit 4.19
11. Adler J, Parmryd I (2012) Colocalization analysis in fluorescence microscopy. Methods Mol Biol 931:97–109
12. Royon A, Papon G (2013) Calibration of fluorescence microscopes. Imag Microsc 3:41–43
13. Royon A, Converset N (2017) Quality control of fluorescence imaging systems. Imag Microsc 12:22–25
14. Zwier JM, Van Rooij GJ, Hofstraat JW et al (2004) Image calibration in fluorescence microscopy. J Microsc 216:15–24
15. Brakenhoff GJ, Wurpel GWH, Jalink K et al (2005) Characterization of sectioning fluorescence microscopy with thin uniform fluorescent layers: sectioned Imaging Property or SIPcharts. J Microsc 219:122–132
16. Model M (2014) Intensity calibration and flatfield correction for fluorescence microscopes. In: Current protocols in cytometry. John Wiley & Sons, Inc., Hoboken, NJ
17. Schindelin J, Arganda-Carreras I, Frise E et al (2012) Fiji: an open-source platform for biological-image analysis. Nat Methods 9:676–682
18. De Chaumont F, Dallongeville S, Chenouard N et al (2012) Icy: an open bioimage informatics platform for extended reproducible research. Nat Methods 9:690–696
19. Matthews C, Cordelières FP (2010) MetroloJ: an ImageJ plugin to help monitor microscopes' health. In: ImageJ User & Developer Conference
20. Matthews C, Cordelières FP (2017) MetroloJ plugin. http://*ImageJ*docu.tudor.lu/doku.php?id=plugin:analysis:metroloj:start
21. Cordelières FP (2018) Co-registration analysis *ImageJ* macro. https://github.com/fabricecordelieres/MiMB2018/blob/master/3-1-3_Co-registration/Macro/3-1-3_Coregistration.txt
22. Ollion J, Cochennec J, Loll F et al (2013) TANGO: a generic tool for high-throughput 3D image analysis for studying nuclear organization. Bioinformatics 29:1840–1841
23. Cordelières FP (2018) Resolution analysis *ImageJ* macro. https://github.com/fabricecordelieres/MiMB2018/blob/master/3-1-4_Resolution/Macro/3-1-4_Resolution.txt
24. Walter J (2006) Spectral unmixing plugins. https://*ImageJ*.nih.gov/ij/plugins/spectral-unmixing.html
25. Gammon S (2006) Spectral unmixing of bioluminescence signals. https://*ImageJ*.nih.gov/ij/plugins/spectral-unmixing-plugin.html
26. Neher R, Theis F, Zeug A (2009) PoissonNMF – blind source separation of fluorescence microscopy data. https://www.mh-hannover.de/cellneurophys/poissonNMF/NMF/
27. Peng T, Thorn K, Schroeder T et al (2017) A BaSiC tool for background and shading correction of optical microscopy images. Nat Commun 8:14836
28. Kirshner H, Aguet F, Sage D et al (2013) 3-D PSF fitting for fluorescence microscopy: implementation and localization application. J Microsc 249:13–25
29. Kirshner H, Sage D (2017) PSF generator plugin. http://bigwww.epfl.ch/algorithms/psfgenerator/
30. Sage D, Donati L, Soulez F et al (2017) DeconvolutionLab2: An open-source software for deconvolution microscopy. Methods 115:28–41
31. Vonesch C, Cristofani RT, Schmit G (2016) DeconvolutionLab1 plugin, http://bigwww.epfl.ch/deconvolution/deconvolutionlab1/
32. Sage D, Vonesch C, Schmit G, et al (2017) DeconvolutionLab2 plugin, http://bigwww.epfl.ch/deconvolution/deconvolutionlab2/
33. Zucker RM, Price OT (2001) Evaluation of confocal microscopy system performance. Methods Mol Biol 44:295–308
34. Zucker R (2004) Confocal microscopy system performance: axial resolution. Microscopy Today 12:38–40
35. Zucker R (2006) Quality assessment of confocal microscopy slide-based systems: instability. Cytometry A 69:677
36. Cole RW, Jinadasa T, Brown CM (2011) Measuring and interpreting point spread functions to determine confocal microscope resolution and ensure quality control. Nat Protoc 6:1929–1941
37. Hng KI, Dormann D (2013) ConfocalCheck – a software tool for the automated monitoring of confocal microscope performance. PLoS One 8:e79879
38. Schneider CA, Rasband WS, Eliceiri KW (2012) NIH Image to *ImageJ*: 25 years of image analysis. Nat Methods 9:671–675

39. Pearson H (2005) Image manipulation: CSI: cell biology. Nature 434:952–953
40. Pearson H (2007) The good, the bad and the ugly. Nature 447:138–140
41. Zimmermann T (2005) Spectral imaging and linear unmixing in light microscopy. Adv Biochem Eng Biotechnol 95:245–265
42. Gammon ST, Leevy WM, Gross S et al (2006) Spectral unmixing of multicolored bioluminescence emitted from heterogeneous biological sources. Anal Chem 78:1520–1527
43. Zimmermann T, Marrison J, Hogg K et al (2014) Clearing up the signal: spectral imaging and linear unmixing in fluorescence microscopy. In: Paddock SW (ed) Confocal microscopy: methods and protocols. Springer New York, New York, NY, pp 129–148
44. Wallace W, Schaefer LH, Swedlow JR (2001) A workingperson's guide to deconvolution in light microscopy. Biotechniques 31:1076–1078. 1080, 1082 passim
45. Sibarita JB (2005) Deconvolution microscopy. Adv Biochem Eng Biotechnol 95:201–243
46. Landmann L (2002) Deconvolution improves colocalization analysis of multiple fluorochromes in 3D confocal data sets more than filtering techniques. J Microsc 208:134–147
47. Landmann L, Marbet P (2004) Colocalization analysis yields superior results after image restoration. Microsc Res Tech 64:103–112
48. Richardson W (1972) Bayesian-based iterative method of image restoration. J Opt Soc Am 62:55–59
49. Lucy LB (1974) An iterative technique for the rectification of observed distributions. Astronomical J 79:745
50. Frigo M, Johnson SG (1998) FFTW: an adaptive software architecture for the FFT. In: Proceedings of 1998 IEEE International Conference Acoustics speech and signal processing. pp 1381–1384 IEEE
51. Frigo M, Johnson SG (2005) The design and implementation of FFTW3. Proc IEEE 93:216–231
52. Weigert M, Schmidt U, Boothe T et al (2018) Content-aware image restoration: pushing the limits of fluorescence microscopy. Nat Methods 15:1090–1097
53. Luisier F, Vonesch C, Blu T et al (2010) Fast interscale wavelet denoising of poisson-corrupted images. Signal Processing 90:415–427
54. Luisier F (2017) PureDenoise plugin. http://bigwww.epfl.ch/algorithms/denoise/
55. Sezgin M, Sankur B (2004) Survey over image thresholding techniques and quantitative performance evaluation. Electron Imaging 13:146–165
56. Sommer C, Gerlich DW (2013) Machine learning in cell biology – teaching computers to recognize phenotypes. J Cell Sci 126:5529–5539
57. Sadanandan SK, Ranefall P, Le Guyader S, et al (2017) Automated training of deep convolutional neural networks for cell segmentation Sci Rep 7:1–7
58. Arganda-Carreras I, Kaynig V, Rueden C et al (2017) Trainable weka segmentation: a machine learning tool for microscopy pixel classification. Bioinformatics 33:2424–2426
59. Arganda-Carreras I, Kaynig V, Rueden C, et al (2016) Trainable segmentation: release v3.1.2. https://doi.org/10.5281/zenodo.59290
60. Carpenter AE, Jones TR, Lamprecht MR et al (2006) CellProfiler: image analysis software for identifying and quantifying cell phenotypes. Genome Biol 7:R100
61. Sommer C, Straehle C, Köthe U, et al (2011) Ilastik: interactive learning and segmentation toolkit. In: 2011 IEEE International Symposium on Biomedical Imaging: From Nano to Macro. pp 230–233
62. Pearson K (1901) On lines and planes of closest fit to systems of points in space. Philos Mag 2:559–572
63. Manders EM, Stap J, Brakenhoff GJ et al (1992) Dynamics of three-dimensional replication patterns during the S-phase, analysed by double labelling of DNA and confocal microscopy. J Cell Sci 103(Pt 3):857–862
64. Manders EMM, Verbeek FJ, Ate JA (1993) Measurement of co-localisation of objects in dual-colour confocal images. J Microsc 169:375–382
65. Dunn KW, Kamocka MM, McDonald JH (2011) A practical guide to evaluating colocalization in biological microscopy. Am J Physiol Cell Physiol 300:C723–C742
66. Spearman C (1904) The proof and measurement of association between two things. Am J Psychol 15:72–101
67. McDonald JH, Dunn KW (2013) Statistical tests for measures of colocalization in biological microscopy. J Microsc 252:295–302
68. Costes SV, Daelemans D, Cho EH et al (2004) Automatic and quantitative measurement of protein-protein colocalization in live cells. Biophys J 86:3993–4003
69. Lachmanovich E, Shvartsman DE, Malka Y et al (2003) Co-localization analysis of complex formation among membrane proteins by computerized fluorescence microscopy:

application to immunofluorescence co-patching studies. J Microsc 212:122–131

70. Neubias Cost Action CA15124 (2018), BISE bio imaging search engine: a bio image information index. http://www.biii.eu/
71. R Core Team (2018) R: a language and environment for statistical computing. https://www.R-project.org
72. Berthold MR, Cebron N, Dill F et al (2009) KNIME – the konstanz information miner: version 2.0 and beyond. ACM 11:26–31
73. Allan C, Burel JM, Moore J et al (2012) OMERO: flexible, model-driven data management for experimental biology, vol 9, pp 245–253
74. Gavrilovic M, Wählby C (2009) Quantification of colocalization and cross-talk based on spectral angles. J Microsc 234:311–324
75. van Steensel B, van Binnendijk E, Hornsby C et al (1996) Partial colocalization of glucocorticoid and mineralocorticoid receptors in discrete compartments in nuclei of rat hippocampus neurons. J Cell Sci 109:787–792
76. Lagache T, Lang G, Sauvonnet N et al (2013) Analysis of the spatial organization of molecules with robust statistics. PLoS One 8: e80914
77. Lagache T, Sauvonnet N, Danglot L et al (2015) Statistical analysis of molecule colocalization in bioimaging. Cytometry A 87:568–579
78. Mutterer J, Zinck E (2013) Quick-and-clean article figures with FigureJ. J Microsc 252:89–91
79. Li Q, Ledoux-Rak I, Lai ND (2015) Influence of incident beam polarization on intensity and polarization distributions of tight focusing spot. Adv Device Mater 1:4–10
80. Schindelin J, Eglinger J, Guizzetti L, et al (2018) Coloc2. https://*ImageJ*.net/Coloc_2
81. Cordelières FP, Bolte S (2018) JaCoP, just another co-localization plugin v2. http://*ImageJ*docu.tudor.lu/doku.php?id=plugin:analysis:jacop_2.0:just_another_colocalization_plugin:start
82. Gilles JF, Dos Santos M, Boudier T et al (2017) DiAna, an *ImageJ* tool for object-based 3D co-localization and distance analysis. Methods 115:55–64
83. Rizk A, Paul G, Incardona P et al (2014) Segmentation and quantification of subcellular structures in fluorescence microscopy images using Squassh. Nat Protoc 9:586–596
84. Lavancier F, Pécot T, Zengzhen L, et al (2018) GcoPS. http://icy.bioimageanalysis.org/plugin/GcoPS
85. Ovesný M, Křížek P, Borkovec J et al (2014) ThunderSTORM: a comprehensive *ImageJ* plug-in for PALM and STORM data analysis and super-resolution imaging. Bioinformatics 30:2389–2390
86. De Chaumont F (2018) Colocalizer. http://icy.bioimageanalysis.org/plugin/Colocalizer
87. Lagache T, Grassart A, Dallongeville S et al (2018) Mapping molecular assemblies with fluorescence microscopy and object-based spatial statistics. Nat Commun 9:698

Chapter 11

Triple-Colocalization Approach to Assess Traffic Patterns and Their Modulation

Daniel Sastre, Irene Estadella, Manel Bosch, and Antonio Felipe

Abstract

Confocal microscopy permits the analysis of the subcellular distribution of proteins. Colocalization between target proteins and specific markers of differential cell compartments provides an efficient approach to studying protein traffic. In this chapter, we describe an automated method to denoise confocal microscopy images and assess the colocalization of their stainings using *ImageJ* software. As a step further from conventional single colocalization measurements, in the proposed method, we analyze stacks of three different stainings using two-by-two comparisons. To demonstrate the reliability and usefulness of our proposal, the method was used to compare the traffic of the voltage-gated Kv1.3 potassium channel, which is a well-defined plasma membrane protein, in the presence and absence of KCNE4, a regulatory subunit that strongly retains the channel intracellularly.

Key words Confocal microscopy, Colocalization, Subcellular distribution, ImageJ, Potassium channel, Protein traffic

1 Introduction

The specific subcellular distribution of a protein is crucial to its normal function. In cell biology, many examples describe that traffic tightly regulates the performance of a protein [1]. One of these examples is the case of ion channels. Ion channels are integral membrane proteins showing differential subcellular localizations. In this context, the localization of a channel is important to regulate its function. Upon activation, these proteins drive specific ions from the intracellular to the extracellular space or vice versa. Thus, ion channels must be located on the membrane to be functional. These proteins are crucial for homeostasis and generation of the action potential [2].

Electronic supplementary material: The online version of this chapter (https://doi.org/10.1007/978-1-4939-9686-5_11) contains supplementary material, which is available to authorized users.

Elena Rebollo and Manel Bosch (eds.), *Computer Optimized Microscopy: Methods and Protocols*, Methods in Molecular Biology, vol. 2040, https://doi.org/10.1007/978-1-4939-9686-5_11,

Kv1.3 is a voltage-gated potassium channel, selective for potassium, which activates upon plasma membrane depolarization. This channel has been related to several human disorders such as obesity, type 1 and type 2 diabetes, autoimmunity, inflammation, neurodegenerative disorders, and certain kinds of cancer [3]. Therefore, Kv1.3 raises interest because it is considered a potential therapeutic target with many pathological implications [4]. Because the channel functions at the plasma membrane, the study of its surface abundance is of considerable interest.

Recent evidences based on confocal microscopy show that the reduction of the Kv1.3 abundance in the plasma membrane triggers a channel downregulation physiological mechanism. This mechanism may be mediated either by internalization [5] or by intracellular retention of Kv1.3 [6]. While the former is a consequence of an augmented endocytosis due to several kinase actions, the latter could be due to the action of modulatory subunits such as KCNE4. As previously shown by co-immunoprecipitation and FRET approaches, this ancillary protein associates to channel Kv1.3 preventing the forward trafficking, thus retaining the channel at the endoplasmic reticulum. This retention effect is specific because it is not caused by any other KCNE-related peptide, such as KCNE2 [6].

In this work, we evaluate how the coexpression of Kv1.3 with different KCNE peptides can modify the channel distribution. To do so, we developed an efficiently automated method on *Fiji* distribution of *ImageJ* freeware [7]. We analyze the colocalization between three components: the plasma membrane, the channel Kv1.3, and the regulatory subunit KCNE4. Images are first processed to reduce the presence of noise. The denoised images are then thresholded to obtain the segmented binary masks which are used to eliminate, from the corresponding original denoised images, all the signal which is not contained inside the mask. The resulting images, containing only the gray-scale signal of interest, are used to extract the colocalization coefficients between each two markers by using the *JACoP* plugin [8]. Both the Manders' [9] and Pearson's [10] colocalization coefficients are obtained per two-by-two comparison (*see* **Notes 1–3**). An overview of the presented method can be seen in Fig. 1.

Kv1.3 is detected intracellularly colocalizing with KCNE and at the plasma membrane colocalizing with the plasma membrane marker. To elucidate where Kv1.3 is more abundant, we have designed a membrane vs. intracellular ratio (M/I ratio; *see* Fig. 2 and **Note 4**) between the Kv1.3 protein fraction that efficiently targets to the membrane (α) and the Kv1.3 protein fraction that colocalizes with KCNE intracellularly (β). To do so, the colocalization of Kv1.3 with both the membrane marker ($\alpha + \gamma$) and KCNE ($\beta + \gamma$) was measured together with the triple-colocalization (Kv1.3-KCNE-membrane, γ) signal fraction. This γ signal fraction

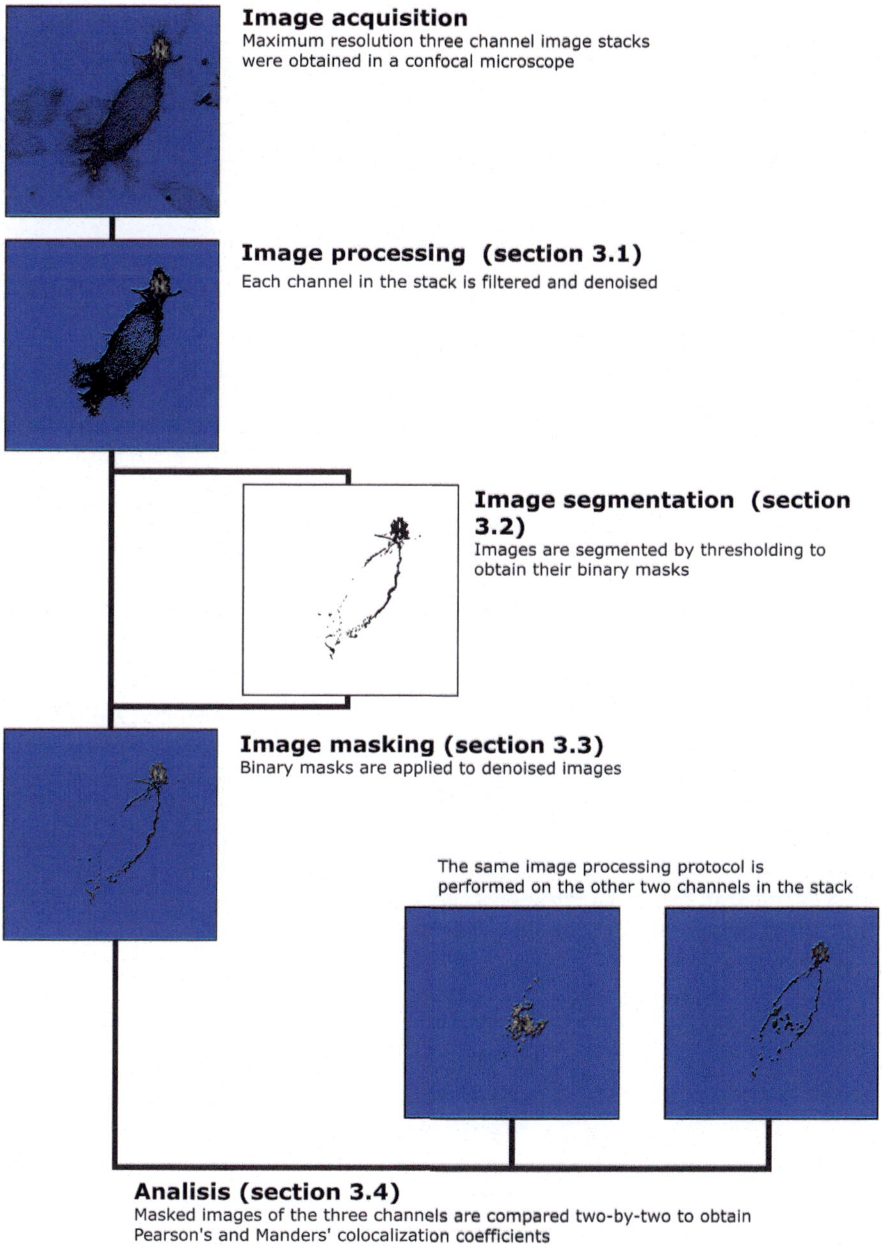

Fig. 1 Overview of the presented method. Acquired images are filtered and denoised (Subheading 3.1). Then they are duplicated and thresholded into binary masks (Subheading 3.2), which are applied to the corresponding original denoised images (Subheading 3.3). Colocalization is assessed (Subheading 3.4) comparing the images obtained in previous section

is subtracted from the double-colocalization signal fractions obtained for Kv1.3 with both the membrane ($\alpha + \gamma$) and the KCNE ($\beta + \gamma$) markers alone. Then the ratio is calculated by dividing α and β. Thus, values over 1 are obtained if Kv1.3 is mainly

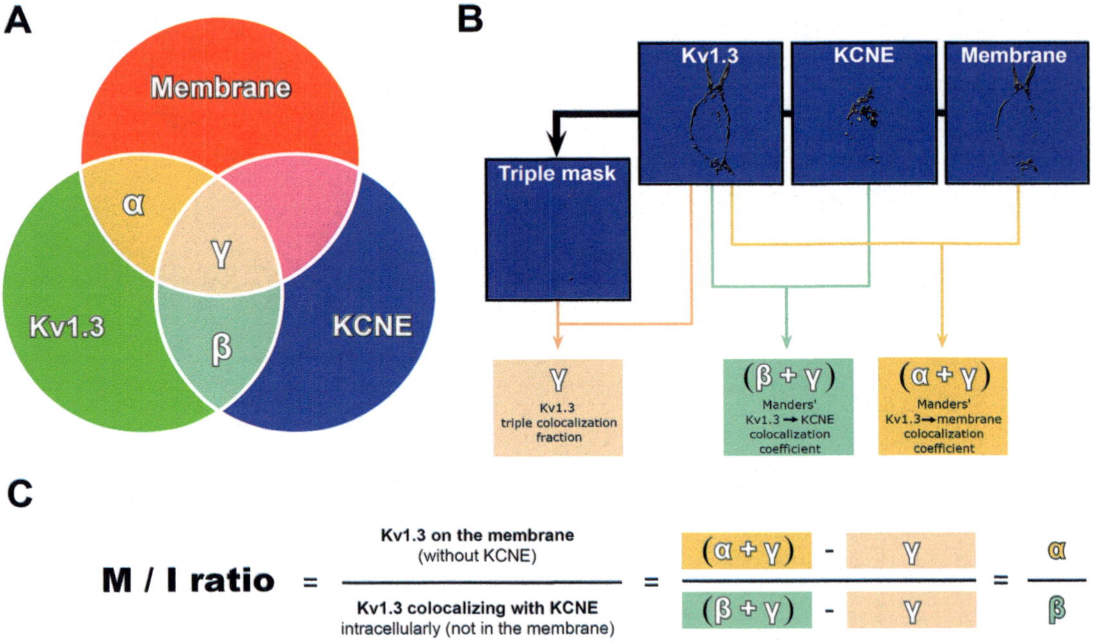

Fig. 2 Explanation of the *M/I* ratio formula. (**a**) Venn diagram of the triple staining. Each color represents a different signal: membrane, in red; KCNE, in blue; and Kv1.3, in green. Different intersections are named α (intersection between membrane and Kv1.3 without KCNE), β (intersection between KCNE and Kv1.3 without plasma membrane), and γ (triple intersection between Kv1.3, KCNE, and membrane). (**b**) These intersections can be measured using Manders' colocalization coefficients for $\alpha + \gamma$ (Manders' coefficient of Kv1.3 over the plasma membrane) or $\beta + \gamma$ (Manders' coefficient of Kv1.3 over the KCNE peptide). Finally, intersection γ can be measured as the fraction of Kv1.3 signal in the triple-colocalization mask ($I_{\mathrm{Kv1.3\_triple}}/I_{\mathrm{Kv1.3}}$). (**c**) Equation that reflects the *M/I* ratio. To obtain the ratio between the intersections α and β, γ ($I_{\mathrm{Kv1.3\_triple}}/I_{\mathrm{Kv1.3}}$) must be subtracted from both Manders' colocalization coefficients ($\alpha + \gamma$ and $\beta + \gamma$)

expressed in the plasma membrane or below 1 if it is retained intracellularly colocalizing with KCNE proteins.

The proposed pipeline can be applied to similar experiments where the colocalization between a subcellular compartment and two proteins needs to be analyzed, for example, in the study of nuclear transport by importins and exportins or to study how ubiquitination affects internalization from the plasma membrane. Thus, the presented method provides an efficient and useful approach to studying differential traffic patterns and evaluating their alteration.

2 Materials

1. Images. The macro works on single-plane composite images with three channels (*see* **Notes 5** and **6** for sample preparation and image acquisition, respectively), acquired in the indicated order (*see* **Note 7**): membrane marker (Channel1 (C1)), the

ancillary subunit KCNE2 or KCNE4 (C2), and the target protein Kv1.3 (C3). Two sample images are provided with this chapter, named "membrane_KCNE2_Kv1.3.tif" and "membrane_KCNE4_Kv1.3.tif," and can be downloaded from the Springer website.

2. Software. The analysis was performed using the *Fiji* distribution of *ImageJ* (version 1.51h) open source program [7].
3. The macro generated in this work, named "TripleColocalizationAnalysis.ijm," is available at [11].
4. Additional tools. Two external plugins and one macro are required. Plugins should be installed according to the instructions on the *ImageJ* website [12].
   (a) *JACoP* (Just Another Colocalization Plugin) version 2.0 is a powerful compilation of colocalization tools developed by Cordelières and Bolte [8, 13]. This plugin should be downloaded from the *JACoP* page on the ImageJ Documentation Wiki [14].
   (b) *BG subtraction from ROI* is an ImageJ plugin which was modified by Collins (McMaster Biophotonics Facility, Hamilton, Ontario). This plugin is gathered with around 200 other plugins in the "MBF ImageJ for Microscopy Collection" [15], which is available online [16].
   (c) Macro for an automatic adjustment of brightness and contrast available at [17]. Its code is included as a function (named *autoAdjust*) in our macro and, thus, requires no further installation. However, if the protocol is followed manually, it can be installed according to the instructions on the *ImageJ* website [18].

3 Methods

The general overview of the protocol is summarized in Fig. 1. Each channel image is processed by filtering and background subtraction to reduce the presence of noise. Then they are segmented by thresholding, and the resulting binary images are used as masks to sample the denoised images, thus creating a background-less image for each channel. Finally, a two-by-two colocalization assay is run between all three background-less channels: (1) the plasma membrane, Kv1.3 colocalization coefficient will be indicative of the Kv1.3 protein fraction distributed at the plasma membrane; (2) the KCNE, Kv1.3 colocalization will show the co-distribution of both molecules anywhere in the cell; and (3) the plasma membrane, KCNE colocalization assay will be indicative of the KCNE distribution at the plasma membrane. In addition to the

```
open();
getDimensions(width, height, channels, slices, frames);
if(!is("composite") || channels != 3){
    exit("The file is not a three-channel composite
image");
}
title = getTitle();
input = File.directory();
//Change names accordingly in the following array
arrayNames = newArray("membrane", "KCNE", "channel");
output = input + "Results" + File.separator;
File.makeDirectory(output);
```

Fig. 3 Macro instructions for the preliminary steps

```
//Function to process each channel image
function processImage(rollingBall){
    roiManager("Select", 0);
    run("BG Subtraction from ROI");
    run("Subtract Background...", "rolling="+rollingBall);
    run("Select None");
    run("Median...", "radius=1");
    run("Gaussian Blur...", "sigma=1");
    run("Select None");
}
```

Fig. 4 User-defined function named *processImage(rollingBall)* used to apply the different processing steps in the macro. When this function is called in the code, it needs the "rollingBall" size passed as an argument of the function (*see* **Note 7**)

colocalization coefficients, the membrane-retention ratio explained in the introduction will be calculated.

The protocol can be manually performed following the step-by-step protocol below, where the menu paths that lead to each corresponding software function are shown. A three-channel image should be first opened as a composite, and the protocol continued in Subheading 3.1. Alternatively, the protocol can be executed automatically using the macro provided. To that aim, drag the macro file to the *Fiji* menu bar and press *Run* (shortcut Ctrl + r). The macro will then ask to open an image, which needs to be a three-channel composite image; otherwise, the macro will exit (*see* macro code in Fig. 3). Once opened, the macro will run the whole protocol. The title of the image and its path will be recorded in two variables, respectively. Additionally, the macro will create an array containing the name of the three proteins under analysis (*see* **Note 8**). Finally, an output directory will be created where the results will be saved. Other functionalities of the macro are also detailed in the protocol.

3.1 Image Processing

In this part, the noise and the background in the images are filtered (*see* the user-defined function *processImage(rollingBall)* illustrated in Fig. 4). An overview of this process is shown in Fig. 5, which illustrates the effect of each individual step.

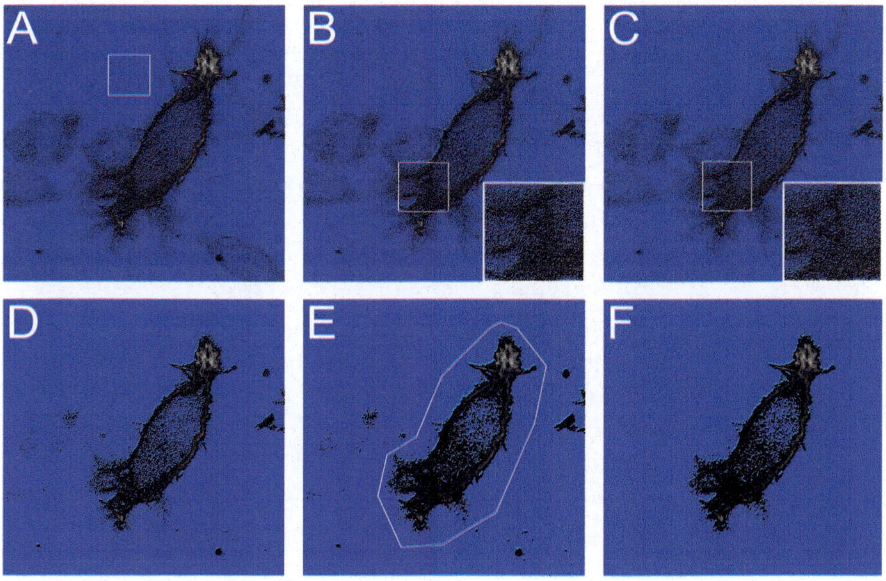

Fig. 5 Processing steps. (**a**) Raw image of the plasma membrane marker shown in HiLo LUT. (**b**) Background subtraction from the white square in (**a**) using the plugin *BG Subtraction from ROI*. (**c**) Background subtraction using the *Subtract Background* function with a *Rolling ball radius* of 50 pixels. A detail of the images **b** and **c** is shown in their respective lower right corners. The image was further filtered with both a *Median* filter (radius = 1 pixel) (**d**) and *Gaussian Blur* (sigma = 1 pixel) (**e**). From the polygonal selection shown in (**e**), image was finally cleaned with the tool *Clear Outside* (**f**)

1. Split the image into its three channels at [Image > Colors > Split Channels]. All the following steps need to be performed for each particular channel.
2. Change the look-up table (LUT) to HiLo at [Image > Lookup Tables > HiLo] (*see* **Note 9**).
3. Draw a region of interest (ROI) on the image background. To use the same ROI size in the different images under analysis, it is recommended to specify the ROI size and position at [Plugins > ROI > Specify ROI]. In this protocol, we used the following parameters to draw a 150 × 150 pixel ROI in the upper left corner: width, 150; height, 150; *X* coordinate, 0; *Y* coordinate, 0; slice, 1; oval, OFF; and centered, OFF. Then the ROI has to be moved to a background area and added to the *ROI Manager* at [Edit > Selection > Add to Manager]; shortcut Ctrl + t. In the programmed macro, the ROI is drawn sizing 15% of the image area. Then the user is asked to move the ROI into a background area. Finally, the ROI is added to the *ROI Manager*.
4. From the *ROI Manager*, select the ROI on each image, and apply the *BG Subtraction from ROI* plugin at [Plugins > Roi > BG Subtraction from ROI]. This plugin successfully eliminates the background from the image (*see* **Notes 10** and **11**).

5. Next, use the plugin *Subtract Background* at [Process > Subtract Background...] to get rid of uneven background levels (*see* **Note 12**). In our example, use a *Rolling ball radius* of 50 pixels for the membrane marker and 500 pixels for the rest of proteins.
6. To minimize the noise in the images, apply both a slight *Median* filter at [Process > Filters > Median...] and a subtle *Gaussian Blur* at [Process > Filters > Gaussian Blur...] using a R*adius* and a S*igma* of 1, respectively (*see* **Note 13**).
7. Finally, in order to suppress either cell debris or other cells that fall into the same image, another ROI needs to be drawn around the cell of interest. Add again the selection to the *ROI Manager*, and select it for each channel to suppress data outside the selection at [Edit > Clear Outside] (*see* the user-defined function *selectSingleCell()* illustrated in Fig. 6). Due to the LUT used in the images, it is important to set the *Background* color to blue at [Edit > Options > Colors...]; otherwise, the cleared area will have an intensity value other than zero.
8. It is recommended to save images obtained after each main step. It allows comparing them with the original ones and to *see* the reliability of the protocol. It can be done manually at [File > Save As > Tiff...]. Inside the macro, this is done using a user-defined function named *saveIntermediate(protein, step)* which requires two arguments: the name of the stained protein and the name of the protocol step (*see* **Note 14**). This function duplicates the image, saves the duplicate in the output directory, and finally closes it (Fig. 7).

```
//Function to select one single cell
function selectSingleCell(){
	roiManager("Select", 1);
	run("Clear Outside");
	run("Select None");
}
```

Fig. 6 User-defined function named *selectSingleCell()* applied to clear outside the cell of interest once a ROI has been drawn and added to the *ROI Manager*

```
//Function to save intermediate files
function saveIntermediate(protein, step){
	run("Duplicate...", "title=duplicate");
	saveAs("Tiff", output + protein + " " + step);
	close();
}
```

Fig. 7 User-defined function named *saveIntermediate()* applied to save intermediate images generated during the processing

3.2 Image Segmentation

In this section, the brightness and contrast of the denoised images from the previous step is optimized. Then images will be segmented and converted to binary masks which will be further fine-tuned by means of binary operations. These steps are performed on duplicate images of each channel to keep a copy of the denoised images from the previous section.

1. Duplicate images at [Image > Duplicate; shortcut Ctrl + Alt + d].
2. Optimize the brightness and contrast of the duplicate image by using the *autoAdjust* function (*see* **Note 15**).
3. Segment images using the *Default Fiji*'s thresholding method at [Image > Adjust > Threshold], ticking the check box *Dark background*. Upon *Apply*, the images will be converted to binary files (*see* the user-defined function *segment()* illustrated in Fig. 8).
4. Clean up the binary masks by means of the binary function *Erode* at [Process > Binary > Erode].
5. To further clean up the binary masks of Kv1.3 and KCNE channels, use the Binary function *Open* at [Process > Binary > Open], which consists in two consecutive operations, erosion and dilation (*see* **Note 16**).
6. In case there are nonspecific intracellular particles in the binary mask of the membrane marker, draw a ROI around and clear them at [Edit > Clear]. If using the macro, it asks to check for such particles and in case there are to specify a single ROI surrounding them (*see* macro code in Fig. 9). Once the ROI is drawn, the selected area is cleared.

```
//Function to segment each channel image
function segment(){
      setAutoThreshold("Default dark");
      run("Convert to Mask");
      run("Erode");
}
```

Fig. 8 User-defined function used to segment the images

```
//Clean intracellular particles
arg = getBoolean("Are there any intracellular particles?");
if (arg==1) {
      setBackgroundColor(255, 255, 255);
      setTool("polygon");
      waitForUser("Add ROI to clear out intracellular
particles");
      run("Clear", "slice");
}
```

Fig. 9 Macro instructions to clean intracellular particles in the binary mask of the plasma membrane channel

```
//OBTAIN & SAVE THE IMAGES WITH THE ORIGINAL INTENSITY
INSIDE MASK
imageCalculator("Min create", "C1-"+title, arrayNames[0]);
rename(arrayNames[0] + " int");
saveIntermediate(arrayNames[0],
"intensityInsideSingleMask");
imageCalculator("Min create", "C2-"+title, arrayNames[1]);
rename(arrayNames[1] + " int");
saveIntermediate(arrayNames[1],
"intensityInsideSingleMask");
imageCalculator("Min create", "C3-"+title, arrayNames[2]);
rename(arrayNames[2] + " int");
saveIntermediate(arrayNames[2],
"intensityInsideSingleMask");
```

Fig. 10 Macro instructions to combine the masks with their respective original denoised images and save the resulting images. The macro renames them with the name of the corresponding protein followed by "intensityInsideSingleMask"

7. Finally, save the binary masks manually at [File > Save As > Tiff...]. The macro will do it by using again the *saveIntermediate(protein, step)* function.

3.3 Image Masking

For each channel image, an operation is performed between the final binary mask (produced in Subheading 3.2) and the previously denoised image (produced in Subheading 3.1). The resulting background-less images containing only the signal inside the masks will be ready for the colocalization analysis (*see* macro code in Fig. 10). An overview of the steps followed in this section is shown in Fig. 11, which illustrates the effect of each individual step. For each channel:

1. Apply the binary masks obtained in Subheading 3.2 to the denoised images obtained in Subheading 3.1. This is done using the *Image Calculator* function at [Process > Image Calculator...]. Select the *Min Operation* between the corresponding pair of images (*see* **Note 17**). Tick *Create new window* to generate a new image.
2. Save each new image as an intermediate result at [File > Save As > Tiff...]. The macro saves them using the *saveIntermediate (protein, step)* function specifying "intensityInsideSingleMask" as the name of the step.

3.4 Analysis

In this part of the protocol, the intensity of the Kv1.3 staining is measured in its corresponding background-less image (named "Kv1.3 intensityInsideSingleMask" in the macro) produced in Subheading 3.3. Moreover, a new mask is generated containing the triple-colocalization fraction, that is, the Kv1.3 segmented signal that is present in both the mask of the KCNE protein and the mask of the plasma membrane marker. This triple-colocalization mask is now applied to the denoised Kv1.3 image, as previously done in Subheading 3.3, and the background-less result used to measure the Kv1.3 signal intensity (*see* macro code in Fig. 12). Both

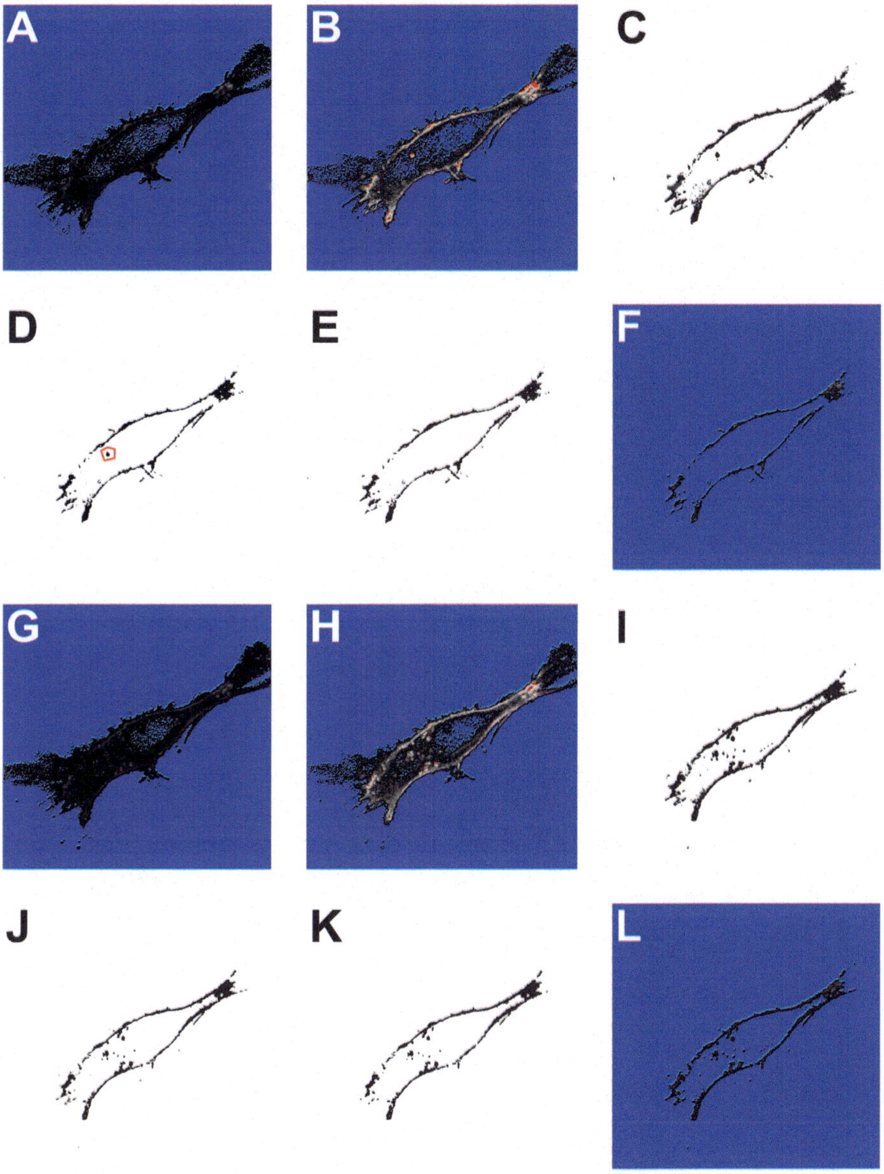

Fig. 11 Main steps during thresholding and the use of the resulting binary images as masks. The process is slightly different for the membrane marker (**a–f**) than for the other channels (of which Kv1.3 is shown as an example in **g–l**). The images have been filtered and denoised (**a**, **g**). The brightness and contrast of the images are adjusted using the *ImageJ*'s *Auto* algorithm (**b**, **h**). The images are thresholded using the *Default* mode (**c**, **i**), and the masks are eroded (**d**, **j**). In the membrane marker image, nonspecific particles are removed (**e**) from a user-defined ROI (shown in red in **d**). In the other channels, the mask is opened (**k**). Finally, the fine-tuned masks (**e**, **k**) are applied to their respective denoised images (**a**, **g**) resulting in the background-less images which are ready to be analyzed (**f**, **l**)

measurements are used to calculate γ (*see* Fig. 2), a quotient that will reflect the fraction of Kv1.3 colocalizing with both KCNE and plasma membrane. This value will be used, together with the

```
//Intensity measurements
//Create triple colocalization image
run("Set Measurements...", "mean display redirect=None
decimal=3");
selectWindow(arrayNames[2] + " int");
run("Measure");
imageCalculator("AND create", arrayNames[2], arrayNames[1])
result1 = getTitle();
imageCalculator("AND create", result1, arrayNames[0]);
result2 = getTitle();
imageCalculator("Min create", "C3-"+title, result2);
rename("intensityInsideTripleMask");
run("Measure");
saveIntermediate(arrayNames[2], "intensityInsideTripleMask"
```

Fig. 12 Macro instructions to generate a triple-colocalization Kv1.3 image (named "Kv1.3 intensityInsideTripleMask") and measure intensity in this and the background-less Kv1.3 image (named "Kv1.3 intensityInsideSingleMask")

colocalization coefficients obtained in **step 8**, to calculate the *M/I* ratio, used in this chapter to quantify the effect of KCNE over the distribution of Kv1.3 in a single value (*see* **Note 4**).

1. Specify *Integrated density* at [Analyze > Set Measurements...] (*see* **Note 18**).
2. Measure the intensity over the background-less Kv1.3 image (named "Kv1.3 intensityInsideSingleMask" in the macro) obtained in Subheading 3.3 at [Analyze > Measure] or using the shortcut Ctrl + m.
3. Combine the three individual binary masks obtained in Subheading 3.2 using the *Image Calculator* tool at [Process > Image Calculator...] (*Operation*: AND, **Note 17**).
4. Save the obtained triple-colocalization mask as an intermediate result.
5. Apply the resulting triple-colocalization mask to the denoised Kv1.3 image obtained in Subheading 3.1 using the *Image Calculator* tool at [Process > Image Calculator...] (*Operation*: Min).
6. Save the new obtained image as an intermediate result. The macro renames the image as "Kv1.3 intensityInsideTripleMask."
7. Measure the intensity at [Analyze > Measure] over the image obtained in **step 5**. Tick the *Summarize* check box.
8. Measure the colocalization between the three background-less images ("membrane intensityInsideSingleMask," "KCNE intensityInsideSingleMask," and "Kv1.3 intensityInsideSingleMask") obtained in Subheading 3.3 using the *JACoP* plugin at [Plugins > JACoP] (*see* macro code in Fig. 13). This analysis is done three times to measure the colocalization between each pair of images: membrane-Kv1.3, KCNE-Kv1.3, and membrane-KCNE. The order in which each pair of images is

```
//Colocalization analysis
run("JACoP ", "imga=[" + arrayNames[0] + " int" +"] imgb=["
+ arrayNames[1] + " int" +"] thra=1 thrb=1 pearson mm");
run("JACoP ", "imga=[" + arrayNames[0] + " int" +"] imgb=["
+ arrayNames[2] + " int" +"] thra=1 thrb=1 pearson mm");
run("JACoP ", "imga=[" + arrayNames[1] + " int" +"] imgb=["
+ arrayNames[2] + " int" +"] thra=1 thrb=1 pearson mm");
```

Fig. 13 Macro instructions to measure colocalization pair-by-pair using the plugin *JACoP*

used in the *JACoP* plugin is important for the final *M/I* ratio measurement (*see* Eq. 1 and **Note 3**). For each measurement, tick *Pearson's coefficient* and *M1 and M2 coefficients* to obtain both Pearson's and Manders' coefficients. Both coefficients are measured because they reflect slightly different properties (*see* **Notes 1–3**). The threshold values can be set to 1 because background pixels have already been removed (*see* Subheading 3.3).

9. Save the intensity measurement results and the colocalization Log window containing the colocalization coefficients.
10. Finally, measure the *M/I* ratio (*see* **Note 4**), using the following formula:

$$\frac{M}{I}\text{ratio} = \frac{\text{Manders M2 (A : membrane B : Kv1.3)} - \frac{I_{\text{Kv1.3}_{\text{triple}}}}{I_{\text{Kv1.3}}}}{\text{Manders M2 (A : KCNE B : Kv1.3)} - \frac{I_{\text{Kv1.3}_{\text{triple}}}}{I_{\text{Kv1.3}}}} \quad (1)$$

Equation 1. Formula to measure the M/I ratio.

where $I_{\text{Kv1.3\_triple}}$ is the intensity of Kv1.3 signal in the "Kv1.3 intensityInsideTripleMask" image, i.e., where the three masks colocalize; $I_{\text{Kv1.3}}$ is the intensity in the background-less image "Kv1.3 intensityInsideSingleMask" (obtained in Subheading 3.3); Manders M2 (A: membrane B: Kv1.3) is the Manders' colocalization coefficient of the Kv1.3 protein over the membrane staining (intersection $\beta + \gamma$ in Fig. 2); and Manders M2 (A: KCNE B: Kv1.3) is the Manders' colocalization coefficient of Kv1.3 over KCNE (intersection $\alpha + \gamma$ in Fig. 2). Both two Manders' coefficients used in this formula are the M2 coefficients because the Kv1.3 image is selected as the image B in the *JACoP* plugin in the two (membrane-Kv1.3 and KCNE-Kv1.3) colocalization analyses. In case the Kv1.3 image was set as the image A in the *JACoP* plugin, the M1 coefficient should be used instead of M2.

11. Close all images and windows. The macro keeps open the final background-less images (named "intensityInsideSingleMask") obtained in Subheading 3.3 for visual inspection and the triple-colocalization "Kv1.3 intensityInsideTripleMask" image obtained in **step 5** of this section.

4 Notes

1. In this colocalization analysis, both Pearson's and Manders' coefficients are measured. While both are standard colocalization parameters, they slightly differ on the properties they are measuring. We used Pearson's coefficient (PC), a correlation-based parameter, as the main colocalization measurement. On the other hand, we used Manders' coefficient, a co-occurrence-based parameter, to calculate the M/I ratio.
2. Pearson's coefficient (PC) is a correlation parameter [10] that describes the relationship between the intensities in two images. It provides the rate of association of two fluorochromes. The value can range from 1 (positive correlation) to −1 (negative correlation), with zero standing for no correlation. PC is especially useful for initial identification of correlations and for examination of complex overlays through ROIs defined but is not enough to evaluate colocalization events rigorously. PC defines the quality of the linear relationship between two signals but does not reflect the slope of such relationship. Moreover, if the sample has two or more different stoichiometry of association, the linear regression will try to fit the segregated dots as one, resulting in a decrease of the PC value [8].
3. Manders' coefficients are an overlap parameter [9] of the proportion of image A signal coincident with image B over the total A intensity (M1) and the proportion of image B signal coincident with image A over the total B intensity (M2). These coefficients will vary from 0 (absence of coincidence) to 1 (both images are identical). Manders' coefficients can be applied even if the intensities in both images are really different from one another. But this is only true if the background is set to zero. Manders' coefficients are very sensitive to noise, and, for that reason, it must be removed. Moreover, they are sensitive at co-occurrence even if it is low. Because of that, it is important to establish a threshold for segmentation. The threshold must be high enough to avoid background and cross talk between fluorophores but at the same time low enough not to lose valuable information [8]. In our case, the threshold was set to 1 because background pixels had already been suppressed from the images used to calculate colocalization.
4. The M/I ratio designed in this chapter is a useful tool to quantify the effect of the coexpressed KCNE peptides over the distribution of the Kv1.3 protein (*see* Eq. 1 and Fig. 2). It represents the fraction of Kv1.3 on the plasma membrane that does not colocalize with KCNE (named α) against the fraction of Kv1.3 that colocalizes with KCNE intracellularly (named β).

To obtain such values, the fraction of Kv1.3 colocalizing with KCNE (intersection $\beta + \gamma$), with the plasma membrane (intersection $\alpha + \gamma$) and with both KCNE and plasma membrane (named γ), must be calculated. Manders' M2 colocalization coefficients between the membrane and Kv1.3 and between KCNE and Kv1.3 reflect precisely the intersections $\alpha + \gamma$ and $\beta + \gamma$, respectively. M2 is defined as the ratio of the summed intensities of pixels from the image B (Kv1.3 in both cases) for which the intensity in the image A is above zero to the total intensity in the image B:

$$\text{Manders M2 (A : membrane B : Kv1.3)} = \frac{\sum \text{Kv1.3}_{\text{coloc\_membrane}}}{\sum \text{Kv1.3}}$$

$$\text{Manders M2 (A : KCNE B : Kv1.3)} = \frac{\sum \text{Kv1.3}_{\text{coloc\_KCNE}}}{\sum \text{Kv1.3}}$$

The intersection γ equals the quotient (named $I_{\text{Kv1.3\_triple}}/I_{\text{Kv1.3}}$ in Eq. 1) between the intensity of Kv1.3 detected inside the triple masked image ("Kv1.3 intensityInsideTripleMask") and the intensity of Kv1.3 detected inside the individual masked image ("Kv1.3 intensityInsideSingleMask"). Thus, γ must be subtracted from both M2 colocalization coefficients to clearly reflect the distribution of Kv1.3 between the membrane and the KCNE positive intracellular compartment:

$$M/I \text{ ratio} = \frac{\frac{\sum \text{Kv1.3}_{\text{coloc}_{\text{membrane}}}}{\sum \text{Kv1.3}} - \frac{\sum \text{Kv1.3\_triple}}{\sum \text{Kv1.3}}}{\frac{\sum \text{Kv1.3}_{\text{coloc}_{\text{KCNE}}}}{\sum \text{Kv1.3}} - \frac{\sum \text{Kv1.3\_triple}}{\sum \text{Kv1.3}}} = \frac{\sum \text{Kv1.3}_{\text{coloc}_{\text{membrane}}} - \sum \text{Kv1.3\_triple}}{\sum \text{Kv1.3}_{\text{coloc}_{\text{KCNE}}} - \sum \text{Kv1.3\_triple}}$$

This was done because some KCNE peptides have better plasma membrane targeting, and even though they may colocalize more with Kv1.3, their coexpression does not induce intracellular retention of the channel.

5. HEK293 cells were incubated on poly-D-lysine-treated coverslips in a six-well culture plate containing 1 mL DMEM supplemented with 10% FBS and 1% penicillin-streptomycin. Twenty-four hours after seeding, cells were transfected with 750 ng Kv1.3-YFP, either 750 ng KCNE2-CFP or 500 ng KCNE4-CFP, and 750 ng membrane marker (Akt-dsRED) using lipotransfectin as a transfection reagent. After 30 h, cells were quickly washed thrice with PBS1x without K^+ and fixed for 10 min with 4% paraformaldehyde (PFA). After washing three times for 5 min with PBS1× without K^+, we used the polyvinyl alcohol Mowiol® as a mounting media. The

coverslips were left to cure at least overnight at room temperature and protected from light.

6. Images were obtained using the confocal microscope Leica TCS SP2 equipped with an Argon and a DPSS 561 nm lasers and a HCX PL APO CS 63× 1.32 OIL objective lens. Images had a pixel size of 116.25 nm and a bit depth of 8-bits. To reduce noise, we used a four-frame average, which consists in rescanning each section four times averaging the intensity of the four scans. DsRed staining (membrane marker) was observed exciting at $\lambda = 561$ nm and detecting between 582 and 666 nm. CFP (KCNE2 and KCNE4) was excited at $\lambda = 458$ nm and detected between 468 and 502 nm. YFP (Kv1.3) was excited at $\lambda = 514$ nm and detected between 524 and 552 nm.
7. Images were sequentially obtained and stacked in the following order: (1) membrane marker (dsRED), (2) KCNE2 or KCNE4 (CFP), and (3) Kv1.3 (YFP). The macro has been designed to work in this same order, and images will be processed and analyzed in consequence.
8. The names of the proteins in the macro array can be manually changed (line 26) according to the current proteins detected and the order in which they were acquired with the microscope.
9. The LUT (look-up tables) refer to the displayed colors per intensity value (from 0 to 255 for an 8-bit image). The *HiLo* LUT presents the image as a gray scale but shows in blue and red the pixels with 0 and 255 values, respectively. This allows for high contrast in the extreme values, being useful for visual evaluation of the background subtraction procedure.
10. To reduce the background in the images, two different plugins are sequentially used to eliminate both the constant and the uneven background signal: the *BG Subtraction from ROI* plugin works better for constant background levels, whereas the *Subtract Background* tool focuses on uneven background levels, i.e., when different sides of the image have different background levels.
11. The *BG Subtraction from ROI* plugin needs a ROI where no specific signal is present. The plugin will analyze the intensity in that region and subtract the mean plus the standard deviation multiplied by an entered factor on the whole image. The effect of this plugin depends on the size of the ROI. Thus, to have consistently identical ROIs in all images, the macro draws a ROI sizing 15% of the image area at the top left corner. Then the user may move it around so that it fits a background region. The program adds the ROI to the *ROI Manager*, selects it, and then runs the *BG subtraction from ROI* plugin.

12. The *Subtract Background* plugin was also used to suppress background from our images. The value chosen for the *Rolling ball radius* is critical in the process and should be kept at least as big as the maximum object size in the field [19]. We chose a value of 50 pixels for the membrane marker and 500 pixels for the probe proteins after a manual inspection of membrane thickness (around 10 pixels) and cell size (500–700 pixels).
13. To denoise the image, we tested different filters on different combinations on a set of images being evaluated visually. Finally, we stayed with the combination of median and Gaussian filters to reduce the two types of noise observed in the images. *Median* filtering replaces each pixel by the median of the surrounding pixels making it especially good at removing "salt-and-pepper" noise while keeping the transients of the image. *Gaussian blur* filter is a low-pass filter that blurs the image using a Gaussian function and thus removes high-frequency noise. Because both filters were applied, we chose subtle parameters: *Radius* = 1 for the *Median* filter, which sets up a 3 × 3 pixel kernel, and *Sigma* = 1 for the *Gaussian blur*, in which 1 pixel is the standard deviation of the filter.
14. Blocs of code that are used several times inside a macro may be wrapped as an internal function and used like the built-in macro functions. When defining the function, use an explicit name to call it and specify its arguments – parameters to use in the internal code—between the parentheses if needed.
15. Images are segmented into a binary mask using *ImageJ*'s *Default* mode, which is a variant from the iterative intermeans *IsoData* method [20]. However, obtaining a consistent mask from an image may require some image preprocessing to segment it effectively. For this purpose, the autoAdjust function from Guimond [21] is used in the macro to automatically adjust the Brightness and Contrast levels in the image. The text used is available at [17], but it can also be done manually at [Image > Adjust > Brightness/Contrast > Auto > Apply]. Because this adjustment alters the image signal, it is only applied on a duplicate to keep the intensity of the source image unaltered. After segmentation, the resulting masks will be combined to the non-altered image duplicates so that colocalization measurements are only performed on unaltered signal levels.
16. The binary masks are further processed with morphology operations such as *Erode*, *Dilate*, and *Open*. Erosion is an operation that deletes pixels from the edges of black objects. On the contrary, Dilation adds pixels to the edges of black objects. Finally, *Open* is a sequence of erosion and dilation to smoothen the mask and remove isolated pixels. These and

other binary functions are specified in the ImageJ Documentation Wiki [22].

17. *Image Calculator* allows to perform several different operations between two images. In the macro, the operations minimum (*Min*) and Boolean conjunction (*AND*) are used. The operation *Min* selects the minimum value of two images. When an image is 8-bit and the other is binary, the resulting image will have the values of the 8-bit image in those pixels whose intensity is 255 in the binary; otherwise, their value will be 0. The number 255 is the value *ImageJ* uses to represent 1 in binary images. On the other hand, the operation *AND* results in an image containing what the two combined images have in common keeping the lower intensity between both. For more information on the *Image Calculator*, please consult the ImageJ Documentation Wiki [23].

18. The *Integrated density* function provides us with two different values: *IntDen* which is the product of the image area and the mean gray value and *RawIntDen* which represents the sum of the gray values of all the pixels in the image. As the intensity of Kv1.3 in the triple-colocalization image (obtained in Subheading 3.4) is divided by the intensity of Kv1.3 in the "Kv1.3 intensityInsideMask" image (obtained in Subheading 3.3), both *IntDen* and *RawIntDen* will provide the same result. On the other hand, the mean intensity could also be chosen, since it is the product of the *Integrated density* and the number of pixels in the image.

References

1. Alberts B, Johnson A, Lewis J, Morgan D, Raff M, Roberts K, Walter P (2015) Molecular biology of the cell, 6th edn. Garland Science, Taylor & Francis Group, LLC, New York
2. Armstrong CM, Hille B (1998) Voltage-gated ion channels and electrical excitability. Neuron 20:371–380
3. Pérez-Verdaguer M, Capera J, Serrano-Novillo C, Estadella I, Sastre D, Felipe A (2016) The voltage-gated potassium channel Kv1.3 is a promising multitherapeutic target against human pathologies. Expert Opin Ther Targets 20:577–591. https://doi.org/10.1517/14728222.2016.1112792
4. Serrano-Albarrás A, Estadella I, Cirera-Rocosa-S, Navarro-Pérez M, Felipe A (2018) Kv1.3: a multifunctional channel with many pathological implications. Expert Opin Ther Targets 22:101–105. https://doi.org/10.1080/14728222.2017.1420170
5. Martínez-Mármol R, Styrczewska K, Pérez-Verdaguer M, Vallejo-Gracia A, Comes N, Sorkin A, Felipe A (2017) Ubiquitination mediates Kv1.3 endocytosis as a mechanism for protein kinase C-dependent modulation. Sci Rep 7:42395. https://doi.org/10.1038/srep42395
6. Sole L, Roura-Ferrer M, Perez-Verdaguer M, Oliveras A, Calvo M, Fernandez-Fernandez JM, Felipe A (2009) KCNE4 suppresses Kv1.3 currents by modulating trafficking, surface expression and channel gating. J Cell Sci 122:3738–3748. https://doi.org/10.1242/jcs.056689
7. Schindelin J, Arganda-Carreras I, Frise E, Kaynig V, Longair M, Pietzsch T, Preibisch S, Rueden C, Saalfeld S, Schmid B, Tinevez JY, White DJ, Hartenstein V, Eliceiri K, Tomancak P, Cardona A (2012) Fiji: an open-source platform for biological-image analysis. Nat Methods 9:676–682. https://doi.org/10.1038/nmeth.2019
8. Bolte S, Cordelières FP (2006) A guided tour into subcellular colocalization analysis in light

microscopy. J Microsc 224:213–232. https://doi.org/10.1111/j.1365-2818.2006.01706.x

9. Manders EMM, Stap J, Brakenhoff GJ, Van Driel R, Aten A (1992) Dynamics of three-dimensional replication patterns during the S-phase, analysed by double labelling of DNA and confocal microscopy. J Cell Sci 103 (Pt 3):857–862
10. Pearson K (1895) Note on regression and inheritance in the case of two parents. Proc R Soc London 58:240–242. https://doi.org/10.1098/rspl.1895.0041
11. (2018) Manel Bosch ijm-Macros. https://github.com/manelbosch76/ijm-Macros. Accessed 5 Jun 2018
12. Rueden C (2013) Plugins—ImageJ. https://imagej.net/index.php?title=Plugins&oldid=24930. Accessed 2 Apr 2018
13. Cordelières FP, Bolte S (2008) JACoP v2.0: improving the user experience with co-localization studies. ImageJ User Dev Conf 174–181
14. Cordelières FP (2008) JACoP [ImageJ Documentation Wiki]. http://imagejdocu.tudor.lu/doku.php?id=plugin:analysis:jacop_2.0:just_another_colocalization_plugin:start. Accessed 2 Apr 2018
15. Collins TJ (2007) ImageJ for microscopy. BioTechniques 43:25–30. https://doi.org/10.2144/000112505
16. Collins TJ (2007) MBF plugin collection. http://imagej.net/plugins/mbf/index.html. Accessed 2 Apr 2018
17. Kota Miura ijm-Macros. https://github.com/miura/IJ_BCautoMacro. Accessed 2 Apr 2018
18. Schindelin J, Cardona A, Eglinger J, Rueden C, Brocher J, Hiner M, Arena ET, Arganda-Carreras I (2010) Introduction to macro programming—ImageJ. https://imagej.net/Introduction_into_Macro_Programming#Installing_macros. Accessed 2 Apr 2018
19. ImageJ Wiki Subtract background [ImageJ Documentation Wiki]. http://imagejdocu.tudor.lu/doku.php?id=gui:process:subtract_background. Accessed 14 Jan 2018
20. Ridler TW, Calvard S (1978) Picture thresholding using an iterative selection method. IEEE Trans Syst Man Cybern 8:630–632. https://doi.org/10.1109/TSMC.1978.4310039
21. Kota Miura ijm-Macros (2014). https://github.com/miura/IJ_BCautoMacro. Accessed 5 Jun 2018
22. ImageJ Wiki (2010) Binary [ImageJ Documentation Wiki]. http://imagejdocu.tudor.lu/doku.php?id=gui:process:binary. Accessed 1 May 2018
23. ImageJ Wiki (2008) Image calculator [ImageJ Documentation Wiki]. http://imagejdocu.tudor.lu/doku.php?id=gui:process:image_calculator. Accessed 1 May 2018

Chapter 12

Photobleaching and Sensitized Emission-Based Methods for the Detection of Förster Resonance Energy Transfer

Timo Zimmermann

Abstract

Förster resonance energy transfer (FRET) is a non-radiative interaction between two molecules that happens at distances in the range of a few nanometers. Using FRET interactions between suitably selected fluorophores allows to study molecular interactions or conformational changes of single molecules on fluorescence microscopes even though the optical resolution of the microscope is limited to distances that are almost two orders of magnitude higher.

In this chapter several variants of FRET detection methods are described that are based either on the targeted photobleaching of one of the participating molecule species or on the direct detection of the fluorescence signal that is created as a result of the FRET interactions.

Key words FRET, Acceptor photobleaching, Donor photobleaching, Ratiometric imaging, Sensitized emission

1 Introduction

1.1 The Principle of FRET

Information on molecular interactions and even on conformational changes can be obtained using fluorescence as a readout of molecular resonance energy transfer processes that were described by Theodor Förster in the middle of the twentieth century [1]. Förster resonance energy transfer (FRET) constitutes a non-radiative form of energy transfer between two molecules; that is, it does not involve the emission and absorption of a photon to transfer the energy between them. The energy is instead transferred between the donor and acceptor molecule through the coupling of the dipole of the excited donor molecule and the dipole of unexcited acceptor molecule. This leads to the de-excitation of the donor molecule and the corresponding elevation of an acceptor molecule electron to an excited state (*see* Fig. 1a, b). The distance in which

Electronic supplementary material: The online version of this chapter (https://doi.org/10.1007/978-1-4939-9686-5_12) contains supplementary material, which is available to authorized users.

Elena Rebollo and Manel Bosch (eds.), *Computer Optimized Microscopy: Methods and Protocols*, Methods in Molecular Biology, vol. 2040, https://doi.org/10.1007/978-1-4939-9686-5_12,

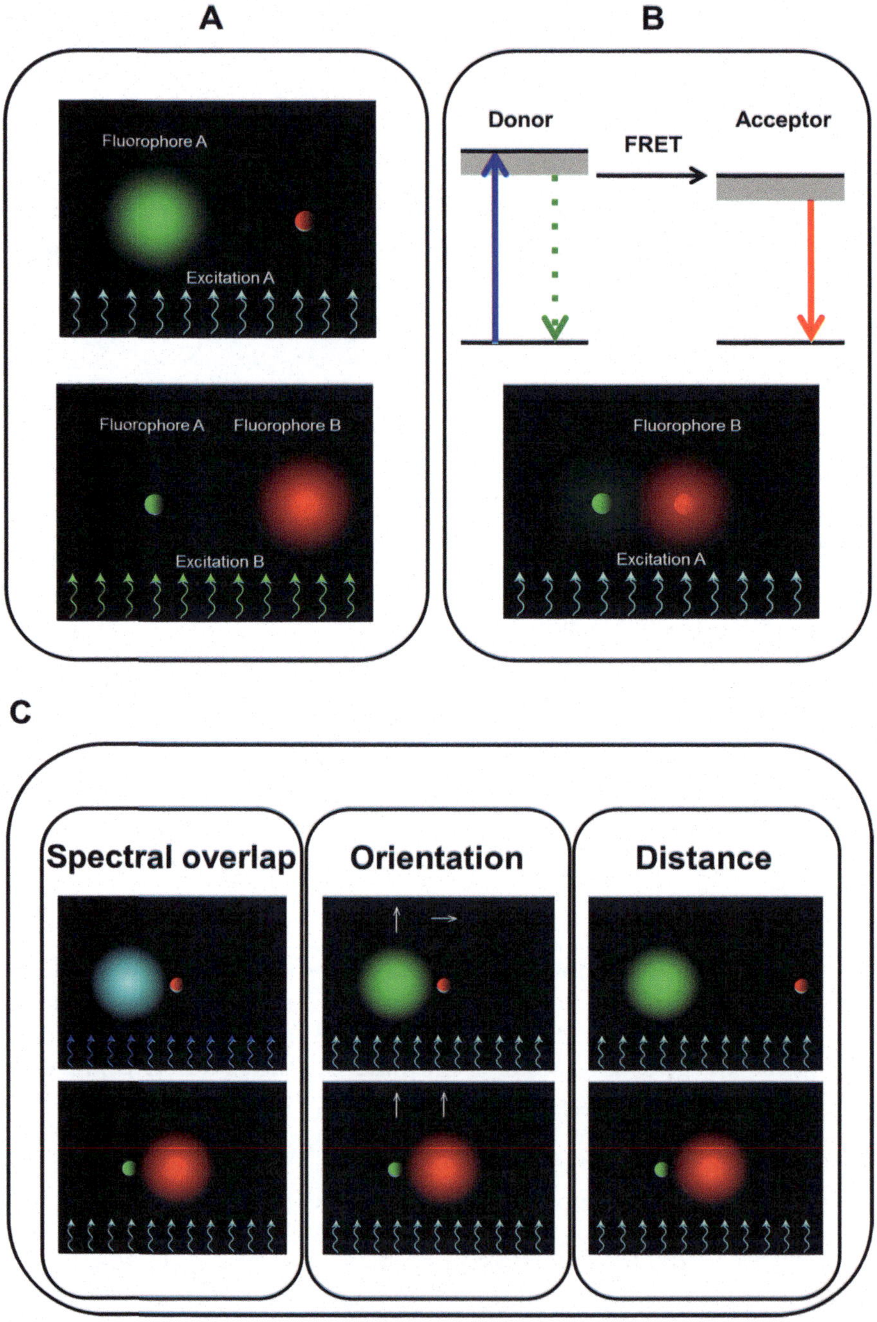

Fig. 1 FRET principle and requirements. (**a**) Shown are schematic representations of two fluorophores (**a**, **b**) with different excitation and emission spectra that can form a FRET pair. The emission of fluorophore A overlaps with the excitation spectrum of fluorophore B. (**b**) Jablonski diagram of the FRET process and illustration of the interaction. In very close proximity, the energy of the excited donor fluorophore can be passed on non-radiatively to a suitable acceptor fluorophore that transitions to the first excited state S1 and emits fluorescence. (**c**) Requirements for FRET interactions consist of partial overlap of the donor emission and acceptor fluorophore excitation spectra, a suitable orientation of the fluorophore dipoles relative to each other and a dependence on close proximity of the two fluorophores

the dipole fields can be coupled for energy transfer is very small (in the range of around 5–7 nm, a small fraction of the wavelength that the emitted light would have) so that the participating molecules have to be in close proximity. This distance constraint constitutes its usefulness for molecular studies as proximity in the nm range is an indication for direct interaction of two molecules (intermolecular FRET). Alternatively, conformational changes within a single molecule can be studied by changes in the efficiency of energy transfer (intramolecular FRET). The principle of FRET lies at the heart of many fundamental biological processes that require the transfer of energy between molecules. Accordingly, it was studied and first described by Förster for the photosynthetic machinery of chlorophyll complexes. FRET can take place between identical molecules (homo-FRET, as studied by Förster) or between different molecules (hetero-FRET) as long as the following requirements are met in forming a FRET pair (*see* Fig. 1c): (1) the energy levels of the excited donor and of the unexcited acceptor molecule must overlap sufficiently to allow the exchange of energy between the molecules; the suitable energy levels are represented by the spectra at which the molecules can absorb and release photons of a certain energy (spectral color); (2) the distance between the molecules must be small enough to allow interaction of the dipole fields that are decaying very rapidly with distance; and (3) the dipole moments of the donor and acceptor molecule need to be oriented relative to each other in a manner that allows energy transfer; at a 90° orientation to each other, no interaction can take place.

These three factors determine the efficiency of the energy transfer between two molecules, i.e., how likely it is that an excited donor will transfer energy to an acceptor molecule instead of emitting a fluorescence photon.

Assuming free rotations of the participating fluorophores (as would be the case for single bond connections) the dipole orientations can often be approximated by a constant that represents an even distribution of all possible orientations. This allows to define the FRET efficiency between donor and acceptor based on one molecular property (spectral overlap) and one external parameter (distance).

The distance at which the FRET efficiency is at 50% (half of the donor excitation cycles de-excite by energy transfer) is called the Förster radius (R_0) of a fluorophore pair. It is characteristic for that combination and is generally in the range of 3–5 nm. The dependence between molecule distance r and FRET efficiency E is defined by the following formula:

$$E = \frac{R_0^6}{r^6 + R_0^6} \quad (1)$$

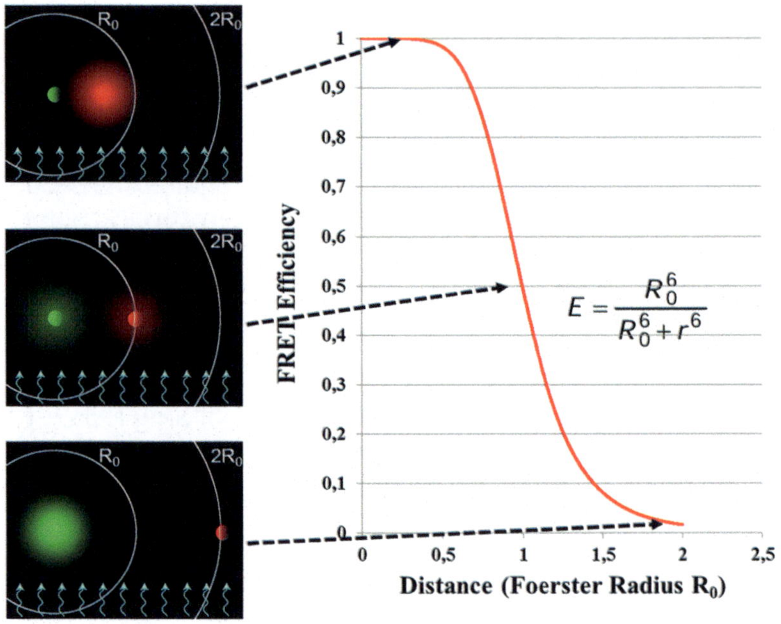

Fig. 2 FRET efficiency *E* represents the likeliness of the donor fluorophore returning to the ground state by resonance energy transfer and is strongly dependent on the distance between the fluorophores of the FRET pair, here defined as the distance at which the transfer efficiency *E* is 50% (Förster radius). Typical Förster radius sizes are between 3 and 5 nm

The dependence of efficiency to distance is such that the transfer efficiency is close to zero already at distances of two times the Förster radius (*see* Fig. 2).

1.2 FRET as a Tool for Molecular Biology

As described above, FRET mechanisms are important parts of many biological processes. The FRET principle has however also found its use in molecular biology studies, as it allows the detection of distance changes in the single nanometer range. Methods that use FRET to study molecular interactions or conformational changes must provide a detectable readout that is based on fluorescence. FRET methods are therefore generally based on a matched pair of distinct fluorophores serving as donor and acceptor that are used to specifically label the molecules (or molecule) under study. The chosen fluorophore combination (referred to as a FRET pair) must fulfill the spectral overlap criteria for FRET interactions. FRET pairs generally consist of a shorter wavelength donor fluorophore and a longer wavelength acceptor molecule.

There are however no absolute prerequisites. Although most methods use distinct fluorescent molecules for FRET interactions (hetero-FRET), applications using only one kind of fluorophore reporter (homo-FRET) also exist. Additionally, a FRET pair for fluorescence readout does not have to consist of a donor and an acceptor fluorophore. In methods detecting changes in donor fluorescence only, the acceptor's fluorescence emission properties are not relevant as long as the donor can be efficiently de-excited (quenched) by the acceptor's energy absorption. In methods specifically detecting the additional FRET-based emission of the acceptor (sensitized emission), the quantum efficiency (proportion of fluorescence emissions to molecule excitations) of the donor can be low as long as the donor efficiently transitions to the excited state from which FRET can take place. In the related technique of bioluminescence resonance energy transfer (BRET), the donor is a bioluminescent molecule instead of a fluorophore that transfers its bioluminescence-created energy to a fluorescent acceptor that is subsequently read out. All this serves to illustrate that fluorescence is only the reporter of the process but does not constitute part of the process itself.

1.3 FRET Detection Methods

Several powerful methods have been described over the years that can be executed on standard biological research microscopes (motorized fluorescence microscopes, confocal microscopes). FRET detection methods can be divided into two main categories (Fig. 3): (1) methods that detect a change in the fluorescence signal distribution upon external disruption of the FRET signal: this category contains the photobleaching methods based on the destruction of either the donor or the acceptor fluorophore, and (2) methods in which the FRET information is contained within the detection channels of the unperturbed measurement. The last category contains the approaches to measure the additional sensitized emission of the acceptor that is created by the FRET interactions in the sample. Currently, the most commonly used FRET technique (ratiometric imaging of intramolecular FRET changes) belongs to the second category and constitutes a reduced variant of the full sensitized emission detection.

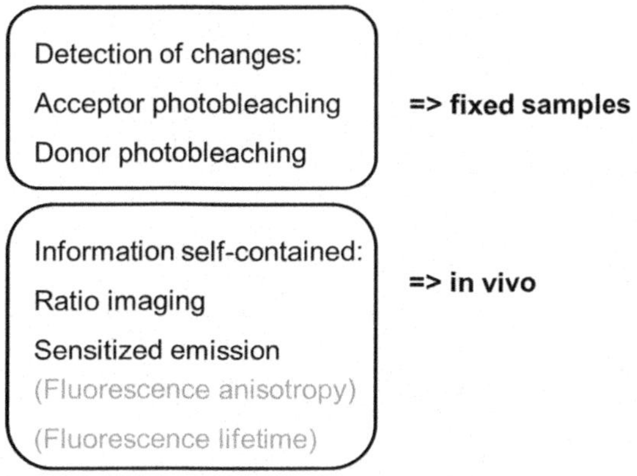

Fig. 3 FRET detection methods fall into two main categories. Based on the approach, the photobleaching methods are better suited for fixed samples

The two categories differ in several aspects (*see* **Note 1**). FRET can in addition be detected through measurements of the excited-state fluorescence lifetime of the donor and acceptor molecules and the comparison of the measured lifetimes with the ones of the fluorophore in non-FRET conditions (*see* **Note 2**).

1.3.1 Acceptor Photobleaching

Acceptor photobleaching [2–4] is a method of FRET detection that can be implemented easily on most confocal microscopes. In its simplest form it is a three-step procedure consisting of a pre-bleach measurement, a bleaching step (destruction of the acceptor fluorophore), and a post-bleach measurement (*see* Fig. 4).

The readout of the method is based on the observation of the donor channel before and after the bleaching step. In case of no FRET interaction, a donor molecule's fluorescence emission will not be affected by acceptor photobleaching and will be the same before and after the bleaching step. In case of FRET interaction between a donor and an acceptor fluorophore, the photodestruction of the acceptor fluorophore can be perceived as an increase in donor fluorescence emission. While donor and acceptor molecules interact through energy transfer, excited donor fluorophores can return to the ground state by transferring the absorbed energy

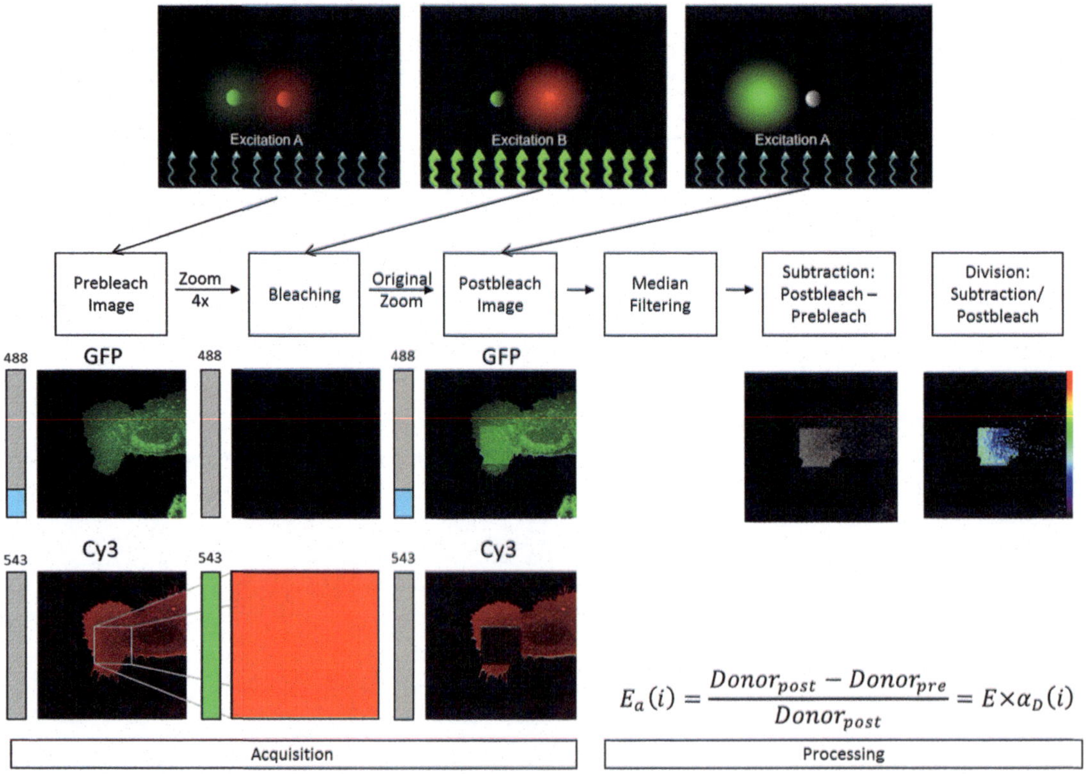

Fig. 4 Schematic of the principle of acceptor photobleaching and of the execution of a one-step acceptor photobleaching experiment

non-radiatively to the acceptor molecule instead of emitting a fluorescence photon. For the same number of donor excitation cycles, less donor fluorescence will therefore be emitted, and they will appear "darker." If transfer is not possible anymore because of the destruction of the accepting partner, donor fluorescence will increase by the amount that was passed on by FRET before (determined by FRET efficiency E).

1.3.2 Donor Photobleaching

Donor photobleaching [5] is possibly the simplest form of FRET detection as it requires only the observation of the donor fluorescence channel over time. It can therefore be implemented even on standard fluorescence microscopes as long as they are capable of acquiring image series.

At the heart of the simplest form of FRET detection lies the same process that also forms the basis for the most advanced form of FRET imaging, fluorescence lifetime imaging microscopy (FLIM). Both methods detect effects that are related to the lifetime of the excited state of the fluorophore and FRET efficiencies derived from these methods can be calculated with basically the same formula.

Donor photobleaching experiments measure changes in the bleaching rate of donor fluorophores in the presence and absence of FRET (*see* Fig. 5). In noninteracting conditions donor molecules will leave the excited state in one of the three ways: (1) returning to the ground state non-radiatively by collisional quenching (creating vibrations and heat); (2) returning to the ground state by fluorescence photon emission; or (3) transitioning to a triplet state from which bleaching can take place with a certain likeliness.

Together, the transition rates of these events will determine the average time a fluorophore stays in the excited state and the relative likeliness of a specific outcome in relation to the other two possibilities. As the fluorophore will be irreversibly bleached with a certain likeliness from the triplet state, the fluorophore population will over time decrease following a kinetic that is determined by the bleaching transition rate.

In FRET interactions, an additional return to the ground state exists for the excited molecule: passing the energy to an acceptor by resonance energy transfer. Accordingly, a fluorophore can in this case de-excite by four instead of three pathways. This has two detectable effects that are both proportional to the efficiency of the FRET process: (1) The average time a fluorophore spends in the excited state decreases as the combined transition rates to the four outcomes make an earlier return from the excited state possible, and (2) the likeliness for a fluorophore to enter the bleaching pathway is proportionally decreased by the proportion of molecules that will enter the FRET pathway. Therefore, both the excited state lifetime of donor molecules and their bleach rate will change in case of FRET interactions.

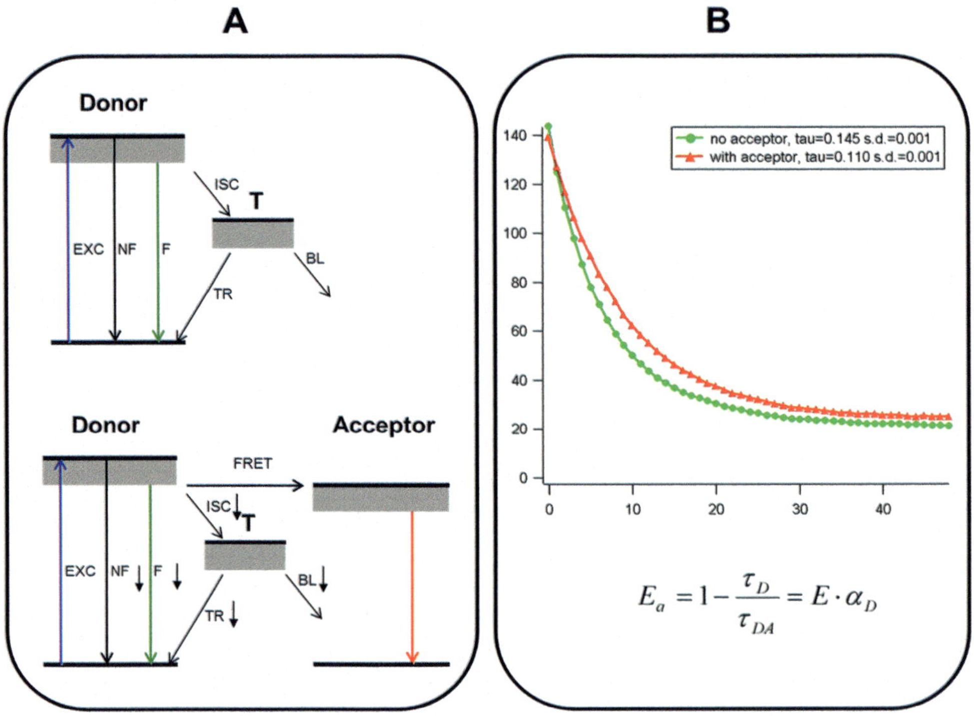

Fig. 5 (**a**) Illustration of the change in the transitions out of the first excited state of a donor fluorophore in the presence or absence of FRET. Introducing transitions by resonance energy transfer will correspondingly lower the time the fluorophore stays in the excited state and the number of transitions through the other pathways. *T* triplet state, *EXC* excitation, *NF* nonfluorescent return to the ground state, *F* fluorescence, *ISC* intersystem crossing, *TR* triplet return, *BL* photobleaching. (**b**) Example of a donor photobleaching experiment, comparing donor bleaching curves in the presence and absence of acceptor

1.3.3 Sensitized Emission-Based Methods

In addition to the methods probing samples for FRET by photobleaching, the FRET effect can also be detected by the outcome of the observed energy transfer, the additional fluorescence that is being emitted by the acceptor fluorophore. This additional signal is referred to as sensitized emission and can be extracted from suitably configured multichannel images. Sensitized emission detection methods do not interfere with the samples (not taking into account possible bleaching and photodamage by the fluorescence observation itself) and can be repeated as image series to observe FRET changes in dynamic samples.

Sensitized emission methods comprise two variants: ratiometric imaging (two channels) and sensitized emission imaging (three channels). Which one is applicable depends on whether the concentration of donor versus acceptor is variable in the sample or not.

Different from photobleaching-based FRET detection methods, sensitized emission methods do not require the strong local illumination provided by laser light sources and can be implemented on both wide-field and confocal instruments. Considerations on fluorophore selection can be found in **Notes 3** and **4**.

In case of a fixed stoichiometry of donor to acceptor, FRET changes can be detected with ratiometric imaging by the observation of both the donor and the acceptor emission channel under donor illumination. A fixed donor-to-acceptor stoichiometry is given in reporter constructs in which the donor and the acceptor fluorophores are permanently connected to the same macromolecule and interact only with their FRET partner on that molecule (intramolecular FRET) but not with fluorophores on other molecules (intermolecular FRET). Changes in the ratio of the two channels are caused exclusively by changes in intramolecular FRET efficiency that reflect changes in the conformation of the molecule affecting the distance and/or the dipole orientation of the fluorophores. Changes in local molecule concentration will not register as this intensity change is reflected in both channels and thus does not alter the ratio between them.

Ratiometric FRET imaging is currently the most commonly used method for FRET detection in biological applications, often using genetically encoded tandem constructs with two fluorescent proteins as a FRET pair [6].

In case of variable donor-to-acceptor stoichiometry, ratiometric FRET imaging is not sufficient for FRET detection. In addition to the sensitized emission created by resonance energy transfer, the acceptor emission channel under donor excitation will contain spectral contributions from the donor fluorophore (the part of its spectrum extending to longer wavelengths) and the acceptor fluorescence (created by direct excitation of the acceptor by the excitation light used for donor stimulation). In ratiometric imaging of fixed stoichiometries, both the donor bleedthrough into the acceptor channel and the amount of directly excited acceptor in the channel are invariable and will contribute to the channel ratio only as fixed offsets that do not change during the measurement. Changes in local concentration of donor and acceptor are always identical and are removed from the ratio by the channel division, which is basically a normalization to the donor channel that removes all concentration information. The ratio can only be altered by changes in FRET as manifested in the sensitized emission of the acceptor and the corresponding quenching of the donor. In a variable stoichiometry setting (fluorophores on separate molecules) changes in FRET and changes in acceptor concentration are indistinguishable from one another if only the acceptor channel under donor excitation is taken into account. A three-channel sensitized emission measurement is therefore a straightforward extension of the ratiometric two-channel approach that takes into account the acceptor concentration by directly measuring it in an additional third channel.

This chapter focuses on the two classes of FRET detection methods described above (photobleaching and sensitized emission-based methods). It provides step-by-step instructions on how to execute the measurements and describes the analysis of the data. To help with understanding and applying the image processing steps on a computer, the analysis procedures are also made available as a set of easy-to-follow macros for the open-source image processing software *ImageJ* which can be run platform-independently on PC, Macintosh, and Linux environments in *Java*.

2 Materials

FRET techniques are detection methods that can be applied to an ever-expanding range of label pairs and sample preparations that cannot be adequately covered in this chapter. The fluorophores have to be selected according to general FRET criteria (*see* Subheading 1). Additional considerations have to be taken according to the selected method groups (*see* **Notes 3** and **4**).

2.1 Instrumentation

1. Confocal microscopy: All methods can be executed on standard confocal instrumentation. This comprises a set of laser illumination lines that are matched to excite the donor and acceptor fluorophores of the selected FRET pair. Current laser light sources are strong enough to efficiently bleach fluorophores in any spectral range and can be aided by objectives with a high numerical aperture. The donor and acceptor fluorescence signals need to be collected in two separate channels. This can be done using defined filters and beam splitters in the light path or using different variants of spectral confocal microscopes. Where possible, channel settings should be defined to maximally separate donor and acceptor signals for the subsequent processing steps. Strongly overlapping signals can be postprocessed by linear unmixing before further analysis. For quantitative analysis, the detectors have to be used in their linear detection range. In case of bleaching methods, damage to very sensitive detection technologies by strong illumination needs to be avoided by deactivating them during the bleaching steps.
2. Wide-field fluorescence microscopy: Wide-field instruments are mainly applicable to the second group of described methods that do not require targeted photobleaching. For efficient fluorescence imaging, a highly sensitive camera is required. The microscope needs to be fully automatized to allow the repeated acquisition of several channels. As this requires the repeated movement of filter components in the light path, separately movable filter components are better suited which can exist in a range of configurations.

2.2 Software and Macros

The processing of FRET images makes use of basic image processing steps (noise filtering, image subtraction, image division) that can be easily implemented in all current image processing software packages. In the context of this chapter, example workflows are available, along with example images, for download as supplementary material for this book as plain-text macros in the widely used open-source Java-based software package *ImageJ*, more specifically its *Fiji* distribution [7].

1. The macros will run in any version of *ImageJ* equal to or above 1.42. *ImageJ* runs in the Java Runtime environment and is therefore not dependent on a specific operating system, but is supported for current Windows versions, Mac OS X, and Linux and can be run more generally on any platform with *Java 8* runtime.
2. The macros only contain the processing steps for the FRET calculations, and background corrections need to be applied separately. In case that the raw data contain background, it may be interesting to run the macros on the uncorrected images as well to understand the influence of background on the efficiency results.
3. The provided example macros are as follows:
   (a) "single step acceptor photobleaching.ijm" for acceptor photobleaching with a single bleaching step
   (b) "linear fit acceptor photobleaching.ijm" for the fitting of gradual acceptor photobleaching data
   (c) "exponential fit donor photobleaching.ijm" for the fitting of donor photobleaching data
   (d) "sensitized emission_median.ijm" for sensitized emission analysis
   (e) "intensity weighing.ijm" for an intensity-weighed image of FRET intensities

2.3 Sample Images

For training purposes, example images are available for download as supplementary material for this book.

1. Two single-step acceptor photobleaching images (*see* Subheading 3.1.1) named "FRET positive_pre_bleach.tif" and "FRET positive_post_bleach.tif."
2. Five sensitized emission detection images (*see* Subheading 3.2.2), named "donor_reference.tif," "acceptor_reference.tif," and "FRET positive_sensitized emission_AA, _DA and _DD."

3 Methods

3.1 Photobleaching Methods

Confocal laser scanning microscopes are very suitable for localized photobleaching methods (*see* **Note 5**). For sample mounting considerations, *see* **Note 6**.

Although acceptor photobleaching experiments are self-calibrating and do not require reference measurements for the execution, control samples containing the labels and fluorophores used in the experiment are essential. These should be samples containing only donor or only acceptor and should be processed and imaged exactly as the samples that contain both and are studied for interactions.

Donor-only samples should remain completely unaffected by the procedure and give exactly the same value before and after acceptor photobleaching. If the donor intensity decreases in the bleached area upon acceptor bleaching the donor is directly affected by the light used for the acceptor and values cannot be compared in a real experiment (*see* **Notes 7** and **8**).

3.1.1 Single-Step Acceptor Photobleaching

Figure 4 shows a generalized workflow for the acquisition and analysis of a single-step acceptor photobleaching measurement in which the acceptor is completely bleached.

1. Acquisition of a pre-bleach image of the donor and acceptor channels: The pre- and post-bleach settings are chosen with similar considerations to those for any clearly separated two channel image of two fluorophores. Especially cross talk of the acceptor into the donor channel needs to be avoided as the analysis depends on the comparison of the donor signals before and after bleaching. A full separation of the acceptor signal is also helpful as it allows to assess the completeness of the bleach and apply a correction in case of incomplete bleaching (*see* **Note 9**).

   The signals in the channels should be in the linear detection range of the detector to allow quantitation using the formula described below. Saturation needs to be avoided. Especially in the pre-bleach image the detected intensities need to be sufficiently below saturation to accommodate a possible signal increase in the donor channel upon photodestruction of the acceptor.

   To be able to distinguish localized changes in FRET, resolution of the image has to be set according to the desired detail level and image noise should be low to allow FRET calculations at the pixel level. While creating a good image is important for this, illumination levels and (in the case of confocal images) pixel dwell time and/or image averaging need to be kept low enough to avoid detectable photobleaching during the acquisition of the pre- and post-bleach images.

2. Bleaching step, in which the acceptor fluorophore is specifically destroyed (for additional considerations *see* **Notes 10** and **11**): Bleaching of the acceptor (in the single-step procedure) should ideally be complete to create a significant change between pre- and post-bleach donor images while making sure that the donor is not directly affected by the bleach illumination. Incomplete bleaching of the acceptor can be taken into account by a straightforward extension of the analysis procedure that will be described below.
3. Acquisition of a post-bleach image of the donor and acceptor channels under the same image settings as the pre-bleach image.
4. FRET efficiency calculation: Assuming complete photobleaching of the acceptor, the apparent FRET efficiency E_a in the sample can be calculated using only the two images of the donor channel pre- and post-bleach using the following formula:

$$E_a = \frac{\text{Donor}_{\text{post}} - \text{Donor}_{\text{pre}}}{\text{Donor}_{\text{post}}} = E \times \alpha_D \tag{2}$$

The measurable E_a is in this case the product of the FRET efficiency between two interacting molecules E and the amount of donor molecules that are interacting with acceptor molecules α_D, expressed as

$$\alpha_D = \frac{[DA]}{[DA] + [D]} \tag{3}$$

Based on acceptor photobleaching experiments alone E_a cannot be further separated into E and α_D.

Furthermore, incomplete bleaching of the acceptor can be taken into account by extending the initial Eq. (2) and taking into account the information from the acceptor channel by creating a correction factor b:

$$b = 1 - \frac{\text{Acceptor}_{\text{post}}}{\text{Acceptor}_{\text{pre}}} \tag{4}$$

$$E_{\text{a corr}} = \frac{E_A}{b} \tag{5}$$

5. Image processing steps for FRET efficiency calculation. The general steps for processing acceptor photobleaching data are represented in "single step acceptor photobleaching.ijm" that is available for download as supplementary material for this book. It basically executes the calculation described in **step 4**

(Eq. 5) on a pre-bleach and a post-bleach image of the donor channel through the steps described below. The provided example images "FRET positive_pre_bleach.tif" and "FRET positive_post_bleach.tif" can be used to try the macro pipeline. In case of nonzero background, execute a background subtraction on both images before macro execution. Further considerations and pitfalls on the quantitation and interpretation of the results can be found in **Notes 12–16**.

6. Pre- and post-bleach images are pre-filtered for noise (using median filtering).
7. An image containing signal increases is created by subtracting the pre- from the post-bleach image.
8. A normalized apparent FRET efficiency image is created by dividing the subtracted image by the post-bleach image.
9. The newly created FRET efficiency image is displayed with a suitable color palette.

3.1.2 Gradual Acceptor Photobleaching

As the apparent FRET efficiency can also be derived from a measurement with incomplete bleaching of the acceptor (*see* Eqs. 4 and 5) acceptor photobleaching experiments can also be executed as a series of partial bleaches at lower illumination intensities [8]. Figure 6 shows a schematic of the approach. *See also* **Note 17**.

1. Acquisition of a pre-bleach image of the donor and acceptor channels. *See also* the instructions of **step 1** of the single-step acceptor photobleaching procedure.
2. Partial bleach of the acceptor fluorophore (for additional considerations *see* **Notes 10** and **11**).
3. Acquisition of a post-bleach image of the donor and acceptor channels under the same image settings as the pre-bleach image.
4. Repeat **steps 2** and **3** until acceptor fluorescence is significantly decreased.
5. Calculate FRET efficiency. The apparent FRET efficiency E_a for every step n $E_a(n)$ is calculated with the same formula (Eq. 2) as in the single-step process. Donor_0 represents the pre-bleach image:

$$E_a(n) = \frac{\text{Donor}_n - \text{Donor}_0}{\text{Donor}_n} \quad (6)$$

$E_a(n)$ is incomplete until all acceptor signal has disappeared but will show a linear increase when plotted against the decreasing amount of acceptor signal Acceptor_n or its normalized equivalent $A(n)$:

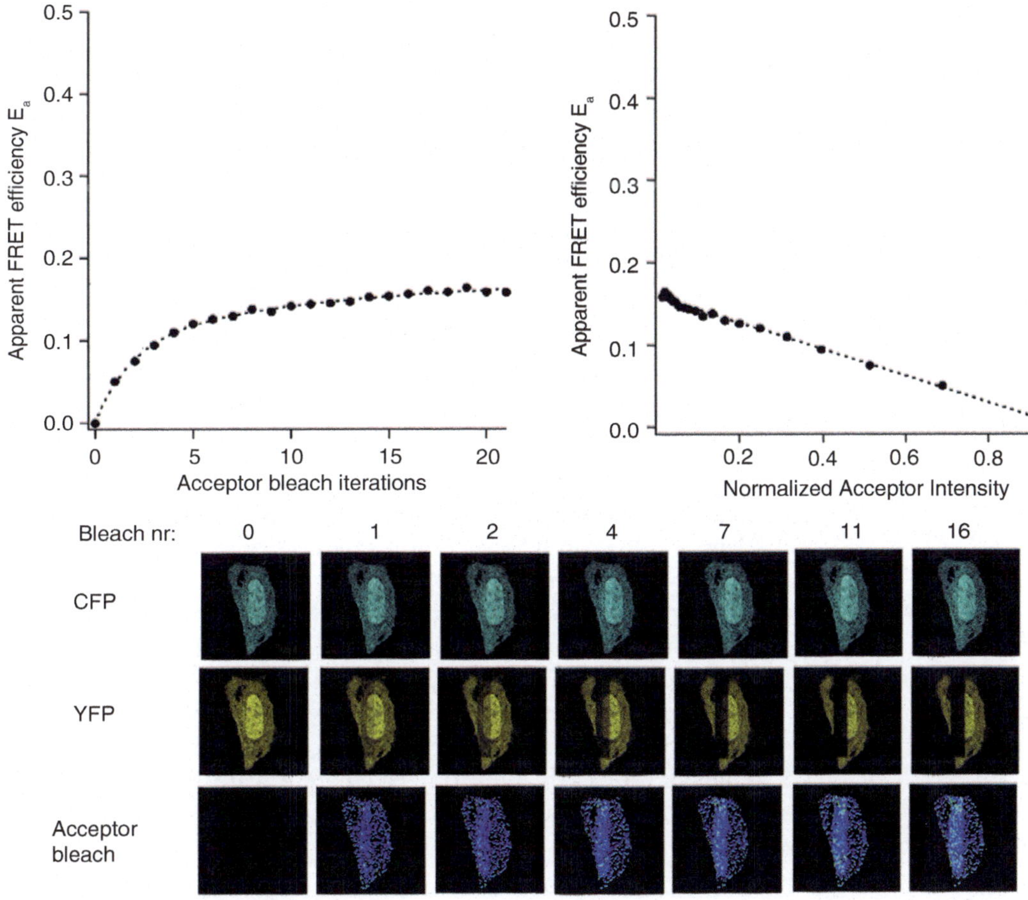

Fig. 6 Example of a gradual acceptor photobleaching experiment. The increase of brightness of the donor is directly correlated with the disappearance of the acceptor signal and the full apparent FRET efficiency can be extrapolated by fitting a line to the FRET efficiency value expected in the complete absence of acceptor signal

$$A(n) = \frac{\text{Acceptor}_n}{\text{Acceptor}_0} \tag{7}$$

Following the linear increase to its intersection with the zero point of the acceptor intensity axis (no acceptor left) will provide the same E_a that can be obtained in a single-step bleaching experiment. This value does not have to be measured, but can be fitted based on the data points of the incomplete bleaching steps.

Using a linear fit a line can be created that intersects with the axis of apparent FRET efficiency ($E_a = 0$) at the pre-bleach image and with the axis for intensity of the acceptor ($A = 0$) at the apparent FRET efficiency of the experiment. Fitting has the benefit that noise on the data points can be better compensated than in a measurement value-based calculation and that acceptor bleaching does not have to be complete to calculate E_a as the value can be extrapolated from the line fit. This makes the method less prone to measurement artifacts as less light is used, and therefore possible direct donor bleaching and acceptor photoconversion effects would be reduced. Such effects would be perceivable as deviations from the linear progression of the data points and would accumulate in later photobleaching steps so that these data points could be left out of the fit. Direct donor photobleaching would lead to a nonlinear decrease of E_a whereas acceptor photoconversion would manifest itself as an additional increase that would deviate from the linear progression. Considerations and pitfalls are the same as for the single-step acceptor photobleaching method (Subheading 3.1.1) and can be found in **Note 16**.

6. The steps to apply a linear regression fit of data values in a comma- or tab-separated text file (∗.csv or ∗.tsv) obtained from gradual photobleaching experiments are presented in the "linear fit acceptor photobleaching.ijm" that is available for download as supplementary material for this book. It uses the curve fitting functions of the *Fiji/ImageJ* distribution to calculate the FRET efficiency at complete bleach of the acceptor. The first column should contain the intensity data from the acceptor channel, and the second column the partial FRET efficiency of each iteration (calculated against the pre-bleach image). To prepare such a file, take in *Fiji* the steps described below.
7. In case of nonzero background, execute a background subtraction on all images or image stacks before macro execution.
8. For the acceptor intensity reference: Collect all acceptor images into an image stack.
9. Select a signal-containing and photobleached area of the sample as a ROI and generate the acceptor values using the stack function *Plot z-axis profile.*
10. Open the values in a text window and copy them into the first column of an Excel file.
11. For the donor channel: Collect all donor images (pre-bleach image + all post-bleach images) into an image stack.
12. Subtract the donor pre-bleach image from that stack using the image calculator function. This creates the numerator part of Eq. (6).
13. Divide the result image stack by the pre-bleach image stack. This completes Eq. (6) by dividing the numerator by the denominator.

14. Apply the same ROI from **step 9** (using, e.g., the *ROI manager*) to the resulting stack and generate the apparent FRET efficiency values using the stack function *Plot z-axis profile.*
15. Open the values in a text window and copy them into the second column of an Excel file. Save the file as a .CSV file.
16. Execute the "linear fit acceptor photobleaching.ijm" macro.

3.1.3 Donor Photobleaching

For this method, image series of donor fluorescence are taken at illumination intensities that are sufficiently high to bleach the donor through the course of the series, ideally to the background level in the image. For experimental considerations, *see* **Notes 18–20**. For donor photobleaching experiments, reference measurements are required. Series of samples with donor only and with donor and acceptor present need to be taken for the analysis. Samples with only acceptor should be taken in addition as experimental control.

1. Take a series of donor images of the sample containing both donor and acceptor at an illumination intensity sufficient to induce gradual photobleaching. Continue acquisition until a background plateau is reached.
2. Repeat the experiment with a sample containing only donor fluorophore.
3. Determine the half-time (*see* **Note 21**) of the donor bleaching curve in the FRET sample and compare it to the sample containing only donor fluorophores (*see* Fig. 5b). In case of FRET interactions, bleaching will be slowed, and the bleaching curve halftime will be increased accordingly. An apparent FRET efficiency E_a can then be determined based on the donor-only halftime (τ_D) and the FRET sample halftime (τ_{DA}):

$$E_a = 1 - \frac{\tau_D}{\tau_{DA}} = E \times \alpha_D \tag{8}$$

   The apparent FRET efficiency is identical to the one that can be obtained with acceptor photobleaching and represents the product of the FRET efficiency E_a at the molecular level and the proportion of interacting donor (*see* Eq. 3). Considerations and pitfalls for the method can be found in **Note 22**.
4. The general steps for fitting an exponential function to donor photobleaching data using the curve fitting functions of the *Fiji/ImageJ* distribution and calculating the halftime τ are illustrated in the *ImageJ* macro "exponential fit donor photobleaching.ijm" that is available for download as supplementary material for this book. The data to apply to the macro can be created in *ImageJ* by following the steps below.
5. Create an image stack of the donor channel images.

6. Measure intensity values in a ROI of the bleaching series using the stack function *Plot z-axis profile*.
7. Save the result as a .CSV file.
8. Execute the macro "exponential fit donor photobleaching. ijm."
9. Creating a FRET efficiency image of a donor photobleaching experiment is not as easy as with other methods, as every pixel in the efficiency image would be the result of a curve fitting operation on the corresponding position of the image stack. This is outside of the capability of most basic processing software and the analysis is better done on the values derived from ROIs of specific image regions. A nonlinear representation (which is not directly corresponding to bleaching halftimes!) of pixel-based bleaching kinetics can be made by creating a sum projection (in ImageJ: [Stacks > Z Project.. > Sum) of all time points of the bleach series and dividing it by the first image of the bleach series (using *Image Calculator*). Higher values in the resulting image would correspond to higher lifetimes (as higher intensities would accumulate in the sum projection compared to faster bleaching regions), but the relation isn't linear and can't be easily converted to halftime values.

3.2 Sensitized Emission-Based Methods

3.2.1 Ratiometric FRET Imaging

Ratiometric FRET imaging in two channels requires wide-field or confocal instrumentation that can create the following channels (*see* **Note 23**): (1) donor emission under donor excitation and (2) acceptor emission under donor excitation.

1. Acquire a time series of the two image channels described above. For the measurement setup refer to **Note 24**.
2. In the processing step, FRET ratios are formed by the division of the acceptor channel (under donor excitation) by the donor channel (under donor excitation). For the processing take into account **Note 25**. The division normalizes the FRET signal to the local concentration of the probe which is effectively removed from the ratio. Changes in FRET are then reflected in the relative changes of the ratio over time or under different experimental conditions.
3. For analysis, the ratio values of the single time points are plotted over time. The ratio values are dependent on the acquisition settings. For an understanding of the values, **Note 26** is very important!

   Using biological controls for the underlying physiological process, maximum and minimum values of intracellular ratios can be obtained. In vitro ratiometric measurements of the reporter construct (under the same channel acquisition settings!) can be used to calibrate the detectable concentration range of a probe ligand (e.g., Ca^{2+}) and the affinity with which it is bound. The ratios can then be used to directly deduce the ligand concentration.

3.2.2 Experimental Procedure for Sensitized Emission FRET Imaging

Sensitized emission measurements require the acquisition of three channels and the acquisition of two calibration samples in addition to the FRET sample to provide the correction factors for subsequent analysis. The acquisition channels are (1) donor emission under donor excitation (F^{DD}); (2) acceptor emission under donor excitation (F^{DA}); and (3) acceptor emission under acceptor excitation (F^{AA}). The calibration samples that need to be acquired in addition to the FRET sample containing donor and acceptor are (*see* **Note 27**) (1) donor-only sample and (2) acceptor-only sample.

An additional important biological control is samples that contain the fluorophores in similar locations, but where no interaction should take place. This should be detectable as a clear negative; if not, the experimental settings and applied corrections will need to be checked. The acquisition can be performed on standard fluorescence systems and confocal systems in different configurations (*see* **Note 28**).

1. Acquire a multichannel image or a time series of the image channels described above using the FRET sample containing donor and acceptor. For the measurement setup refer to **Note 29**.
2. Take a multichannel image of the donor-only sample under the same imaging conditions.
3. Take a multichannel image of the acceptor-only sample under the same imaging conditions.
4. For the subsequent processing and analysis of the FRET sample data calculate the donor crosstalk correction factor R_D from the donor-only sample data (apply background correction as described in **Note 25**):

$$R_D = \frac{F^{DA}}{F^{DD}} \tag{9}$$

5. Calculate the direct acceptor excitation correction factor R_A from the acceptor-only sample data (apply background correction as described in **Note 25**):

$$R_A = \frac{F^{DA}}{F^{AA}} \tag{10}$$

6. R_D and R_A define the contributions of the donor and acceptor fluorescence to the F^{DA} channel. In case of FRET, this channel would additionally contain the sensitized emission signal of the acceptor. Removing these contributions from the channel

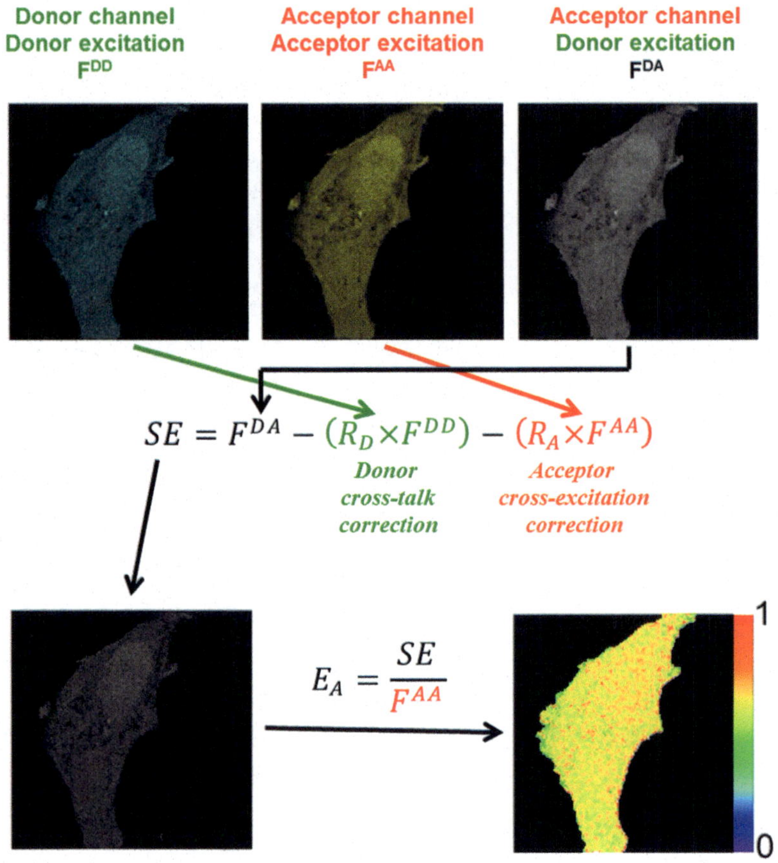

Fig. 7 Schematic of the acquisition and analysis of sensitized emission measurement data

allows to identify the sensitized emission contribution (SE) by executing the following calculation, using the three channels of the FRET sample measurement [9]:

$$\mathrm{SE} = F^{\mathrm{DA}} - \left(R_{\mathrm{D}} \times F^{\mathrm{DD}}\right) - \left(R_{\mathrm{A}} \times F^{\mathrm{AA}}\right) \tag{11}$$

By removing the donor fluorescence cross talk from the channel F^{DA} with the first correction ($R_{\mathrm{D}} \times F^{\mathrm{DD}}$) and the directly excited acceptor with the second ($R_{\mathrm{A}} \times F^{\mathrm{AA}}$), any remaining contribution in that channel would constitute the sensitized emission FRET signal SE (*see* Fig. 7).

7. The detection of the SE signal would be a qualitative proof of FRET interactions, but the value cannot be directly used for quantitative analysis as it is still affected by the concentrations of the interacting fluorophores. To obtain an apparent FRET efficiency, the SE value can be normalized to the acceptor concentration [10] (*see* Fig. 7):

$$E_a = \frac{F^{DA} - (R_D \times F^{DD}) - (R_A \times F^{AA})}{F^{AA}} = \frac{SE}{AA} = E \times \alpha_A \times C \quad (12)$$

$$\alpha_A = \frac{[DA]}{[DA] + [A]} \quad (13)$$

The apparent FRET efficiency E_a is the product of the FRET efficiency E at the molecular level, the amount of interacting acceptor α_A, and a measurement setting-dependent constant C. For the interpretation of the values and processing requirements, *see* **Notes 30–32**. Alternative analysis approaches for sensitized emission measurements are covered in **Notes 33** and **34**.

8. The general steps for extracting the sensitized emission signal from the FRET-containing channel F^{DA} and normalizing it to the acceptor concentration are presented in the *ImageJ* macro "sensitized emission_median.ijm" that is available for download as supplementary material for this book, along with example images. It basically executes the calculation described in **step** 7 (Eq. 12) on an image set consisting of the three channels required for sensitized emission through the steps below.
9. In case of nonzero background, execute a background subtraction on both images before macro execution.
10. The (separately measured) correction factors for donor cross talk and direct acceptor excitation are provided in input dialog windows.
11. The three input images are pre-filtered (median filtering).
12. The cross talk and direct excitation images are created based on the F^{DD} and F^{AA} reference channels and subtracted from the F^{DA} FRET channel image.
13. The resulting image of the sensitized emission is normalized to the acceptor channel by image division.
14. The resulting FRET efficiency image is displayed with a suitable false color scheme.

3.3 Displaying FRET Efficiency Images

All FRET detection methods allow the creation of images that show the local distribution of the chosen FRET index values (referred to as apparent FRET efficiency in the above sections) in the sample. FRET indices represent the strength of FRET interactions, not the accumulated amount of FRET signals in a specific location. Taking the subtraction of the pre- from the post-bleach donor image (in acceptor photobleaching) will highlight areas with unquenched donor signal, but areas of high signal could represent the accumulation of many weak interactions or represent few strong interactions. The apparent FRET efficiencies only become distinguishable after normalizing to the concentration of fluorophores in the location. FRET efficiency images therefore don't contain information on the sample intensities anymore and it becomes difficult to distinguish signal-bearing areas from low-intensity background in the sample. As both presence and absence of FRET are biologically meaningful, displaying the images with a linear color *LookUp-Table* (LUT) going from dark to bright values is not helpful, as it highlights only one part of the values while leaving the other part "in darkness." This can be overcome by creating false color images using multicolor LUTs that are representing the whole value range in bright colors (e.g., following the distribution of the visible light spectrum). Such a false color display helps in seeing all values well and is needed for a proper display of the FRET efficiency values. It makes it however even harder to distinguish relevant signals from background areas.

After assigning a suitable false color LUT to the FRET efficiency image, it is possible to reintroduce local concentration information by intensity weighing the image using one of the fluorophore channels of the FRET dataset. Like this, the color hue is preserved to represent FRET efficiency, but the color brightness reflects the local concentration.

1. Select the fluorescence channel (donor or acceptor) to be used for the intensity weighing and normalize it by dividing by the maximum value of its bit range (e.g., 255 for 8-bit images). The resulting image contains values between 0 and 1 and needs to contain floating point values.
2. Transform the FRET efficiency image with its false color LUT into a color image of RGB type, which has red, green, and blue intensity values for every image pixel. Such a color image does not contain the FRET efficiency value anymore, but its convolution with the color values of the false color LUT.
3. Split the RGB image into the three-color channels (red, green, and blue) and multiply each channel with the normalized intensity image created in **step 1**.
4. Merge the altered color channels into a new, intensity-weighed RGB image.
5. The *ImageJ* macro "intensity weighing.ijm" does this job for a FRET efficiency image created from sensitized emission data. The macro can be downloaded as a supplement to the book chapter. The images used are the acceptor channel image "FRET positive_sensitized emission_AA" and the apparent FRET efficiency image that was created with the macro "sensitized emission_median.ijm" for sensitized emission imaging. This approach is mainly useful for FRET methods where the

information is self-contained and would provide highly misleading images when applied to photobleaching methods that bleach only subregions of the image.

3.4 Comparison of FRET Data Across Acquisition Methods

Both described photobleaching methods and procedures for sensitized emission can provide linear apparent FRET efficiencies. With the proper correction [11], donor-normalized sensitized emission results can be directly compared with photobleaching results as the apparent FRET efficiency $E_{\text{app donor}}$ can for all of them be defined as

$$E_{\text{app donor}} = E \times \alpha_{\text{D}} \tag{14}$$

(*see also* Eq. 2)with α_{D} defined by Eq. (3). In Fig. 8, the gradual acceptor photobleaching experiment from Fig. 6 is extended for the sensitized emission channel and the results are identical. These values are independent from the measurement conditions and the instrument with which they were taken.

A second equally correct apparent FRET efficiency with a different value can be taken from the sample when normalizing to the acceptor channel. Both contain the molecular FRET efficiency E (defined by the distance and orientation of the fluorophores) but $E_{\text{app acceptor}} = E \times \alpha_{\text{A}} \times C$. C is a measurement-dependent factor representing the relative brightness of the FRET-containing channel to the acceptor detection channel. It can be defined as

$$C = \frac{\varepsilon_{\text{A}}}{\varepsilon_{\text{D}}} \times R_{\text{A}} \tag{15}$$

where R_{A} is defined by Eq. (10) and $\frac{\varepsilon_{\text{A}}}{\varepsilon_{\text{D}}}$ is the ratio of the acceptor and donor extinction coefficients at donor illumination. An instrument-independent $E'_{\text{app acceptor}}$ can then be defined as

$$E'_{\text{app acceptor}} = \frac{E_{\text{app acceptor}}}{C} \tag{16}$$

It is important to understand that $E_{\text{app donor}}$ and $E'_{\text{app acceptor}}$ are distinct parameters and will in most cases provide different values as the local concentrations of donor and acceptor are frequently unconnected. The values are not only not identical, but they can also change independent from one another during the measurement (*see* **Note 35**).

If $E_{\text{app donor}}$ and $E'_{\text{app acceptor}}$ can be expected to be different for intermolecular FRET experiments, they should be identical in intramolecular FRET measurements (ratiometric imaging). Here, a 1:1 stoichiometry could be expected with α_{D} and α_{A} at 100% and no unbound fluorophores present. While the stoichiometry for the ratiometric construct can normally be considered to be fixed inside an experiment, it will very often deviate from the expected 1:1 distribution. Ratiometric probes based on tandem constructs of fluorescent proteins are clearly defined by their genetic sequence

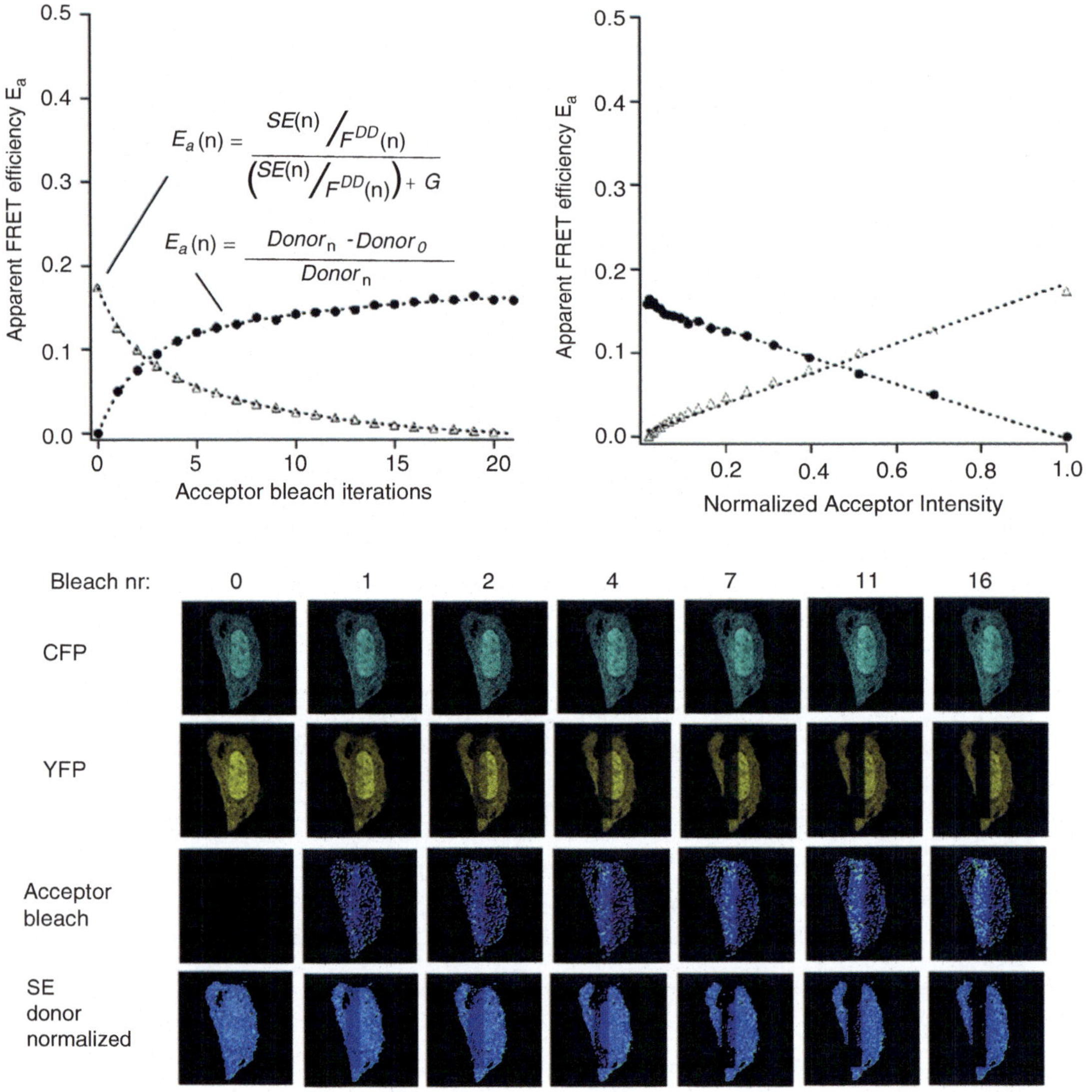

Fig. 8 Comparison of apparent FRET efficiencies calculated by gradual acceptor photobleaching and by applying donor-normalized sensitized emission with linear correction (formula according to [11]) using the experiment from Fig. 6. The FRET values are reciprocally matched as one method (sensitized emission) measures the presence and the other method the disappearance of the FRET effect. *n* represents the bleach iteration used for the calculation

as an unseparable couple, but the biology of fluorescent proteins is complex enough to allow for significant differences in folding efficiency and in reactions to the fluorophore environment. This means that a certain portion of constructs will only have one functional fluorophore (or even none). *See* **Note 36**.

4 Notes

1. Bleaching-based methods compare measurements before and after a targeted perturbation (photodestruction of a FRET partner) to provide an apparent FRET efficiency (*see* Eqs. 2 and 8) that is not affected by the imaging conditions and the instrument that were used. It would however be affected by molecule exchanges between bleached and unbleached areas as can be expected in living samples or dynamic in vitro systems. It is therefore most suited to fixed samples that have no turnover of molecules to provide an accurate measurement value. The bleaching step also alters the sample permanently as it changes the fluorophore concentrations in the sample. Measurements can't be repeated under the same conditions so that it is not suited for observations over time even if turnover in the sample during the bleaching measurement could be neglected.

   Sensitized emission-based methods do not directly interfere with the sample and do not require before/after comparisons that could be affected by dynamical processes. They are therefore suited to the study of living or dynamic samples. These methods require calibration measurements of samples containing only donor or acceptor for FRET analysis and the apparent FRET efficiencies they provide contain instrument- and acquisition-dependent scaling factors. Direct comparisons of experiments on different instruments are not possible unless these scaling factors are fully defined.

2. Donor de-excitation through FRET interactions can then be measured as a decrease in donor lifetime and an ingrowth of acceptor lifetime. This approach can provide the most complete quantitative information (FRET efficiency and interacting molecule fractions) by fitting the curves of the participating fluorophore states to the measured decay curve. Lifetime imaging can be considered as belonging to the second category of non-perturbing FRET measurements but will not be considered further in the scope of this chapter as lifetime measurements require specific instrumentation and the execution of the measurement and repeatability for time series in in vivo samples deviate significantly from the other methods in that category.

3. Spectral separability in itself is not fundamental factor of a FRET pair's efficiency. In homo-FRET the interacting fluorophores are identical and spectral separability is accordingly zero. It is however an important factor in its detectability. FRET pair selection favors not the most efficient FRET pairs, but those that perform well and at the same time provide clear separation for detection and bleaching. In the case of homo-FRET between identical fluorophores, overlap is not optimal as

a fluorophore normally only has limited overlap between the "red" edge of its excitation spectrum and the "blue" edge of its emission spectrum. Hetero-FRET interactions with the best overlap between donor emission and acceptor excitation have a high efficiency but significant detection limitations as there is channel cross talk and fluorophore cross-excitation so that in viable combinations efficiency is "sacrificed" for separability. Such limitations could largely be overcome by linear unmixing of the acquired channels, but in the case of photobleaching applications this is not possible as bleaching of one partner of the FRET pair would inevitably also affect the other due to large overlap of their excitation spectra. Suitable FRET pairs for photobleaching require either limited overlap or long Stokes shifts (spectral separation of excitation and emission spectra of a fluorophore).

4. As for photobleaching methods, the selection of a suitable fluorophore pair for sensitized emission measurements has to take into account both the efficiency of the FRET pair and the detectability of the FRET interaction. Whereas photobleaching methods mainly focus on reading out the donor, sensitized emission methods (including ratiometric constructs) require a clearly detectable acceptor that actually creates sensitized emission photons. In photobleaching methods, the absorption properties of the acceptor are important to allow efficient energy uptake by resonance energy transfer, but they do not require good acceptor fluorescence emission properties as represented by the quantum efficiency of the molecule. In case of acceptor photobleaching, the acceptor must be bleachable, and for donor photobleaching acceptor emission could be fully neglected and even a "dark" quencher would be sufficient.

   For sensitized emission methods the requirements are inverted. Acceptor fluorophores should have a high quantum efficiency to create a clearly detectable sensitized emission signal. Donor fluorescence emission is less important as the FRET efficiency of a FRET pair is not affected by how a molecule returns to the ground state from the excited state. For sensitized emission, the donor molecules have to act as light harvesters for the acceptor molecules. Whether the proportion of excited states that do not lead to resonance energy transfer return to the ground state by fluorescence photon emission or by collisional quenching and heat generation is irrelevant. Even donor fluorophores with a low quantum efficiency can therefore be used. This is evidenced in some of the ratiometric fluorescent protein constructs where a donor with low quantum efficiency like blue fluorescent protein "feeds" a much brighter acceptor (green or yellow fluorescent protein) and most of the ratiometric change is caused by the increase or decrease of the acceptor fluorescence which is disproportionately higher than that of the donor.

5. Confocal laser scanning microscopes are very suitable for localized photobleaching methods since powerful laser illumination is focused into a diffraction-limited spot as part of their operating principle. Together with rapid intensity regulation this allows to efficiently apply high illumination intensities in designated areas of the sample as is required both for FRAP and for FRET photobleaching experiments. In modern confocal microscopes, highly automated acceptor photobleaching workflows are either implemented as dedicated modules or easily implemented as part of a more general photobleaching module.
6. Acceptor photobleaching is normally performed in fixed samples that can be mounted on microscope slides like standard fluorescence samples. Many conventional mounting media can be used, but it is important that they don't have properties or have chemicals (antifade agents) added that impede photobleaching. This is normally desirable to keep samples stable for a long time during observation, but in the case of photobleaching methods it interferes with the experimental result, often in complex and unforeseeable manners.
7. Bleaching is complex and dependent on the chemical environment. The linearity of bleaching phenomena is still poorly understood. Using lower absolute intensities for bleaching, possibly in conjunction with more bleach repetitions, may alleviate the problem, as maybe more light than needed to efficiently bleach the acceptor is used and the acceptor can be bleached at lower intensities while not affecting the donor fluorophore.
8. Bleaching the acceptor in the absence of donor is important to rule out photoconversion of the acceptor into a photoproduct that is detectable in the donor channel. This would create false positives in the analysis. Photochemistry of fluorophores can be very complex and photoconversions can be observed with many fluorescent proteins [12, 13] and also other fluorophore classes. In these experimental controls, more important than proving the absolute absence of such phenomena it is important to understand whether they affect the experiment in the conditions used. Yellow fluorescent protein (YFP) can be photoconverted into a shorter wavelength photoproduct that becomes detectable in the spectral range used for cyan fluorescent protein detection [13]. The effects of this on the accuracy of YFP acceptor photobleaching experiments were controversially discussed in several scientific publications as some groups considered it problematic whereas others could not see it in

their measurements. The photoconversion clearly exists, but is much less bright than the original acceptor signal [14]. Groups working with samples of approximately equal donor-to-acceptor concentrations do not detect it in their measurements whereas results where the acceptor concentration is significantly higher than that of the donor may be affected.

9. In practice, separate channel settings for the two fluorophores should be easily achievable, as FRET pair fluorophores with significant overlap are not a good choice for photobleaching as explained in **Note 3**.
10. Although there is no absolute requirement for this, constraining the bleaching step to a subregion of the image is beneficial, as it creates internal control regions in the dataset that allow to compare areas that were affected or unaffected by the bleach. This allows to detect movement artifacts between pre- and post-bleach images, as well as observe bleaching during the acquisition of pre- and post-bleach images.
11. To minimize sample stability problems bleaching should be executed at high intensities with the smallest possible interval between pre- and post-bleach images. In case of limited laser power, selection of an image subregion for bleaching increases the bleach efficiency as, in addition to intensity upregulation of the laser line, the complete illumination of the image frame is concentrated into a smaller area.
12. As they are self-normalizing to values between 0 (no FRET interactions) and 1 (FRET efficiency of 100%), acceptor photobleaching results do not depend on the measurement conditions of the experiment and can be directly compared even when taken on different instruments and without any additional calibration data.
13. Done correctly (i.e., including the proper experimental controls), acceptor photobleaching experiments with a positive outcome provide a qualitative answer (FRET takes place) and a semiquantitative result (the apparent FRET efficiency) while not providing a direct readout of the FRET efficiency at the molecular level.
14. The calculations for acceptor photobleaching can be applied to regions of interest or on every pixel of the same image. As this includes a division of very similar values (in the pre- and post-bleach image), the intensity differences between image objects (information bearing) and background areas in the image are removed, and noise (especially on the low-intensity background) is pronounced. Moderate image filtering and thresholding (using a very low background threshold) can improve the image of apparent FRET efficiency.

15. As with all image divisions, it is essential to remove image background on both images before division to avoid intensity-dependent deviations of the efficiency values.
16. The sample-based controls for acceptor photobleaching experiments ensure that the fluorophore properties do not affect the outcome of the measurements. There are however additional factors that can create false results and that need to be taken into account. In image-based calculations, values at the specific pixel positions of the image are used in the calculation and allow to measure and compare FRET efficiencies in different regions of the sample as seen in the image. Pixel-based calculations work only correctly when the images used for the calculation are perfectly registered to each other. Even slight mismatches, caused, e.g., by sample drift between the pre- and post-bleach images, will cause artifacts in the image. In the given calculation the edges of structures would give false readouts as they are transitions between areas of high and low intensities and their presence or absence in the second image would lead to edge effects. A clear indication of shift-related errors in a FRET efficiency image would be the presence of strong values at the edges of the objects, both inside and outside the bleached image area. As any area outside of the bleached region should be the same before and after bleaching, any E_A values there will indicate problems with the experiment, as those areas serve as internal controls. The problems could be noise related, a technical problem with the imaging settings (images are not taken with the same settings or the light source or detector is unstable), or shift related. Shift-related errors mainly highlight structural edges and can thus be disambiguated from the other two possibilities. If the shift is lateral, the images can be easily registered to one another by cross-correlation and the edge artifacts will disappear. Due to the optical sectioning capability of confocal microscopes, even slight changes of focus can severely affect the pre- and post-bleach images. Such errors cannot be corrected at the level of the single images as the additional z-information is missing. In case of series of z-sections (3D stacks) mismatches can also be corrected in z using the same cross-correlation approach.

    To minimize translational mismatches between pre- and post-bleach images by image drift, the time for bleaching and image acquisition should be generally kept as low as possible. Using strong lasers and automatized (instead of manual) acquisition routines, this is quite possible.

    In certain cases, measured FRET efficiencies may vary (decrease) over time even in fixed (and positionally stable!) samples. The photochemistry of many fluorophores is sufficiently complex to contain reversible fluorophore dark states, in addition to the irreversibly bleached state normally associated with photobleaching. Such transient and reversible dark states play a role in fluorophore blinking at the single-molecule level, thus being fundamentally important for many localization-based super-resolution microscopy methods. Fluorophores will return from such transient dark states to an excitable state either according to the intrinsic lifetime of the dark state or by illumination with specific shorter wavelengths (photoactivation or back pumping). In case that a significant portion of interacting acceptor molecules populates a reversible dark state, their return to the ground state will reestablish their participation in FRET interactions and can therefore change the amount of detectable donor emission over time [15]. In such cases the time to readout after bleaching should be minimized and not be varied. Ideally, fluorophores with significant reversible dark-state populations should not be used as acceptors or only imaged in chemical conditions that do not favor the formation of reversible states.

The most fundamental consideration of any acceptor photobleaching experiment is to minimize the amount of illumination light needed to achieve an accurate result. Using too high intensities or too many scan repetitions to create the pre- and post-bleach images can lead to good-looking images with low noise levels but can induce observation bleaching that will affect the readout of the experiment and could mask the fluorescence increase in the donor channel. Using too much light for acceptor bleaching may affect the donor fluorophores in the area and could also mask any donor fluorescence increase caused by the destruction of FRET interactions. In case of evenly distributed signals in the sample, using "unnatural" angular bleach regions may be helpful, as fuzzy bleach borders that extend beyond the straight delineated bleach area in the acceptor channel may be an indication of too much light, as the Gaussian distribution of the scanning beam extended significantly into areas that should not have been affected. In the donor channel, an absence of FRET signals in the middle of a small bleach area surrounded by an unsharp halo of positive FRET signals can also indicate the same, i.e., too much intensity in the designated bleach area and better signals in areas that were reached only by outer regions of the Gaussian-shaped bleach beam.

17. Whereas single-step acceptor photobleaching cannot be efficiently implemented on normal wide-field fluorescence microscopes with additional strong bleaching sources, gradual acceptor photobleaching procedures are possible and have been described [16].

18. Similar to FRAP experiments, imaging the sample until a plateau is reached (in the case of FRAP the recovery endpoint, in the case of donor photobleaching the image background) will help in creating an accurate exponential curve fit that is needed for the analysis.
19. Images of the acceptor channel are not necessary for the analysis but are needed to check for bleaching effects on the acceptor fluorophore.
20. The following conditions should be established: (1) The donor channel has to be set up without saturation of the donor signal to allow subsequent fitting and quantitation. (2) The intensity decrease should be perceptible between images so that background intensity is reached in reasonable time, thus avoiding problems with image drift. (3) The background plateau should not be reached after a few frames already as this does not provide a good representation of the bleaching curves and may not allow to accurately detect differences in the signal decay. The intensity decrease should be stretched over the image series with a clear beginning and endpoint.
21. Bleach curves of fluorophores measured on the microscope follow an exponential decay kinetic that is the direct representation of the bleach rate of the fluorophore (here stated for the simplest case of a single bleach pathway and disregarding additional reversible dark states). They can thus be characterized by the half-time of their signal decay.
22. In addition to the potential FRET sample containing donor and acceptor, a sample containing only donor is required for the analysis as a reference value for no FRET interaction. It can also serve as a control for other photochemical artifacts like increases by photoactivation that would increase donor signal upon observation. Such behaviors may be rare but can be observed in fluorescent proteins like Kindling and also in other fluorophores. The acceptor-only control is necessary to check against direct photobleaching of the acceptor by the donor illumination. In case photobleaching can be detected under the chosen observation conditions, the experiment should not be executed with the chosen FRET pair because a mixture of donor and acceptor photobleaching renders the results unusable. In case more than one excitation wavelength is available to image the donor fluorophore, a shorter excitation wavelength can be tried to overcome the problem of direct acceptor bleaching.

    Even though donor photobleaching experiments are very easy to execute, this is the method with the most possible and encountered artifacts and therefore the least recommended approach. The acceptor fluorophore has to be significantly more photostable than the donor for this method to work.

As excitation spectra are generally very wide and tend to extend significantly into shorter wavelengths relative to the excitation maximum, acceptor fluorophores are often excited by the wavelengths that are used to efficiently image the donor fluorophore. In multichannel imaging experiments this problem of cross-excitation is overcome by imaging the shorter wavelength fluorophore through a suitably chosen bandpass filter that rejects emission cross talk, but in bleaching experiments the cross-excitability cannot be ignored. Therefore, donor photobleaching experiments work best for donors with long Stokes shifts. Even then, the photostability of the acceptor is essential, as in the case of FRET interactions the donor acts as a light harvester for the acceptor. So even when the acceptor is completely insensitive to the donor excitation light, it will be stimulated through FRET and can thus be induced to bleach.

When acceptor bleaching (direct or through FRET) happens, donor photobleaching experiments become extremely complex and unquantifiable [15]. In the case of no FRET interaction, the donor bleaching curve will not be affected by additional acceptor bleaching. In the case of FRET however, the reduced photobleaching caused by the energy transfer will be overlaid by the unquenching of donor fluorescence caused by the disappearance of acceptor which will at the same time reduce the bleach protection by resonance energy transfer. In cases of the acceptor bleaching more efficiently than the donor (supported by the donor bleach protection through FRET and by the donor acting as energy collector for the acceptor) the donor intensities in the sample can even transiently increase as the acceptors in the FRET pairs bleach away before decreasing with the same decay rate as a donor-only sample since the FRET interactions have been destroyed. Such curves are unusable for quantitation and can only act as a qualitative proof of interaction when compared to donor-only samples.

23. These channels can be created in many different configurations: (1) First, in wide-field systems with a single camera, dedicated fluorescence filter cubes with the required specificity can be alternated in the light path to take the two channels one after the other. This is relatively slow as all fluorescence components have to be moved as a whole and is not suitable for the observation of fast and transient processes. (2) Second, by placing the excitation, beam-splitting and emission components into separate devices (e.g., filter wheels), the acquisition can be made more efficient for single-camera systems. In that case, the excitation (donor fluorescence) and beam-splitting components (single dichroic for donor excitation or double

dichroic that will only be used for donor excitation) can remain in place and only the emission filter is alternated for donor and acceptor fluorescence detection. This way, less components are moved, and significantly faster processes can be observed with several double-frame pairs per second. (3) Third, as the excitation light is the same for both channels, they can be acquired simultaneously. This can be achieved by introducing an image splitter into the emission light path that separates the donor and acceptor emission and projects them on distinct areas of the camera to create a spectrally split image side by side on a single-camera chip. Instead of a beam splitter and a single camera, a more dedicated setup can contain an emission dichroic that splits the signal onto two cameras, one for each fluorescence channel. Ratiometric acquisitions are then only limited by the frame rate of the camera. (4) Last, confocal microscopes are intrinsically capable of simultaneous multichannel imaging as they contain several detectors for different emission ranges. These can be configured so that the donor and acceptor fluorescence channels are read out at the same time under donor excitation. Ratiometric acquisitions are thus only limited by the frame acquisition speed which can be adapted over a wide range of speeds on confocal microscopes.

24. The following conditions need to be established: (1) First, the channels should be set up in a manner that prevents signal saturation taking into account that FRET changes will affect the signal intensity in the images. (2) As the same excitation is used, the channels should be taken in parallel, when possible, or sequentially (one after the other) with minimal time delay to avoid movement artifacts in the subsequent image calculation. (3) In the case of acquisition of time series of transient FRET changes, the frame rate has to be matched to the required time resolution for the events. (4) Last, as in all in vivo imaging experiments, light exposure to the sample must be minimized to avoid bleaching and photodamage during observation.

25. To create an accurate and concentration-independent ratio, it is important to remove any image background in the channels before the image division. Lack of or incomplete background correction will lead to different ratios for different concentrations, as the sums of a fixed (background) and a variable component (signal) in the channels will give, upon image division, different values for the same intensity-based signal distribution, thus creating intensity-dependent artifacts. Insufficiently corrected ratio images often have graded ratio distributions from the center to the periphery of cells, giving these a "fried egg" impression where normally a completely even distribution throughout the whole cell would be expected.

26. The absolute value of the ratio is defined by the spectral range and the intensity settings of the detection channels and cannot be directly compared between different instruments and image settings. As the values of the image ratio depend on the chosen acquisition settings, a single FRET ratio observation provides little to no qualitative information on FRET presence and no quantitative information on FRET efficiency. The ratio values obtain relevance according to the observed variations inside the time course of a measurement which reveal the relative strength and the kinetics of FRET changes.
27. These calibration samples should ideally be the same donor and acceptor constructs used in the FRET sample and be provided in concentrations comparable to the ones in the sensitized emission measurement so that they are imaged in the same intensity ranges as the real sample.
28. These channels can be created in the following manners: (1) First, using specific filter cubes that match the requirements for the F^{DD}, F^{DA}, and F^{AA} channels: As mentioned for ratiometric imaging, acquiring the channels like this is relatively slow. (2) Alternatively, using separately controllable components (e.g., filter wheels) and a double-dichroic beam splitter, the emission filters should be first switched under donor excitation (creating channels F^{DD} and F^{DA}), and then the excitation filters (creating the additional channel F^{AA}). This allows efficient and fast imaging of sensitized emission. (3) Another solution would be based on image splitter and double camera, as described for ratiometric imaging. Sequential imaging of donor excitation (creating channels F^{DD} and F^{DA}) and acceptor excitation (for channel F^{AA}) is required. (4) Last, on confocal microscopes, channels F^{DD} and F^{DA} can be acquired simultaneously. Excitation then is switched for channel F^{AA}.
29. The following conditions need to be established: (1) First, saturation of the acquisition channels needs to be avoided to allow subsequent FRET analysis. (2) Then, as the subsequent analysis includes image subtraction and division steps, movement artifacts between the channels need to be avoided. This is easiest achieved in confocal microscopes where $F^{DD} + F^{DA}$ and F^{AA} channel acquisition can be alternated for every single scan line. Differences between the channels are thus reduced to the millisecond range. (3) Additionally, in case of dynamic processes, the intervals between time points have to be matched to the speed of the process under observation. (4) Last, as in all in vivo imaging experiments, light exposure to the sample must be minimized to avoid observation bleaching and photodamage during observation.

30. The values for both E and α_A can range from 0 to 1, but C is not limited to that range and E_a can therefore be >1 without implying a molecular FRET efficiency above 100%. C depends on how the fluorescence signals are represented in the channels F^{DA} and F^{AA}. If the sensitized emission signal in F^{DA} is displayed with significantly higher values than the acceptor channel reference F^{AA}, C is increased. If F^{AA} is represented higher, C is decreased. A C of 1 implies that the channel sensitivity for the sensitized emission signal is equal to the sensitivity for directly imaging the acceptor.
31. To provide accurate results, the images must be background corrected before creating the correction values (Eqs. 9 and 10) and the FRET calculations (Eq. 12). Incomplete background corrections would result in intensity-dependent artifacts as already described for ratiometric FRET imaging.
32. As two different excitation settings are used to image the three channels, the donor and acceptor illumination fields may not be fully identical, especially toward the periphery of the images. This would cause different ratios for the same FRET value in different regions of the images when the sensitized emission derived from the donor excitation-based F^{DA} channel is divided by the acceptor illumination-based F^{AA} channel (Eq. 12). In cases of uneven illumination fields, a flat-field correction has to be applied to the image channels before the calculation. The necessity for such a correction can be tested by imaging structures of the acceptor-only sample in different regions of the image and calculating the acceptor excitation correction R_A (Eq. 10). If different values are created, a flat-field correction based on imaging evenly distributed fluorescence needs to be performed.
33. The described procedure for sensitized emission analysis is the simplest possible approach. It is helpful for an initial understanding of the method and the kind of calculations that need to be performed. It corrects for the expected major contributions of donor cross talk and direct acceptor excitation and should work for samples that are spectrally well separated and stretched over a limited range of intensities. Other, more elaborate methods have been described that (1) take into account the additional crosstalk factors of acceptor fluorescence in the donor channel and direct excitation of the donor by the acceptor [17–19] which allows accurate FRET detections for low efficiencies and (2) calibrate the channel-specific instrument response over the total range of concentrations (intensities) to allow accurate detection of a wide range of efficiencies at different fluorophore concentrations [20].

Normalization of the sensitized emission signal to the label concentration can be done in different forms and with different FRET indices as outcomes (overviewed in [11]). Normalization to the donor channel [11, 20–22], the acceptor channel [10, 19, 23], or a product of donor and acceptor signals [17, 24] will give different values and differ in linearity. Acceptor-normalized methods represent FRET interactions relative to the amount of bound acceptor (Eq. 13) and give linear values as the directly measured acceptor channel is not affected by FRET effects. Donor channel-normalized methods represent FRET interactions relative to the amount of bound donor (Eq. 3). They should provide results similar to those obtained with acceptor or donor photobleaching methods, but the sensitized emission results are affected by the quenching effect of FRET on the donor fluorescence signal used for normalization which renders them nonlinear [21, 22], unless corrected [11, 20, 25, 26]. Normalization to a product of the donor and acceptor signals [17, 24] does not directly follow the interaction with either donor or acceptor populations.

The apparent FRET efficiency values are directly affected by the presence of pools of unlabeled donor and acceptor molecules that interact with their labeled counterparts. Even if the initial pool of unlabeled molecules is negligible, photobleaching will inevitably create pools of undetectable molecules that have the same effect. Most sensitized emission measurement values will therefore be affected by progressive photobleaching during the measurement unless this is specifically corrected for [11].

34. As mentioned before, the spectral separability of a FRET fluorophore pair is not directly linked to its FRET efficiency but is important for the detectability of FRET interactions. Because of the width of most fluorophore spectra, the FRET pairs with best overlap between donor emission and acceptor excitation and therefore the highest FRET efficiency will almost inevitably suffer from significant cross talk between donor and acceptor emission and significant cross-excitation. Cross-excitation can be reduced by the selection of a donor with a long Stokes shift between excitation and emission, but it was shown that high FRET efficiency will not provide a good ratiometric construct if the emissions are not well separated [27]. CFP/YFP tandem constructs will therefore perform better with a lower FRET efficiency than a combination of Sapphire (a UV-excitable GFP) with YFP that has excellent spectral overlap, but poor separability.

The described sensitized emission methods clean up the channel that contains the additional FRET signal by removing emission cross talk by a straightforward subtraction of the donor contribution. Some methods also take into account (minor) acceptor contribution to the donor channel [17–19], but subtraction-based methods are not performing well in the case of significant overlaps as the remaining "pure" signal fractions are small in comparison to the crosstalk contributions and a lot of the total signal is rejected in the subtraction steps. Linear unmixing methods [28] are an elegant solution to separate strongly overlapping signals into their single components and reassign the total signal into the unmixed channels instead of rejecting large parts of it. Linear unmixing can be applied in a variety of manners, distributing the signal over multiple narrow channels [28] or using fewer but wider channels [29]. Linear unmixing can be directly applied to ratiometric FRET imaging and will provide improved ratios as the baselines created by emission cross talk are removed. As a consequence, established FRET pairs will display greater ratiometric changes and hitherto unused highly overlapping combinations with high FRET efficiency become available [30, 31]. A range of spectral approaches for FRET imaging have been described [32–34], also [35]. Spectral methods perform comparable to other methods [36] and have the advantage of allowing the use of more efficient FRET pairs, removal of autofluorescent background, and collection of the full emission spectra [37]. The methods may however be affected by lower sensitivity of the spectral detector [37] and unfavorable error propagation during the unmixing calculations in case of noisy data [34].

35. This can be demonstrated by considering a gradual acceptor photobleaching experiment like the one in Figs. 6 and 8. Whereas $E_{\mathrm{app\ donor}}$ disappears proportional to the unquenching of the donor fluorescence, $E'_{\mathrm{app\ acceptor}}$ remains unchanged while the signal disappears (smaller changes occur due to limited sensitivity of the detection channel).

    Acceptor photobleaching will decrease the pools of bound and unbound acceptor with the same likeliness. For an acceptor signal decrease of x (defined as current acceptor value divided by initial value) the calculation would be as follows:

    For the donor channel:

$$\alpha'_{\mathrm{D}} = \frac{(1-x)\times[\mathrm{DA}]}{((1-x)\times[\mathrm{DA}])+[\mathrm{D}]} \tag{17}$$

    For the acceptor channel:

$$\alpha'_{\mathrm{A}} = \frac{(1-x)\times[\mathrm{DA}]}{(1-x)\times([\mathrm{DA}]+[\mathrm{A}])} = \frac{[\mathrm{DA}]}{[\mathrm{DA}]+[A]} = \alpha_{\mathrm{A}} \tag{18}$$

So α'_D will change according to x whereas α'_A will remain equal to α_A and therefore be independent of x.

36. In the case of the commonly used combination of cyan (CFP) and yellow fluorescent proteins (YFP) for ratiometric imaging, unenhanced versions of YFP are known to not fold efficiently into their fluorescent configuration. The CFP donor can therefore exist uncoupled from a working acceptor fluorophore. Doing quantitative sensitized emission measurements on such constructs reveals different occupation ratios α_X for donor and acceptor, with those of the acceptor higher, as they are more often coupled to a working donor than the other way around. Using enhanced variants of the initial protein version helps minimize such effects. A common combination of enhanced fluorescent proteins of the CFP/YFP categories is Cerulean (a brighter and more stable CFP variant) and Venus (YFP with improved folding capabilities at the temperatures required for mammalian cell cultures).

Incomplete acceptor folding would add a donor baseline (and vice versa in the case of the donor) and therefore decrease the dynamic range of the ratio readout. Different maturation rates of donor and acceptor could lead to different ratios over time, especially in experiments that cover days [38].

Fluorescent proteins react differently in different environments and for YFP-like proteins pH and chloride sensitivity [39] and signal loss during PFA fixation [25] can significantly change the measurable FRET signal and cause misinterpretations.

Performing full sensitized emission measurements instead of ratiometric imaging allows to assess such effects and should be considered as additional controls to avoid the misinterpretation of ratiometric FRET measurements alone.

In case of protein relocations from the cytosol (three-dimensional environment) to a membrane (two-dimensional environment) intramolecular FRET can be overlaid with intermolecular FRET which would affect the measured FRET values. Also, in FRET between separate molecules, a FRET partner may have more than one interaction partner.

Acknowledgments

The fitting macros “linear fit acceptor photobleaching.ijm” and “exponential fit donor photobleaching.ijm” were kindly provided by Raul Gomez, Centre for Genomic Regulation, Barcelona.

References

1. Förster T (1948) Zwischenmolekulare energiewanderung und fluoreszenz. Ann Phys 437 (1–2):55–75
2. Bastiaens P, Majoul IV, Verveer PJ, Söling H-D, Jovin TM (1996) Imaging the intracellular trafficking and state of the AB5 quaternary structure of cholera toxin. EMBO J 15 (16):4246–4253
3. Karpova T, Baumann C, He L, Wu X, Grammer A, Lipsky P, Hager G, McNally J (2003) Fluorescence resonance energy transfer from cyan to yellow fluorescent protein detected by acceptor photobleaching using confocal microscopy and a single laser. J Microsc 209(1):56–70
4. Wouters FS, Bastiaens PI, Wirtz KW, Jovin TM (1998) FRET microscopy demonstrates molecular association of non-specific lipid transfer protein (nsL-TP) with fatty acid oxidation enzymes in peroxisomes. EMBO J 17 (24):7179–7189
5. Jovin TM, Arndt-Jovin DJ (1989) FRET microscopy: digital imaging of fluorescence resonance energy transfer. Application in cell biology. In: Kohen E, Ploem JS, Hirschberg JG (eds) Cell structure and function by microspectrofluorometry. Academic Press, Orlando, pp 99–117
6. Miyawaki A, Llopis J, Heim R, McCaffery JM, Adams JA, Ikura M, Tsien RY (1997) Fluorescent indicators for Ca 2+ based on green fluorescent proteins and calmodulin. Nature 388 (6645):882
7. Schindelin J, Arganda-Carreras I, Frise E, Kaynig V, Longair M, Pietzsch T, Preibisch S, Rueden C, Saalfeld S, Schmid B, Tinevez JY, White DJ, Hartenstein V, Eliceiri K, Tomancak P, Cardona A (2012) Fiji: an open-source platform for biological-image analysis. Nat Methods 9(7):676–682. https://doi.org/10.1038/nmeth.2019
8. Amiri H, Schultz G, Schaefer M (2003) FRET-based analysis of TRPC subunit stoichiometry. Cell Calcium 33(5–6):463–470
9. Youvan DC, Silva CM, Bylina EJ, Coleman WJ, Dilworth MR, Yang MM (2003) Calibration of fluorescence resonance energy transfer in microscopy using genetically engineered GFP derivatives on nickel chelating beads. Biotechnology 3:1–18
10. Wouters FS, Verveer PJ, Bastiaens PI (2001) Imaging biochemistry inside cells. Trends Cell Biol 11(5):203–211
11. Zal T, Gascoigne NR (2004) Photobleaching-corrected FRET efficiency imaging of live cells. Biophys J 86(6):3923–3939
12. Kremers G-J, Hazelwood KL, Murphy CS, Davidson MW, Piston DW (2009) Photoconversion in orange and red fluorescent proteins. Nat Methods 6(5):355
13. Valentin G, Verheggen C, Piolot T, Neel H, Coppey-Moisan M, Bertrand E (2005) Photoconversion of YFP into a CFP-like species during acceptor photobleaching FRET experiments. Nat Methods 2(11):801
14. Seitz A, Terjung S, Zimmermann T, Pepperkok R (2012) Quantifying the influence of yellow fluorescent protein photoconversion on acceptor photobleaching-based fluorescence resonance energy transfer measurements. J Biomed Opt 17(1):011010
15. Sinnecker D, Voigt P, Hellwig N, Schaefer M (2005) Reversible photobleaching of enhanced green fluorescent proteins. Biochemistry 44 (18):7085–7094
16. Van Munster E, Kremers G, Adjobo-Hermans-M, Gadella TW (2005) Fluorescence resonance energy transfer (FRET) measurement by gradual acceptor photobleaching. J Microsc 218 (3):253–262
17. Gordon GW, Berry G, Liang XH, Levine B, Herman B (1998) Quantitative fluorescence resonance energy transfer measurements using fluorescence microscopy. Biophys J 74 (5):2702–2713
18. Nagy P, Vámosi G, Bodnár A, Lockett SJ, Szöllősi J (1998) Intensity-based energy transfer measurements in digital imaging microscopy. Eur Biophys J 27(4):377–389
19. van Rheenen J, Langeslag M, Jalink K (2004) Correcting confocal acquisition to optimize imaging of fluorescence resonance energy transfer by sensitized emission. Biophys J 86 (4):2517–2529
20. Elangovan M, Wallrabe H, Chen Y, Day RN, Barroso M, Periasamy A (2003) Characterization of one-and two-photon excitation fluorescence resonance energy transfer microscopy. Methods 29(1):58–73
21. Vanderklish PW, Krushel LA, Holst BH, Gally JA, Crossin KL, Edelman GM (2000) Marking synaptic activity in dendritic spines with a calpain substrate exhibiting fluorescence resonance energy transfer. Proc Natl Acad Sci 97 (5):2253–2258

22. Żal T, Żal MA, Gascoigne NR (2002) Inhibition of T cell receptor-coreceptor interactions by antagonist ligands visualized by live FRET imaging of the T-hybridoma immunological synapse. Immunity 16(4):521–534
23. Erickson MG, Alseikhan BA, Peterson BZ, Yue DT (2001) Preassociation of calmodulin with voltage-gated Ca2+ channels revealed by FRET in single living cells. Neuron 31(6):973–985
24. Xia Z, Liu Y (2001) Reliable and global measurement of fluorescence resonance energy transfer using fluorescence microscopes. Biophys J 81(4):2395–2402
25. Chen H, Puhl HL 3rd, Koushik SV, Vogel SS, Ikeda SR (2006) Measurement of FRET efficiency and ratio of donor to acceptor concentration in living cells. Biophys J 91(5): L39–L41
26. Hoppe A, Christensen K, Swanson JA (2002) Fluorescence resonance energy transfer-based stoichiometry in living cells. Biophys J 83 (6):3652–3664
27. Heim R (1999) Green fluorescent protein forms for energy transfer. Methods Enzymol 302:408–423
28. Dickinson M, Bearman G, Tille S, Lansford R, Fraser S (2001) Multi-spectral imaging and linear unmixing add a whole new dimension to laser scanning fluorescence microscopy. BioTechniques 31(6):1272–1279
29. Zimmermann T, Rietdorf J, Pepperkok R (2003) Spectral imaging and its applications in live cell microscopy. FEBS Lett 546 (1):87–92
30. Schleifenbaum A, Stier G, Gasch A, Sattler M, Schultz C (2004) Genetically encoded FRET probe for PKC activity based on pleckstrin. J Am Chem Soc 126(38):11786–11787
31. Zimmermann T, Rietdorf J, Girod A, Georget V, Pepperkok R (2002) Spectral imaging and linear un-mixing enables improved FRET efficiency with a novel GFP2–YFP FRET pair. FEBS Lett 531(2):245–249
32. Chen Y, Mauldin JP, Day RN, Periasamy A (2007) Characterization of spectral FRET imaging microscopy for monitoring nuclear protein interactions. J Microsc 228 (2):139–152
33. Wlodarczyk J, Woehler A, Kobe F, Ponimaskin E, Zeug A, Neher E (2008) Analysis of FRET signals in the presence of free donors and acceptors. Biophys J 94 (3):986–1000
34. Woehler A, Wlodarczyk J, Neher E (2010) Signal/noise analysis of FRET-based sensors. Biophys J 99(7):2344–2354
35. Megias D, Marrero R, Martinez Del Peso B, Garcia MA, Bravo-Cordero JJ, Garcia-Grande-A, Santos A, Montoya MC (2009) Novel lambda FRET spectral confocal microscopy imaging method. Microsc Res Tech 72 (1):1–11. https://doi.org/10.1002/jemt.20633
36. Domingo B, Sabariegos R, Picazo F, Llopis J (2007) Imaging FRET standards by steady-state fluorescence and lifetime methods. Microsc Res Tech 70(12):1010–1021
37. Pietraszewska-Bogiel A, Gadella T (2011) FRET microscopy: from principle to routine technology in cell biology. J Microsc 241 (2):111–118
38. Nagai T, Ibata K, Park ES, Kubota M, Mikoshiba K, Miyawaki A (2002) A variant of yellow fluorescent protein with fast and efficient maturation for cell-biological applications. Nat Biotechnol 20(1):87
39. Griesbeck O, Baird GS, Campbell RE, Zacharias DA, Tsien RY (2001) Reducing the environmental sensitivity of yellow fluorescent protein mechanism and applications. J Biol Chem 276(31):29188–29194

Chapter 13

In Vivo Quantification of Intramolecular FRET Using RacFRET Biosensors

Manel Bosch and Elena Kardash

Abstract

Genetically encoded FRET biosensors are powerful tools to visualize protein activity and signaling events in vivo. Compared with a biochemical approach, FRET biosensors allow a noninvasive spatial-temporal detection of signaling processes in live cells and animal tissues. While the concept of this technique is relatively simple, the experimental procedure is complicated and consists of several steps: (1) biosensor optimization; (2) data acquisition; and (3) image processing with each step posing its own challenge. In this chapter, we discuss steps (2) and (3) with the emphasis on the intramolecular RacFRET biosensor. We describe the design principle of the biosensor, the experimental imaging setup for acquiring FRET data in zebrafish embryos expressing the RacFRET biosensor, and the step-by-step ratio image generation protocol using *Fiji* software. We discuss important considerations during FRET data acquisition and analysis. Finally, we provide a macro code for the automated ratio image generation.

Key words FRET biosensors, Sensitized emission, FRET ratio imaging, Zebrafish, Rac, Rho GTPases, Fiji/ImageJ, Macro programming

1 Introduction

Förster resonance energy transfer (FRET) is an electrodynamic phenomenon of a non-radiative energy transfer that occurs between two fluorophores called donor and acceptor. FRET requires a spectral overlap between donor emission and acceptor absorption and occurs at a distance between donor and acceptor ranging from 10 to 100 Å. Additional conditions may apply such as certain relative orientations between the emission dipole of the donor and the absorption dipole of the acceptor but these are usually negligible when biological samples are concerned [1]. During FRET, the donor in the excited state transfers the energy to the acceptor in the ground state. As a result, the emission from the

Electronic supplementary material: The online version of this chapter (https://doi.org/10.1007/978-1-4939-9686-5_13) contains supplementary material, which is available to authorized users.

Elena Rebollo and Manel Bosch (eds.), *Computer Optimized Microscopy: Methods and Protocols*, Methods in Molecular Biology, vol. 2040, https://doi.org/10.1007/978-1-4939-9686-5_13,

donor molecule is reduced in the process called quenching while the emission from the acceptor is increased. These changes can be detected in a fluorescence microscope. The relative emission levels from donor and acceptor molecules upon donor excitation reflect FRET efficiency, which in turn can be used to measure the distances between molecules [1, 2].

One of the most common biological applications of FRET is to measure distances between molecules [3, 4]. FRET sensitivity allows to detect minute changes in macromolecules directly in living tissues otherwise not accessible by conventional imaging techniques such as those involving regular fusion proteins. This can be very useful when studying protein-protein interactions and protein conformational changes in living cells. Intramolecular FRET-based biosensors were specifically designed for investigating signaling pathways in real time. Since the creation of calmodulin—the first FRET biosensor for measuring calcium signaling in living cells [5]—multiple biosensors were designed aimed at deciphering a range of signaling pathways and protein activities in tissues and cells. Today we have access to numerous biosensor modules, among which are those to study kinases, Rho GTPases, neuronal activity, PH levels, and calcium signaling—just to name a few [4, 6–9]. In this chapter, we focus on the intramolecular FRET biosensors for measuring the activity of Rho GTPases [8, 10]; specifically we consider the RacFRET biosensor [11, 12].

Rac protein belongs to a family of Rho GTPases, which is a family of small proteins (about 21 kDa) that control multiple cellular processes ranging from cell migration and differentiation to endocytosis and cell division [10]. Rho GTPases exist in two main states: the active, GTP bound, and the inactive, GDP bound. In the active state, Rho GTPases form complexes with their effector proteins to induce signaling events controlling cellular physiology and behavior. Rac protein is one of the prominent Rho GTPases controlling actin polymerization during cell shape changes such as those involving cell migration [10]. There are several isoforms of Rac proteins, which may only differ in a few amino acids yet perform diverse functions. A number of FRET-based sensors were developed to study the activity of Rho GTPases in living cells [8, 11, 13, 14]. In the past decade, several modifications were introduced to Rac and RhoA biosensors to improve their sensitivity, accuracy, and dynamic range [12, 15–17]. It is important to remember that intramolecular FRET biosensors for Rho GTPases reflect the balance between GEF and GAP activities that would activate the respective endogenous Rho GTPase. These biosensors, while being unique to the specific Rho GTPase, might not differentiate between the closely related isoforms of the specific protein.

A typical FRET biosensor for Rho GTPase is a single-chain macromolecule consisting of two fluorescent proteins serving as donor and acceptor during the energy transfer linked together with

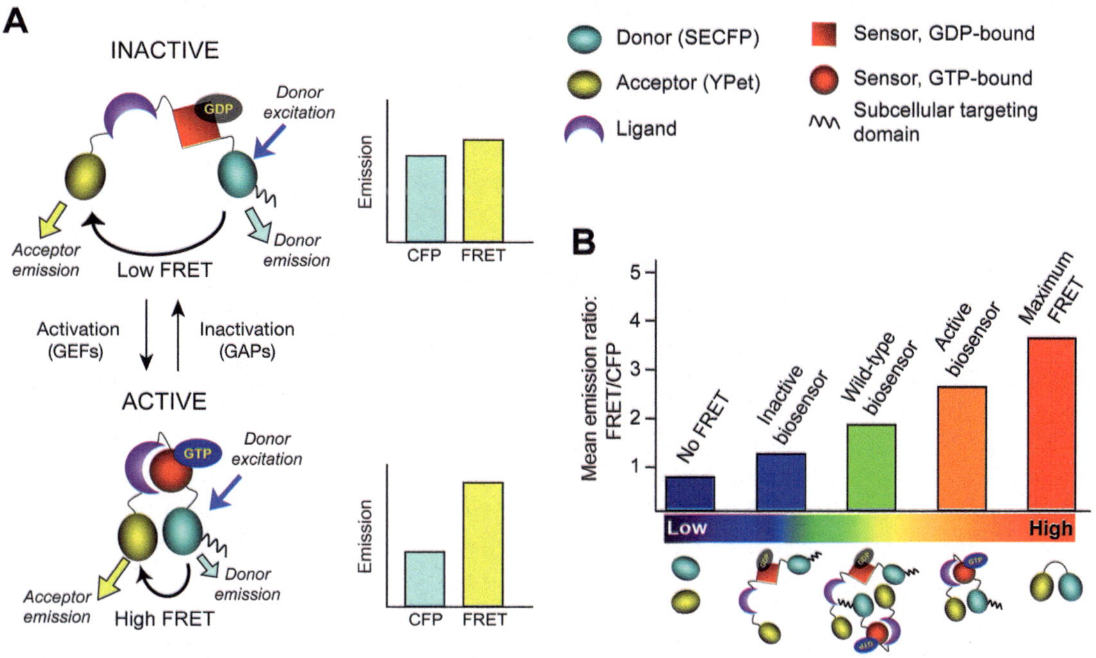

Fig. 1 The principal design of an intramolecular FRET biosensor for Rho GTPases. (**a**) The scheme shows a molecular chain consisting of two fluorophores, the YPet and the CFP fused to the sensor and ligand domains by flexible linkers. Sensor domain binds GDP or GTP moiety, which is controlled by the GAP and GEF groups of proteins, respectively. In the GTP-bound form, the sensor interacts with the ligand. The graphs show relative emission values for the inactive and the active biosensor states detected in the CFP and FRET channels when the CFP excitation wavelength is used. (**b**) The graph shows a schematic representation of the relative ratio values for various controls during biosensor optimization. The actual values would depend on the specific biosensor and the microscopy setup. From left to right: the "no FRET" situation when CFP and YPet proteins are expressed in equal amounts; the inactive biosensor with its sensor domain permanently bound to GDP (insensitive to GEFs); the wild-type biosensor existing in two different states: active and inactive; the active biosensor with its sensor domain permanently bound to GTP (insensitive to GAPs) and therefore bound to the ligand domain; the permanent FRET between CFP and YPet linked with a short linker

two additional protein domains: a ligand and a sensor that can interact when the sensor is activated (Fig. 1a). In the inactive state of a biosensor, its sensor domain is bound to the GDP and there is no interaction between the ligand and the sensor domains, which keeps two fluorophores apart resulting in low FRET values. Upon activation caused by the exchange of GDP with GTP, the GTP-loaded sensor binds to the ligand, which results in a conformational change within the biosensor bringing donor and acceptor closer to one another that causes an increase in FRET signal (Fig. 1a) [4, 11]. The most commonly used donor and acceptor pair in a FRET biosensor is cyan (CFP) and yellow (YFP) variants of the green fluorescent protein; however other fluorophore combinations are also possible [4, 18]. The position of the fluorophores with respect to the ligand and sensor domains can vary depending on the specific biosensor design [15, 16].

What would be the best combination for CFP and YFP variants in a biosensor? One of the most important properties of intramolecular FRET biosensors is their dynamic range, which is defined as the difference between the active and inactive states of the biosensor, which in turn relies on the effectiveness of energy transfer between the donor and acceptor. Dynamic range is directly related to the sensitivity of the biosensor. Several molecular modifications have improved significantly the yellow and cyan fluorophores in terms of their brightness, stability, and the capacity to participate in energy transfer [19, 20]. We have tried different combinations of CFP and YFP variants in our FRET experiments using Rac biosensor backbone to find the pair exhibiting the best dynamic range. While the combination YPet-CyPet displayed the strongest FRET efficiency [12, 20], we do not advise using this pair in the intramolecular FRET biosensor because of a strong reduction in signal intensity in the CFP channel during the energy transfer, which makes it difficult to obtain sufficient signal level for the ratio imaging. When signal intensity is below certain level, artifacts may be caused during ratio image generation. In our hands, the best results are obtained with SECFP-YPet combination [12, 17] while other research groups use CyPet-YPet combination [20, 21].

There are several important aspects to consider when using FRET biosensor in living cells and tissues. A FRET biosensor must meet the following criteria: (1) neutrality: a good FRET biosensor should not interfere with the normal cell physiology and endogenous signaling pathways; (2) availability: the biosensor must be accessible to the activation/deactivation machinery of the cell the same way as the endogenous protein is; (3) localization: the biosensor must localize within the cell similarly to the endogenous protein; (4) dynamic range: it is important to detect significant difference between inactive and active states of the biosensor while the activity measured with the wild-type biosensor should fall between the values obtained when using the inactive and active states (Fig. 1b, *see* **Notes 1–5**).

There are multiple strategies to measure FRET efficiency that include acceptor photobleaching, sensitized emission, fluorescence lifetime measurement, and other techniques [1, 22, 23]. Here, we focus on the sensitized emission as one of the most common methods used to measure FRET when imaging live samples in a fluorescence microscope. When measuring sensitized FRET emission, the sample expressing FRET biosensor is excited at the CFP excitation wavelength and two emission channels are acquired: the CFP emission and the FRET emission. In the FRET channel, the emission measured is composed of three components: the CFP emission that bleeds through the YFP channel, the YFP emission due to a non-direct excitation, and the YFP emission caused by the energy transfer. In case of intermolecular FRET, additional controls are required to estimate the nonspecific signal from both the donor and the acceptor in the FRET channel [24]. For intramolecular FRET biosensor, the stoichiometric ratio between the CFP and YFP fluorophores is 1; therefore the only controls needed are those for the active and the inactive states of the biosensor to define the range between the low and the high activation levels [12, 16] (Fig. 1b, *see* **Note 6**).

The main challenge when using FRET biosensors in live cells and particularly in live animals is to achieve sufficient level of signal-to-noise (S/N) ratio in the images. The expression level of a biosensor in live embryos might be low due to the toxicity or insufficient time for fluorophore synthesis and folding, which is the case when studying early developmental stages. To overcome this, several options are available. When using a camera to acquire images, use higher binning to increase the signal level; when using a confocal microscope, use higher pixel size and slower speed of scanning. In case of the early stages of zebrafish embryonic development, we suggest delaying the desired stage by incubating the embryos at lower temperature (25 °C instead of 28 °C)—that would delay the required stage allowing more time for fluorophore production/folding (*see* **Note 7**).

Here we describe the protocol for FRET ratio imaging in live zebrafish embryos using FRET biosensor for visualizing Rac activity. First, we provide a brief overview of the experimental setup and working conditions when using FRET biosensors in live embryos; then we present the ratio generation algorithm and discuss the possible pitfalls during this process; finally, we provide a macro, which can be used for analyzing large sets of data in an automated fashion. This protocol can also be used for ratio imaging in other applications since the principles of dividing one image by another are the same. While here we focus on working with RacFRET biosensor in zebrafish embryos, the same principles will apply when using other FRET biosensors in different animal models or in cell culture conditions.

2 Materials

Depending on the developmental stage, the experiment may take between several hours and several days. For early developmental stages (until 6 h post fertilization), it would be ideal to use a transgenic line, in which the FRET biosensor is maternally expressed. That would allow enough time to accumulate the sufficient protein level in the embryo for ratio imaging. Injecting a purified protein into one-cell-stage fertilized embryo is another option for working with the early stages.

2.1 Zebrafish Work

1. Zebrafish embryos: transgenic embryos expressing the biosensor or wild-type embryos for injection with sense mRNA are collected in the morning and used for further manipulations.
2. Embryo medium: zebrafish embryos are maintained in the embryo medium [25, 26].
3. Custom-made ramps to make agarose molds for imaging: for upright microscope setups, we use custom agarose molds adapted to the embryo size.
4. Ultrapure-grade agarose to make mounting molds.
5. Low-melting agarose: used in certain mounting setups when the embryo needs extra fixation.
6. Injection needles are made by pulling glass capillaries 1.0 mm OD.
7. Dissecting microscope equipped with the fluorescence source, i.e., Leica MZ10F.

2.2 Molecular Biology

1. DNA encoding RacFRET biosensors: wild-type, positive, and negative controls relevant for the specific biosensor and fluorophores. For RacFRET, we use RacV12FRET as a positive control for high level of activation and RacN17 as a negative control for low level of activation. RacN17 mutant is expressed together with the wild-type biosensor [12].
2. mMessage machine for synthesizing capped mRNA with sp6 or T3 polymerases depending on the minimal promoter used.
3. Phenol red (optional).

2.3 FRET Ratio Imaging Setup Using Wide-Field Illumination

1. Fluorescence microscope able to acquire two images simultaneously: The DualView Imager [27] (*see* **Note 8**) and a cooled CCD camera were used to acquire the sample images provided in this chapter.
2. Optical configurations for the filter wheel and the DualView (*see* **Note 8**):
   - (a) YFP filter cube for identifying cells expressing FRET biosensor: exciter 500AF25, dichroic 525DRLP, and emitter 545F35.
   - (b) FRET filter cube used in the microscope filter wheel in a combination with the DualView for obtaining CFP and FRET emission values: exciter 440AF21 and dichroic 455DRLP; the emission filter is removed.

(c) DualView is fitted with 505dcxr beam splitter, ET 480/40 nm and ET 535/40 emission filters for CFP and YFP channels respectively.

3. Fluorescence excitation source for wide-field microscopy (*see* **Note 9**).
4. Objective lens: Plan-Apochromat water-dipping lens, e.g., 20x or 40x (1.0 NA).

2.4 FRET Ratio Imaging Setup Using Confocal Fluorescence Microscopy

The confocal microscope should be equipped with:

1. A laser line required to excite the donor: 458 nm for CFP excitation.
2. At least two detectors (PMT or GaAsP) to detect CFP and YFP emissions simultaneously.
3. An objective: the same as in Subheading 2.3, **step 5**.

2.5 Software and Macro for Ratio Image Generation

The present algorithm is based on *Fiji* version 1.52 g [28, 29] and requires the following plug-ins:

1. NucMed [30].
2. TurboReg [31, 32].
3. MultiStackReg (normally provided with the original *Fiji* package) [33].
4. Ratio Plus [34].
5. The macro containing the ratio algorithm can be downloaded from GitHub [35]. This algorithm is adapted for either single plane or time series images acquired in a wide-field microscope equipped with the DualView.

2.6 Sample Images

Four sets of raw data are provided for practising the ratio image generation protocol and can be downloaded from the Springer website.

1. Single-frame images showing wild-type polarized germ cells in zebrafish (shown in Figs. 2, 3, and 4):
   "Wild_Type_CFP.tif" and "Wild_Type_FRET.tif".
2. Time-lapse movie containing a wild-type zebrafish migrating germ cell, separated in two channels: "Wild_Type_Movie_CFP.tif" and "Wild_Type_Movie_FRET.tif".
3. Single frames of zebrafish somatic cells expressing the activated form of the RacFRET biosensor, to be used as positive control (*see* Fig.4): "Active_CFP.tif" and "Active_FRET.tif".
4. Single frames of zebrafish somatic cells expressing the negative form of the RacFRET biosensor, to be used as negative control (*see* Fig. 4): "Inactive_CFP.tif" and "Inactive_FRET.tif".

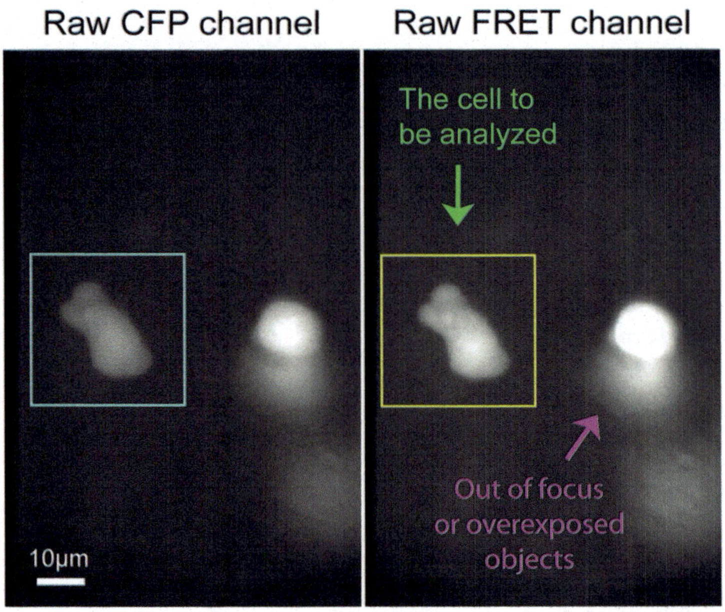

Fig. 2 Image processing **steps 1–4**. The original raw data corresponding to CFP and FRET channels are shown. These images were acquired simultaneously on the Zeiss Axioplan microscope equipped with the DualView camera set at binning 4. Images used: Files "Active_CFP.tif" and "Active_FRET.tif". ROI selection suggestion, dimensions: 65 × 65 (px). Scale bar: 10 μm

3 Methods

3.1 FRET Biosensor Expression in Zebrafish Embryos

1. Zebrafish work should be carried out according to the instructions in [25, 26].
2. Inject embryos with mRNA for RacFRET biosensor into a single cell at 8- or 16-cell stage to create mosaic labeling or use transgenic embryos expressing a biosensor. A detailed protocol for zebrafish embryo injection is described elsewhere [36].
3. Raise embryos to the desired stage (typically between 8 and 40 h post fertilization).
4. Select positively labeled embryos using a dissecting microscope equipped with the fluorescent excitation source. We prefer to use YFP filter because YFP signal is the brightest. Important: Use the lowest illumination intensity possible because FRET is highly sensitive to the bleaching (*see* **Note 9**).
5. Remove the chorion using forceps.
6. Mount the embryos in the orientation desired for imaging.

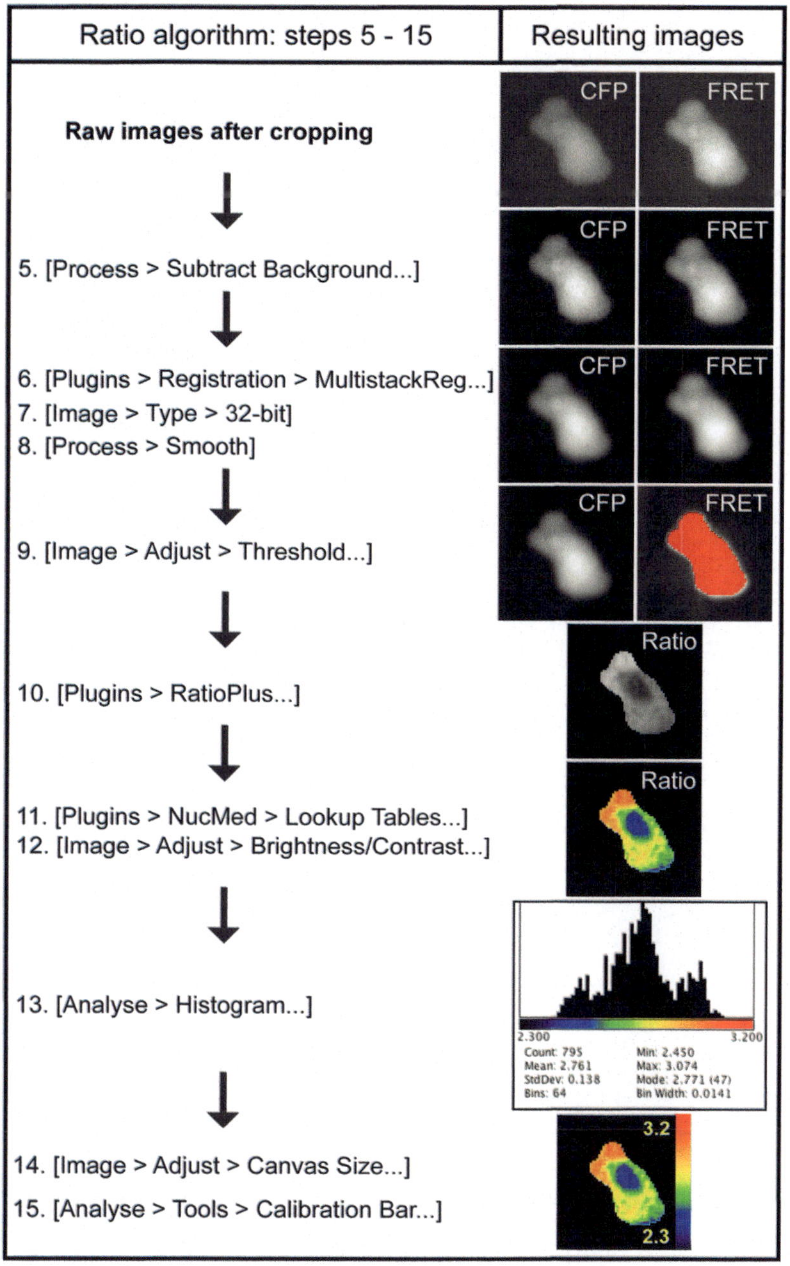

Fig. 3 Image processing **steps 5–15**. The chart shows steps in the ratio image generation algorithm after the ROI selection and cropping of the images. Resulting images for each step are shown

3.2 Image Acquisition on a Wide-Field Illumination Microscope

1. Place the mounted embryos under the microscope.
2. Use the lowest intensity illumination possible to locate the labeled cells expressing the biosensor suitable for imaging (using low-intensity illumination extremely important because

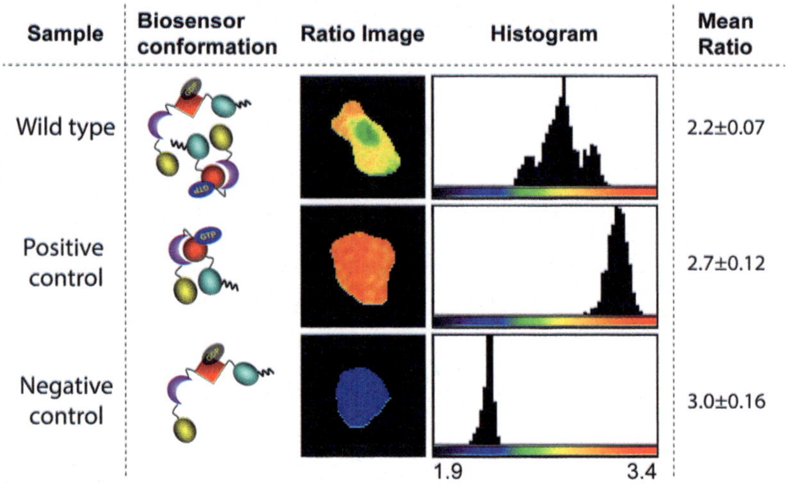

Fig. 4 Controls for RacFRET biosensor. The controls used to define the range between the active and the inactive states of RacFRET biosensor are shown. Wild type shows a migrating primordial germ cell expressing RacFRET biosensor; the positive control shows a somatic cell in a zebrafish embryo expressing RacV12FRET biosensor. V12 mutation in the Rac domain results in the biosensor that is insensitive to the GAP (GTPase-activating proteins); the negative control shows a somatic cell in the zebrafish embryo expressing a wild-type biosensor together with RacN17 mutant. The RacN17 mutant inhibits the GEF (guanine nucleotide exchange factors) preventing biosensor activation. For each control, a scheme shows the expected folding of the biosensor accompanied with the ratio image and a histogram

over-illumination would destroy FRET) as in Subheading 3.1, **step 3** (*see* **Note 9**).

3. Acquire data under the following conditions: Use FRET filter cube for CFP excitation and simultaneous acquisition of CFP and YFP emission channels in the DualView to generate CFP and FRET images, respectively. Make sure that there is no saturation in the recorded channels (*see* **Note 10**). When optimizing imaging conditions for the first time, acquire samples with the positive and negative controls to establish the dynamic range of the biosensor.
4. Acquire sufficient number of events at each developmental stage for making statistical analysis (*see* **Notes 11** and **12**).

3.3 Image Acquisition on a Confocal/Two-Photon Microscope

1. Place the sample under the microscope. In case of using fluorescence source for locating the labeled cells, use the lowest possible illumination to locate the object to be imaged.
2. When optimizing imaging conditions, we recommend using low-resolution scanning format (256 × 256 pixels) and trying pinhole apertures higher than 1 AU as well as different

acquisition speeds, in order to find the best combination that achieves the best S/N ratio.

3. Acquire data following the same principles as in **step 4**, Subheading 3.2.

3.4 Image Processing and Ratio Image Generation Algorithm

During image processing, the FRET/CFP ratio image is produced from the raw data images in a series of processing steps using *Fiji* software. The steps listed below can be executed manually or automatically using a macro provided in Subheading 2.5. Each step during image processing and ratio generation algorithm is aimed at improving the image quality as well as correcting for possible artifacts. Wrongly executed, these can cause more artifacts. We advise first-time users to run the algorithm manually step by step before applying the accompanying macro for automated ratio image generation. This exercise will help to understand the function of each step as well as to detect possible errors that may arise during this process.

1. Launch *Fiji.*
2. Open the image files corresponding to the FRET and CFP channels or the respective stack images in case of a time lapse at [File > Open...] or drag the image files directly into *Fiji.*
3. Select the appropriate region of interest (ROI) containing the cell or group of cells to be analyzed. Avoid overlapping, saturated, or out-of-focus cells next to the one to be analyzed (Fig. 2). Use the FRET image to generate the ROI using [Edit > Selection > Specify...]. Select the *Width* and the *Height* to include the object to be analyzed and to ensure sufficient area around the cell for background correction. For the images provided here, anything between 65 × 65 and 75 × 75 (in pixels) would be suitable. To store the ROI, select [Analyze > Tools > ROI Manager...], press *Add*, and the respective ROI will appear in the *ROI Manager* window. A quicker way to activate the *ROI Manager* window is by pressing *t* on the keyboard. Switch to the CFP image and select the ROI within the *ROI Manager* box to activate the ROI at the same location as in the other image. Another way to place the ROI at the same location in the other image is to press *shift e* on the keyboard after switching to the second image.
4. Crop the images: [Image > Crop] (*see* **Note 13**).
5. Correct the background for uneven illumination. It should be applied to both images. In case of acquiring images in the wide-field microscope, uneven illumination is very common. To correct it, we suggest using the rolling ball algorithm implemented in *Fiji* at [Process > Subtract Background...]. The typical range for the *Rolling ball radius* is between 50 and 200 px (*see* **Note 14**). Press *OK.*

6. Register the two channels. This step is necessary when two channels are misaligned, which is often the case when using the DualView on a wide-field microscope (*see* **Note 15**). *MultistackReg* works for both single images and stacks. This can be done at [Plugins > Registration > MultiStackReg...]. Select the CFP image or stack as *Stack 1* and set *Use as Reference* in *Action 1*. Select the FRET image or stack as *Stack 2* and set *Align to First Stack* in *Action 2*. Select *Rigid Body* for transformation.
7. Convert images to 32 bits at [Image > Type > 32-bit] (*see* **Note 16**). This step should be applied to both images.
8. Filter the images using [Process > Smooth]. Filtering is often used during image processing to reduce the background noise. We use a smooth filter, which is a mean filter of radius 1 (*see* **Note 17**).
9. Segment the images. This step is necessary to avoid artifacts in the background during ratio image generation at a later step. During segmentation, the minimum and the maximum intensity values are set and the values outside the threshold are ignored. This is done at [Image > Adjust > Threshold...]. Select the *Default* algorithm and tick the options *Red* and *Dark Background*. In case of working with stacks, select *Stack Histogram*, which is necessary to avoid artifacts caused by the intensity variations due to bleaching. Then press *Apply* and tick *Set background pixels to NaN* when prompted to convert all background pixels to NaN (not a number). NaN pixels will be discarded in any image calculation (*see* **Note 16**).
10. Generate the ratio image using [Plugins > RatioPlus...]. Choose the FRET image as *Image 1* and the CFP image as *Image 2*. Press *OK*. We recommend saving this image as *.tif* file and perform further manipulations (next steps) such as adjusting the intensity range and preparing the image for presentation on a duplicate of this ratio image.
11. Assign a lookup table (LUT) to the ratio image using the *NucMed* plug-in, which offers a large variety of LUTs in addition to those already integrated in *Fiji*. This plug-in can be found at [Plugins > NucMed > Lookup Tables]; choose *Blue Green Red* from the list under *LUT name*.
12. Adjust the intensity range of the ratio image using [Image > Adjust > Brightness/Contrast...]. Press *Set* in the "B&C" window and type in the desired values. Here we used 2.3 and 3.2 as low and high values, respectively (*see* **Note 18**). Duplicate the ratio image, convert it to RGB, and save it for presentation purposes.
13. Show the histogram of the ratio image. The histogram is a useful tool to obtain the mean and the max/min values, and

to visualize the noise in the images by the spread of the intensity values. It also serves for visual representation of the biosensor range (Figs. 3 and 4). This function is found at [Analyze > Histogram]. In the *Histogram* window, use 100 bins, unselect the *Use pixel value range*, and set up the same *X min* and *X max* values as chosen for the range in the previous step.

14. Increase the canvas or image size. This step will increase the background space in the image to prepare it for use in presentations and figures. This extra space is necessary for instance to add a calibration bar (added in the next step) to fit into the image or any additional information such as the specific condition. There are two options to do it, either resizing the canvas of the image using [Image > Adjust > Canvas Size...] or resizing the entire image at [Image > Adjust > Size...] using *Constrain aspect ratio*, *Average when downsizing*, and *Interpolation: None.* In both cases, it is required to type in the new parameters for *Width* and *Height.* In this protocol, the image was resized to 400 px width.

15. Add a calibration bar to the ratio image. This step would place a colored calibration bar showing the ratio intensity differences on the image. This function can be found at [Analyze > Tools > Calibration Bar...]. The options used in this protocol are: *Location: Upper Right*; *Fill color: Black*; *Label color: White*; *Number of Labels: 3*; *Decimal places: 1*; *Font size: 14*; and *Zoom factor: 1.5.*

3.5 Macro for the Automated Processing

All the steps described in Subheading 3.4 can be easily automated by using the macro recorder in *Fiji* at [Plugins > Macros > Record...]. The recorder keeps track of all the functions called and allows for creating a script that automatically repeats the whole procedure over the same set of images. Such basic script needs to be further edited to make it usable with other sets of images and, to that aim, some new functionalities may be required. This section describes step by step all such functionalities that have been added to the recorded script in order to create the provided macro. Although no programming detail is particularly granted, the specific blocks of the code are shown in Figs. 5, 6, 7, and 8 to complement the information given at each step, and to be used as example in other programming pipelines.

1. Open the macro or drag and drop it to the *Fiji* bar and press *Run* in the script editor window.

2. Before opening an image, the macro pops up a dialog box (Fig. 5a) where the minimum and maximum ratio values should be introduced (*see* **Note 18**). These values will be stored in two variables and will be used in the latest macro steps to adjust the brightness and contrast of the ratio image. The

A

```
//Create the dialog box asking for ratio values
Dialog.create("FRETRatioAnalysis");
Dialog.addNumber("Minimum value for the ratio", 1.9);
Dialog.addNumber("Maximum value for the ratio", 3.4);
Dialog.show;
minRatio = Dialog.getNumber();
maxRatio = Dialog.getNumber();
```

B

```
//Open FRET image, collect data and open its corresponding CFP image
showMessage("Open a FRET image when asked");
open();
title = getTitle();
fretID = getImageID();
input = File.directory;
experiment = substring(title, 0, lengthOf(title)-8);//Removing FRET and file extension
cellType = substring(title, lengthOf(title)-8, lengthOf(title)-4);//Looking for FRET word
extension = substring(title, lengthOf(title)-4);//Looking for file extension
getDimensions(width, height, channels, slices, frames);
if(cellType=="FRET"){
  open(experiment + "CFP" + extension);//Pair of FRET image previously open
  cfpID = getImageID();
}else{
  exit("There is no FRET image open");
}
```

C

```
//Calculate the ratio
run("Ratio Plus", "image1=[FRET] image2=[CFP] background1=0 clipping_value1=0 background2=0 clipping_value2=0 multiplication=1");
ratioID = getImageID();
setMinAndMax(minRatio, maxRatio);
```

Fig. 5 Macro blocks, part 1. (**a**) This set of commands creates a dialog box to ask for the minimum and maximum values of the ratio range. These values are stored in two variables, "minRatio" and "maxRatio", respectively. (**b**) This code block asks to select a FRET image to be opened and then stores its dimensions, name, path, and ID in variables. Afterwards, the macro checks whether this image corresponds to the FRET channel and opens the matching image corresponding to the CFP channel; otherwise, an error message appears and the macro stops. (**c**) This set of commands creates a ratio image with the range of values predetermined in (**a**)

default values shown are fine for the current sample images. Once the values are entered, press the *OK* button to proceed.

3. The next macro block (Fig. 5b) is aimed to open the FRET image. A message is shown warning about this. After pressing *OK* the macro forces the computer browser to pop up for the FRET image to be selected. Once opened, the file's general information (dimensions, name, image ID, and path to the input folder) is collected. The macro then checks whether the

A

```
//Add a calibration bar
selectImage(ratioID);
run("Lookup Tables", "resource=Lookup_Tables/ lut=[Blue Green Red]");
run("Duplicate...", "title=copy duplicate");
resizeImage();
run("Calibration Bar...", "location=[Upper Right] fill=Black label=White number=3 decimal=1 font=14 zoom=1.5 overlay");
run("Flatten", "stack");
calBarID = getImageID();
```

B

```
//Function to resize an image and increase its canvas by 100px width
function resizeImage(){
  if(width<400){
    run("Size...", "width=400 height=400 constrain average interpolation=None");
    height = getHeight();
  }
  width = getWidth() + 100;
  run("Canvas Size...", "width="+width+" height="+height+" position=Center-Left zero");
}
```

Fig. 6 Macro blocks, part 2. (**a**) This portion of macro applies a LUT to the ratio image, duplicates and resizes it, and adds a calibration bar. (**b**) This user-defined function named *resizeImage()* is used to resize and modify the canvas size of the active image

image contains the word "FRET" in its name and then for its partner (CFP image) in the same input folder. In case the selected image is not a "FRET" one, the macro shows an error message and stops (*see* **Note 19**).

4. Both of the open images are renamed as "FRET" and "CFP" to facilitate further operations with them inside certain *Fiji* functions (i.e., **step 5**).
5. The macro pops up a message window asking the user to draw a ROI around the cell of interest in the "FRET" image. Once drawn, press *OK* in the message window and the macro will draw the same ROI in the "CFP" image.
6. **Steps 4–9** in Subheading 3.4 are then executed and the macro controls the action at every step and the specific image it is applied to.
7. The macro calls the *Ratio Plus* function using the "FRET" and "CFP" names generated earlier in **step 3** (Fig. 5c, line 84).
8. The brightness and contrast of the ratio image (obtained in **step 5**) are adjusted using the minimum and maximum values defined at the beginning of the macro execution (Fig. 5c, line 86).

A

```
//Obtain histogram
selectImage(ratioID);
rename("Ratio");
if(slices>1){
  stackHistogram();
}else{
  run("Histogram", "bins=100 x_min=" + minRatio + " x_max=" + maxRatio + " y_max=Auto");
}
```

B

```
//Function to create a stack of histograms obtained from each slice in a stack
function stackHistogram(){
  setBatchMode(true);
  for(i=1; i<=slices; i++){
    selectImage(ratioID);
    setSlice(i);
    run("Histogram", "bins=100 x_min=" + minRatio + " x_max=" + maxRatio + " y_max=Auto");
    rename("frame " + i);
  }
  run("Images to Stack", "name=histoStack title=frame use");
  rename("Histogram of Ratio");
  setBatchMode(false);
}
```

Fig. 7 Macro blocks, part 3. (**a**) This set of commands creates the histogram with the range of values defined in Fig. 5a, lines 25 and 26. (**b**) This user-defined function named *stackHistogram()* is used to obtain a histogram from each slice of a stack and then combine all histogram images into a new stack

A

```
//Create Results folder inside the input folder
output = input + "Results" + File.separator;
File.makeDirectory(output);
if(!File.exists(output)){
  exit("Unable to create the folder");
}
```

B

```
//Save images and histogram
selectImage(calBarID);
saveAs("Tiff", output + experiment + " Ratio_RGB");
selectImage(ratioID);
saveAs("Tiff", output + experiment + " Ratio");
selectWindow("Histogram of Ratio");
saveAs("Tiff", output + experiment + " Ratio_Histogram");
if(isOpen("copy")){
  selectWindow("copy");
close();
```

Fig. 8 Macro blocks, part 4. (**a**) This code creates an output folder named "Results" inside the input folder selected in Fig. 5b. (**b**) This set of commands saves the original ratio image, its copy with the calibration bar, and the histogram inside the "Results" folder created in (**a**)

9. After applying the LUT, the ratio image is duplicated and then resized using a user-defined function named *resizeImage* (Fig. 6a, line 92). This function is defined at the end of the macro (Fig. 6b, lines 129–136) and it is used to resize an image in case it is smaller than 400 px width and finally enlarge the width of the canvas by 100 px (*see* **Note 20**).
10. The calibration bar (Fig. 6a, line 93) is accommodated in the upper right corner of this resized image. First it is added as an overlay to the ratio image and then both calibration bar and image are flattened to obtain an RGB image.
11. The histogram for the ratio image is generated using the minimum and maximum values entered at the beginning of the execution of the macro (Fig. 7a). If the ratio image is a stack the macro calls another user-defined function named *stackHistogram()* (Fig. 7a, line 101). This function is defined at the end of the macro (Fig. 7b, lines 139–150) and it is used to generate a histogram of each slice and finally to convert all histograms into a stack.
12. The macro creates an output folder named "Results" inside the input folder where the original images are located (Fig. 8a).
13. Finally, the ratio image, its duplicate with the calibration bar created in **step 8**, and the histogram are saved inside the output folder created in **step 10** (Fig. 8b).

4 Notes

1. FRET biosensors are unique tools that allow obtaining instant information about protein activity and signaling events in living organisms. Experiments involving FRET biosensors are complex and require multiple steps, each step involving another discipline: molecular biology and animal work to optimize the biosensor expression in a living tissue; microscopy for observation and image acquisition; and computer-assisted image processing for analyzing the raw data. Each of these steps brings specific challenges and here we describe the most common ones and offer possible solutions. We do not discuss the principles of the molecular design for FRET biosensors because these are far beyond the scope of this protocol.
2. Toxicity: The biosensor should not cause any adverse effects such as cell death, abnormal shape, interference with proper cell cycle, and other normal cellular functions. Therefore, the very first experiment when working with biosensors involves expressing it in the embryo either globally or in the population of cells and monitoring proper development. Ideally, there should be no difference between embryos expressing the

biosensor and the embryos expressing a control protein that is known to be nontoxic.

3. Neutrality: The biosensor should not interfere with the endogenous signaling processes within the cells. While one can never be 100% sure that no other signaling pathways are affected by the biosensor overexpression, a simple observation of cellular shapes, rate of cell division, and proper animal development should indicate if the biosensor might be causing developmental abnormalities due to signaling interference. For example, in case of migrating cells, the proper migration should be a good indication that the biosensor is not interfering with important signaling events. Other examples might include a proper shaping of the targeted tissues and the expression of relevant markers in the presence of a biosensor.
4. Subcellular targeting: The biosensor should be localized in the same subcellular areas as the corresponding endogenous protein. This can be tested by comparing the subcellular distribution of the overexpressed biosensor with the expression pattern of the endogenous protein probed by an antibody staining. If the antibody is not available, a GFP fusion of the corresponding protein can be used as an alternative. In case of Rho GTPases, the C-terminal farnesylation site of the respective Rho GTPase fused to the 5′ of the biosensor was used to ensure the proper localization to the membrane where these proteins normally carry out their work [11]. Depending on the tissue and cell types, the biosensor localization might require further optimization. For example, in zebrafish germ cells, the original FRET biosensor for Rac protein had shown the predominant accumulation in the nucleus and several modifications such as eliminating the nuclear localization sequence in the Rac farnesylation site or a deleting this C-terminal domain were made to ensure a proper targeting [17]. Additional considerations for optimizing biosensor for subcellular targeting and proper functioning include changes in the order of different modules in a biosensor backbone [16, 21].
5. Availability and responsiveness: The biosensor should be accessible and responsive to the endogenous activation and deactivation machinery that acts on the corresponding protein. This can be tested by measuring FRET in the presence of one of the known activators of the protein studied. In case of RacFRET biosensor, we have used the active domain of Tiam1, one of the Rac activators [17].
6. Controls: For each biosensor, the controls for the active and inactive state will help defining the range between high and low FRET levels. Even in the inactive state of the biosensor, a low FRET level will be detected in many cases; therefore a sample

expressing CFP and YFP fluorophores cannot serve as a negative control (Fig. 1b). For the very first optimization of FRET measurements when using new equipment/conditions, it is recommended to use additional controls that will allow distinguishing between no FRET and a constant FRET (Fig. 1b).

7. Signal to noise level: When dividing one image by another, high signal intensity in the images is crucial. When signal intensity drops below certain level, which is at least double of the background noise, artifacts may arise during the division. Typically, the CFP channel is expected to have the lowest intensity level because CFP variants in general have lower quantum yield as compared with YFP variants and in addition the CFP signal will be reduced due to the energy transfer. It can also be difficult to achieve high signal intensity in the embryos because of the possible toxic effects of the high number of biosensor molecules to the embryo. To overcome these, we recommend several strategies:
   - (a) Use brighter CFP and YFP variants. Many CFP and YFP variants were created showing an improved brightness and energy transfer [20]. These were successfully utilized in biosensors.
   - (b) Delay embryonic development by lowering the temperature. In case when the observation takes place at the early developmental stages (before 8 h post fertilization), the shorter incubation time may compromise the biosensor brightness due to the time it takes for the fluorophore folding and maturation. Keeping the embryos at a lower temperature (24 °C instead of 28 °C for zebrafish embryos) would delay the embryonic development and add the time required for biosensor signal accumulation.
   - (c) When possible, use transgenic animals instead of mRNA injection for a transient expression. In transgenic animals, the signal-to-noise ratio is often higher as compared to mRNA injection. Also, transgenic technology allows targeting specific tissues and cell populations, which can be helpful in improving the signal.
8. DualView is an optical device that allows simultaneous acquisition of two spectrally distinct images, therefore avoiding any lag between the acquisition of the two images and minimizing the photo-destruction of fluorophores during the acquisition process. The filter configurations proposed here are optimal for the CFP/YFP FRET pair. If different fluorescent proteins are used as FRET pair in a biosensor, these configurations must be adjusted using the respective excitation/emission parameters for specific fluorophores [18].

9. Bleaching is the worst enemy of FRET because strong illumination destroys energy transfer between two fluorophores; therefore samples expressing FRET biosensors should never be exposed to a strong fluorescence source. Avoid bleaching by reducing the intensity of the fluorescence excitation source. We recommend using about 25% or less of the available fluorescence intensity. This can be achieved by reducing the power of the fluorescence or by fitting optical density filters in front of the fluorescence source.
10. Saturation: Another enemy of FRET is signal saturation. Saturated images cannot be used for quantitative data analysis because they do not contain the correct intensity values. Saturation normally occurs in the FRET channel and can be detected by the apparent uniform signal in the image or using pseudo color lookup tables in the image acquisition software that present saturated pixels during data acquisition.
11. Number of cells to be analyzed: We recommend measuring at least 30 cells for each condition, especially when optimizing imaging conditions and establishing the range for biosensor activity (*see* Subheading 3.4 and **Note 18** for further details on image processing). This would allow to account for data variability when establishing the dynamic range of the biosensor and comparing between different cell types, developmental stages, or other conditions such as mutants or drug treatments.
12. When working with live embryos, we observed an increase in average FRET level in biosensors during time. This increase is not biosensor related because we observe it as well in the SECFP-YPet fusion module, which cannot depend on the external activators [12]. Therefore, it is important to perform every set of measurements at the same developmental stage when working with embryos.
13. After the images were cropped, it is not possible to go back to the original image except for opening it again. In a multistep protocol such as ratio image generation, it can be useful to duplicate the region of interest instead of cropping the images and work on a copy. That would allow going back to the original images and start over in case there was an error at one of the steps.
14. Background correction: Uneven illumination in a wide-field microscope will result in variations in intensity across the image. This noise can be corrected with the *Subtract Background* function implemented in *Fiji*. This function employs a rolling ball algorithm [37]. The *Rolling ball radius* (in pixels) should be larger than the largest object in the image that is not part of the background. In case of the confocal imaging, the background noise in the detectors can be corrected by

subtracting a fixed value from each pixel of the image corresponding to the background signal level: [Process > Math > Subtract. . .].

15. Registration: During registration, two images or two stacks are aligned and shifts along *x*- and *y*-axes are corrected. In a wide-field microscope, it is often the case that the CFP and FRET channels are shifted relative to one another, especially when the DualView is used. When two images are misaligned, the ratio image will contain artifacts, which might be misinterpreted as real data. A typical artifact in the ratio image caused by the division of two misaligned images would manifest as abnormally high and low values at the opposite edges of the cell.
16. 32-bit conversion: Original raw data generated in a microscope would typically be in 8-bit or 16-bit format. The conversion to 32 bits allows conversion of the background pixels to NaN (not a number) during the segmentation step and retaining the intensity information within the object. This allows saving time instead of creating a mask separately and multiplying it to the original images. More importantly, the 32-bit conversion generates floating numbers, which is essential when dividing one image by another as it allows decimal numbers.
17. Filters are used to reduce the noise and to improve the image quality. However, it may introduce artifacts or result in data loss. We prefer using the *Smooth* filter in this protocol for its mild effects on averaging the data during filtering. It is possible to use other filters such as *Gaussian Blur* and *Median* if necessary; however we do not recommend applying radius larger than 2 px to avoid artifacts.
18. Defining the dynamic range between the active and inactive states in the RacFRET biosensor: The range values are defined based on the measured ratio values in each specific experiment using positive and negative controls that should provide low FRET and high FRET values. For the visual representation, it is recommended to set the range at values slightly lower than the minimal and slightly higher than the maximum values measured for the inactive and the active biosensor states. In that case, the cell expressing the inactive biosensor form will be clearly visible on the dark background (Fig. 4). It is expected that the active and inactive controls would have a very narrow range shown in their histograms as compared with the wild-type situation. The range in the controls reflects the noise in the biosensor and should be smaller as compared to the wild-type situation (Fig. 4).
19. The ratio image in this protocol is performed dividing a pair of images. To avoid mistakes when opening the two required images, we programed the macro to ask for a FRET image

and then the macro itself opens its partner, i.e., the CFP image. To be able to do so, the name of the images must end in "FRET" and "CFP", respectively (Fig. 5b): for example "MyControl_FRET.tif" and "MyControl_CFP.tif".

20. Cropping the images to select only the cell or cells of interest performed in this protocol results in images sizing less than 400 px.

Acknowledgments

Elena Kardash is grateful to Dr. Nadine Peyriéras at CNRS/BioEmergences and her laboratory for providing support during the creation of this chapter. E.K. was supported by the ANR-10-INBS-04 through the National Infrastructure France-BioImaging supported by the French National Research Agency and by the 2017-ITN-721537 as part of the ITN ImageInLife Marie Skłodowska-Curie Actions when working on this chapter.

References

1. Lakowicz JR (2007) Principles of fluorescence spectroscopy. Springer, New York
2. Clegg RM (2009) Förster resonance energy transfer—FRET what is it, why do it, and how it's done. In: FRET and FLIM techniques, Laboratory techniques in biochemistry and molecular biology, 33, 1st edn. Elsevier, Amsterdam
3. Okamoto K, Sako Y (2017) Recent advances in FRET for the study of protein interactions and dynamics. Curr Opin Struct Biol 46:16–23
4. Ni Q, Mehta S, Zhang J (2018) Live-cell imaging of cell signaling using genetically encoded fluorescent reporters. FEBS J 285:203–219
5. Miyawaki A, Llopis J, Heim R, McCaffery JM, Adams JA, Ikura M, Tsien RY (1997) Fluorescent indicators for Ca2+ based on green fluorescent proteins and calmodulin. Nature 388:882–887
6. Kurokawa K, Ohba Y, Nagal T, Miyawaki A, Matsuda M (2001) Spatio-temporal images of growth-factor-induced activation of Ras and Rap 1. Nature 411:1065–1068
7. Miyawaki A (2003) Visualization of the spatial and temporal dynamics of intracellular signaling. Dev Cell 4:295–305
8. Kiyokawa E, Aoki K, Nakamura T, Matsuda M (2011) Spatiotemporal regulation of small GTPases as revealed by probes based on the principle of Förster Resonance Energy Transfer (FRET): implications for signaling and pharmacology. Annu Rev Pharmacol Toxicol 51:337–358
9. Shimozono S, Iimura T, Kitaguchi T, S-i H, Miyawaki A (2013) Visualization of an endogenous retinoic acid gradient across embryonic development. Nature 496:363–366
10. Haga RB, Ridley AJ (2016) Rho GTPases: regulation and roles in cancer cell biology. Small GTPases 7:207–221
11. Itoh RE, Kurokawa K, Ohba Y, Yoshizaki H, Mochizuki N, Matsuda M (2002) Activation of Rac and cdc42 video imaged by fluorescent resonance energy transfer-based single-molecule probes in the membrane of living cells. Mol Cell Biol 22:6582–6591
12. Kardash E, Bandemer J, Raz E (2011) Imaging protein activity in live embryos using fluorescence resonance energy transfer biosensors. Nat Protoc 6:1835–1846
13. Pertz O, Hahn KM (2004) Designing biosensors for Rho family proteins—deciphering the dynamics of Rho family GTPase activation in living cells. J Cell Sci 117:1313–1318
14. Nakamura T, Aoki K, Matsuda M (2005) Monitoring spatio-temporal regulation of Ras and Rho GTPases with GFP-based FRET probes. Methods 37:146–153
15. Pertz O, Hodgson L, Klemke RL, Hahn KM (2006) Spatiotemporal dynamics of RhoA activity in migrating cells. Nature 440:1069–1072

16. Fritz RD, Letzelter M, Reimann A, Martin K, Fusco L, Ritsma L, Ponsioen B, Fluri E, Schulte-Merker S, van Rheenen J, Pertz O (2013) A versatile toolkit to produce sensitive FRET biosensors to visualize signaling in time and space. Sci Signal 6:rs12
17. Kardash E, Reichman-Fried M, Maître J-L, Boldajipour B, Papusheva E, Messerschmidt E-M, Heisenberg C-P, Raz E (2010) A role for rho GTPases and cell-cell adhesion in single-cell motility in vivo. Nat Cell Biol 12:47–53
18. Bajar BT, Wang ES, Zhang S, Lin MZ, Chu J (2016) A guide to fluorescent protein FRET pairs. Sensors (Basel):16, 1488
19. Sawano A, Miyawaki A (2000) Directed evolution of green fluorescent protein by a new versatile PCR strategy for site-directed and semi-random mutagenesis. Nucleic Acids Res 28: e78
20. Nguyen AW, Daugherty PS (2005) Evolutionary optimization of fluorescent proteins for intracellular FRET. Nat Biotechnol 23:355–360
21. Komatsu N, Aoki K, Yamada M, Yukinaga H, Fujita Y, Kamioka Y, Matsuda M (2011) Development of an optimized backbone of FRET biosensors for kinases and GTPases. Mol Biol Cell 22:4647–4656
22. Jares-Erijman EA, Jovin TM (2003) FRET imaging. Nat Biotechnol 21:1387–1395
23. Jares-Erijman EA, Jovin TM (2006) Imaging molecular interactions in living cells by FRET microscopy. Curr Opin Chem Biol 10:409–416
24. Xia Z, Liu Y (2001) Reliable and global measurement of fluorescence resonance energy transfer using fluorescence microscopes. Biophys J 81:2395–2402
25. Kimmel CB, Ballard WW, Kimmel SR, Ullmann B, Schilling TF (1995) Stages of embryonic development of the zebrafish. Dev Dyn 203:253–310
26. Westerfield M (2000) The zebrafish book: a guide for the laboratory use of zebrafish (*Danio rerio*). http://zfin.org/zf_info/zfbook/cont.html
27. DualView. https://www.photometrics.com/products/multichannel/dv2.php
28. Schneider CA, Rasband WS, Eliceiri KW (2012) NIH image to ImageJ: 25 years of image analysis. Nat Methods 9:671–675
29. Fiji. https://imagej.net/Fiji/Downloads
30. Parker JA NucMed. http://www.med.harvard.edu/JPNM/ij/plugins/NucMed.html
31. Thévenaz P, Ruttimann UE, Unser M (1998) A pyramid approach to subpixel registration based on intensity. IEEE Trans Image Process 7:27–41
32. TurboReg. http://bigwww.epfl.ch/thevenaz/turboreg/
33. Busse B MultistackReg. http://bradbusse.net/downloads.html
34. Magalhães P (2003) Ratio Plus. https://imagej.nih.gov/ij/plugins/ratio-plus.html
35. Bosch M (2018) FRET ratio. https://github.com/manelbosch76/ijm-Macros
36. Rosen JN, Sweeney MF, Mably JD (2009) Microinjection of zebrafish embryos to analyze gene function. J Vis Exp 25:1115
37. Rolling ball. http://imagejdocu.tudor.lu/doku.php?id=gui:process:subtract_background

Chapter 14

Cell Proliferation High-Content Screening on Adherent Cell Cultures

Pau Carrillo-Barberà, Jose M. Morante-Redolat, and José F. Pertusa

Abstract

Pulse-chase experiments using 5-bromo-2′-deoxyuridine (BrdU), or the more recent EdU (5-etynil-2′-deoxyuridine), enable the identification of cells going through S phase. This chapter describes a high-content proliferation assay pipeline for adherent cell cultures. High-throughput imaging is followed by high-content data analysis using a non-supervised *ImageJ* macroinstruction that segments the individual nuclei, determines the nucleoside analogue absence/presence, and measures the signal of up to two additional nuclear markers. Based upon the specific combination with proliferation-specific protein immunostaining, the percentage of cells undergoing different phases of the cell cycle (G0, G1, S, G2, and M) might be established. The method can be also used to estimate the proliferation (S phase) rate of particular cell subpopulations identified through labelling with specific nuclear markers.

Key words Bioimage analysis, Cell proliferation, High-content, Ex vivo, Neural stem cells

1 Introduction

Scoring cell proliferation has become a routine technique in many laboratories. This can be achieved by different approaches that are based on quantifying the cell DNA content, labelling newly synthetized DNA, measuring cell metabolism, or detecting cell cycle-specific proteins. Among them, one of the most popular and reliable methods consists in incubating the cells for a short period of time with halogen-containing pyrimidines, such as BrdU (5-bromo-2′-deoxyuridine) [1] or the more recently developed thymidine analogue EdU (5-etynil-2′-deoxyuridine) [2], in order to label the genomic DNA of cells undergoing S phase of the cell cycle.

Furthermore, these DNA synthesis-based methods can be combined with the detection of proliferation-specific proteins to

Electronic supplementary material: The online version of this chapter (https://doi.org/10.1007/978-1-4939-9686-5_14) contains supplementary material, which is available to authorized users.

Elena Rebollo and Manel Bosch (eds.), *Computer Optimized Microscopy: Methods and Protocols*, Methods in Molecular Biology, vol. 2040, https://doi.org/10.1007/978-1-4939-9686-5_14,

estimate the percentage of cells in other cell cycle phases, thus obtaining a more detailed analysis of the culture proliferation. One of the most commonly used markers is Ki67, which is present within the nucleus of cycling cells during G1, S, G2, and M phases, but not during quiescence (G0) [3, 4]. Additionally, phospho-histone 3 (PHH3) can be used to identify those cells that are specifically undergoing mitosis (M phase) [5]. By combining a short nucleoside analogue pulse (S phase) with immunocytochemical detection of Ki67 (cycling cells) and PHH3 (M phase), the entire range of cell cycle phases in the sample can be determined.

Alternatively, nucleoside pulse-chase may be combined with the detection of other nuclear markers, e.g., antigens associated to specific subpopulations present in the culture, which would allow to estimate the proliferation (S phase) rate of each individual subpopulation.

Our main goal here was to develop a protocol for non-supervised, high-content image analysis of ex vivo cell proliferation assays based on nucleoside analogue pulse alone or in combination with other nuclear markers that, although developed and tested with murine neural stem cells (NSCs), could be applied to any kind of cell type growing on adherent conditions. We here provide a step-by-step method on how to create the macroinstruction, programmed in *ImageJ* Macro language (IJM), along with the user instructions. The method takes into account the issues described below.

First, in order to achieve a non-supervised analysis, the macroinstruction must manage automatically all the images of the experiment, which entails (i) finding them in the proper directory, (ii) opening them in the correct sequence, and (iii) systematically saving the corresponding data so they can be correctly identified in the end of the process.

Second, the macroinstruction must be compatible with image datasets captured with different imaging systems. Since we worked with the automated imaging system IN Cell Analyzer 2000 (GE Healthcare) that allows fast capturing of high-content assays performed with cells cultured in multiwell plates, the macroinstruction has been initially programmed to recognize the IN Cell Analyzer file naming convention (*see* **Note 1**). In order to make it compatible with image datasets captured with different imaging systems, we have included a "Filename Transformation" mode that transforms the input image dataset into a recognizable IN Cell Analyzer-like naming scheme (*see* **Note 2**). Therefore, the script can be run with images taken from any kind of experimental setup (i.e., other cell types, other kinds of culture plates, etc.) or obtained with other types of automated imaging systems or even conventional microscopes, as long as certain conditions are met (*see* Subheading 2).

Third, the use of automated imaging systems, such as IN Cell Analyzer, facilitates the analysis of a high number of experimental conditions and the acquisition of hundreds of images in a short period of time. However, one common drawback of such systems is that a certain number of images might not contain any relevant biological information (i.e., images of regions with no cells present or taken out of focus). Segmenting these images by means of a filtering approach will probably generate fake results. Accordingly, the irrelevant images must be discarded to avoid unreliable results. Therefore, the script includes an automatic quality checkpoint aimed to identify and discard nonrelevant images.

Fourth, each individual experiment, even following the same protocol and experimental conditions, generates an image dataset with unique singularities (depending on the research specimen, the sample fixation and labelling process, the image acquisition setup, etc.). As a consequence, the application of the exact same parameters that worked with one image set might go awry with a different one. In order to overcome this limitation, the macroinstruction includes a Pre-Analysis mode which allows the user to tune up the parameters that will be applied to the workflow and visualize the resulting segmentation outcome before running the complete analysis.

Fifth, each nucleus in the image must be identified and analyzed as an individual object, based on the general nuclear staining included in the experimental design. By operating on this particular image, the macroinstruction obtains a binary image where each individualized nucleus is used as a mask to interrogate the rest of channels starting with the nucleoside analogue image. It is important to take into account that cells undergoing S phase might incorporate variable amounts of the nucleoside analogue during the pulse. Seemingly, depending on the chase duration and the cell proliferation rate, the labelling can be "diluted" to a different extent. Consequently, the nucleoside analogue staining pattern is usually heterogeneous ranging from an irregular punctate pattern to a full nuclear staining of variable intensity (or a combination of both) (*see* Fig. 1). However, in order to calculate the proliferation rate, the analysis must include all nucleoside-positive cells, regardless of their pattern or intensity. For this reason, the macroinstruction works with the segmented nucleoside analogue image. This way every time that a nuclear mask redirected to this image co-localizes with one or more particles, the macro allocates the nucleus as positive. Additionally, in case other markers were included in the experiment, the nuclear mask can be redirected to the corresponding channel images but this time to the original grayscale ones in order to retrieve the mean gray level, so the labelling of other markers in each nucleus can be further evaluated.

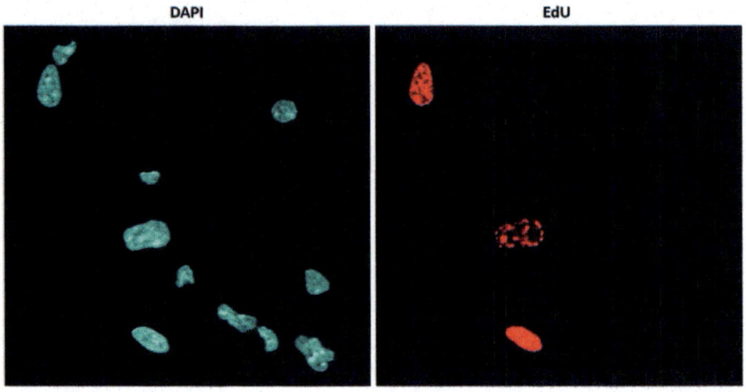

Fig. 1 Representative images of a general nuclear staining (left) and the corresponding nucleoside analogue (right) extracted from a proliferation experiment performed with adherent ex vivo cultures of murine neural stem cells. After a 1 h pulse of EdU, a few cells have incorporated the nucleoside analogue and, therefore, were undergoing S phase of the cell cycle during the pulse. Note that DNA-synthetizing cells incorporate/dilute the nucleoside analogue to different extents and, as a consequence, a variety of EdU-staining patterns can be found among the nuclei (right)

Finally, the ultimate results table must contain the measurements retrieved from all the processed images, properly identifying each nucleus by its well and field codes. This will allow a posterior user-adapted data analysis.

2 Materials

2.1 Image Dataset

The present bioimage protocol has been developed to analyze high-content image datasets from ex vivo cell proliferation experiments (*see* **Note 3**). As already mentioned, image sets acquired from sources other than IN Cell Analyzer (GE Healthcare) can be used, as long as they are transformed to an IN Cell Analyzer-like naming pattern (*see* **Notes 1** and **2**). The dataset must consist of monochromatic images (never RGB) of at least the general nuclear staining and the nucleoside analogue channels. Images of up to two extra nuclear markers might also be included in the analysis. All of them must be saved as Tagged Image Format (TIF) files in a unique folder with no subfolder system.

In order to test the present protocol, this chapter provides an example image dataset that consists of fluorescence microscopy images of a proliferation assay performed with a short EdU pulse on primary cultures of murine NSCs obtained from the adult subependymal zone seeded in matrix-coated 96-well culture plates (*see* **Note 4**). Although the chapter is mainly dedicated to image processing and analysis, some details on how to culture NSCs and perform the nucleoside analogue pulse and detection are provided in **Notes 5** and **6**.

2.2 Fiji Distribution [6] of ImageJ Software

ImageJ is an imaging software created by Wayne Rasband at the National Institutes of Health (Bethesda, Maryland, USA) in 1987 and originally named NIH Image. It is an open-source platform supported by a large user community that continuously implements the program in a cooperative effort [7, 8]. The *ImageJ* version used in this work is 1.52f.

2.3 ImageJ Macro

The macro "Cell_proliferationHCS.ijm" is provided as supplementary material and can be downloaded from GitHub [9], along with the group of images to serve as example dataset. Updated versions will be progressively uploaded. It has been developed using the *ImageJ* Macro Language (NIH) as explained below. A description of all the necessary variables, operators, conditional and looping statements, and user-defined functions, as well as all built-in macro functions, can be found at [10].

3 Methods

Figure 2 shows a schematic representation of the macroinstruction dialog boxes in the form of a decision tree. The macro starts with a dialog box (Fig. 3) where the user must (i) choose one of the available analysis options and (ii) set the path of the directory that contains the image dataset.

The macro offers two different analysis modes ("Analysis" and "Pre-Analysis (parameter tweaking)") along with a complementary mode called "Pre-Analysis (visualization)" that displays graphically the result of the segmentation of previous pre-analyses. Additionally, it includes an optional mode called "Filename Transformation" that recodifies the filenames of a provided image dataset to make them recognizable by the macro, in case of having captured it with an imaging system different than IN Cell Analyzer.

In order to obtain a successful object extraction, we have designed an image analysis workflow that is applied to both the general nuclear staining and the nucleoside analogue images. First, the grayscale images are subjected to enhance and restoration

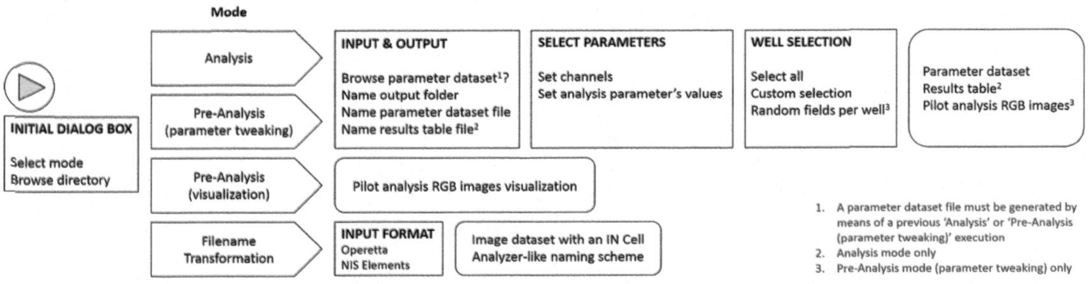

Fig. 2 Schematic representation of the macroinstruction dialog boxes decision tree

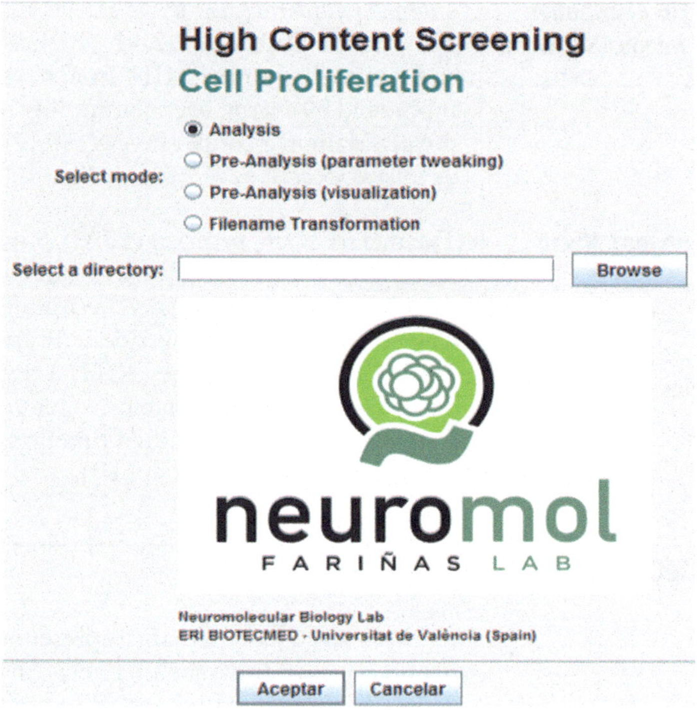

Fig. 3 Initial dialog box

operations that facilitate subsequent segmentation. These include background subtraction, contrast enhancement (normalization), and Gaussian filtering. Then, segmentation is achieved by a thresholding algorithm. Finally, a series of binary operations is applied: *Convert to Mask* to obtain a new set of binary images, *Fill Holes* to restore incomplete objects, *Erode* and *Open* to eliminate noise particles, *Watershed* to individualize clumped objects, and a final selection by size to remove any background particles that may remain. This workflow has been optimized for the experimental design described above. However, different image datasets might require some adjustments in order to get a successful analysis. For that purpose, the "Pre-Analysis (parameter tweaking)" mode performs a pilot analysis on manually selected wells, randomly picking a few fields of each of them and allowing the user to adjust the parameters of the analysis workflow. This Pre-Analysis generates and saves a text file (TXT) that contains the applied parameter settings so that it can be loaded in further Pre-Analysis or Analysis runs. This mode also generates and displays at the end of the process an image stack file composed of RGB images of each of the pre-analyzed fields with the resulting nuclear and nucleoside analogue segmentation outlined with colored lines (Fig. 4), so the user can visually check whether the applied parameters achieved a satisfactory segmentation of the objects of interest or not. Note

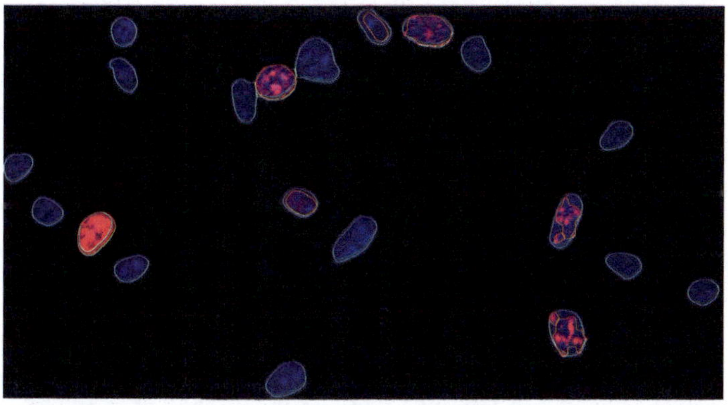

Fig. 4 The "Pre-Analysis (parameter tweaking)" mode creates RGB images of the fields submitted to the pilot analysis. In order to visually check the resulting segmentation, the outlines of the nuclei (cyan) and nucleoside analogue (orange) are drawn on the images

that the RGB image stack is generated and saved in the output folder only during "Pre-Analysis (parameter tweaking)" mode. However, once generated, it can be visualized anytime by means of the "Pre-Analysis (visualization)" mode in the initial dialog box.

Finally, the "Analysis" mode analyzes all the fields in the selected wells by means of a cell by cell algorithm. The macro includes a default dataset of values for all the parameters to be applied during segmentation of nuclei and nucleoside analogue images, although they can be manually tuned up by the user or substituted by loading a pre-established parameter dataset file. Once the analysis is completed, the obtained data is stored in a results table, which is saved as a TXT file.

3.1 How to Run the Macroinstruction

1. Drag and drop the "Cell_proliferationHCS.ijm" file to the *ImageJ* menu bar. Thereby, *ImageJ* will display the script into the macro editor window.
2. Press the *Run* button to initialize the macro. Alternatively, it is also possible to install the macro and even create a shortcut to run it by pressing a single key [10].

3.2 Initial Dialog Box

The macro starts with a dialog box generated using a universal #@parameter notation [11]. Depending on the selected mode, other dialog boxes are generated, this time with the *ImageJ* dialog functions (*see* **Note 7**). The user must follow these steps:

1. Choose the macro mode {Analysis, Pre-Analysis (parameter tweaking), Pre-Analysis (visualization), or Filename Transformation}.
2. Browse a directory to work to. The selected mode and the directory path are then assigned to different variables.

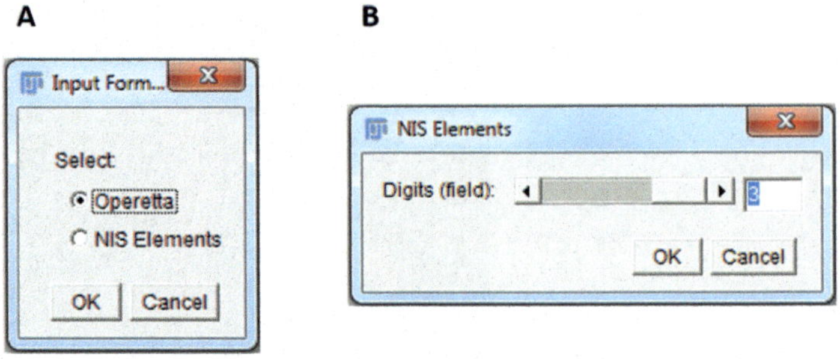

Fig. 5 Having selected the "Filename Transformation" mode, the macroinstruction displays a dialog box asking for the image dataset origin (**a**). In case of working with NIS-Elements images, the macroinstruction also asks the number of digits encoding the field in the filename (**b**)

3.3 Filename Transformation Mode

At the moment, the macro can recognize and rename images captured with Operetta high-content microscope (PerkinElmer) or with NIS-Elements software (Nikon). IN Cell Analyzer and Operetta systems have their own default fixed pattern for encoding the image files, whereas NIS-Elements software filenames consist of a user customizable prefix followed by certain information added automatically by the program. For this reason, we have included in **Note 2** a series of recommendations on how to pattern the filename prefix to make the resulting image datasets transformable by the macro. The Filename Transformation contains the following steps:

1. A dialog box generated with the *ImageJ* dialog functions (*see* **Note 7**) prompts the user to select the source of the image dataset to transform, either Operetta or NIS-Elements (Fig. 5a). Future macro updates will include compatibility with other imaging devices and software.
2. One specific requirement of the macro is that the field number in the filename be encoded by three digits. Operetta-acquired images already meet this condition, but, in the case of NIS-Elements images, the user has to enter the number of digits (1–3) that the current dataset actually contains in the field code (Fig. 5b). This way the macro will automatically fill the missing positions with zeros, if less than three are selected.
3. Once all the files have been renamed, the macro prints on screen an "end of process" message and the path of the folder that contains the transformed image dataset.

3.4 File Management

An important part of the script is dedicated to the automated management of files, which allows a non-supervised analysis. In case that "Analysis" or "Pre-Analysis (parameter tweaking)" mode has been selected, the macro checks if the directory contains TIF files to work to. Then, the macro extracts the required information (well, field, and channel) from the TIF filenames. Both aims are achieved as follows:

1. An array containing the names of the files in the directory path is created using the *getFileList()* function. The array is sorted using the *Array.sort()* built-in function. Based on the "tif" extension at the end of the filename, the number of TIF files in the directory is counted by means of the *endsWith()* built-in function placed into a *for* loop. If there are no TIF files, an "error message" is displayed, and the macro execution terminates by means of the *exit()* built-in function. Or else, the macro creates an array containing only the names of the TIF files in the directory path.
2. The macro creates (i) two arrays that will progressively store the corresponding well and field information of each TIF filename and (ii) two variables, one to count the number of wells and the other to count the number of fields. Regarding the arrays, this information is obtained from each filename by the *substring()* built-in function and stored by means of a *for* loop. Meanwhile, into the same loop, every time that one field code (three digits number) is different to the previous one, the number of fields is increased by one. The same happens with wells. Then, a series of simple calculations are performed in order to obtain:
   (a) The number of images per well = the number of images divided by the number of wells
   (b) The number of images per field = the number of images divided by the number of fields
   (c) The number of fields per well = the number of fields divided by the number of wells
3. The channel information is also extracted from the images' filenames. The macro creates an array that will contain the names of all the channels used in the experiment. Since the already calculated number of images per field equals the number of channels, this number is used to set the size of the array (adding one extra space to include an "Empty" value). By using the built-in macro functions *indexOf()* and *lastIndexOf()* into a *while* loop, the starting and ending indexes of the channel names are determined and used to store those strings in the array by means of the *substring()* function. The string "Empty" is stored in the last string space.
4. Another array is created to extract only the channel names (omitting the "Empty" string) by using the *Array.slice()* built-in function. It will be used in the "Select Parameters" dialog box (*see* Subheading 3.5). As mentioned before, the macro offers the possibility to analyze up to four channels (nucleus, nucleoside analogue, and two optional extra

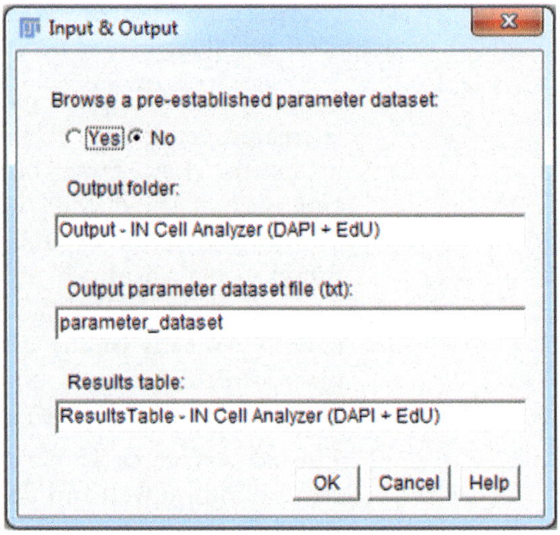

Fig. 6 Input & Output dialog box

markers). In case no extra markers are included, the "Empty" option in the "Select Parameters" dialog box prevents the macro to search for those images during the analysis.

3.5 "Pre-Analysis (Parameter Tweaking)" and "Analysis" Parameterization

The "Analysis" and "Pre-Analysis (parameter tweaking)" modes start with a series of dialog boxes generated with the *ImageJ* dialog functions (*see* **Note** 7) that prompt the user to provide the required information to perform the analysis. First, the macro displays the "Input & Output" dialog box (Fig. 6) where the user has to (i) decide between uploading a parameter dataset file generated in a previous run and not, (ii) name the output folder, and (iii) name the parameter dataset file that will be generated in that specific run. Additionally, the user also has to name the results table file (only in "Analysis" mode). In case of having decided to load a preexistent parameter dataset, the macro will open a window to browse the TXT. Thereafter, the macro displays the "Select Parameters" dialog box (Fig. 7) that has three different sections. In the "Channel Selection" section, the user has to match the nuclear staining, the nucleoside analogue, and the two optional nuclear markers with the correspondent channel that was used to capture them. A list of the available channel names is displayed as a drop-down menu in each case. Since the nuclei and nucleoside analogue images are essential for the analysis, their drop-down menus do not include an "Empty" option, whereas the optional markers can be excluded from the analysis by choosing "Empty" as channel. The other two sections present a list of the parameters that will be applied along the analysis workflow for nuclei and nucleoside analogue segmentation, respectively. The dialog box already displays an established value for each of the parameters, either default or retrieved from the selected parameter dataset file. At this point these values can be

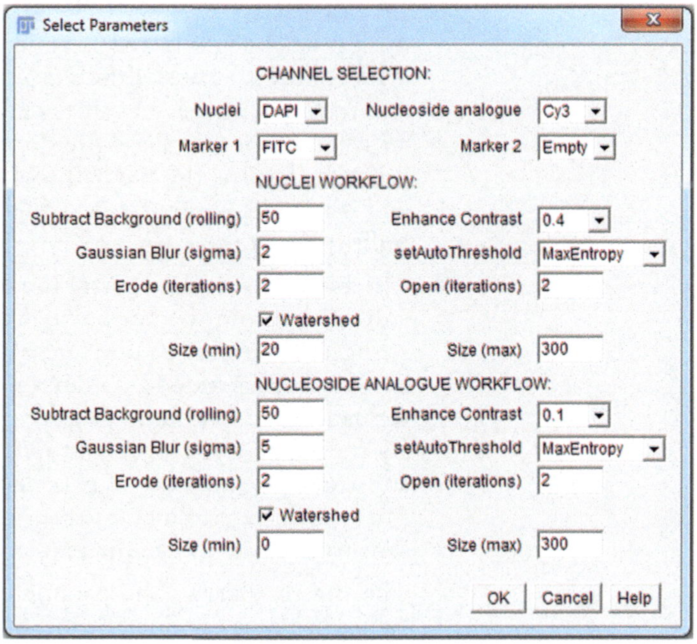

Fig. 7 Select Parameters dialog box

readjusted manually in case applying them to a specific image set does not yield a satisfactory segmentation output. A further explanation of these parameters and how to adjust them can be found in **Note 8**.

The macro saves in each run the applied parameter values in a parameter dataset file stored in the output folder, both in "Pre-Analysis (parameter tweaking)" and "Analysis" modes. In case of saving this file with the same name than a pre-existing file in the folder, the macro will replace it without asking. For this reason, it is advisable to create different output folders for each run. This part of the script consists of the following steps:

1. Before displaying the dialog box, the original folder name is obtained by using the *File.getName()* built-in function, which returns the last name (the string after the last backslash) of the directory path name sequence.
2. The "Input & Output" dialog box asks the user to choose between browse a previously created parameter dataset file and not.
3. It is necessary to name the output folder (original folder name preceded by the string "Output" by default) and the parameter TXT ("parameter_dataset" by default).
4. If the "Analysis" mode has been selected, the user also has to name the results table. By default, the name starts with the string "ResultsTable," followed by the name of the original folder.

5. In case that "Browse a pre-established parameter dataset" option has been selected, a conditional statement (*if*) asks for a TXT file, by means of the *File.openDialog()* built-in function, and retrieves the preset values of all the parameters (otherwise, an *else* statement assigns the default parameters' values). This function returns the file path, which is used to open the TXT and obtain its content as a string with the *File.openAsString()* built-in function.
6. The built-in function *split()* breaks the string into an array of substrings using the newline delimiter ("\n"), where each substring corresponds to one line of the TXT. Each row contains two values, separated by a tab: (a) the first value is the parameter name, and (b) the value after the tab is the parameter value. By means of a *for* loop, the *split()* function breaks each row substring into an array of two elements using the tab delimiter ("\t"), and the parameter value (second element of the array) is sent to an array of parameters.
7. The macro displays the parameter dataset (either the default values or those obtained from a browsed parameter dataset file) in the "Select Parameters" dialog box, so they can be edited.
8. The user has to select the name of the channel corresponding to the nuclei, nucleoside analogue, and additional marker (Marker_1 and Marker_2) images, which are displayed as drop-down menu options. If they are set as "Empty," they will be excluded from the analysis.
9. The user has to determine the parameters that will be used for the image analysis workflow. The same sequence of operations will be applied to transform both nuclei and nucleoside analogue images into binary images, but different values for each parameter can be set to each of them (*see* **Note 6**).
10. Once the user has set the parameters, the macro verifies that the same channel has not been selected twice. Or else, an error message is displayed, and the macro execution is terminated by using the *exit* function.
11. An output folder is created into the original directory with the name indicated by the user by means of the *File.makeDirectory()* built-in function.
12. A parameter dataset TXT is also created (*see* **Note 7**). Generated as a blank table [Plugins > New > Table...], the TXT displays the name and the value of each parameter separated by a tab delimiter ("\t") in different lines. This format allows the extraction of all the parameters values. It is saved as a TXT in the output folder by using the built-in function *saveAs*.

3.6 Well Selection

Once all the parameters have been set up, a "Well Selection" dialog box (Fig. 8a) allows the user to choose between including all the

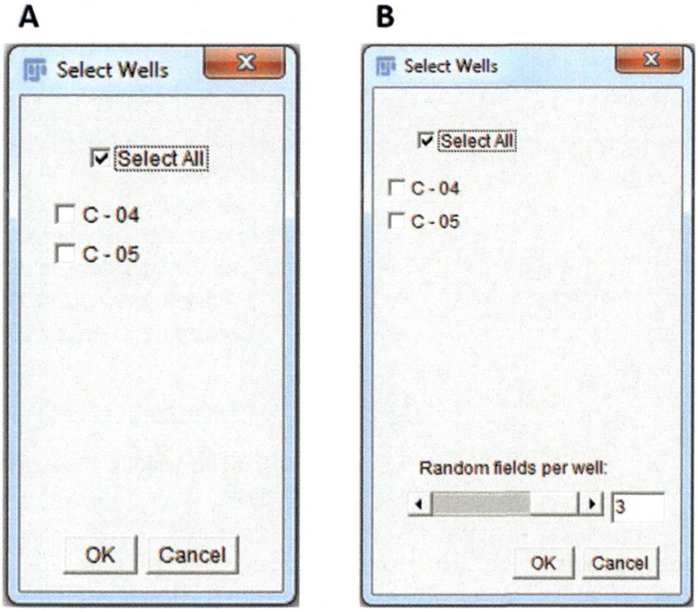

Fig. 8 Select Wells dialog box in "Analysis" (**a**) and "Pre-Analysis (parameter tweaking)" (**b**) modes

wells in the analysis ("Select all") and just a few selected ones (note that unchecking the "Select all" box is essential to carry out a custom selection; otherwise the macro will ignore it). In the "Pre-Analysis (parameter tweaking)" mode, the user can also set the number of random fields per well (between 1 and 10) that will be processed by means of the bottom slider (Fig. 8b). The Well Selection displays the following steps:

1. The "Well Selection" dialog box is generated with several checkboxes: (i) a top checkbox to "Select all" the wells founded in the folder and (ii) a checkbox corresponding to each of the wells labelled with the well code. It allows to customize the Well Selection, but the "Select all" checkbox is active by default.
2. Additionally, when the "Pre-Analysis (parameter tweaking)" mode has been selected, only a few fields per well are necessary to perform a pilot analysis. The range of the allowed number of fields (up to 10) is displayed as a slider. The specific fields to be analyzed will be selected randomly (*see* Subheading 3.7.2).
3. Finally, clicking *OK* in the "Well Selection" box will initiate the analysis, but, before progressing to this step, the macro verifies that at least one well has been selected. If there are no wells selected, an error message is displayed, and the macro execution is terminated by using the *exit()* built-in function.

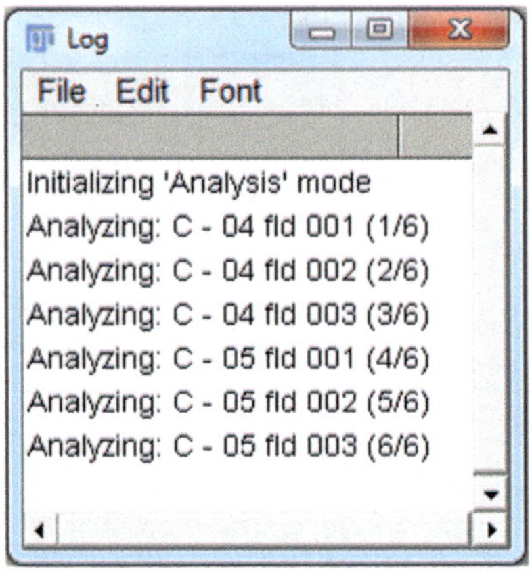

Fig. 9 Example of the Log window displayed in "Analysis" mode showing the progress of the analysis

3.7 Pre-Analysis Workflow

3.7.1 Batch Mode

In order to reduce the processing time, the macro runs as if in background mode (called *batch mode*), i.e., the images that are being analyzed and the intermediate images resulting from the segmentation workflow are not displayed on screen. Therefore, to help the user follow the analysis process, the macro progressively prints in a log window: an initialization message, the well and field that are being currently analyzed, and a final "end of process" message once all the images have been processed (Fig. 9).

1. The batch mode is turned on by using the *setBatchMode(true)* built-in function (*see* **Note** 7).
2. From this point, the script is written into a *for* loop whose iterations are equal to the number of wells included in the analysis.
3. Before starting the Pre-Analysis of each well, the macro checks if the well has been included in the assay by the user. Otherwise, it jumps to the next well without performing the Pre-Analysis.

3.7.2 Random Selection of Images for Pre-Analysis

The user has already chosen the well/s and the number of field/s per well that will be pre-analyzed, but the macro is programmed to select randomly which fields will be included:

1. A new array is created containing a number of elements equal to that of random fields by using the *newArray()* function.
2. Then, a *while* loop starts, running until the array is completely filled. The statement starts generating a random integer number (between zero and the fields per well number minus one). This is done by combining the *random()* and *round()* built-in functions.

3. After that, by means of a *for* statement, the macro checks that the produced number has not been already included in the array.
4. Then the corresponding images are opened and processed.
5. The macro includes two quality checkpoints to ensure that (i) all the channel images associated to a well are present (*see* Subheading 3.7.3) and (ii) the nuclei image actually contains analyzable objects (*see* Subheading 3.7.5). In case both checkpoints are passed successfully, the generated random number is added to the array, and the macro will try to fill the next element. Conversely, if any of them fails, the *while* statement will be repeated until the number of random fields specified by the user passes the quality checkpoints.

3.7.3 "All Channels Present" Quality Checkpoint and Channel Identification

For each randomly picked field, the macro opens all the corresponding channel images and confirms that all the expected channels have a matching open image. In case any of them is missing, the macro excludes the incomplete field from the analysis.

1. All the images associated to each randomly selected field are opened using the *open()* built-in function.
2. The number of opened images (*nImages* built-in function) is compared to the number of calculated images per field (*see* Subheading 3.4). If there is a difference, all of them are closed by means of the *cleanUp()* user-defined function (*see* **Note 9**), and the program proceeds to the next field.
3. Then, each image is identified according to the channel name selected by the user in the "Select Parameters" dialog box (*see* Subheading 3.5) using the *channelIdentification()* user-defined function. This is achieved by checking whether the string corresponding to the channel name is present in each image filename (obtained with the *getTitle()* built-in function) by means of the *indexOf()* function.
4. Each image is then renamed only with the retrieved channel name by using the *rename()* built-in function (since the macro opens the images field by field, the channel name is the only relevant data necessary to continue the bioimage analysis workflow).

3.7.4 8-Bit Transformation

IN Cell Analyzer 2000 generates 12-bit images. However, as it will be explained in the following lines, both the nuclei and the nucleoside analogue images will be converted into binary masks during the analysis workflow, so there is no need for such high bit depth.

1. The images corresponding to the nuclei and nucleoside analogue channels are converted to 8-bits [Image > Type > 8-bit].

2. The additional marker images remain with their original bit depth, since they will be used to obtain the nuclear mean gray value of each of the cells.

3.7.5 "Maxima Filter" Quality Checkpoint

As already discussed, one frequent problem associated with the computerized imaging is the possibility of capturing images with nonrelevant information. We have developed a simple test, built as a user-defined function (*maximaFilter()*), to detect such images, so that they can be removed from the analysis. The following steps are performed using the images that contain the general nuclear staining:

1. Select the image by means of the *selectImage()* built-in function.
2. *Duplicate* the selected image [Image > Duplicate]. From this point, the *maximaFilter()* function always uses the duplicated image.
3. *Subtract Background* [Process > Subtract Background...] with a 50 pixel *Rolling ball radius.*
4. *Enhance Contrast* [Process > Enhance Contrast... (*Normalize* option checked)] allowing a 0.4% of *Saturated pixels.* As a result, the signal to noise ratio (S/N) is increased in the images with relevant content (in-focus nuclei). Conversely, the nonrelevant images, whose pixels display an entropic distribution of gray levels, maintain the heterogeneity.
5. *Find Maxima* at [Process > Find Maxima...] with a Noise Tolerance of 100 and selecting the option "Count" at the "Output type" drop-down menu. Relevant images have approximately one local maxima per nucleus (*see* **Note 3** about recommended cell density), whereas thousands of maxima are found in images where no nucleus has been captured (*see* Fig. 10). However, images containing no cells but large amounts of saturated debris usually yield low number of local maxima ($\leq$10). The count of local maxima is displayed into the *Results* window.
6. The result is obtained by using the *getResult()* built-in function.
7. The *Results* table is erased at [Analyze > Clear Results] (this command is also available by right-clicking on the *Results* window).
8. The duplicated image is closed by means of the *close()* function.
9. The "localMaxima" variable returns the count of local maxima. As will be explained afterward, the macro performs a nucleus-by-nucleus analysis based on the use of the *createCountMasks()* user-defined function, which unavoidably restricts the maximum number of nuclei that can be analyzed in a single image to 255 (*see* Subheading 3.8.1). Therefore, to ensure that the macro processes just relevant and analyzable images, by means

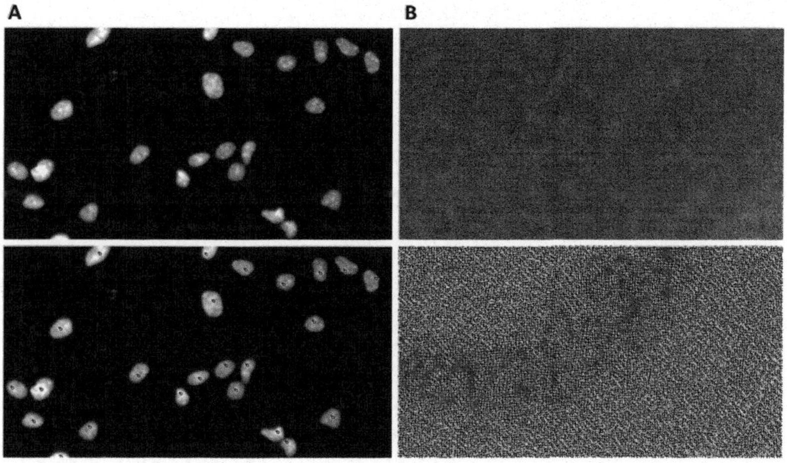

Fig. 10 The "*maximaFilter()* checkpoint" is based on the number of local maxima found in an image [Process > Find Maxima. . .] of DAPI-stained nuclei after having subtracted the background [Process > Subtract Background. . .] and enhanced the contrast [Process > Enhance Contrast. . .]. Applied to focused image containing nuclei (**a**, top), the number of local maxima (shown as a point selection) should be very similar to the number of nuclei in the image (**a**, bottom). Applied to an image without biological relevant content (**b**, top), the number of local maxima that can be found usually is hundreds of times greater (**b**, bottom)

of a conditional *if* statement, it surveys the result of the "localMaxima" variable excluding from the analysis all the images with less than 11 or more than 255 local maxima. Otherwise, images are closed by means of the *cleanUp()* user-defined function (*see* **Note 9**).

3.7.6 Image Processing and Segmentation

Once all the random images in each well have been selected, they are subjected to the image processing pipeline to identify and retrieve the objects of interest: all the individual nuclei on the general nuclear staining image and the positive nuclei undergoing S phase during the pulse on the nucleoside analogue image.

The script calls a *segmentationPreAnalysis()* user-defined function that includes all the necessary steps to convert both the nuclei and the nucleoside analogue 8-bit images into binary images. The function takes as arguments the parameters' values established by the user during parameterization (*see* Subheading 3.5 and **Note 8**) and applies them to both nuclei and nucleoside analogue images separately.

1. *Subtract Background* at [Process > Subtract Background. . .] applying the *Rolling ball radius* established by the user.
2. *Enhance Contrast* at [Process > Enhance Contrast. . .], with the *Normalize* check, is executed if the user has established a value

different from "None." The user must specify a value between 0 and 0.4 as the maximum number of allowed *Saturated pixels.*

3. A *Gaussian Blur* filter is applied [Process > Filters > Gaussian Blur...] to smooth the image according with the *Sigma (Radius)* value established by the user.
4. An auto-thresholding algorithm (found at [Image > Adjust > Threshold...]) is then applied by using the method specified by the user and the *Dark background* check.
5. Images are segmented to obtain binary masks by means of *Convert to Mask* at [Process > Binary > Convert to Mask]. Then, a series of binary operations are performed to improve the obtained binary masks explained in the subsequent steps.
6. *Fill Holes* at [Process > Binary > Fill Holes] is used to restore those nuclei that might have been left with internal pixels with background value (resembling holes) as a result of the thresholding algorithm.
7. *Erode* at [Process > Binary > Erode] is applied if the number of iterations established by the user is greater than zero.
8. *Open* at [Process > Binary > Open] is applied if the number of iterations established by the user is greater than 0.
9. *Watershed* at [Process > Binary > Watershed] if the user selected it, to further separate contiguous objects.
10. "Size selection" at [Analyze > Analyze Particles...]. With the aim of removing from the analysis any remaining debris or image artifacts consequence of the preprocessing, a minimum and maximum size are established in order to consider a particle as a nuclear mask.

3.7.7 Generation of an Output Image Stack for "Pre-Analysis (Visualization)" Mode

Taking into account that the aim of the "Pre-Analysis (parameter tweaking)" mode is not to quantify but to set the specific parameter values that would yield a proper image segmentation, the Pre-Analysis does not analyze the images. Instead, it generates a RGB stack that includes all the processed images, overlaying the outlines of the generated binary masks for both nuclei and nucleoside analogue. This way the user can visually check the result of the segmentation. The macro includes a "Pre-Analysis (visualization)" mode that retrieves and displays the output stacks of previous Pre-Analysis runs, so that they can be revised by the user anytime.

1. First, both nuclei and nucleoside analogue initial grayscale images are merged at [Image > Color > Merge Channels...] keeping the original images. The nuclei image is selected at the *C3 (blue)* drop-down menu, while the nucleoside analogue image is selected at the *C1 (red)* drop-down menu.

2. The merged channels are converted to an RGB image at [Image > Color > Stack to RGB].
3. Then, a thresholding lookup table (LUT) is generated over the segmentation resulting images that contain the binary masks by means of the *setThreshold()* function. To this aim, both the lower and upper threshold levels are set to 255.
4. A selection is created at [Edit > Selection > Create Selection] and added to the *ROI Manager* at [Analyze > Tools > ROI Manager...] by using the *roiManager("Add")* built-in function.
5. The RGB image is selected by the *selectImage()* function.
6. The foreground color is set by the *setForegroundColor(r, g, b)* built-in function. When the *segmentationPreAnalysis()* function is executed for the nuclei image, the foreground color is set to cyan (0, 255, 255) and for the nucleoside analogue image to orange (255, 105, 0).
7. The selection is drawn in the foreground color on the RGB image by means of the *roiManager("draw")* function.
8. The selection is deleted by using the *roiManager("delete")* function and deactivated at [Edit > Selection > Select None]. The resulting images are saved as TIF files at the output directory by using the *saveAs* function.
9. At this point, the macro invokes the *cleanUp()* user-defined function (*see* **Note 9**).
10. The obtained image dataset is displayed by means of the *setVisualization()* user-defined function. The images are imported as an *Image Sequence* at [File > Import > Image Sequence...] using the output directory path. Since the folder would also contain the parameters' TXT (and eventually other files added by the user), only the TIF files are imported by forcing that their filename contains a "tif" string.
11. An image frame is added to the entire stack by using the *Canvas Size* command at [Image > Adjust > Canvas Size...] including an extra space at the bottom to show the outline mask color legend (cyan for "Nuclei segmentation outlines" and orange for "Nucleoside analogue segmentation outlines").
12. Legend lines are drawn using the *drawLine()* built-in function (it can be tuned up with *setForegroundColor()* and *setLineWidth()* functions), and text is added by means of the *drawString()* built-in function (it can be tuned up with *setForegroundColor()*, *setFont()*, and *setJustification()* functions), and this information is added to all of the slides in the stack by a *for* loop.
13. Finally, the batch mode is turned off by means of the *setBatchMode(false)* function, allowing the interpreter to display the

stack, which is saved as a TIF file named "Multi-image" in the output folder (*saveAs* function). Additionally, the macro initial dialog box (*see* Subheading 3.2) includes a "Pre-Analysis (visualization)" mode that allows the user to check anytime the RGB images produced by the "Pre-Analysis (parameter tweaking) mode." In this case, the "Multi-image.tif" file is opened using the directory path selected by the user.

3.8 Analysis Workflow

The initial part of the script is very similar to that of the "Pre-Analysis mode (parameter tweaking)," but the workflow has several differences. It also starts turning on the batch mode by means of the *setBatchMode(true)* function. Moreover, as for the Pre-Analysis script, the analysis is performed well by means of a *for* loop. However, there are several specific changes: (i) all the fields belonging to a specific well will be analyzed once this well is checked in the "Well Selection" dialog box. (ii) Grayscale images are not merged to obtain an RGB image nor saved for visualization. (iii) A quantitative nucleus-by-nucleus analysis is performed, and the results are stored and saved in a results table.

3.8.1 General Analysis Structure

The analysis script is performed as follows taking into account that, for each well, a *while* loop will repeat the workflow for each of the fields in the well.

1. First, like in the Pre-Analysis, the well checkbox selection (*see* Subheading 3.6) is taken into account, and the corresponding channel images of the first field are opened.
2. Then, the macro performs the "all channels present" checkpoint (*see* Subheading 3.7.3). The selected open images are then preprocessed, segmented (as explained for Pre-Analysis mode), and analyzed (*see* flowchart in Fig. 11).
3. Channel identification, 8-bit conversion, and *maximaFilter()* checkpoint are performed as explained in the Pre-Analysis workflow (*see* Subheadings 3.7.3, 3.7.4, and 3.7.5).
4. The *segmentation()* user-defined function performs the image preprocessing and all the segmentation steps that have been described in the *segmentationPreAnalysis()* function, except the size selection. This is because apart from obtaining binary mask images (the output the *segmentationPreAnalysis()* function), the "Analysis" workflow requires a "Count Masks" image from the general nuclear staining channel to count the number of particles and perform the subsequent nucleus-by-nucleus analysis.
5. The command *Analyze Particles* at [Analyze > Analyze Particles...] is used to perform a size selection of the objects in the nucleoside analogue binary mask in order to discard small background particles or clumped cells remaining after preprocessing the image. The size range established by the

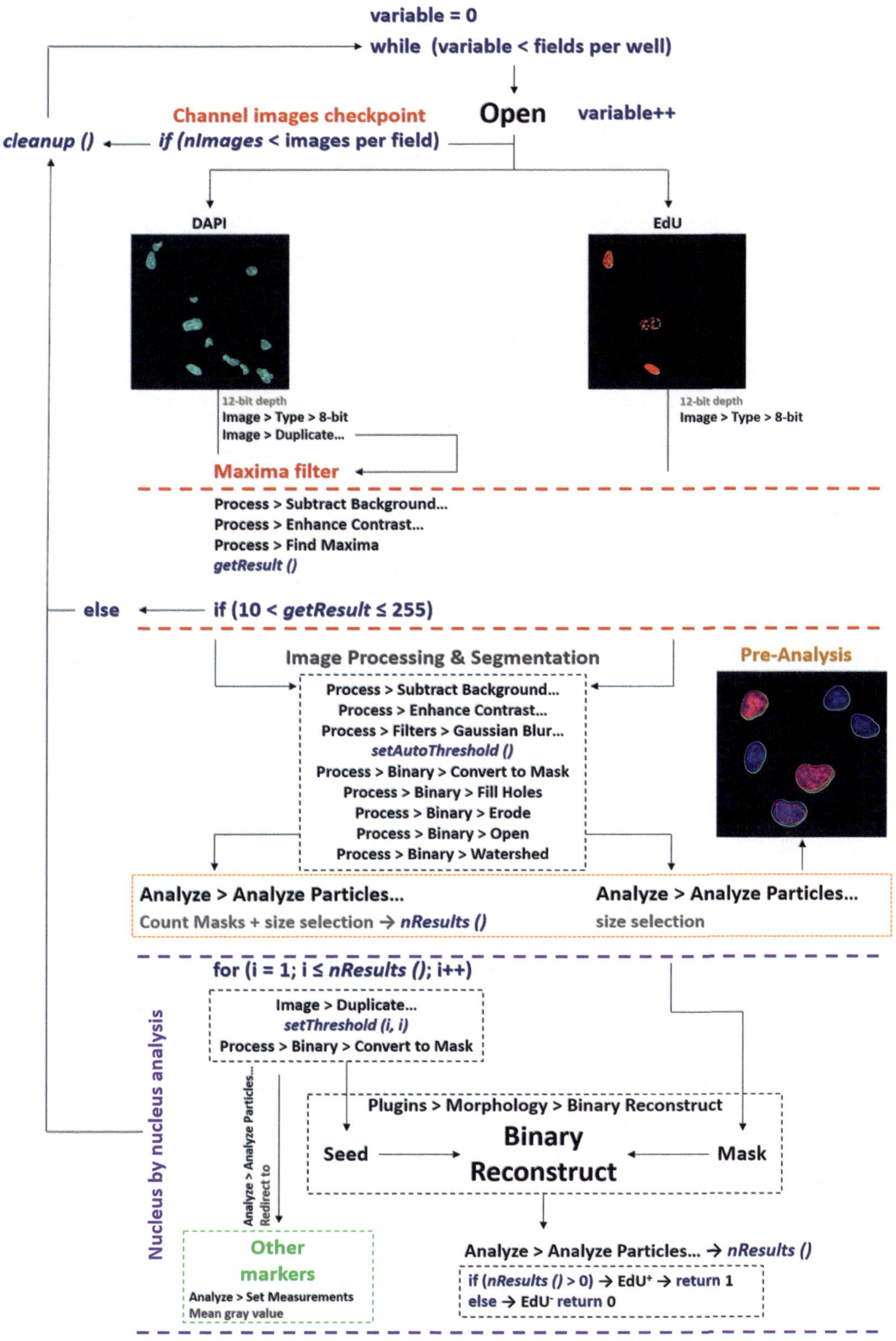

Fig. 11 Schematic representation as a flowchart of the image analysis workflow described in this chapter

user is applied at the *Size* option. A new image containing the proper size objects is generated by selecting the option *Masks* at the *Show* drop-down menu.

6. The *createCountMasks()* user-defined function creates a "Count Masks" image using the binary mask image containing all the nuclei particles and returns the number of particles in the image. This can be done at [Analysis > Analyze Particles. . .] by selecting the option *Count Masks* at the *Show* drop-down menu. A size selection is also performed applying the values established by the user. As a result, in the "Count Masks" image, the background remains zero, whereas each nuclear particle is assigned a different gray level starting from 1 to 255 (*see* Fig. 12). As explained before, this imposes a limit to the maximum number of nuclei that can be analyzed in a single image and explains why images with more than 255 local maxima are discarded by the *maximaFilter()* checkpoint.

3.8.2 Nucleus-by-Nucleus Analysis

From this moment on, nuclei are analyzed one by one within a *for* loop by means of the segmentation of each individual nucleus. At the end of each round, the nucleoside analogue results, the additional markers results, and the image information (well code and field number) of the analyzed nuclei in the field are added to different arrays. The main steps are performed by two user-defined functions described below. The analysis of each single nucleus is performed before the initiation of a new loop.

1. First, the nucleoside analogue analysis is performed by means of the *nucleosideAnalogue()* user-defined function. The "Count Masks" image is used to obtain a mask of each single nucleus within a *for* loop. At the beginning of each loop, the "Count Masks" image is duplicated at [Image > Duplicate], keeping the original intact for the following loops.
2. Segmentation is then performed using the *setThreshold()* built-in function, setting the actual loop counter value as both the lower and upper threshold levels (from 1 to the number of total nuclei particles). This way, only one nucleus is segmented at a time at [Process > Binary > Convert to Mask].
3. The segmented nucleus is used as a "seed" to interrogate the presence of any object in the corresponding nucleoside analogue binary mask by means of the *Binary Reconstruct* plugin [12], at [Plugins > Morphology > BinaryReconstruct]. It yields a new binary mask that retrieves all particles co-localizing with the "seed," if there is any.
4. Then, the function *Analyze Particles* at [Analysis > Analyze Particles. . .] displays one result for each "reconstructed" particle from the "mask" (nucleoside analogue) image.

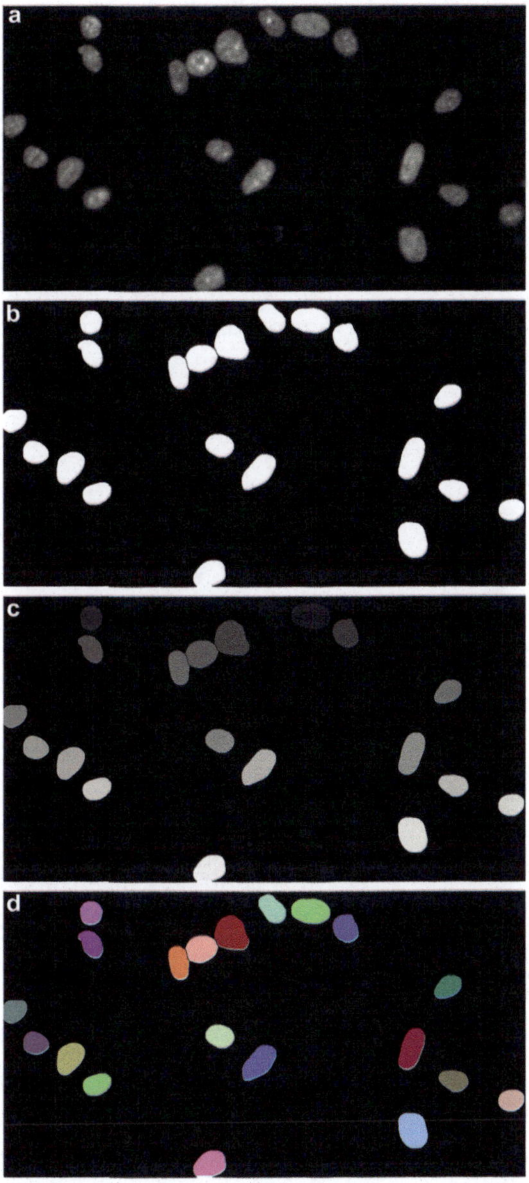

Fig. 12 Image processing steps to obtain individualized nuclei particles from the general nuclear staining image. (**a**) Original 12-bit depth image showing all nuclei stained with DAPI. (**b**) Binary image in which each nucleus has been individualized as a single particle. (**c**) By using the *Analyze Particles* command at [Analyze > Analyze Particles. . .], if the *Count Masks* option is selected at the drop-down menu *Show*, the binary image is turned to an 8-bit depth image where all the background pixels are set to zero (black), and a different gray level, from 1 till 255, is assigned to each particle pixels. (**d**) Although this step is not included in the script, the gray level differences at the "Count Masks" image are better appreciated applying the *glasbey on dark* LUT [LUT > glasbey on dark], which assigns a different color to each gray intensity in the image, while the background remains black

5. The *nResults* function returns the number of results. Then, if the seed retrieves one or more particles (*nRsesults* value is greater than zero), the nucleus is scored as nucleoside analogue-positive, and the function returns "1." Conversely, if the number of particles is zero, the nucleus is considered nucleoside analogue-negative, and the function returns "0."
6. The *Results* table is erased at [Analyze > Clear Results].
7. The additional markers analysis is performed by means of the *markerAnalysis()* user-defined function. This function is used in the same loop of the *nucleosideAnalogue()* function in order to perform a nucleus-by-nucleus analysis. The macro can analyze up to two additional markers, but the function is the same for both, just changing the channel name, given as an argument. It is necessary to use the nucleus mask image obtained at the beginning of the loop that analyzes the nucleoside analogue (during the *nucleosideAnalogue()* function), which has not been closed.
8. The analysis is performed as a conditional (*if*) statement when the channel name has not been set as "Empty" in the "Select Parameters" dialog box. Otherwise (*else*), the function returns "0."
9. The nucleus mask image is activated with the *selectImage()* function.
10. The measurements that will be recorded are specified at the *Set Measurements* window [Analyze > Set Measurements]: *Mean gray value* is checked, and the marker image (channel name) is selected at the *Redirect to* drop-down menu.
11. The result is recorded by *Analyze Particles* at [Analyze > Analyze Particles...] and returned by the *getResult* function.
12. The partial *Results* table is erased at [Analyze > Clear Results].
13. Having analyzed the additional marker/s by the *markerAnalysis* function, the nucleus mask and the nucleoside analogue "reconstructed" images are closed by using the *close* function.
14. At the end of each field loop, the results of all the retrieved nuclei are transferred to a series of arrays ("nucleoside analogue," "marker 1," "marker 2," "image") designed to house the whole screening. In order to do so, in each round the macro uses four temporal arrays that will store just the results of the present round. The size of these temporal arrays depends on the number of cells in the image, so, in each round, it is set by the number of nuclei retrieved by *Analyze Particles*. Because the total number of nuclei that will be found in the whole image dataset remains unknown until the end of the analysis, the size of the final storage arrays cannot be set beforehand. To solve this, the results of the first round are directly transferred

to the final arrays, but the following ones are added to the previous stored data by using the *Array.concat()* built-in function.

15. Then, the *cleanUp()* function (*see* **Note 7**) is also used here to close the remaining images after the analysis of each field.

3.8.3 Generation of the Results Table

Once all the selected wells have been processed, the batch mode is turned off by means of the *setBatchMode(false)* function, and the macro generates a custom results table that will contain all the analysis data, so far stored in arrays. It is created and filled by means of the *resultsTable()* user-defined function (*see* **Note 5**).

1. A blank table is generated [Plugins > New > Table...].
2. The table headings are printed: n, image, S phase, marker 1, and marker 2 channel names.
3. By means of a *for* loop, each piece of data is printed in a row separating each column data by tab delimiters ("\t"). The "n" column displays just the nuclei numeration. The "image" column contains a string formed by the well code plus the field number of the corresponding image, both elements separated by the string "fld" . The "S phase" column indicates the nucleoside analogue presence/absence (0/1). If additional markers have been selected in the "Select Parameters" dialog box, the corresponding columns will show the measured mean gray value (bit depth would depend on the original image: 8-bit, 12-bit, 16-bit, etc.). Conversely, when an additional marker has been excluded of the analysis, its column is headed by "Empty" and filled with zeros.
4. The table is saved as a TXT into the output folder, named as the user previously established, and closed at [File > Close]. It contains all the analyzed nuclei and for each of them the well code and field number of origin (image), the presence (1) or absence (0) of nucleoside analogue staining (S phase), and, in case any additional marker was included, the corresponding intensity measurement (Fig. 13).

4 Notes

1. The IN Cell Analyzer acquisition software automatically saves the whole image dataset in a unique folder, without any subfolder system, storing each image as a TIF file and generating a filename that contains the following information:
   (a) The well alphanumeric code, consisting of a 6-character string placed between the filename indexes 0 and 5, both included. The string is formed by a capital letter (row information) and a two-digit number (column

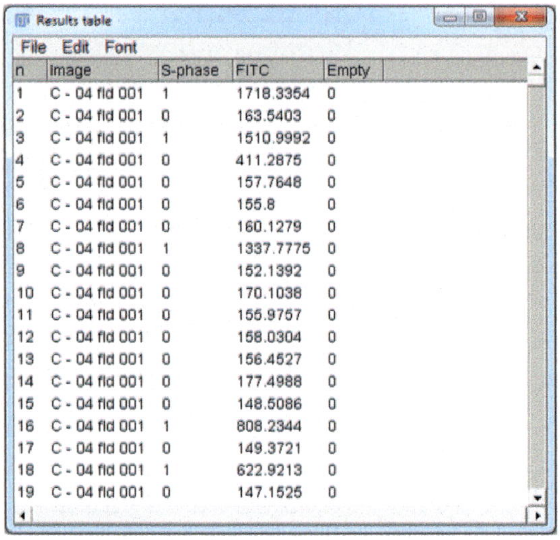

n	Image	S-phase	FITC	Empty
1	C - 04 fld 001	1	1718.3354	0
2	C - 04 fld 001	0	163.5403	0
3	C - 04 fld 001	1	1510.9992	0
4	C - 04 fld 001	0	411.2875	0
5	C - 04 fld 001	0	157.7648	0
6	C - 04 fld 001	0	155.8	0
7	C - 04 fld 001	0	160.1279	0
8	C - 04 fld 001	1	1337.7775	0
9	C - 04 fld 001	0	152.1392	0
10	C - 04 fld 001	0	170.1038	0
11	C - 04 fld 001	0	155.9757	0
12	C - 04 fld 001	0	158.0304	0
13	C - 04 fld 001	0	156.4527	0
14	C - 04 fld 001	0	177.4988	0
15	C - 04 fld 001	0	148.5086	0
16	C - 04 fld 001	1	808.2344	0
17	C - 04 fld 001	0	149.3721	0
18	C - 04 fld 001	1	622.9213	0
19	C - 04 fld 001	0	147.1525	0

Fig. 13 Example of a results table

information) separated by a hyphen escorted by two blank spaces (e.g., A - 01).

(b) The corresponding field number, expressed as a three-digit number placed between the filename indexes 11 and 13, both included (e.g., 001).

(c) The fluorophore, wavelength, or channel name expressed as a string of alphanumeric characters between the former substring "wv" and a subsequent substring composed by a blank space and a hyphen (e.g., the DAPI image corresponding to the first field captured in the well in row C, column 4 will be named: C - 04(fld 001 wv DAPI - DAPI).tif).

Therefore, the macroinstruction has been programmed so as to recognize this naming convention and extract the necessary information, without which the analysis cannot be performed.

2. In case of having acquired the image dataset with an imaging system different than IN Cell Analyzer, the image filenames will be unrecognizable by the macro, so it will be necessary to recode them to meet the macro requirements. To achieve this, the macro includes a "Filename Transformation" mode that recognizes the file naming pattern of Operetta high-content microscope (PerkinElmer) as well as of the NIS-Elements software (Nikon), extracts the necessary information (well code, field number, and channel), and renames the file in an IN Cell-like fashion. In order to preserve the original data, the macro creates a subfolder in the original directory that will store the renamed files. Filenames of images

acquired with Operetta are typically saved following the pattern, r01c01f01p01rc1-ch1sk1fk1fl1.tiff, so the macro searches inside this sequence and extracts the values found immediately after the strings: (i) "r" = row and "c" = column (well code) and (ii) "f" = field and (iii) "ch" = channel. IN Cell Analyzer registers the field number in the filename with three digits, so an extra "0" is added before the extracted value. The renamed file is saved with a new pattern (i.e., r01c01(fld 001 wv ch1—ch1).tif). Regarding NIS-Elements, in this case the software allows the customization of some of the elements included in the filename, but it cannot completely mimic the IN Cell Analyzer pattern. However, certain rules can be taken into account to help the "Filename Transformation" mode recognize and rename the images: (i) set a six-character *Prefix* string (e.g., A - 01 following the IN Cell Analyzer scheme, r01c01 following the Operetta scheme, s01c01 sorting by slide and coverslip, etc. Note that the *ImageJ* Macro language is case and blank space sensitive when defining strings). Since the well code string cannot exceed six characters, if the *Prefix* string exceeds that number, only the first six will be considered. If the string is smaller, it will be filled with blank spaces to meet the six-character requirement. (ii) Set three *Digits* for the field number. However, if just one or two digits were used for the field code, the macro offers the possibility of filling the missing spaces with zeros, once the user has introduced the number of digits that was used in the filename. (iii) Export the images at [File > Import/Export > Export Multiple ND Files. . .] as TIF files checking the "Mono image for each channel" option, so the NIS-Elements software will automatically add the channel (c1, c2, c3. . .) between the field number and the format extension. Thereby, a filename like s01c01001c1.tif will be generated, and the macro could transform it into the IN Cell-like s01c01(fld 001 wv c1 - c1).tif.

3. The experimental design consists of cells growing in adherent conditions, either attached to the plastic or to coated surfaces, at a cell density that yields images with up to 255 nuclei per image. Proliferation is assessed by detection of traceable modified nucleoside analogues that are incorporated to the newly synthesized DNA of cells undergoing S phase during a short pulse with the molecule. After that, cells are washed and fixed, and the presence of the incorporated halogenated nucleosides (such as BrdU) is revealed by fluorescent immunocytochemistry with specific antibodies or, in the case of EdU, by click chemistry using a fluorophore-conjugated azide. Finally, a general nuclear staining such as DAPI (4′,6-diamidino-2-phenylindole) or Hoechst is used to identify all the nuclei present in each sample, regardless of its proliferative state. The extra

nuclear markers can be detected by conventional immunocytochemistry using specific primary antibodies combined with fluorophore-conjugated secondary antibodies.

4. Using an IN Cell Analyzer 2000, we acquired images of 64 fields per well with a 40× objective adjusting the camera focus by means of a *laser autofocus* method. Captured fields were homogenously distributed among the well surface to ensure a representative sampling of the cells in each condition, applying these settings to all the wells in the plate.

5. NSCs are cultured in serum-free defined medium and are usually grown in suspension where they proliferate and form spherical aggregates called neurospheres. Alternatively, they might be cultured on adherent conditions over extracellular matrix-coated surfaces. Detailed protocols for isolation, culture, and handling of these cells have been previously described [13]. Nevertheless, as already mentioned, this image analysis protocol might be adapted to different types of cells growing on adherent surfaces. Specific details for handling other cell types should be found elsewhere. The high-throughput imaging protocol described in this chapter has been performed as follows: (1) Individual NSCs, obtained from freshly disaggregated neurospheres, were seeded at a density of 2×10^4 cells per well in matrix-coated 96-well culture plates. (2) Plates were pre-treated for 24 h with Matrigel® (BD) to ensure cell adhesion. (3) Matrigel® was removed, and wells were washed with water before seeding the cells. (4) Cells were cultured in NSC medium containing a defined hormone cocktail and epidermal growth factor (EGF) and basic fibroblast growth factor (bFGF) as mitogens [13, 14]. (5) Cultures were incubated for 24 h at 37 °C temperature and 5% CO_2 atmosphere.

6. BrdU, CldU, or IdU might be used to perform a nucleoside analogue pulse-chase. In that case, their presence must be revealed by immunocytochemistry with specific antibodies including an initial step of DNA hydrolysis for antigen retrieval [1]. Additionally, other proliferation proteins, such as Ki67 or PHH3, or population specific nuclear markers present only in a subset of cells in the culture might be included to achieve a more detailed analysis of the culture. The presence of these markers could be revealed by immunocytochemistry with specific antibodies following standard procedures. In order to perform the nucleoside analogue pulse in the example dataset, the culture medium was carefully removed to prevent the release of the cells from the adhesive matrix. EdU was prepared according to the manufacturer's instructions (Thermo Fisher Scientific), diluted in NSC medium, and incubated with cells for 1 h. After the pulse, cells were fixed and permeabilized using *cytoskeletal buffer* [15] at 37 °C during 20 min. The

presence of EdU was revealed by click chemistry combined with Alexa Fluor® azide following the manufacturer's instructions. Finally, nuclei were stained with DAPI (2 mg ml^{-1} in ddH_2O) for 5 min at room temperature.

7. Some macro examples to see how to display a dialog box can be checked in DialogDemo [16] or DialogDemo2 [17]. To see how to open a blank table and write to, you can check SineCosineTable2 [18]. To see an example of how to process images in batch mode and display only the resulting images, you can check BatchModeTest [19], BatchMeasure [20], BatchSetScale [21], or ReplaceRedWithMagenta [22]. To see an example of how to close windows, you can check CloseAllWindows [23]. Note that the *cleanUp* user-defined function described in this chapter is adapted from this macro example. The *Close All* command at [File > Close All] can also be used to close just image windows.

8. Both "Analysis" and "Pre-Analysis" workflows submit the images to the same image processing and segmentation steps meeting the parameter criteria set by the user: (1) *Subtract Background* [Process > Subtract Background...] applies the *Rolling ball radius* set by the user. It is recommended to not use a rolling ball radius smaller than the largest object in the image. (2) *Enhance Contrast* [Process > Enhance Contrast...] applies the percentage of pixels allowed to become saturated set by the user. Normally, 0.1% saturation is enough (excessive number of saturated pixels is not desirable), although low-contrast image datasets would benefit from higher percentage. Since *Normalize* is checked, the image pixel values are recalculated to meet the maximum range of the image type (e.g., 0–255 for 8-bit images). In order to skip the *Enhance Contrast* algorithm, the drop-down menu includes a "None" option. (3) *Gaussian Blur* [Process > Filters > Gaussian Blur...] applies the *Sigma (Radius)* set by the user to smooth the image in order to make both background and objects more homogeneous, hence facilitating segmentation. Usually slight blurring (1–2 radius) would be enough. A radius of "0" will skip this step. (4) Segmentation is performed by *Threshold [Image > Adjust > Threshold...]* automatic methods. All the algorithms available can be obtained by using the *getList ("threshold.methods")* built-in function. The choice of algorithm would have a great impact on the workflow output, and different algorithms might work better than others for a specific image dataset. (5) *Convert to Mask* [Process > Binary > Convert to Mask] and (6) *Fill Holes* [Process > Binary > Fill Holes] do not need any user input and are always applied in order to perform the segmentation and fill the hollow particles, respectively. (7) *Erode* [Process > Binary > Erode] and

(8) *Open* [Process > Binary > Open] are performed applying to the binary mask the iterations (repetitions of the same algorithm) set by the user, so can be skipped setting them as "0." Both options are useful to clean the binary mask of unwished small particles, but *Open* is preferable if the user wants to affect the bigger particles to a lesser extent. It is really important to clean the nucleoside analogue image as much as possible from background particles since the size of the objects of interest may be smaller than in the nuclei image (small spotted pattern). (9) *Watershed* [Process > Binary > Watershed] can be useful to separate convex touching particles. Since applying it to extremely clumped objects can be counterproductive, it is offered as an optional step (checkable box). (10) "Size selection" is performed according to the limits established by the user by using *Analyze Particles* [Anayze > Analyze Particles...] and selecting *Mask* at the *Show* drop-down menu. In the case of the general nuclear staining, the size limits must include the expected nuclei size in the image. However, since nucleoside analogue particles might be much smaller, we recommend setting a much lower limit. The values are taken as size-squared units for scaled images or as pixel units for images without spatial scale.

9. The *cleanUp()* user-defined function closes the images that remain open and the *Results* and *Threshold* windows. By means of the *isOpen* built-in function, placed as an *if* condition, the macro checks if the *Threshold* and *Results* windows are open and, if true, closes them at [File > Close]. Additionally, a combination of the built-in function *nImages* and a *while* statement closes all open images: as long as *nImages* function returns a result greater than zero, the statement is repeated until there are no remaining open images.

Acknowledgments

This work was supported by Spanish Ministerio de Ciencia, Innovación y Universidades (project SAF2014-54581-R) and cofinanced by the European Social Fund.

References

1. Eidinoff ML, Cheong L, Rich MA (1959) Incorporation of unnatural pyrimidine bases into deoxyribonucleic acid of mammalian cells. Science 129(3362):1550–1551
2. Chehrehasa F, Meedeniya AC, Dwyer P, Abrahamsen G, Mackay-Sim A (2009) EdU, a new thymidine analogue for labelling proliferating cells in the nervous system. J Neurosci Methods 177(1):122–130
3. Scholzen T, Gerdes J (2000) The Ki-67 protein: from the known and the unknown. J Cell Physiol 182(3):311–322
4. Silvestrini R, Costa A, Veneroni S, Del Bino G, Persici P (1988) Comparative analysis of

different approaches to investigate cell kinetics. Cell Tissue Kinet 21(2):123–131

5. Tapia C, Kutzner H, Mentzel T, Savic S, Baumhoer D, Glatz K (2006) Two mitosis-specific antibodies, MPM-2 and phospho-histone H3 (Ser28), allow rapid and precise determination of mitotic activity. Am J Surg Pathol 30(1):83–89
6. Schindelin J, Arganda-Carreras I, Frise E, Kaynig V, Longair M, Pietzsch T, Preibisch S, Rueden C, Saalfeld S, Schmid B, Tinevez JY, White DJ, Hartenstein V, Eliceiri K, Tomancak P, Cardona A (2012) Fiji: an open-source platform for biological-image analysis. Nat Methods 9(7):676–682
7. Schneider CA, Rasband WS, Eliceiri KW (2012) NIH image to ImageJ: 25 years of image analysis. Nat Methods 9(7):671–675
8. Schindelin J, Rueden CT, Hiner MC, Eliceiri KW (2015) The ImageJ ecosystem: an open platform for biomedical image analysis. Mol Reprod Dev 82(7–8):518–529
9. https://github.com/paucabar/Cell_proliferation_assay. Accessed 5 Oct 2018
10. https://imagej.nih.gov/ij/developer/macro/macros.html. Accessed 13 Sep 2018
11. https://imagej.net/Script_Parameters. Accessed 13 Sep 2018
12. Landini G (2008) Advanced shape analysis with ImageJ. Proceedings of the second ImageJ user and developer conference, Luxembourg, 6--7 November, p 116–121. ISBN 2-919941-06-2. Plugins available from http://www.mecourse.com/landinig/software/software.html. Accessed 13 Sep 2018
13. Belenguer G, Domingo-Muelas A, Ferron SR, Morante-Redolat JM, Farinas I (2016) Isolation, culture and analysis of adult subependymal neural stem cells. Differentiation 91 (4–5):28–41
14. Ferron SR, Andreu-Agullo C, Mira H, Sanchez P, Marques-Torrejon MA, Farinas I (2007) A combined ex/in vivo assay to detect effects of exogenously added factors in neural stem cells. Nat Protoc 2(4):849–859
15. Hua K, Ferland RJ (2017) Fixation methods can differentially affect ciliary protein immunolabeling. Cilia 6:5
16. https://imagej.nih.gov/ij/macros/DialogDemo.txt. Accessed 13 Sep 2018
17. https://imagej.nih.gov/ij/macros/DialogDemo2.txt. Accessed 13 Sep 2018
18. https://imagej.nih.gov/ij/macros/SineCosineTable2.txt. Accessed 13 Sep 2018
19. https://imagej.nih.gov/ij/macros/BatchModeTest.txt. Accessed 13 Sep 2018
20. https://imagej.nih.gov/ij/macros/BatchMeasure.txt. Accessed 13 Sep 2018
21. https://imagej.nih.gov/ij/macros/BatchSetScale.txt. Accessed 13 Sep 2018
22. https://imagej.nih.gov/ij/macros/ReplaceRedWithMagenta.txt. Accessed 13 Sep 2018
23. https://imagej.nih.gov/ij/macros/CloseAllWindows.txt. Accessed 13 Sep 2018

Chapter 15

HCS Methodology for Helping in Lab Scale Image-Based Assays

Joaquim Soriano, Gadea Mata, and Diego Megias

Abstract

High-content screening (HCS) automates image acquisition and analysis in microscopy. This technology considers the multiple parameters contained in the images and produces statistically significant results. The recent improvements in image acquisition throughput, image analysis, and machine learning (ML) have popularized this kind of experiments, emphasizing the need for new tools and know-how to help in its design, analysis, and data interpretation. This chapter summarizes HCS recommendations for lab scale assays and provides both macros for HCS-oriented image analysis and user-friendly tools for data mining processes. All the steps described herein are oriented to a wide variety of image cell-based experiments. The workflows are illustrated with practical examples and test images. Their use is expected to help analyze thousands of images, create graphical representations, and apply machine learning models on HCS.

Key words HCS, *ImageJ* macro language, Data mining, Automated analysis, Machine learning

1 Introduction

High-content screening (HCS) technology can be resumed as the combination of automated microscopy and image analysis to extract quantitative data from cell populations [1]. Pharmaceutical companies have been using HCS in drug discovery for many years [2, 3]. However, over the past decade, academic environments have also incorporated this kind of assays into basic research experimental approaches, increasing the number of biological applications and their complexity [4–6]. The latest technological advances have generated a fast evolution of this technology, and nowadays it is used for testing thousands of potential therapeutic compounds, as well as for cell perturbation treatments, such as RNAi and gene mutagenesis [7–9].

The use of conventional automated microscopes combined with open-source image analysis software has greatly reduced

J. Soriano and G. Mata contributed equally to this work.

Elena Rebollo and Manel Bosch (eds.), *Computer Optimized Microscopy: Methods and Protocols*, Methods in Molecular Biology, vol. 2040, https://doi.org/10.1007/978-1-4939-9686-5_15,

HCS costs, popularizing this technology either for large- or medium-scale assays. This makes it really important to define clear workflows and robust standardization strategies.

HCS experiments can be divided into at least four steps: (i) assay design, (ii) image acquisition, (iii) image analysis, and (iv) results review and data mining. The blind confidence in HCS results obtained from image analysis platforms with limited revision or without considering possible sources of variability is a high risk that is addressed in this chapter.

In order to facilitate HCS implementation, we have developed a set of tools that includes a specific routine on *ImageJ* [10], at present one of the most widespread scientific image analysis software. This routine does not need installation and is easy to use, since graphical user interfaces guide the user through all the process. In addition, we provide workflows, developed in *Orange* [11] for data mining tasks. *Orange* is an open-source software devoted to graphical visualization and machine learning (ML), with an intuitive and user-friendly environment. For better comprehension, the purposed HCS workflow will be illustrated step by step with real examples. The materials used for these examples can be applied, with minor adjustments, to a wide variety of image cell-based experiments.

2 Materials

We provide all the materials needed to implement the forthcoming protocol:

1. Representative images obtained from HCS assay [12]. Images belong to a cell-based assay aimed at determining whether different drugs induce nuclear translocation of a fluorescent marker (assay information can be found in [13]). Images were acquired on a HCS equipment Opera LX system (Perkin Elmer) using a 20× Dry 0.5 NA objective. Ten fields were captured on each well of a 96-well plate (micro-clear Greiner-Bio one). Two images were taken per field (Fig. 1): one registering DNA staining (channel 1: DAPI) and a second containing a green fluorescent protein signal (channel 2: GFP). Each well contained adherent cultured cells that were either untreated (control samples), treated with a drug that induces GFP nuclear translocation (stimulated control samples), or treated with different drugs whose effect on GFP nuclear translocation is unknown (experimental samples).
2. Image J Macro. Two macros are provided as supplementary material. The first macro, "HCS_Analysis.ijm," is intended to (i) detect and measure objects, (ii) compute the experiments Z factor and Z scores, and (iii) check whether the objects

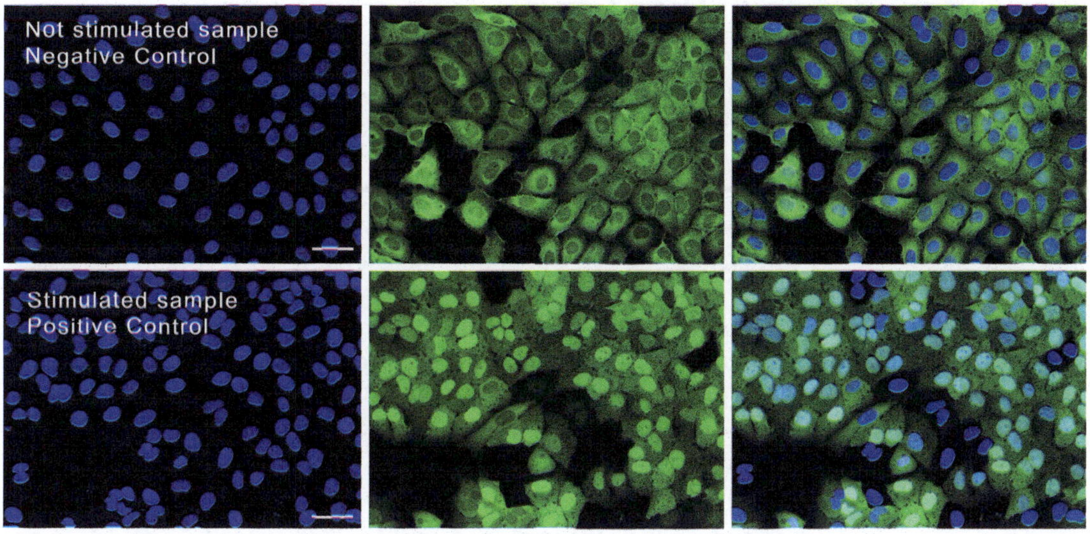

Fig. 1 Representative images of assay control samples. Two channels and merged images for untreated (top) and treated (bottom) samples with the nuclei staining in blue and showing the changes in green fluorescent protein localization. Scale bar: 50 μm

detection is properly done. The second macro, "detection_macro.ijm," can be used as a template to design a more complex object detection. This last macro can be automatically called by "HCS Analysis" (*see* **Step 9** in Subheading 3.2.1) and is configured to detect nuclei on the images provided. In order to use the first macro, copy the files "HCS_Analysis.ijm" and "HCS-Check results.ijm" to the *ImageJ*'s plugins folder. All files are available at [12]. *ImageJ (Fiji)* software can be downloaded from [14].

3. Data results. The files "ex_plate_data.csv" and "ex_plate_data_class.csv" are included for testing data mining processes. The results file obtained from the example images can also be used, but the provided ones contain more information for a better user's experience [12].
4. *Orange* workflows. Files to load pre-configured workflow examples for data review are included. These files will help create heat maps, establish features ranking, make variables reduction, and build supervised or unsupervised machine learning classifications; all workflows and data files are available at [12]. The *Orange* software can be downloaded from here [15].

3 Methods

3.1 Recommendations for HCS Assays

The design of HCS assays has been addressed by many authors [16–19]; reviewing this topic is out of the scope of this chapter. In any case, the following subsections cover the most relevant tips, according to our expertise.

3.1.1 Assay Design

1. Define the cell phenotype beforehand. In automated assays, it is critical to have a clear cell phenotype and a set of variables to describe it numerically. For example, in the provided case, fluorescent signal translocation can be determined by measuring the mean nuclear green intensity. In the non-treated samples, most of the signal is located at the cytoplasm, but the nuclear signal increases when a cell is stimulated.
2. Mind the controls. One of the differences between conventional and automated assays is that, usually, the latter are repeated in time, which requires a higher level of standardization. While in conventional assays control samples are used to test staining specificity using different combinations of the primary and secondary antibodies, in automated assays the possible variability of staining intensity or changes in treatment activity through the different samples or days should be considered. When repeating an experiment, tested samples may not be in the same growth rate, reagents dispensing could vary, and their efficacy may change with different production batches [20] or storage time. In order to solve these issues, it is important to add to this kind of assays activity controls, i.e., samples treated with a reagent known to stimulate the effect under study. The comparison between this activity control and a non-treated control will define the assay window, showing the minimum and maximum effect we should expect in our assay (*see* Table 1 and **Note 1**).
3. Standardize staining and treatments. It is important to check all your materials and reagents and standardize your experiment. The staining conditions should be checked beforehand, making sure cell morphology is unaltered, especially when performing in vivo experiments. Keep staining concentrations to a minimum in order to avoid signal cross-talk and reduce the experiment cost. Since fixation is usually a critical step in most of the procedures, fixation artifacts must be always carefully checked.
4. Plan your replicates. The number of technical replicates of the controls and experimental samples depends on the variability of the assay. There is no fixed rule to define a general number of replicates. As a guideline, if the amount of samples is huge and the differences between stimulated and non-stimulated samples

Table 1
Control samples recommended for HCS assays

Recommended controls	
Sample	**Purpose**
Not stained	Autofluorescence levels Extremely important in tissue samples
Sample with only secondary antibodies (Abs)	Test specificity/cross-reaction of secondary Abs
In double-stained samples, prepare two samples combining first primary Ab with secondary Ab for the second primary Ab (and the other way around)	In cases where Ab are developed in similar species (such as rat and mouse), you can observe cross-reaction between Abs
Full staining not treated	Basal level of activity
Full staining treated known stimulus	To know the induced level of activity
Not stained treated known stimulus	Is rare but sometimes the treatment increases autofluorescence or background

are clear, it is possible to work without replicates, keeping in mind that this strategy will increase the number of samples that will require further confirmation. It is usual to work with triplicates when the assay is based on a complex cell phenotype or when it has intrinsic sources of variability. These three values per sample create a trend in the results reducing the variability and the possibility of false categorization. Adding more technical replicates should be always carefully considered, since each additional one increases the costs to benefits relationship of the experiment.

5. Watch the cell confluence. This is an important variable. The correlation of assay results needs to systematically reproduce cell confluency throughout the different samples. Comparing biological effects from samples with variability in confluency is complex, since they usually have different growth rates, but additionally many cells will create clumps that will make it difficult to get data from single cells, and too few cells per field will require the acquisition of more images per sample. Some experiments require using specific cell models, while others allow for choosing the cell line. In the latter, it is important to run a pilot experiment to compare expression and activity levels between different cell lines.

3.1.2 Image Acquisition

Image capturing should balance image quality and acquisition throughput. The following tips are meant to that aim.

1. Mind the number of images per sample. This can vary depending on the cell confluence and the frequency of the studied

event on each random picture. For example, assays that check cell division aberrations need to quantify mitotic cells, and these are usually a small percentage of the total population. To estimate the minimum number of pictures needed, automated acquisitions with different number of images on a control sample can be used. At a certain number of images, values should reach an equilibrium.

2. Adjust exposure times. In order to increase acquisition throughput and avoid possible signal saturation, firstly set up the exposure times on the stimulated control sample, and then check the untreated ones. The signal-to-noise ratio should always be considered.
3. Minimize the number of channels. This depends on the cell phenotype observed. Each additional channel in the capture delays the acquisition and increases its cost. Sometimes, it is better to add channels in a further experiment with only the candidate samples obtained from a more simple experiment. Whenever possible, it is important to use different channels for image segmentation (in our example DAPI to find the nuclei) and signal quantification (GFP in the example).
4. Choose the proper imaging technology. Both confocal and wide-field microscopy are used in automated captures. Usually, confocal images have higher contrast and therefore better signal-to-noise ratio, facilitating segmentation and quantification steps. However, confocal acquisition is time-consuming and expensive, so it can be replaced by wide field for applications where three-dimensional structures are not critical, such as nucleus quantification in cell monolayers, number of cells, etc.

3.2 HCS Processing

The following subsections cover the three main steps regarding HCS experiments processing: (i) Image Analysis and (ii) Results Review.

3.2.1 Image Analysis

Objects identification is one of the most critical steps in image analysis; therefore, the automation of this process should be done very carefully. First, it is important to assume that, due to the high number of images, it will not be possible to manually review most of the identified objects; alternatively, quality checks should be added to the quantification protocol to make sure the objects are properly detected. This quality checks are useful to distinguish possible side effects and outliers from positive results and can be defined based on parameters such as the number of detected objects, its size, or any morphological criteria. These should always be measured on control and experimental samples, keeping the values from controls as a reference to detect and evaluate any possible deviations. Once the quality criteria are defined, any sample not meeting them

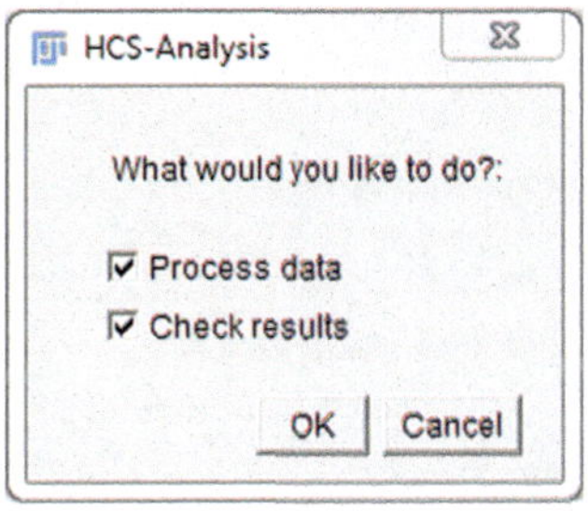

Fig. 2 Dialog to start analysis process and/or check results

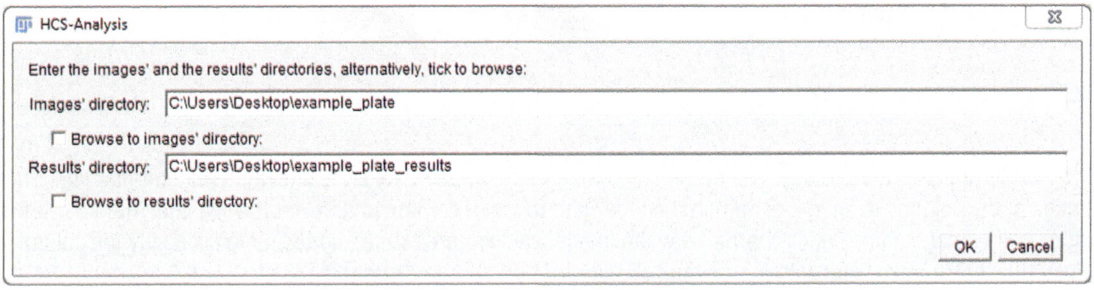

Fig. 3 Dialog to define directory paths for images and results.

should be observed, preferably on the original images, in order to discriminate between true object variability and artifacts, for the latter to be excluded from the dataset.

"HCS Analysis" provides a means to automatically detect and measure objects of interest as well as compute Z factor and Z scores, values that will help evaluate the screening quality and find possible hits. The following steps are meant to help through all the steps involved in an HCS assay, including the screening quality, the image processing, and the reviewing of the results.

1. Start *ImageJ*. Choose "HCS Analysis" in the Plugins menu.
2. On the following dialog (Fig. 2), choose *Process data* in order to detect and measure objects of interest. Choose *Check results* if you wish to observe detected objects on image sets (*see* **Note 2**).
3. Set the paths to the folder containing the images to be analyzed and to the folder where to save the results (Fig. 3, *see* **Note 3** for more information and file format requirements).
4. Enter the digits of the images' names that codify the sample's row and column; the number of positive controls, negative controls, and experimental samples; and the number of images that were captured per field (Fig. 4). More information and a second example can be found in **Note 4**.

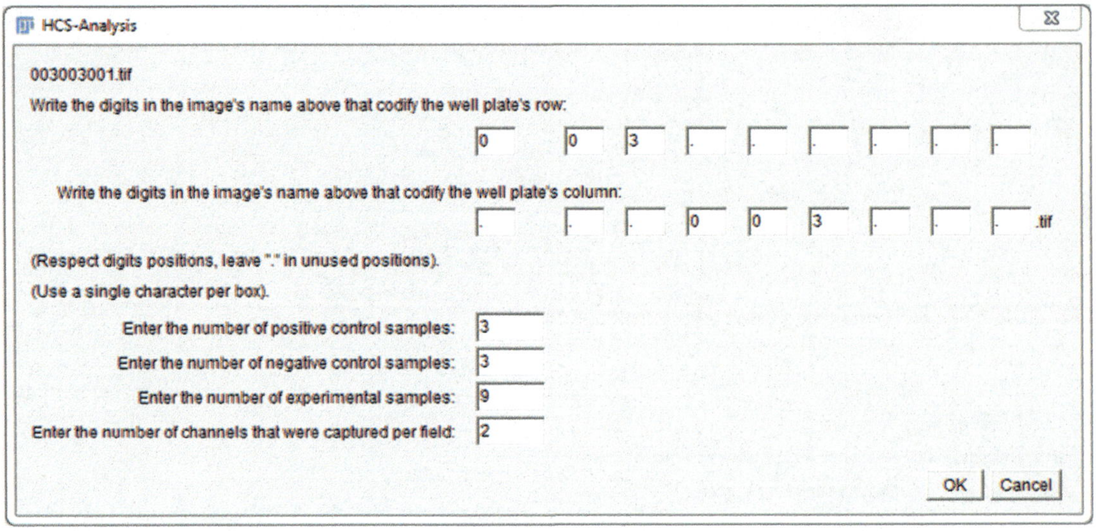

Fig. 4 HCS analysis third dialog. The first image name in the images' folder is shown in the first text line. The image name belongs to an image captured on the third row and column of a 96-multiwell plate. The first three digits of the image name codify the well row the image was captured from; digits four to six codify the column. Numerical fields were filled with the number of samples and captured channels (3, 3, 9, and 2, respectively, in the ongoing example)

5. Choose the variables which are going to be measured (Fig. 5a). Detailed information on these features is shown by clicking on the *Help* button (*ImageJ*'s wiki will be open).
6. Choose the channels where objects of interest are and which one you want to measure (Fig. 5b).
7. Choose a thresholding method. Mark *Select objects that fulfill* if you want to set all parameters now. Mark *Apply a pre-programmed thresholding routine* if you have a designed macro and jump to **Step 9**. Only one option should be selected (unmarking the other option).
8. Configure the parameters for the thresholding method as shown in Fig. 5c, where objects brighter than 35 pixel values, bigger than 190 pixels, smaller than 1700 pixels, and whose circularity is bigger than 0.6 get selected; after implementing a watershed algorithm, the objects are extracted for quantification.
9. This option is chosen by simply selecting the checkbox *Apply a pre-programmed thresholding routine* in Fig. 5d and entering the macro's path either manually or using a browser. For more information about object detection routines, *see* **Note 5**.
10. Enter the name of an image belonging to each positive and negative controls and choose whether to compute the *Z* factor (Z') and/or *Z* scores (Fig. 6). Names of images can be

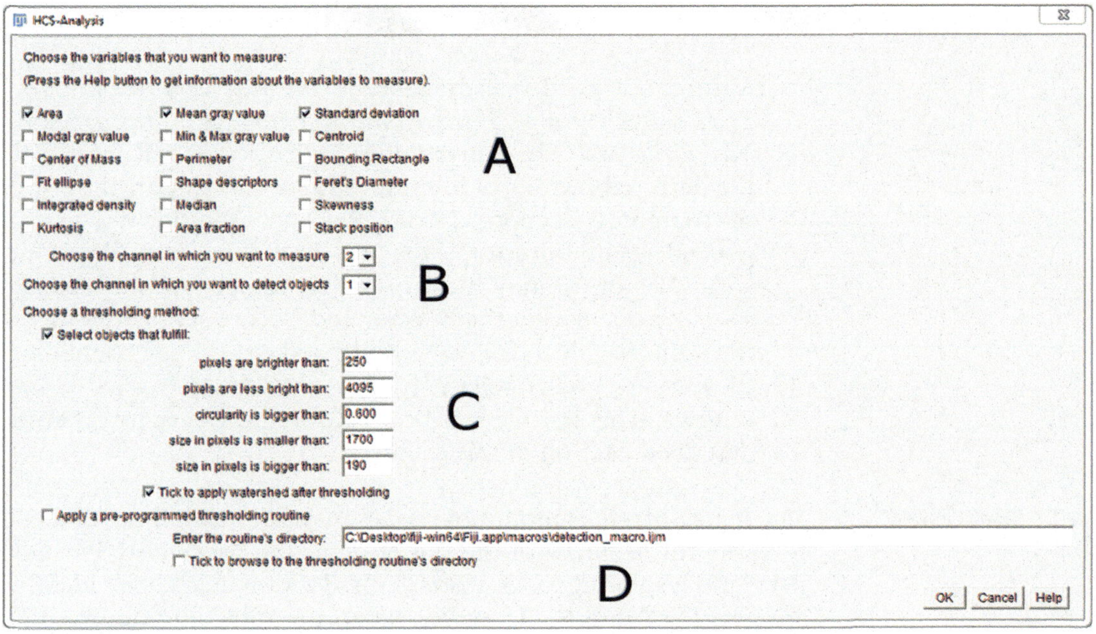

Fig. 5 Dialog for image analysis options selection. (**a**) Objects variables to be measured. (**b**) Channels of interest. (**c**) Object detection parameters, (**d**) Alternative detection macro option

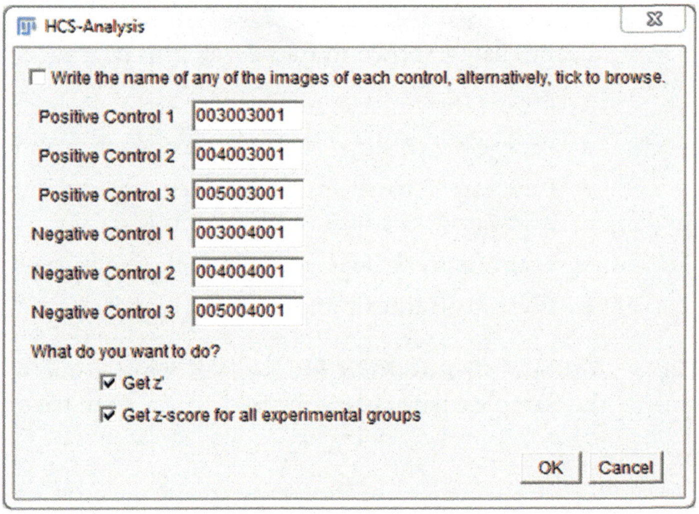

Fig. 6 Dialog for controls selection and *Z* factor calculations

introduced either manually or using a browser. For more details and the formulas description, *see* **Notes 6** and 7.

11. After pressing the *OK* button (*see* Fig. 6) on the graphical user interface, "HCS Analysis" will process all images according to the introduced parameters. As a result, four folders will be created on the results directory. Three of them will be for experimental samples and for negative and positive controls—

Experimental conditions, *Negative Controls*, and *Positive Controls*. They will contain text files with the numeric results of the features selected for analysis—*Experimental wells resume/Negative controls resume/Positive controls resume* and *Measurements Well nameWell*. The fourth folder (called *ROIs*) will contain all the detected regions of interest (ROIs), allowing further revision (*see* next section, Results Review). Additionally, in the general results directory, you will find three text files: One with *Z*-factor (prime) calculations (file named *z-prime resume*), a second one having the *Z*-factor and z-score summaries for all the samples (*HCS Analysis resume*) and a third one containing the mean per well of all the features analyzed (file called *means resume*). This last file can be used for the *Orange* workflows described later on in this chapter.

3.2.2 Results Review

The increase in the amount of images makes it much more difficult to verify the accuracy of the object detection procedure. Plotting data and exploring its statistical distribution may help identify artifacts (*see* **Note 8**). However, when the variability observed in the results file is moderate, it is not easy to determine whether the result is real or an error produced in a certain step of the image analysis. To address this problem, the "HCS Analysis" macro provides a means to identify the detected objects on the original images. When the option *Check results* is selected (Fig. 2, *see* **Note 2**), the dialog shown in Fig. 7 will pop up.

1. Press *OK*.
2. Press *key 1* to show all detected objects in an image (*see* Fig. 8).
3. Press *key 2* to show all detected objects in a stack of images belonging to a well or piece of tissue.
4. You can press *key 1* or *2* in any moment to check the images.
5. Press *key 3* to exit and stop the macro.

3.3 HCS Data Mining

Although theoretically HCS experiments come to an end when all the samples have been imaged, once you have the results, it is

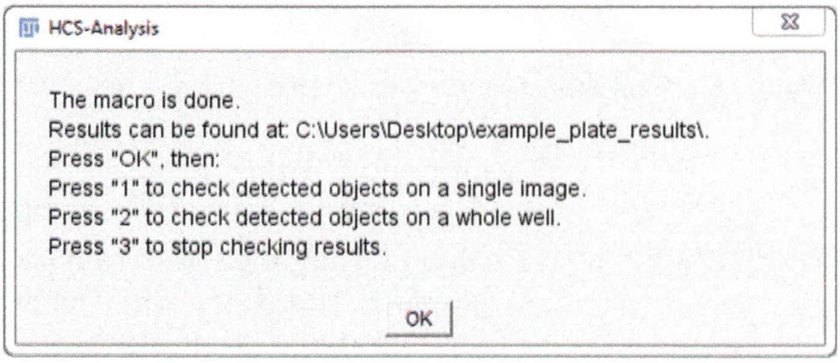

Fig. 7 Dialog window for *Check results* options

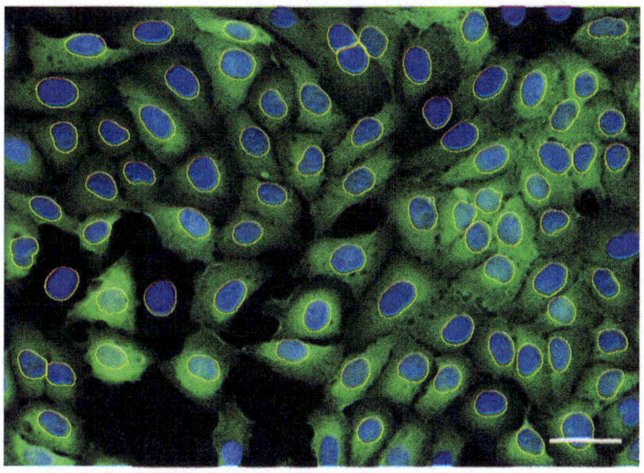

Fig. 8 Example of nuclei segmentation review. Yellow circles mark the objects detected on the blue channel. Scale bar: 50 μm

critical to have tools available for their visualization, interpretation, and manual or automated classification [21–23]. The following subsections cover a series of protocols directed to these aims. First, one of the most useful ways of presenting HCS data is introduced (Subheading 3.3.1): heat maps. Then, several classification methods are provided. The most common way to classify data is to determine manual thresholds using the variables of interest or by using the z-score. However, every time a threshold is set up, there is a probability that a false positive or a false negative is classified. A good threshold minimizes the number of false values and usually requires not only one variable but a combination of them. The multiparametric results associated with HCS assays make it more difficult to establish valid thresholds using only one feature. To help to this purpose, machine learning [21–23] classification solutions can be used (Subheading 3.3.2), as well as principal components analyses (PCA; Subheading 3.3.3). Also, supervised and unsupervised classification methods are provided. Unsupervised learning (the dataset is not previously categorized or classified) is here approached by cluster analysis, a strategy that groups a set of data in such a way that instances (or samples) from the same cluster are more similar to each other than to those in the other clusters. We use mainly two methods of unsupervised learning, k-Means [24] (Subheading 3.3.4) and hierarchical clustering [25] (Subheading 3.3.5). In cluster analyses, the dataset is not categorized or classified previously. Rather, a strategy is used to group a set of data in such a way that instances (or samples) from the same cluster are more similar to each other than to those in the other clusters. Additionally, supervised classification is approached (Subheading 3.3.6). Finally, the HCS output interpretation is covered in Subheading 3.3.7. All the procedures provided are based in *Orange*. Once this software is executed, it can open and run all the examples provided in this chapter. Each of the topics described

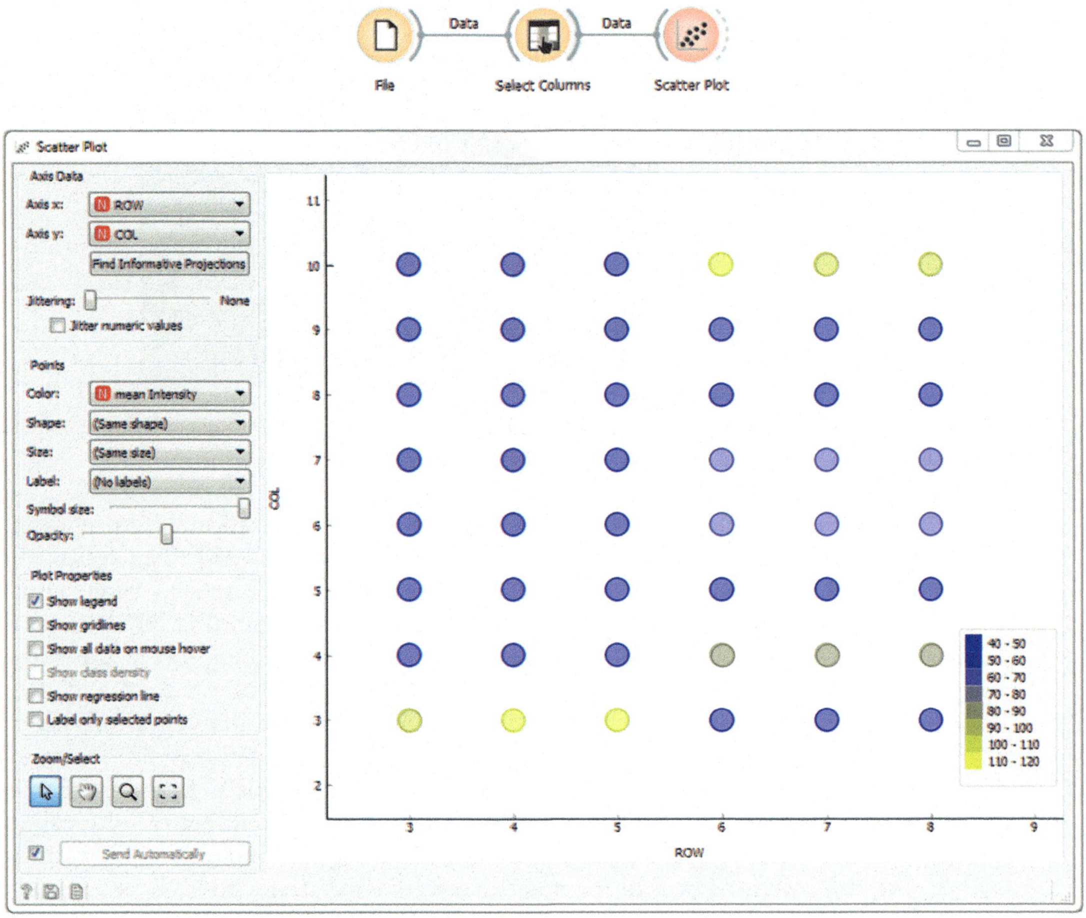

Fig. 9 Heat map example containing average intensities of the nuclei green mean intensity

below has a specific template for testing purposes. For some tips about using *Orange*, *see* **Note 9**. In addition, there are some details of some widgets used in the workflow explained in **Note 10** for better understanding.

3.3.1 Results Visualization: Heat Maps

Heat maps code the results from the different samples and assign them colors, making it easier to visualize the differences between samples. To generate a heat map in *Orange*, we use a *File* and a *Scatter Plot* widget as explained in the following steps (*see* Fig. 9):

1. Open the *Orange* file "HCS Heat Map.ows."
2. Click on the widget called *File* and load the file: "ex_plate_data.csv." In this example, the widget called *Select Columns* allows you to check the variables whether they are placed correctly.
3. Click on the *Scatter Plot* widget. It shows a heat map for each feature of the sample. You can change the feature in the *Color* option.

3.3.2 Results Classification: Machine Learning

Humans since babies are trained routinely to classify the surrounding objects into categories and perform this kind of task as something natural. The problem arises when the number of objects to be classified increases, resulting in a tedious job, or when the dimension exceeds the human capabilities. Machine learning is a field of artificial intelligence (AI) that tries to solve this kind of problems. Specifically, the purpose of machine learning is to develop techniques allowing the computer to adapt its behavior from examples [24]. There are three main parts in the process of ML, input, algorithm, and output, and each separate part is complex, requiring special attention. For more information about these parts, please *see* **Note 11**.

Results from HCS assays gather all the features obtained from an image. Selecting the most useful features is essential for success using ML. The *feature selection* strategies are aimed at reducing the number of features (a number known usually as dimension), thus removing those that are either redundant or irrelevant without losing too much information. These methods simplify the models, making the interpretation of the results easier, shortening the training time, and avoiding high dimensionality. The algorithms compute the features correlation and score them according to their matching with the class. In order to add the class, the file with the data ("ex_plate_data.csv") can be edited in Microsoft Excel—or in a similar program. Then a column with a label for the positive and negative controls and the rest of the wells labeled as "experimental" should be added. The following workflow (Fig. 10) is an example:

1. Open the file "HCS Feature Selection.ows."
2. Load the file called "ex_plate_data_class.csv" using the *File* widget.
3. Use the *Select Columns* widget to choose the feature that determines the class; in the current example, this feature is "class." It is necessary to have a dataset with the control samples manually classified.
4. Using the *Select Rows* widget, you choose only the samples that are classified as positive or negative controls. The class of the remaining samples is named as "experimental," since we do not know if they are going to be active or not.

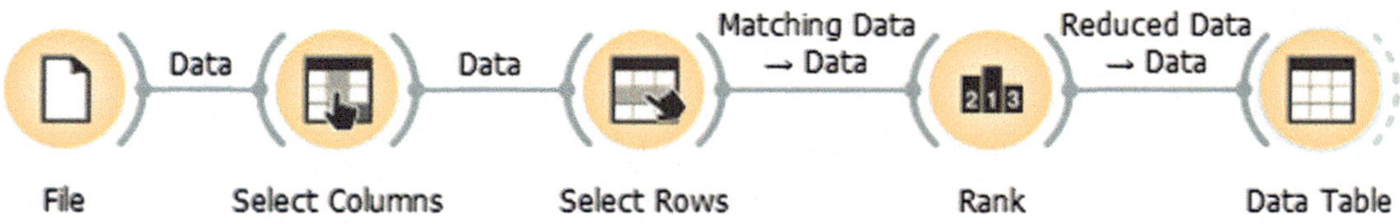

Fig. 10 *Orange* workflow example where the widget called *Rank* is used to obtain fewer features

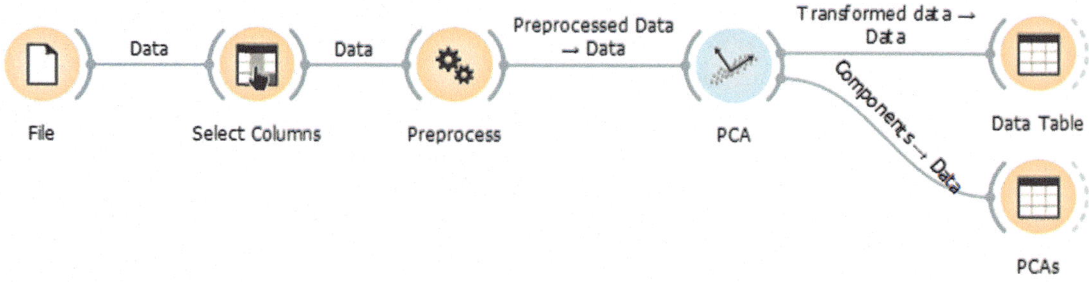

Fig. 11 *Orange* workflow example where the widget called PCA is used to reduce the number of features

5. The *Rank* widget shows the relevance of the features, according to different algorithms.
6. Use *Data Table* widget to check the relevance of the features for each instance or sample.

3.3.3 Results Classification: PCA

Principal component analysis [26] is another method that reduces the dimensionality of the samples. This procedure converts a set of possibly correlated features into a set of linearly uncorrelated new features called principal components. The components are sorted according to their original variance, that is, the first component has more variability in the data than the rest. The following steps (*see* Fig. 11) explain how to work on it:

1. Open "HCS Principal Component Analysis.ows" file.
2. Use the *Select Columns* widget to put the features that are not relevant for the classification as *Meta Attributes*. In the example, these features are "ROW" and "COL."
3. Use the *Preprocess* widget to normalize the data.
4. Click on the *PCA* widget for choosing the number of components. The provided example is small; however, it is enough to understand this concept. In this case, two components were used; therefore, the dimension is reduced to two, and the variance covered is high.
5. See the new features named components in the *Data Table* widget. In addition, the *PCA* widget shows the relation among to features to create the components.

3.3.4 Unsupervised Sample Classification: k-Means

The following workflow describes how to use the unsupervised method called k-Means using the data provided (*see* Fig. 12 and **Note 12** for more information):

1. Open "HCS k-Means Unsupervised Learning.ows" file.
2. As in the previous example, use the *Select Columns* widget to put the features which are not relevant for the classification as *Meta Attributes*. In our example, these features are "ROW" and "COL."

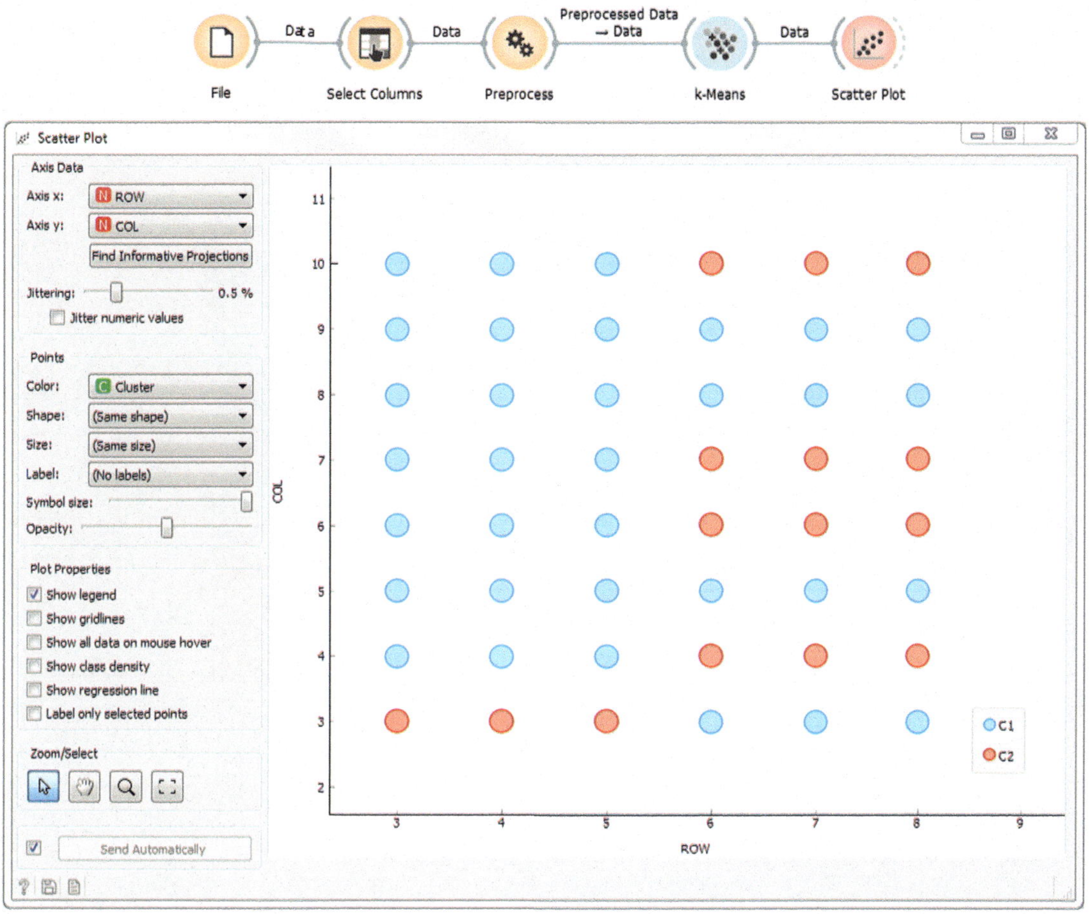

Fig. 12 *Orange* workflow example to work with a k-Means methods (upper panel); below, the results obtained using two clusters to classify the sample

3. Use the *Preprocess* widget to normalize the data.
4. Use the *k-Means* widget to fix the number of clusters to divide the dataset. In the example, the number of clusters is set to two; the option to check what happens with a higher number of clusters can be also tested.
5. Check the results in the *Scatter Plot* widget. Use the option *Color* to show the results according to their cluster. In addition, it is required to fill out the option called *Axis Data* with "ROW" and "COL" in order to show the plot as in Fig. 12.

3.3.5 Unsupervised Sample Classification: Hierarchical Clustering

Using one learning method does not mean that other analyses are not possible. The following workflow (Fig. 13) shows how to perform unsupervised cluster analyses using a different approach, the hierarchical clustering [27] (*see* **Note 13**).

1. Open the file called "HCS Hierarchical Clustering.ows."

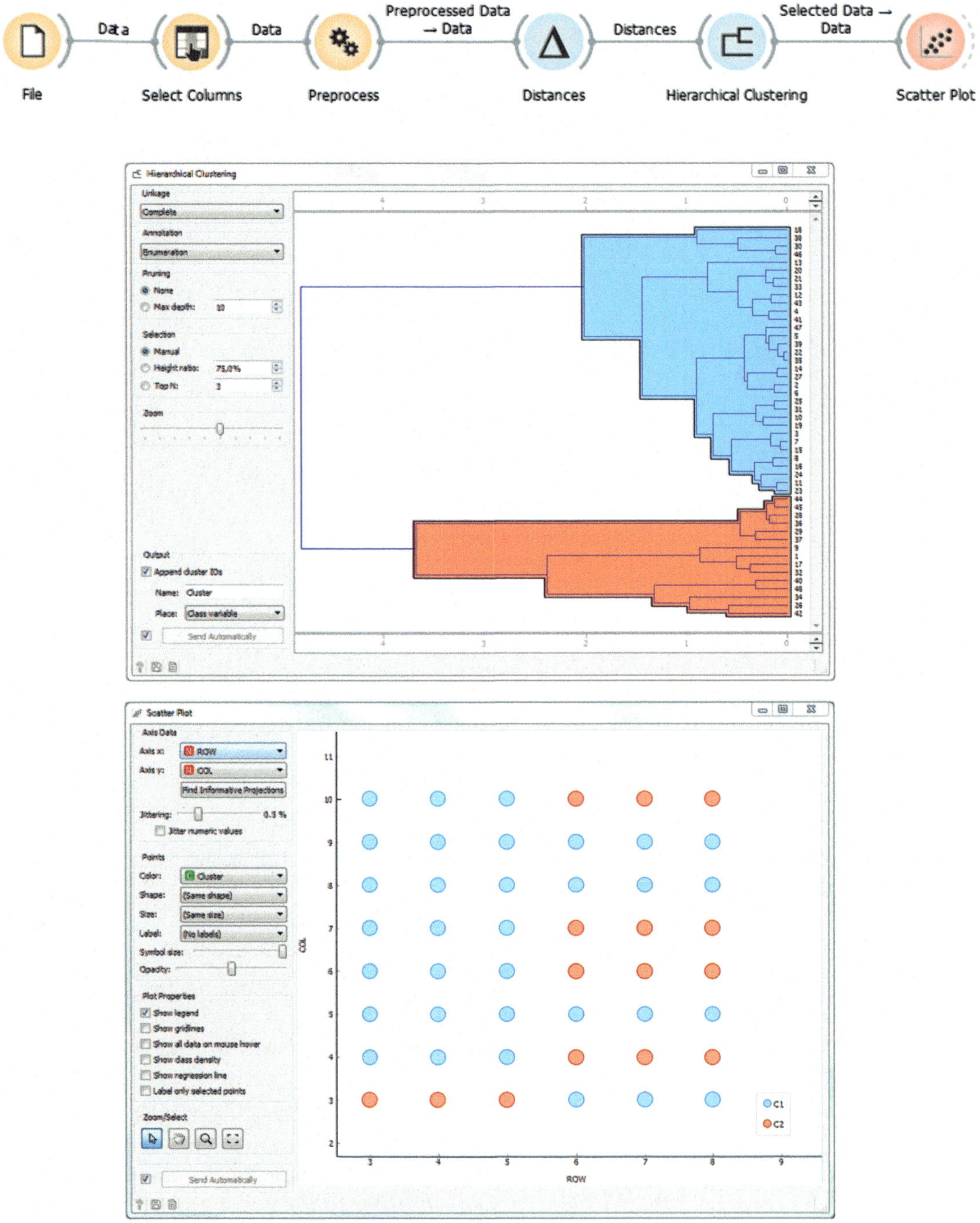

Fig. 13 Hierarchical clustering example. *Orange* workflow (top), dendrogram obtained for the sample (middle), and visualization of the results obtained using a scatter plot (bottom)

2. This step is the same as the previous example. Use the *Select Columns* widget to put the features which are not relevant for the classification as *Meta Attributes*. Use the *Preprocess* widget to normalize the data.
3. The *Distances* widget computes the distance between instances. In our example, this distance is computed between rows.
4. The result is shown using a dendrogram, which is visible by clicking on the *Hierarchical Clustering* widget.
5. Select the two groups that appear in the dendrogram, and use the *Scatter Plot* widget to show the results. These groups have to be selected manually; the two groups with the highest distance between them can be observed in Fig. 13.

3.3.6 Supervised Sample Classification: k-Nearest Neighbors

Supervised methods require training the computer by giving all the information obtained from an input of data (mean intensity, size, etc.), manually classified in the different classes. Based on this information, the algorithm learns and finds patterns, which will be used to classify other data without manual classification. The result is the data classification or the prediction of a number, in this case, called regression. In order to apply a supervised method, three datasets are necessary: a training dataset, a validation set, and a test dataset (*see* **Note 14** for more information). There are different methods of supervised learning; an example is an algorithm known as k-nearest neighbors (kNN) [19] (*see* **Note 15**). The following workflow example (Fig. 14) shows how to work on a supervised method:

1. Open file called "HCS kNN Supervised Learning.ows."
2. Load "ex_plate_data_class.csv" file, since in this case, we need to train the model knowing the class of some samples. In this example, we have manually classified the control samples as

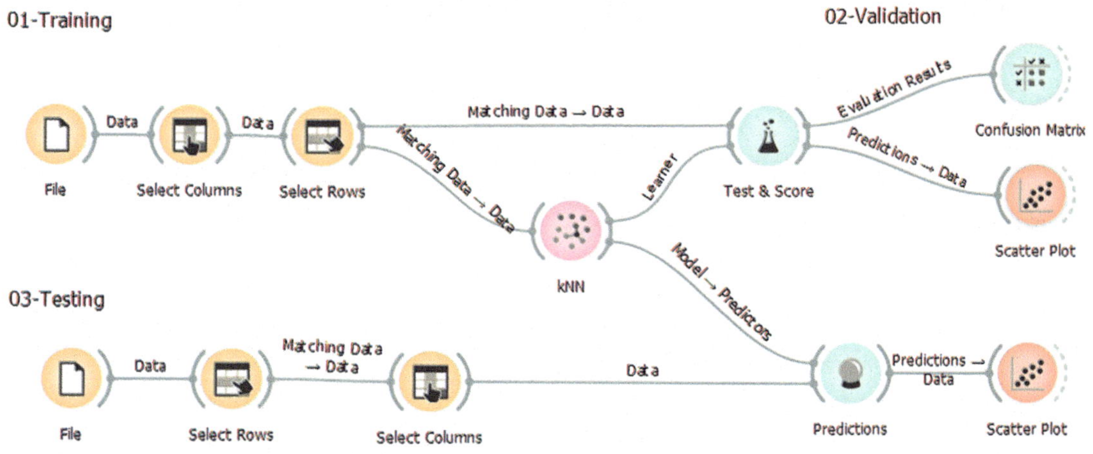

Fig. 14 The top workflow is designed to obtain a model from the dataset controls. The bottom one shows how to use this model to classify the whole dataset

negative and positive controls. The control samples are used as training and validation dataset; the rest are the testing dataset.

3. Use the *Select Columns* widget to put in *Meta Attributes* those features that are not relevant in the analysis as "ROW" and "COL." In addition, you have to choose the *Target Variable*, which is the "class" feature.
4. The *Select Rows* widget is used to choose only the samples for which we know their class. In our case, we choose the rows whose class is defined by "positive control" or "negative control." This subset is used for the training of the model.
5. The *Test & Score* widget needs an algorithm to be able to compute a model and validate the results. In our example, the model shown is *kNN* (other methods can be used in *Orange* software), and the method known as *Cross Validation* is used to validate the model, on the *Test & Score* widget. In addition, this widget shows some parameters of the validation as the precision and the recall; *see* **Notes 16** and **17** for more details.
6. The *Confusion Matrix* shows a table with the evaluation results.
7. Use the *Scatter Plot* widget to check that the data subset is correct.
8. To test the rest of the samples with this model, you have to load the same file called "ex_plate_data.csv" in the second *File* widget.
9. Use the *Select Rows* widget to select all rows classified as "experimental."
10. Use the *Select Columns* widget to empty the *Target Variable*, and drag and drop the irrelevant features for the analysis in the box for *Meta Attributes*.
11. The *Predictions* widget shows the result of the testing with the previous model.
12. To see the results in a graph, you can use the *Scatter Plot* widget and choose *kNN* result in the *Color* option.
13. The last step is the interpretation of the results. You can find more information about it in **Notes 16** and **17**.

4 Notes

1. The stimulus used for induction in activity controls should be chosen considering the maximum reaction expected in the test, a too exaggerated reaction renders an unrealistic difference between the treated and non-treated sample.
2. Choosing *Check results* will launch "HCS-Check Results.ijm." If only the *Check results* option is selected, results have

supposedly been obtained in a previous analysis session. An interface similar to that of Fig. 3 will pop up to enter images and results directories (go to **Step 3** on Subheading 3.2.1 Image Analysis section). If both *Process data* and *Check results* are selected, "HCS Check Results" will be automatically installed after object detection and measurement will be performed.

3. The "HCS Analysis" macro works under the assumption that all the images that are going to be analyzed are contained in the same folder (no images should be kept in any subfolder). Images creation date is supposed to be kept unaltered (so the first image in this folder is the first image that was captured and so on). All images on the same well or piece of tissue and images on the same field are supposed to be captured sequentially. Images with more than one channel must be a composite or a stack of images with a channel per slice. "HCS Analysis" only reads *.tif* format files. Not meeting this criterion may cause the macro to fail. Paths can either be manually introduced in a text field or specified at browser that will pop up after choosing a checkbox.
4. Alternative use of the third "HCS Analysis" dialog with a different well code for a sample on the third row and column of a 96-multiwell plate (*see* Fig. 15). The sixth digit of the image name codifies the row that contains the well where the image was captured; the eighth and ninth digits codify the column. "HCS Analysis" assumes that all image names have the same number of characters. If this were not the case, images should be renamed to fulfill this requirement (e.g., a collection of two images such as "well B-3.tif, well B-11.tif" should be renamed as "well B-03.tif, well B-11.tif").

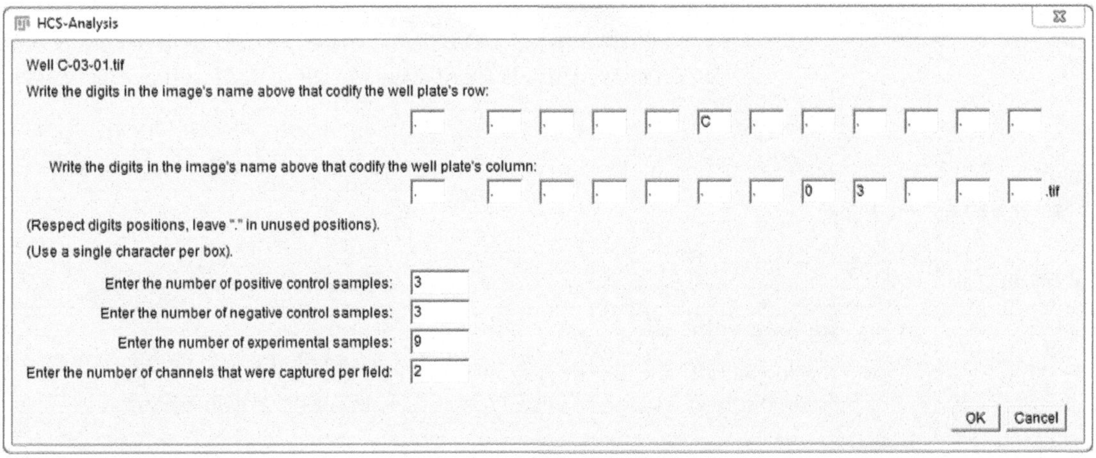

Fig. 15 Alternative use of the third "HCS Analysis" dialog with a different well code for a sample

5. Object detection routines should be designed to work using the object detection channel image as input and delivering a binary image as output. The output image background pixel value should be 0, and the detected objects pixel value should be 255. No global commands that affect non-selected images (e.g., such as a command closing all opened images) should be used. A simple object detection routine ("detection_macro.ijm") that detects objects based on the parameters shown in Fig. 5c is provided as a starting template to routine's design. To use a detection routine, select *Apply a pre-programmed thresholding routine*, and check that *Select objects that fulfill* option is unselected.

6. *Z*-factor and z-scores are the most widely accepted parameters used for evaluation of reliability and suitability of screening assays and for hit identification [16, 21]. Both *Get z* and *Get z-score for all experimental group* can be calculated with the "HCS Analysis" macro provided. The *Z*-factor considers in its calculation both the assay variability and the dynamic range, determining if there is enough difference between treated and non-treated controls. This will assess the assay quality and if it is going to be possible to distinguish the response from a candidate sample from the controls. The z-score is an index to identify possible hits from a screening [27]. This value quantifies the difference between samples and represents the number of standard deviations that separates a given sample from the mean of the data population. This score can be applied when the number of samples is high and most of the tested samples are inactive. *See* **Note 7** for more information and formulas description. When the number of samples tested is low or most of the samples show some activity but at different levels, other statistical tests such as *t*-test for pairs or Mann-Whitney and Dunnett's tests can be applied. The latter one makes several samples comparison against a control group.

7. The *Z*-factor is used to measure how reliable is the difference between our controls (*see* Table 2). It is calculated according to

Table 2
***Z* factor values' meaning**

Z factor	
>0.5	Assay with robust differences
0–0.5	Poor reproducibility check for alternatives to increase the value. It is accepted for complex phenotypic assays but needs a careful review of the possible results
Rest of the values	It is not likely to determine the differences for the assay to be useful

the formula below, where σ_{pos} is the standard deviation of technical replicates for the treated control sample, σ_{neg} is the standard deviation of technical replicates for the non-treated control, μ_{pos} is the average of technical replicates for the treated control sample, and μ_{neg} is the average of technical replicates for the non-treated control sample.

$$Z' = 1 - \frac{3(\sigma_{pos} + \sigma_{neg})}{|\mu_{pos} - \mu_{neg}|}$$

The *Z*-factor writes in the numerator the standard deviation of both control samples multiplied by three, so it penalizes the variability for both distributions and normalizes it with the difference between mean values in the denominator. In short, the *Z*-factor measures how tight is the distribution of control values, how far apart are the distributions of both positive and negative controls, and therefore if it is going to be possible to distinguish a positive effect in our test samples. The *Z*-factor should be tested on a pilot experiment with a small number of selected control samples before running a large-scale assay.

The *z*-score is calculated per each sample and is defined according to the formula below, where x_i is the sample result (e.g., mean cell intensity),

μ corresponds to the mean of all the samples (except controls), and σ is the standard deviation of all the samples (except controls).

$$z - \text{score} = \frac{x_i - \mu}{\sigma}$$

A higher z-score (e.g., above 2) means a strong difference with the rest of the samples.

8. Within a specific sample population, data values above or below two to three times the standard deviation of the mean in normally distributed populations are usually good candidates to be discarded as atypical data values.

9. We describe some useful tips if you are a beginner user in *Orange*. To use any widget and create or continue with a workflow, click on the widget and drag and drop on the right side of the screen. In addition, if you double click on a widget, a new window is opened with the configuration or the information of that widget. For example, if you are using a widget called *Scatter Plot*, when you double click on it, a window with a scatter plot of the data is shown. The widgets are linked among them; therefore, the previous widget supplies the next one. Moreover, some widgets as *Select Column* or *Select Rows* can be used to check that the variables are in the correct place before to use them to analyze.

10. Although you can find the information using the *Help* button inside each widget, we describe some details of some widgets as follows:

    Use the widget *Select Columns* to select and check the features. The features which are going to be used in the analysis have to be in the panel called *Features*. If the file contains at least a column with the class of the features, that feature has to be in the panel called *Target Variable*. The rest of the features which you want to use but they do not participate in the analysis will be in the panel called *Meta Attributes*.

    Rank widget is used to know the relevance features according to the correlation of the features with the class feature. There are different algorithms to be applied on the left side of the window.

    Preprocess widget is used to normalize the dataset among other aims. In our example, we normalize the features using a center by median and scaled by standard deviation. Therefore, the features are in the same range of the values.

11. There are three main parts in the process of machine learning, the input, the algorithm, and the output. Input refers to the data we want to analyze. It could be a set of images or data. It is important to know what we want to analyze, what information we can obtain from the data, and how we obtain it. This information is referred to as the attributes. In the case of an image, we need to transform the information found in the image into data. There are different kinds of features in an image, some of which are based directly on its intensity, for instance, the mean, the median, the skewness, etc. Other features include the perceived texture, which is quantified using a set of measurements computed from the image. Some texture features are an aggregation of simple features that give us information about the spatial arrangement of color or intensity in an image, such as its coarseness, contrast, directionality, and line-likeness, among others. Regarding algorithms, these can be classified into different types, being *Supervised* and *Unsupervised* the most important, both of which include different methods [19]. Finally, the output gives a solution to problems and reports whether our model is viable or not. A model is the result obtained from a set of samples used to train an algorithm. This result will classify additional samples. The final step is to understand and interpret the output data.

12. K-Means clustering [19] is a popular method for cluster analysis in data mining. It divides a set of data with n instances into k-clusters. Each instance belongs to the cluster which is closest, that is, the cluster which has the nearest mean. K-Means classify a dataset into k-clusters; the instances in each cluster are similar.

13. Hierarchical clustering [21] is a method that creates a hierarchy of clusters. The key point in this method is to use an appropriate metric since it will determine the shape of the clusters. According to the metric, the instances may be close to other ones or vice versa. In fact, a cluster is formed when all the dissimilarities among the instances in the cluster are below a particular level. To visualize the results, a dendrogram is used.
14. There are three datasets needed for supervised learning: a training dataset, a validation set, and a test dataset. The first contains all available information, including the features and the class of each sample. A model is obtained from this training dataset and then tested using a validation dataset. This model should be able to determine the class of each sample, and these results are then compared to those of the training dataset.

    This step allows improving the model by repeating the process until the obtained output reaches a certain degree of precision with respect to the annotated input.

    There are different techniques to obtain a set of validation from the set of training. A popular method is known as cross-validation. It consists of doing k-subsets of the data to train with ($k-1$)-subsets and validate with one subset. This process is repeated k-times, and the obtained result is the average of all results.

    The last step runs a test dataset. Again, this set of data only contains the features of the samples, and the algorithm will determine the class.
15. kNN method is used for classification and for regression. The output dataset consists of the k-nearest samples in the feature space. For classification purposes, the algorithm assigns a class to each sample based on the classes of its k-nearest neighbors (k is a positive integer). For regression, the average of the values of its k-nearest neighbors is the value of the output.
16. Output interpretation is the last step to finish the analysis. If we know the class of the instances we can compare them with the predicted results (this is only possible if we know previously the data as in the validation dataset). The outcomes are formulated in a matrix, called a confusion matrix or contingency matrix (Fig. 16). The confusion matrix allows the visualization of the performance of an algorithm. Each column of the matrix depicts the predictions of the algorithm, while each row represents the "actual" instances—those defined by the expert—or vice versa.

    In addition, terms and formulas can be derived as a result of this matrix. The most commonly used parameters are *recall* and *precision*, which are also analyzed for each image. For definitions of these parameters, *see* **Note 17**.

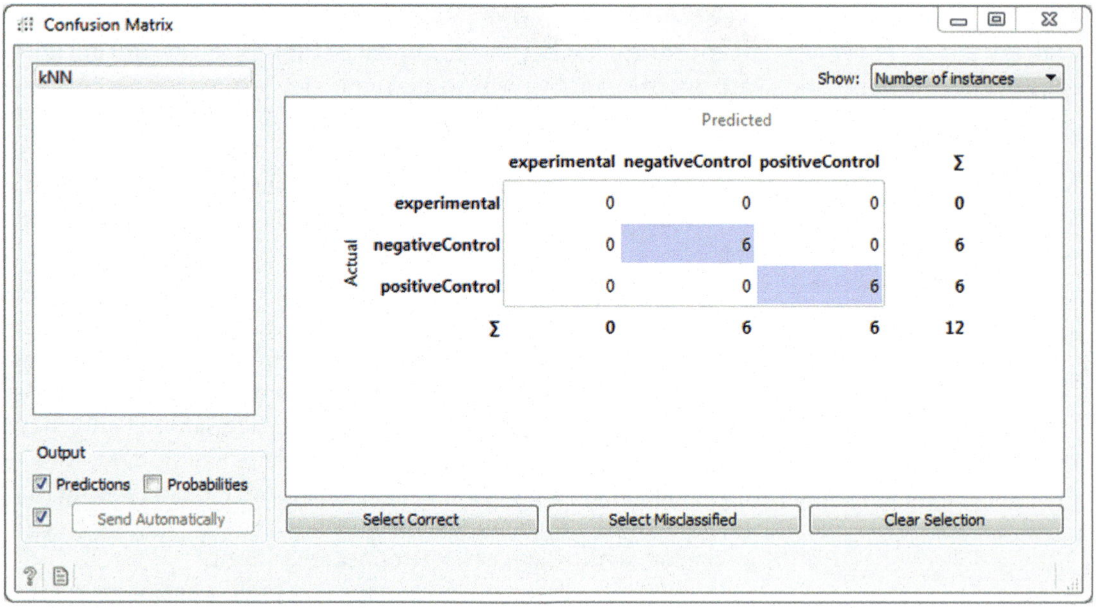

Fig. 16 Confusion matrix for the sample controls

17. *Recall*, also known as sensitivity, is a parameter for a true positive rate (TPR). It measures the proportion of positives which are correctly identified.

$$\text{Recall} = \text{TPR} = \frac{\text{TP}}{\text{P}} = \frac{\text{TP}}{\text{TP} + \text{FN}}$$

(i.e., the percentage of instances correctly identified as meeting a criterion, being TP the true positive instances, FN false negative, and P the actual positive instances).

Precision, also called positive predictive value (PPV), is the fraction of relevant instances among the ones classified with the same predicted condition

$$\text{Precision} = \text{PPV} = \frac{\text{TP}}{\text{TP} + \text{FP}}$$

(i.e., the percentage of instances which are correctly identified as positive (TP) among all the instances classified with the same condition, being FP the false-positive instances).

Acknowledgments

We acknowledge Pablo J. Fernandez-Marcos, PhD, Madrid Institute of Advanced Studies IMDEA-Food, for his scientific contribution. We are grateful to Elena Rebollo and Manel Bosch for their help and critical reading.

References

1. Hartig SM, Newberg JY, Bolt MJ, Szafran AT, Marcelli M, Mancini MA (2011) Automated microscopy and image analysis for androgen receptor function. Methods Mol Biol 776:313–331. https://doi.org/10.1007/978-1-61779-243-4_18
2. Starkuviene V, Pepperkok R (2007) The potential of high-content high-throughput microscopy in drug discovery. Br J Pharmacol 152 (1):62–71. https://doi.org/10.1038/sj.bjp.0707346
3. Martinez NJ, Titus SA, Wagner AK, Simeonov A (2015) High-throughput fluorescence imaging approaches for drug discovery using in vitro and in vivo three-dimensional models. Expert Opin Drug Discov 10(12):1347–1361. https://doi.org/10.1517/17460441.2015.1091814
4. Westhoff JH, Giselbrecht S, Schmidts M, Schindler S, Beales PL, Tonshoff B, Liebel U, Gehrig J (2013) Development of an automated imaging pipeline for the analysis of the zebrafish larval kidney. PLoS One 8(12):e82137. https://doi.org/10.1371/journal.pone.0082137
5. Garvey CM, Spiller E, Lindsay D, Chiang CT, Choi NC, Agus DB, Mallick P, Foo J, Mumenthaler SM (2016) A high-content image-based method for quantitatively studying context-dependent cell population dynamics. Sci Rep 6:29752. https://doi.org/10.1038/srep29752
6. Fang Y, Eglen RM (2017) Three-dimensional cell cultures in drug discovery and development. SLAS Discov 22(5):456–472. https://doi.org/10.1177/1087057117696795
7. Henser-Brownhill T, Monserrat J, Scaffidi P (2017) Generation of an arrayed CRISPR-Cas9 library targeting epigenetic regulators: from high-content screens to in vivo assays. Epigenetics 12(12):1065–1075. https://doi.org/10.1080/15592294.2017.1395121
8. de Groot R, Luthi J, Lindsay H, Holtackers R, Pelkmans L (2018) Large-scale image-based profiling of single-cell phenotypes in arrayed CRISPR-Cas9 gene perturbation screens. Mol Syst Biol 14(1):e8064
9. Boutros M, Bras LP, Huber W (2006) Analysis of cell-based RNAi screens. Genome Biol 7(7): R66. https://doi.org/10.1186/gb-2006-7-7-R66
10. Schindelin J, Arganda-Carreras I, Frise E, Kaynig V, Longair M, Pietzsch T, Preibisch S, Rueden C, Saalfeld S, Schmid B, Tinevez JY, White DJ, Hartenstein V, Eliceiri K, Tomancak P, Cardona A (2012) Fiji: an open-source platform for biological-image analysis. Nat Methods 9(7):676–682. https://doi.org/10.1038/nmeth.2019
11. Demsar JCT, Erjavec A, Gorup C, Hocevar T, Milutinovic M, Mozina M, Polajnar M, Toplak M, Staric A, Stajdohar M, Umek L, Zagar L, Zbontar J, Zitnik M, Zupan B (2013) *Orange*: data mining toolbox in python. J Mach Learn Res 14:2349–2353
12. Materials can be downloaded from. https://github.com/ConfocalMicroscopyUnit/HCSmaterials
13. Link W, Oyarzabal J, Serelde BG, Albarran MI, Rabal O, Cebria A, Alfonso P, Fominaya J, Renner O, Peregrina S, Soilan D, Ceballos PA, Hernandez AI, Lorenzo M, Pevarello P, Granda TG, Kurz G, Carnero A, Bischoff JR (2009) Chemical interrogation of FOXO3a nuclear translocation identifies potent and selective inhibitors of phosphoinositide 3-kinases. J Biol Chem 284 (41):28392–28400. https://doi.org/10.1074/jbc.M109.038984
14. Fiji download website. https://*ImageJ*.net/Fiji/Downloads
15. *Orange* download website. https://*Orange*.biolab.si/download/
16. Bray MA, Carpenter A (2004) Advanced assay development guidelines for image-based high content screening and analysis. doi: NBK126174 [bookaccession]
17. Nierode G, Kwon PS, Dordick JS, Kwon SJ (2016) Cell-based assay design for high-content screening of drug candidates. J Microbiol Biotechnol 26(2):213–225. https://doi.org/10.4014/jmb.1508.08007
18. Johnston PA, Shinde SN, Hua Y, Shun TY, Lazo JS, Day BW (2012) Development and validation of a high-content screening assay to identify inhibitors of cytoplasmic dynein-mediated transport of glucocorticoid receptor to the nucleus. Assay Drug Dev Technol 10 (5):432–456. https://doi.org/10.1089/adt.2012.456
19. Jackson D, Lenard M, Zelensky A, Shaikh M, Scharpf JV, Shaginaw R, Nawade M, Agler M, Cloutier NJ, Fennell M, Guo Q, Wardwell-Swanson J, Zhao D, Zhu Y, Miller C, Gill J (2010) HCS road: an enterprise system for integrated HCS data management and analysis. J Biomol Screen 15(7):882–891. https://doi.org/10.1177/1087057110374233
20. Voskuil JL (2017) The challenges with the validation of research antibodies. F1000Res

6:161. https://doi.org/10.12688/f1000research.10851.1

21. Singh S, Carpenter AE, Genovesio A (2014) Increasing the content of high-content screening: an overview. J Biomol Screen 19 (5):640–650. https://doi.org/10.1177/1087057114528537
22. Sommer C, Hoefler R, Samwer M, Gerlich DW (2017) A deep learning and novelty detection framework for rapid phenotyping in high-content screening. Mol Biol Cell 28 (23):3428–3436. https://doi.org/10.1091/mbc.E17-05-0333
23. Fuchs F, Pau G, Kranz D, Sklyar O, Budjan C, Steinbrink S, Horn T, Pedal A, Huber W, Boutros M (2010) Clustering phenotype populations by genome-wide RNAi and multiparametric imaging. Mol Syst Biol 6:370. https://doi.org/10.1038/msb.2010.25
24. Jones TR, Carpenter AE, Lamprecht MR, Moffat J, Silver SJ, Grenier JK, Castoreno AB, Eggert US, Root DE, Golland P, Sabatini DM (2009) Scoring diverse cellular morphologies in image-based screens with iterative feedback and machine learning. Proc Natl Acad Sci U S A 106(6):1826–1831. https://doi.org/10.1073/pnas.0808843106
25. Rijsbergen CJV (1979) Information retrieval. Butterworth-Heinemann, Newton, MA
26. Jolliffe IT (1986) Principal component analysis and factor analysis. In: Principal component analysis. Springer New York, New York, NY, pp 115–128. https://doi.org/10.1007/978-1-4757-1904-8_7
27. Brideau C, Gunter B, Pikounis B, Liaw A (2003) Improved statistical methods for hit selection in high-throughput screening. J Biomol Screen 8(6):634–647. https://doi.org/10.1177/1087057103258285

Part III

Methods Based on ImageJ Plugin Development

Chapter 16

Filopodia Quantification Using FiloQuant

Guillaume Jacquemet, Hellyeh Hamidi, and Johanna Ivaska

Abstract

Filopodia are fingerlike membrane protrusions that are extended by cells in vitro and in vivo. Due to important roles in sensing the extracellular microenvironment, filopodia and filopodia-like protrusions have been implicated in numerous biological processes including epithelial sheet zippering in development and wound healing and in cancer progression. Recently, there has been an explosion in the number of software available to analyze specific features of cell protrusions with the aim of gaining mechanistic insights into the action of filopodia and filopodia-like structures. In this methods chapter, we highlight an open-access software called *FiloQuant* that has been developed to specifically quantify the length, density, and dynamics of filopodia and filopodia-like structures from in vitro and in vivo generated samples. We provide step-by-step protocols on (i) how to install *FiloQuant* in the *ImageJ* platform (Fiji), (ii) how to quantify filopodia and filopodia-like protrusions from single images using *FiloQuant*, and (iii) how to track filopodial protrusions from live-cell imaging experiments using *FiloQuant* and *TrackMate*.

Key words Filopodia, Filopodia-like protrusions, FiloQuant, ImageJ, Fiji, Filopodia properties

1 Introduction

Filopodia are "antenna-like" protrusions, which contain cell-surface receptors such as integrins and cadherins, and are responsible for constantly probing the cellular environment [1–3]. Filopodia have been implicated in multiple cellular processes including single- and collective cell migration [4–6], wound healing [7], extracellular matrix (ECM) remodeling [8], and the capture of exosomes [9]. Over the last 5 years, multiple tools have been designed to analyze and quantify filopodial protrusions from microscopy images, and these have emphasized the growing demand for automated quantitative approaches. To the best of our knowledge, these tools include *FiloDetect* [10], *CellGeo* [11], *ADAPT* [12], *Filopodyan* [13], and *FiloQuant* [6], each with unique strengths and

Electronic supplementary material: The online version of this chapter (https://doi.org/10.1007/978-1-4939-9686-5_16) contains supplementary material, which is available to authorized users.

Elena Rebollo and Manel Bosch (eds.), *Computer Optimized Microscopy: Methods and Protocols*, Methods in Molecular Biology, vol. 2040, https://doi.org/10.1007/978-1-4939-9686-5_16,

limitations [6, 13]. Despite some of these tools being readily and freely available, step-by-step instructions on how to use the software are often lacking, thus limiting usability by others. In this methods chapter, we provide step-by-step protocols on (i) how to install *FiloQuant* in *Fiji* (Subheading 3.1) [14], (ii) how to quantify filopodia and filopodia-like protrusions from single images using *FiloQuant* (Subheading 3.2), and (iii) how to analyze filopodia dynamics using *FiloQuant* and *TrackMate* [15] (Subheading 3.3).

FiloQuant is a user-friendly and modifiable tool for automated detection and quantification of filopodia properties such as length and density. *FiloQuant* works by first defining the cell or colony edge in an input image using intensity-based thresholding (Fig. 1). Long and thin protrusions, such as filopodia, are then removed from the cell or colony edge using a binary operation such as *open* or *erode* (Fig. 1, image A). In parallel, the same original input image is separately enhanced to optimize filopodia detection (Fig. 1, image B). Enhancements available include a convolution kernel (found to be very effective in improving detection of faint filopodia) and/or the *Enhanced Local Contrast* function [16]. Image A and B are then superimposed to specifically isolate filopodia located at the cell boundary (Fig. 1, image C). Using the *skeletonize* and *AnalyzeSkeleton* algorithms [17], the number and length of these cell-edge filopodia are then automatically measured. Filopodia density can also be determined by calculating the ratio of filopodia number to edge length (extracted by *FiloQuant* from the edge detection image, i.e., image A).

To detect the filopodia-free edge (Fig. 1, image A), *FiloQuant* first uses intensity-based thresholding to create a binary image (image where each pixel is either black or white). To generate this image, the user needs to input a number that corresponds to the highest intensity value where the cell/colony is fully thresholded (i.e., the entire cell/colony is represented with intact cell edges and protrusions but without excess background noise). On this binary image, *FiloQuant* erases fingerlike protrusions using the open and/or erode mathematical morphology operations (operations that consider only the shape of an object and not its intensity). The operation *open* creates gaps between objects that are connected by only thin lines of pixels, while *erode* shrinks an image by stripping away the most outward layer of pixels. We found that the operation *open* is especially well-suited to erase filopodia-like protrusions as these are thin structures consisting of only a few lines of pixels from tip to base.

Improved filopodia detection (Fig. 1, image B) can be achieved through a convolution kernel and/or the *Enhanced Local Contrast* function. Convolution is an image processing method that allows a pixel's intensity to be defined by the intensity of neighboring pixels. The kernel provides the necessary information that dictates the contribution (weight) of each neighboring pixel to the convolution

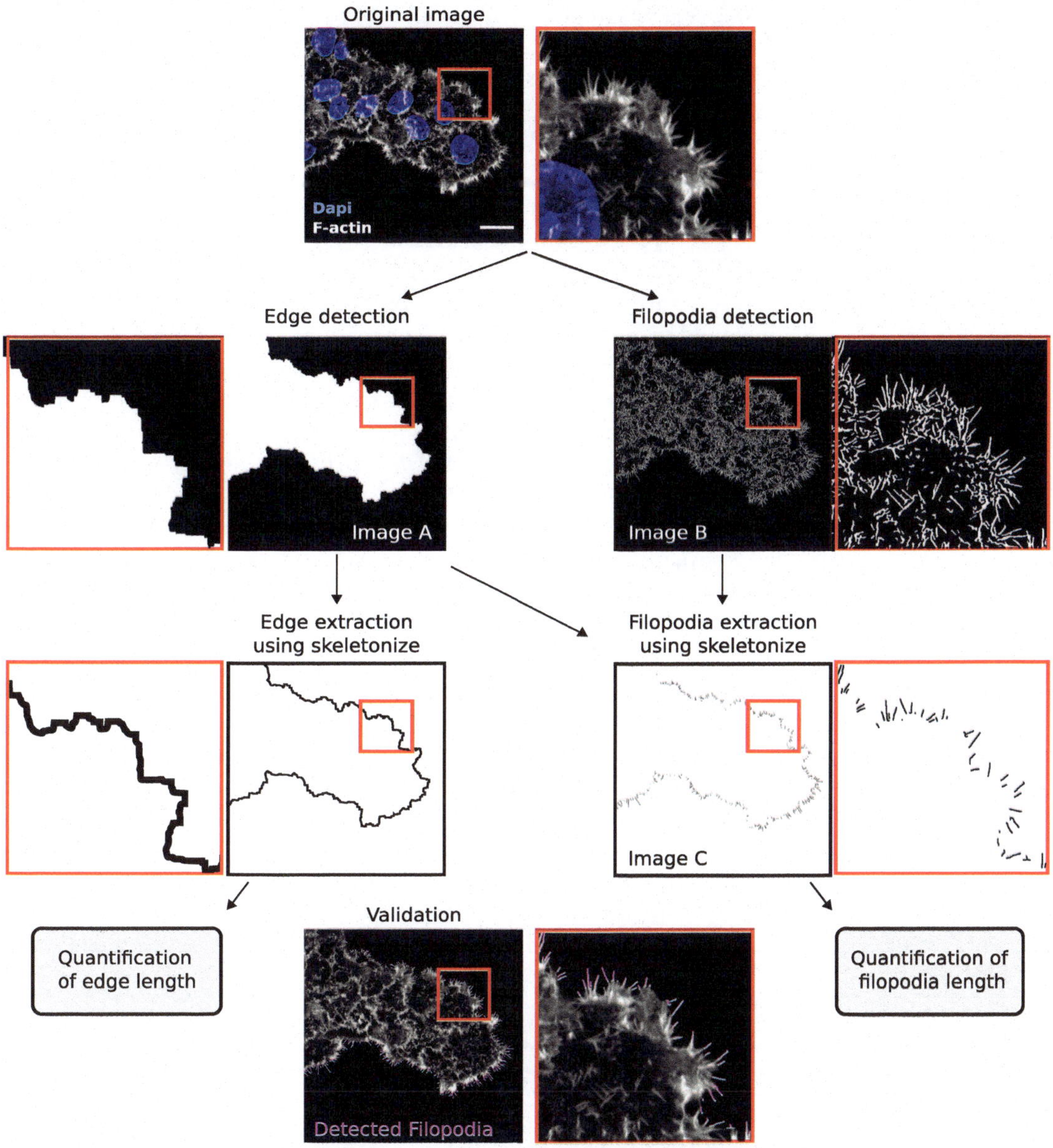

Fig. 1 FiloQuant workflow. Representative images outlining the different steps in *FiloQuant* used to detect and quantify filopodia-like protrusions. In brief, the original image (input) undergoes two parallel processing steps. In the left panel, the cell edge is defined and detected by intensity-based thresholding and by "erasing" the filopodia, using binary operations such as *open* or *erode* (edge detection, image A). In the right panel, the image is enhanced to optimize detection of faint filopodia without introducing noise (filopodia detection, image B). The resulting images are subtracted to isolate edge filopodia (filopodia extraction, image C), and filopodia number and length are automatically analyzed using the *Skeletonize3D* and *AnalyzeSkeleton* algorithms. Detected filopodia are highlighted in magenta in the final image. Filopodia density can be also quantified by determining the ratio of filopodia number to edge length (extracted from the edge detection image). The original image shows MCF10A ductal carcinoma in situ cells invading collectively through a fibrillar collagen gel, stained for actin and DAPI and imaged using a spinning disk confocal microscope. Red insets denote magnified regions. Scale bars, 20 μm. From Jacquemet et al. (2018)

process. The convolution kernel is then performed for every pixel in the image. The *Enhanced Local Contrast* function is an *ImageJ* plug-in that employs the adaptive histogram equalization image processing technique [16]. This strategy divides the image into multiple sections and computes the best contrast for each section, leading to improved intensities in faint areas. We found that the convolution kernel, coded within *FiloQuant*, is particularly suited for enhanced filopodia detection.

FiloQuant is a plug-in for the freely available *Fiji* platform with interoperating systems compatibility [14, 18, 19]. It can be easily installed in *Fiji* using the *FiloQuant ImageJ* update site. *FiloQuant* contains step-by-step user validation to help users achieve optimal settings for filopodia detection. As it was designed to simplify and speed up the analysis of filopodia properties, semiautomated and fully automated versions of *FiloQuant* are also available. In the semiautomated version, users can rapidly analyze a large number of images while keeping control over the settings used to analyze each image and modify these settings on the fly to improve the accuracy of detection. In the automated version, users can choose the settings for analyzing a large number of images at once. This latter version of *FiloQuant* is especially useful for screening purposes or to analyze filopodia properties and dynamics from live-cell imaging data. It is important to note that *FiloQuant* is currently limited to the detection and quantification of filopodia and other fingerlike protrusions that extend out from the cell edge and is not applicable for studying filopodia localized at cell-cell junctions, or for detecting dorsal and ventral filopodia. *FiloQuant* currently works only on 2D images and does not yet support 3D analyses.

2 Materials

1. Images to be analyzed. This protocol assumes that the user has already acquired images of cells displaying filopodia or filopodia-like protrusions for quantification. Filopodia or filopodia-like protrusions are rich in filamentous actin (f-actin) and are therefore visualized by f-actin markers/probes/stains such as phalloidin, SiR-actin, or lifeact. **Notes 1** and **2** contain tips on how to prepare samples (*see* **Note 1**) and acquire images (*see* **Note 2**) to optimize filopodia detection by *FiloQuant.* Test images are available as supplementary documents.
2. A computer running Windows, macOS, or Linux.
3. The latest version of *Fiji.* We recommend the *Fiji* distribution in [20] as it already contains all the necessary dependencies required by *FiloQuant.* However, *FiloQuant* can also be run in *ImageJ* [18] with the installation of extra plug-ins (*see* **Note 3**).

3 Methods

3.1 Installation of FiloQuant in Fiji Using the FiloQuant ImageJ Update Site

1. In *Fiji*, click on [Help > Update].
2. Then click on *Manage update sites.*
3. Find *FiloQuant* from the list and tick the box.
4. Close the *Manage update sites* window.
5. In the *ImageJ* Updater window, click on *Apply* changes.
6. Restart *Fiji.*
7. The three versions of *FiloQuant* can now be found under [Plugins > FiloQuant].

3.2 FiloQuant: Step-by-Step Instructions for Single-Image Analysis

The single-image or semiautomated version of *FiloQuant* contains step-by-step user validation of the various image processing steps to help users achieve optimal settings for filopodia detection. Below you will find detailed instructions related to the various steps for the single-image analysis version. The settings are recapitulated in Table 1. The analysis process in the semiautomated version of *FiloQuant* is identical to the one described below with the exception that the user is prompted to choose the location of the folder containing the images to be analyzed and a folder where the results can be saved (*see* **Note 4**):

1. Open the image to be analyzed by drag and drop on the *Fiji* bar.
2. In *Fiji*, start *FiloQuant* by clicking on [Plugins > FiloQuant > Single image FiloQuant].
3. Choose a region of interest (ROI) to be analyzed by drawing a square or other shape around the ROI on the initial image, and then click *OK*. If you want to analyze the entire image, and not an ROI, just click *OK*.
4. To calculate optimal brightness/contrast settings, you can select an ROI by drawing a square and then clicking *OK*. Alternatively, to use the whole image for auto contrast calculations, just click *OK*.
5. Input the parameters that *FiloQuant* will use to generate the filopodia-free cell edge. In this step, *FiloQuant* will threshold the cell edges (using intensity-based thresholding), and erase filopodia-like protrusions using the operation(s) *open* and/or *erode* (Fig. 1, image A). Information about the parameters can be found in Table 1. To control the parameters used by *FiloQuant* to measure the length of the cell or colony edge at a later stage (*see* **step 10**), tick the box *Input the parameters to modify edge length measurement?* If the box remains unticked, default settings are used instead.

Table 1
Summary of the various adjustable parameters available in FiloQuant

Parameters that FiloQuant will use to generate the filopodia-free cell edge	
Threshold for cell edge	Intensity value to be used for the intensity-based thresholding (input range, 1–255). This number should correspond to the highest intensity value where the cell/colony is fully thresholded
Number of iterations for Open	Open is the process used by FiloQuant to shave off the filopodia. Input here the number of times you want to perform this operation (input range, 1–100)
Number of cycles for Erode Dilate	Alternative method to shave off filopodia. Input here the number of times you want to perform this operation (input range, 1–100). We recommend to use Open first
Fill holes on edges?	Tick this box if you want FiloQuant to fill the holes at the image boundary
Fill holes?	Tick this box if you want FiloQuant to fill the holes in the middle of the image
Input the parameters to modify edge length measurement?	Tick this box if you want to control how the edges are measured
Parameters to detect filopodia	
Threshold for filopodia	Intensity value to be used for the intensity-based thresholding (input range, 1–255)
Filopodia minimum size	Input the minimum size (in pixels) of structures to be considered further for analysis
Filopodia repair cycles	Input the number of times you want to perform the "close" operation to try to restore broken filopodia. We recommend 0 or 1. Inputting 0 will disable the option
Use convolve to improve filopodia detection?	Tick this box if you want to use a standard convolution kernel to help with filopodia detection. We recommend this option as we found it to be very powerful in extracting faint filopodia
Use local contrast enhancement to improve filopodia detection?	Tick this box if you want to use this option to improve the detection of faint filopodia. This option needs to be disabled if the image is noisy
Filopodia detection: maximum distance from the cell edge?	Input the maximum distance (in pixels) that filopodia are allowed to be from the cell edge to be considered further for analysis. Inputting 0 will disable this option
Contour measurement	
Number of iterations for Close	Input here the number of times you want to perform this operation (input range, 1–100)
Number of iterations for Erode	Input here the number of times you want to perform this operation (input range, 1–100)

(continued)

Table 1
(continued)

Number of iterations for Dilate	Input here the number of times you want to perform this operation (input range, 1–100)
Batch mode:	
Batch mode: stack analysis?	Tick this box if the folder you want to analyze contains stacks of images rather than single images (*see* **Note 10**)

The Batch mode parameter is only available in the automated version of FiloQuant

6. Validate filopodia-free cell-edge detection; if you are happy with the threshold, click *OK*. If you are not happy with the thresholding, untick the box and click *OK*. You will then be able to remodify the parameters to detect the filopodia-free cell edge. Repeat until you are satisfied with the parameters (*see* **Note 5** for tips on difficult thresholding).
7. Input the parameters that *FiloQuant* will use to detect filopodia. In this step, *FiloQuant* will optimize the input image to improve filopodia detection (Fig. 1, image B). Information about the parameters can be found in Table 1.
8. Validate filopodia detection; if you are happy with the threshold, click *OK*. If you are not happy with the thresholding, untick the box and click *OK*. You will then be able to remodify the parameters for filopodia detection. Repeat until you are satisfied with the parameters. This step allows you to optimize the parameters to improve filopodia detection. Filopodia filtering in function of size happens at a later stage and cannot be validated at this step.
9. Validate filopodia detection (final validation). This step allows you to verify if the entire workflow is optimal for filopodia detection. If you are happy with filopodia detection (in magenta), click *OK*. If you are not happy with the filopodia detection, untick the box and click *OK*. You will be able to restart the analyses from **step 3**. Repeat until you are satisfied with the parameters.
10. Set the parameters to measure the length of the cell edge. This window becomes available only if the box *Input the parameters to modify edge length measurement?* is ticked in **step 5**. Otherwise, default parameters are used. The options that can be used to smooth the edges, before measurements are taken, can be found in Table 1.
11. Validate contour (cell edge) detection. If you are happy with the contour detection, click *OK*. If you are not happy with the contour detection, untick the box and click *OK*. You will then be able to remodify the parameters for contour detection. Repeat until you are satisfied with the parameters.

12. The results tables and processed images can be found at the same location as the original image. For more information on the results file generated by *FiloQuant*, *see* **Note 6**. Example images generated by *FiloQuant* are displayed in Fig. 2. Errors in detection and quantification, resulting in approximate values, can arise when filopodia density is too high (*see* **Note 7**), when the intensity of individual filopodia is too low (*see* **Note 8**), or when multiple filopodia intersect each other (*see* **Note 9**).

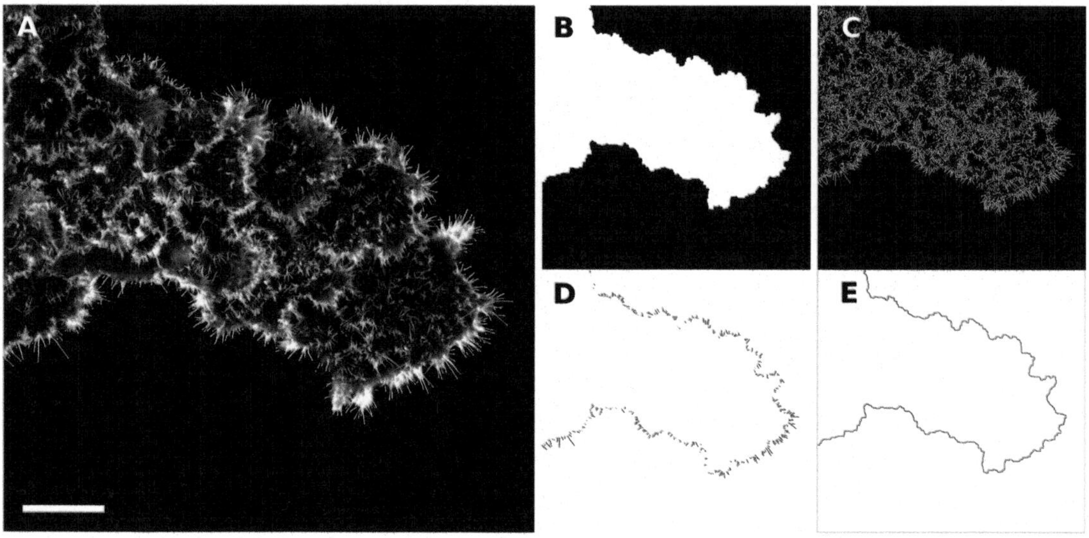

F

	A	B	C	D	E
1		coordinate x	coordinate y	Filopodia length µm	Edge length µm
2	1	14.196	60.332	1.097	383.014
3	2	15.337	61.917	0.791	0
4	3	14.893	62.297	0.28	0
5	4	14.893	62.677	0.153	0
6	5	15.146	59.572	0.933	0
7	6	15.78	57.481	1.473	0
8	7	16.034	62.867	0.813	0
9	8	16.287	44.109	0.791	0
10	9	16.287	46.453	0.586	0
11	10	16.541	55.96	0.063	0
12	11	16.541	56.974	0.063	0
13	12	16.604	31.244	1.32	0
14	13	17.301	46.96	1.362	0
15	14	17.491	43.982	1.355	0
16	15	17.808	42.524	0.739	0
17	16	18.125	38.785	0.686	0
18	17	18.125	39.799	1.552	0
19	18	18.188	41.003	0.433	0
20	19	18.505	14.513	1.635	0
21	20	18.505	54.058	0.306	0
22	21	18.632	33.208	0.802	0
23	22	18.632	55.516	0.433	0
24	23	18.759	33.842	1.287	0
25	24	18.886	53.235	0.19	0
26	25	18.949	12.421	0.776	0

Fig. 2 Images generated by FiloQuant. (**a–e**) In addition to results tables, *FiloQuant* generates multiple images including a "validation image," where the detected filopodia are highlighted in magenta (**a**). *FiloQuant* also saves an image of the cell/colony edge (**b**), as well as the image used to detect filopodia (**c**), and a mask of the images containing the detected filopodia (**d**) and the detected contour (**e**). A representative results table generated by *FiloQuant* analysis is shown. From Jacquemet et al. (2018)

3.3 Analysis of Filopodia Dynamics Using FiloQuant and TrackMate

In the automated version of *FiloQuant*, users can choose the settings for analyzing a large number of images, or a stack of images, at once (*see* **Note 10**). This version of *FiloQuant* is especially useful for screening purposes and/or to analyze filopodia properties and dynamics from live-cell imaging data. When analyzing live-cell imaging data, in addition to the outputs described previously (*see* Subheading 3.2 and **Note 6**), *FiloQuant* will also generate a time projection of detected filopodia and a tracking file (Fig. 3). The time projection of the detected filopodia is generated by the *ImageJ* plug-in *Temporal-Color Code* [21]. The tracking file consists of an image where each of the detected filopodia is reduced to a single dot which can be easily detected and tracked using automated tracking software such as *TrackMate* (Fig. 3a). The tracking file is generated from the detected filopodia using the binary operation *ultimate (eroded) points.* Below you will find detailed instructions on how to automatically analyze filopodia dynamics using *FiloQuant* and *TrackMate* (Fig. 3) [17].

3.3.1 Filopodia Identification Using FiloQuant

1. Save the movie(s) to be analyzed as a stack(s) in a folder while ensuring that there are no spaces in the file name(s) or in the file path(s). Image stacks need to be in .tiff or .tif format.
2. In *Fiji*, start *FiloQuant* by clicking on [Plugins > FiloQuant > Automated FiloQuant].
3. Choose the input folder and click *Select.*
4. Choose the output folder where the results will be saved and click *Select.*
5. Input all the parameters to be used by *FiloQuant* for the analysis. Ensure that the option *Batch mode: stack analysis?* is enabled (*see* **Note 10**). Click *OK* to start the analysis. Details of the various parameters available in *FiloQuant* can be found in Table 1. Suitable *FiloQuant* parameters could be determined beforehand by analyzing one image of the movie(s) using the single-image *FiloQuant* protocol (Subheading 3.2). Click *OK* to start the analysis.
6. Review the *FiloQuant* analysis by checking (1) the time projection of detected filopodia (Fig. 3c) and (2) the movie where the detected filopodia are highlighted in magenta (Fig. 3b; to open the movie, drag the folder called "Tagged_skeleton_RGB" into *Fiji*).

3.3.2 Filopodia Tracking Using TrackMate

1. Open the tracking file (stack) of the detected filopodia for further analysis in *TrackMate* (*see* **Note 11**). To do so, drag the folder called "Tracking_file" into *Fiji*. When prompted click *yes* for *Open all images as stack?*
2. In *Fiji*, start *TrackMate* by clicking on [Plugins > Tracking > TrackMate].

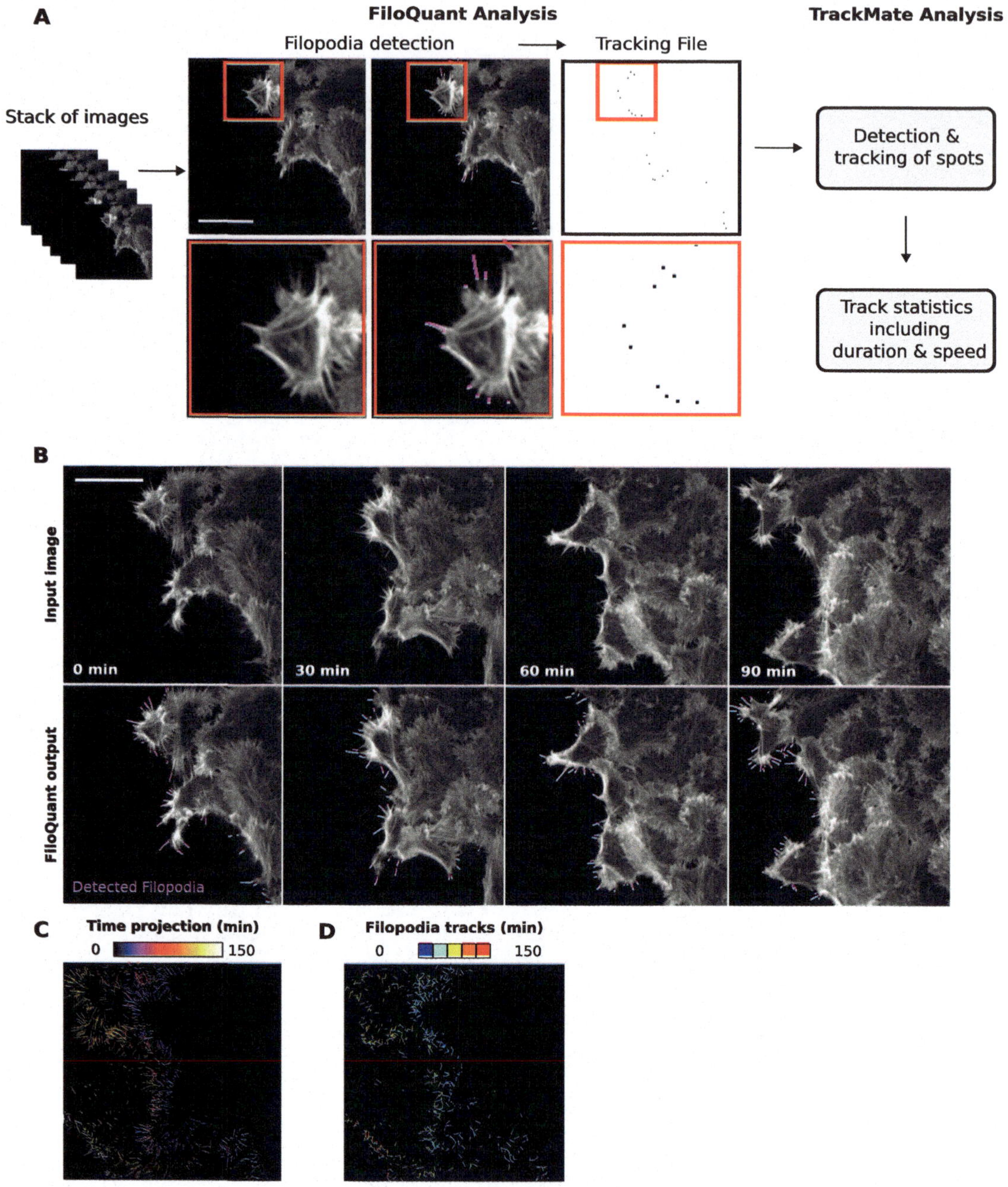

Fig. 3 Filopodia dynamics analyzed using FiloQuant and TrackMate. (**a**) Workflow explaining how the automated version of *FiloQuant*, with the *Batch mode: stack analysis?* option enabled, can be connected to *TrackMate* to analyze filopodia dynamics. For each image of the stack, *FiloQuant* will detect filopodia (as in Fig. 1) but also generate a tracking image containing a single dot locating each filopodium (generated using the binary option *Ultimate Points*). These dots can then be automatically detected and tracked over the timeframe of the movie using *TrackMate*. Filopodia track properties such as speed and duration can also be generated within *TrackMate*. (**b–d**) Example analysis of filopodia dynamics using *FiloQuant* and *TrackMate*. Filopodia detected by *FiloQuant* (magenta) are displayed (scale bar, 25 μm) at different time points (**b**). In

3. When prompted click *yes* for *Z/T swapped*.
4. Ensure that the calibration settings are correct, and then click *Next*.
5. Select *LoG detector* and click *Next*. The *LoG detector* was found to work well with *FiloQuant* output data.
6. Choose an *estimated blob diameter* of 0.1 μm and a *threshold* of 10. *Enable subpixel localization*. Click on *Preview* to assess the detection parameters. If suitable, click *Next*. In the tracking file generated by *FiloQuant*, each detected filopodium is reduced to a single dot, which can be easily detected using these parameters.
7. At the end of the detection, click *Next*.
8. Click *Next*.
9. Choose *HyperStack Displayer* and click *Next*.
10. Click *Next*.
11. Choose the *simple LAP tracker* and click *Next*. The *simple LAP tracker* is ideal for tracking filopodia as it deals only with gap-closing events and prevents the detection of splitting and merging events.
12. Choose the appropriate *linking max distance* (i.e., 1 μm), *gap-closing max distance* (i.e., 1 μm), and *gap-closing max frame gap* (i.e., 1). These values should reflect the timeframe and the scale of the imaging performed. Then click *Next*.
13. At the end of the analysis, click *Next*.
14. To display statistical information on the filopodia tracks (such as track duration or track speed), click on *Analysis*. Results are now ready to be further analyzed.
15. To generate an image of the filopodia tracks (Fig. 3d), first choose the appropriate parameters for *Track display mode* and the *Set color by* options, and then click *Next*. At the following step, click *Next*. Then click *Execute*.

Fig. 3 (continued) addition, a time projection of detected filopodia (**c**) and filopodia tracks are shown (**d**). The time projection was directly generated by *FiloQuant* and is color coded as a function of time (**c**). The filopodia tracks (**d**) were generated by *TrackMate* and are color coded as a function of the track starting time (track index). The images used are MCF10A ductal carcinoma in situ cells stably expressing LifeAct-mRFP invading through fibrillar collagen I and imaged live using a spinning disk confocal microscope (one picture every 3 min). Red insets denote magnified regions

4 Notes

1. One way to improve filopodia segmentation by *FiloQuant* is to, when possible, improve sample preparation. (i) For fixed samples, in our hands, best results were obtained using the following protocol: cells were plated on high-quality coverslips (#1.5) or on high-quality MatTek dishes (#1.5 or #1.7). Cells were fixed using 4% paraformaldehyde (PFA) for 10 min at room temperature. Cells were permeabilized using 0.25% Triton X-100 for 3 min at room temperature and blocked with 1 M glycine for 30 min. For short-term storage (weeks, recommended protocol): Cells were incubated with Alexa Fluor 488 phalloidin (F-actin binding probe to visualize filopodia) at 4 °C (1/100 dilution in PBS) until imaging (minimum length of staining: overnight at 4 °C). Just before imaging, the sample was washed three times in PBS and mounted in Vectashield media (or other soft mounting media). For longer-term storage (months): Cells were incubated overnight with Alexa Fluor 488 phalloidin at 4 °C (1/100 dilution in PBS) and washed three times with PBS. Coverslips were then mounted on slides using Mowiol (or other hard mounting media). (ii) For live-cell imaging, we recommend the use of a bright green fluorescent protein (e.g., mEmerald, mClover3, or mNeonGreen) that is membrane targeted or tagged to an actin probe (e.g., Lifeact or Utrophin) to visualize filopodia or filopodia-like protrusions.
2. Another way to improve filopodia segmentation by *FiloQuant* is to use high-end and/or super-resolution microscopes. However, this is also the less practical solution if such instruments are not available. Below is a list of imaging setups we found to work best and to produce images that are efficiently segmented using *FiloQuant*. (i) Structured illumination microscopy provides excellent image resolution enabling beautiful filopodia segmentation in *FiloQuant*. (ii) Another option would be to acquire images using a microscope system (e.g., TIRF, spinning disk) equipped with a high magnification and high NA objective (i.e., 100×) and a low-noise camera (Orca Flash 4 camera). The low-noise camera generates ideal images for the efficient segmentation of filopodia. (iii) A third possibility would be to use an imaging system (e.g., TIRF, spinning disk) equipped with a high magnification and high NA objective (i.e., 100×) and an EMCCD camera. If the image is too noisy and/or the resolution too low, following image acquisition, images can then be further processed using freely available software that improve image resolution and decrease noise. We found that the *SRRF ImageJ* plug-in [22] works very well for such purposes.

3. To run *FiloQuant* in *ImageJ*, users need to install the following dependencies: *Enhanced Local Contrast* (*CLAHE.class* [16]), *Skeletonize3D.jar* [23], *AnalyzeSkeleton.jar* [24], and *Temporal-Color Code* [21]. To run *FiloQuant* in the *Fiji* version of *ImageJ*, no additional dependencies are needed.
4. In the semiautomated version of *FiloQuant*, users can analyze, rapidly, a large number of images while keeping control over the settings used to analyze each image and modify these settings on the fly to improve the accuracy of detection. The analysis process is very similar to the one described for the single-image analysis except that the user is prompted to choose the location of the folder containing the images to be analyzed and a folder where the results can be saved at the beginning of the analysis. Importantly, the images to be analyzed need to be in a .tiff or .tif format.
5. The initial thresholding of the cell edge can sometimes be challenging. It is possible, during **steps 1** and **2**, to manually outline the cell using the *ImageJ freehand tool* and then use the *fill* or *clear* functions.
6. Results include a "results.csv" file, which contains (i) *x*, *y* coordinates of each detected filopodium (in the specified image calibration unit), (ii) the length of the detected filopodia (in the specified image calibration unit, as indicated in the heading), and (iii) the length of the detected edge (in the specified image calibration unit, as indicated in the heading). Results also include a "settings.csv" file containing all the settings used for the analysis of this particular image. This file can be found in the folder "intermediate files." Multiple images are also saved during the analysis process as highlighted in Fig. 2.
7. When filopodia density is very high (maximal density depends on the imaging modality and resolution), two problems may arise: (i) individual filopodia can still be detected but can no longer be thresholded along their entire length (*FiloQuant* analyses will underestimate the filopodium's length) and (ii) individual filopodia can no longer be detected (*FiloQuant* will fail to count or detect filopodia properties). High filopodia density may be resolved by improving the image resolution (*see* **Note 2**).
8. When the labeling intensity of a filopodium is extremely weak, it can create problems in thresholding the entire protrusion and lead to the detection of multiple fragmented filopodia instead of a single protrusion. This will artificially increase the number of filopodia detected by FiloQuant and result in an underestimation of the true length of the filopodium. As the software also provides the coordinates of each filopodium counted and/or measured, users can re-evaluate the data and surmise

if a filopodium has been incorrectly measured. The broken filopodia phenotype can be improved by tuning the *Threshold for filopodia* parameter or by enabling *Use convolve to improve filopodia detection* or *Use local contrast enhancement to improve filopodia detection*. The *Filopodia repair cycles* may also fix filopodia that are broken (only the filopodia broken by one or two pixels). However, the best way to resolve the "broken filopodia" issue is to acquire images with higher signal to noise ratio (*see* **Notes 1** and **2**).

9. *FiloQuant* can detect and quantify branching and crossing filopodia. However, in the present version of the plug-in, each branch will be counted and measured as independent filopodia. This may result in an overestimation of filopodia numbers and an underestimation of true filopodia length. However, as the software also provides the coordinates of each filopodium measured, it is recommended to re-evaluate the data and to surmise if a filopodium has been incorrectly measured.
10. The *Batch mode: stack analysis?* option allows the analysis of stacks of images. Tick this box if the folder you want to analyze contains stacks of images rather than single images, e.g., live-cell imaging data. This option will (i) organize the results data differently (more suitable for stacks), (ii) generate a time projection of the detected filopodia, and (iii) generate a tracking file of the detected filopodia that can be further analyzed using automated tracking software such as *TrackMate*. If your input folder contains single images, untick the box.
11. *TrackMate* was chosen over other available *ImageJ* tracking plug-ins because of its user-friendly interface and high flexibility. However other automated tracking software can also be used. *TrackMate* is a very versatile tool and can generate many outputs including statistical track features (track length, object speed), images, and movies. For more information on how to use *TrackMate*, visit [15, 25].

References

1. Arjonen A, Kaukonen R, Ivaska J (2011) Filopodia and adhesion in cancer cell motility. Cell Adhes Migr 5:421–430
2. Biswas KH, Zaidel-Bar R (2017) Early events in the assembly of E-cadherin adhesions. Exp Cell Res 358:14–19
3. Jacquemet G, Hamidi H, Ivaska J (2015) Filopodia in cell adhesion, 3D migration and cancer cell invasion. Curr Opin Cell Biol 36:23–31
4. Jacquemet G, Green DM, Bridgewater RE et al (2013) RCP-driven α5β1 recycling suppresses Rac and promotes RhoA activity via the RacGAP1-IQGAP1 complex. J Cell Biol 202:917–935
5. Paul NR, Allen JL, Chapman A et al (2015) α5β1 integrin recycling promotes Arp2/3-independent cancer cell invasion via the formin FHOD3. J Cell Biol 210:1013–1031
6. Jacquemet G, Paatero I, Carisey AF et al (2017) FiloQuant reveals increased filopodia density during breast cancer progression. J Cell Biol 216:3387–3403. https://doi.org/10.1083/jcb.201704045

7. Wood W, Jacinto A, Grose R et al (2002) Wound healing recapitulates morphogenesis in Drosophila embryos. Nat Cell Biol 4:907–912
8. Sato Y, Nagatoshi K, Hamano A et al (2017) Basal filopodia and vascular mechanical stress organize fibronectin into pillars bridging the mesoderm-endoderm gap. Development 144:281–291
9. Heusermann W, Hean J, Trojer D et al (2016) Exosomes surf on filopodia to enter cells at endocytic hot spots, traffic within endosomes, and are targeted to the ER. J Cell Biol 213:173–184
10. Nilufar S, Morrow AA, Lee JM et al (2013) FiloDetect: automatic detection of filopodia from fluorescence microscopy images. BMC Syst Biol 7:66
11. Tsygankov D, Bilancia CG, Vitriol EA et al (2014) CellGeo: a computational platform for the analysis of shape changes in cells with complex geometries. J Cell Biol 204:443–460
12. Barry DJ, Durkin CH, Abella JV et al (2015) Open source software for quantification of cell migration, protrusions, and fluorescence intensities. J Cell Biol 209:163–180
13. Urbančič V, Butler R, Richier B et al (2017) Filopodyan: an open-source pipeline for the analysis of filopodia. J Cell Biol 216:3405–3422. https://doi.org/10.1083/jcb.201705113
14. Schindelin J, Arganda-Carreras I, Frise E et al (2012) Fiji: an open-source platform for biological-image analysis. Nat Methods 9:676–682
15. Tinevez J-Y, Perry N, Schindelin J et al (2017) TrackMate: an open and extensible platform for single-particle tracking. Methods (San Diego, Calif) 115:80–90
16. Enhance Local Contrast (CLAHE)—ImageJ. https://imagej.net/Enhance_Local_Contrast_(CLAHE)
17. Arganda-Carreras I, Fernández-González R, Muñoz-Barrutia A et al (2010) 3D reconstruction of histological sections: application to mammary gland tissue. Microsc Res Tech 73:1019–1029
18. Rueden CT, Schindelin J, Hiner MC et al (2017) ImageJ2: ImageJ for the next generation of scientific image data. BMC Bioinformatics 18:529
19. Schneider CA, Rasband WS, Eliceiri KW (2012) NIH image to ImageJ: 25 years of image analysis. Nat Methods 9:671–675
20. Fiji is just ImageJ. https://fiji.sc/
21. Temporal-Color Code. https://imagej.net/Temporal-Color_Code
22. Gustafsson N, Culley S, Ashdown G et al (2016) Fast live-cell conventional fluorophore nanoscopy with ImageJ through super-resolution radial fluctuations. Nat Commun 7:12471
23. Skeletonize3D. http://imagej.net/Skeletonize3D
24. AnalyzeSkeleton. http://imagej.net/AnalyzeSkeleton
25. TrackMate. https://imagej.net/TrackMate

Chapter 17

Coincidence Analysis of Molecular Dynamics by Raster Image Correlation Spectroscopy

David F. Moreno and Martí Aldea

Abstract

The dynamics of cellular processes is a crucial aspect to consider when trying to understand cell function, particularly with regard to the coordination of complex mechanisms involving extensive molecular networks in different cell compartments. Thus, there is an urgent demand of methodologies able to obtain accurate spatiotemporal information on molecular dynamics in live cells. Different variants based on fluorescence correlation spectroscopy have been used successfully in the analysis of protein diffusion and complex or aggregation status. However, the available approaches are limited when simultaneous spatial and temporal resolutions are required to analyze fast processes. Here we describe the use of raster image correlation spectroscopy to analyze the spatiotemporal coincidence of collaborating proteins in highly dynamic molecular mechanisms.

Key words Protein dynamics, Diffusion, Aggregation, Brightness, Correlation microscopy

1 Introduction

While our understanding of most aspects of cellular physiology has reached impressive heights, very little is known about the dynamics at the molecular level of most processes within the cell, and researchers are becoming increasingly interested in methodologies that provide precise information on molecular concentration, dynamics, and organization at high temporal and spatial resolution in live cells. In this regard, fluorescent proteins have allowed the analysis of molecular dynamics by different methods based on fluorescence wide-field, total internal reflection fluorescence (TIRF) or confocal microscopy. Fluorescence correlation spectroscopy (FCS) is perhaps the best known of these methodologies and provides high temporal resolution information about concentration and diffusion properties of the target fluorescent fusion

---

Electronic supplementary material: The online version of this chapter (https://doi.org/10.1007/978-1-4939-9686-5_17) contains supplementary material, which is available to authorized users.

Elena Rebollo and Manel Bosch (eds.), *Computer Optimized Microscopy: Methods and Protocols*, Methods in Molecular Biology, vol. 2040, https://doi.org/10.1007/978-1-4939-9686-5_17,

protein within the cell [1, 2]. However, this technique can only obtain data from a very small cellular region (the focal volume) at a given time and, for this reason, cannot provide the spatiotemporal information needed to understand molecular processes that simultaneously involve cellular components at large distances. To solve this caveat, a number of methods have been developed that extract spatial information on molecular dynamics from time-resolved image data, each of them with advantages and drawbacks. Time-lapse image correlation analysis (ICS) can be used to analyze protein complex status with high spatial discrimination [3, 4], but the temporal resolution is limited by the frame rate, usually in the order of seconds, and it is unsuited to analyze the much faster molecular dynamics inherent to most cellular processes. On the other hand, raster image correlation spectroscopy (RICS) exploits the temporal dimension inherent to images created by confocal microscopy [5, 6]. Since it analyzes the spatial correlation between adjacent pixels in the microsecond timescale, RICS is able to analyze fast dynamic processes such as molecular diffusion and, more important, provides with spatial information at high resolution. We have applied both FCS and RICS methods to the analysis of chaperone dynamics in live yeast cells, and, in our experience, FCS is superior to RICS with regard to their sensitivity to changes in the coefficient of diffusion of the target protein. However, RICS proved very robust at determining the number of moving particles at high spatial resolution, a very useful parameter to analyze protein complex or aggregation status. Based on this observation, we have developed a method for spatiotemporal analysis of pairs of proteins that physically and functionally interact in a highly dynamic manner, which we call coincidence analysis. Briefly, pixels along the raster line are used to obtain autocorrelation functions and produce high-resolution maps of the particle brightness (B), i.e., the number of fluorescent molecules per moving particle (Fig. 1a–c). These maps can be obtained from time-lapse stacks to follow also slower dynamics of the target protein (Fig. 1d). Then, B maps of each protein are compared by correlation analysis to obtain a coincidence index, i.e., a value that reports the level of spatial correlation of the brightest complexes for both proteins within a narrow time window. Figure 2a shows representative B maps of Ydj1-GFP and Ssa1-mCherry, two chaperones that transiently interact during folding of client proteins [7], or GFP and mCherry as control. As observed in the merged images, B maps produced by the two chaperones are strikingly similar, showing a much higher level of correlation compared to controls (Fig. 2b). While RICS cross-correlation estimates the degree of co-occurrence of two proteins in the moving particles, the approach proposed here not only accounts for this situation but also considers the spatiotemporal coincidence of moving particles as a function of their brightness, i.e., their complex or aggregation status. Although photon-

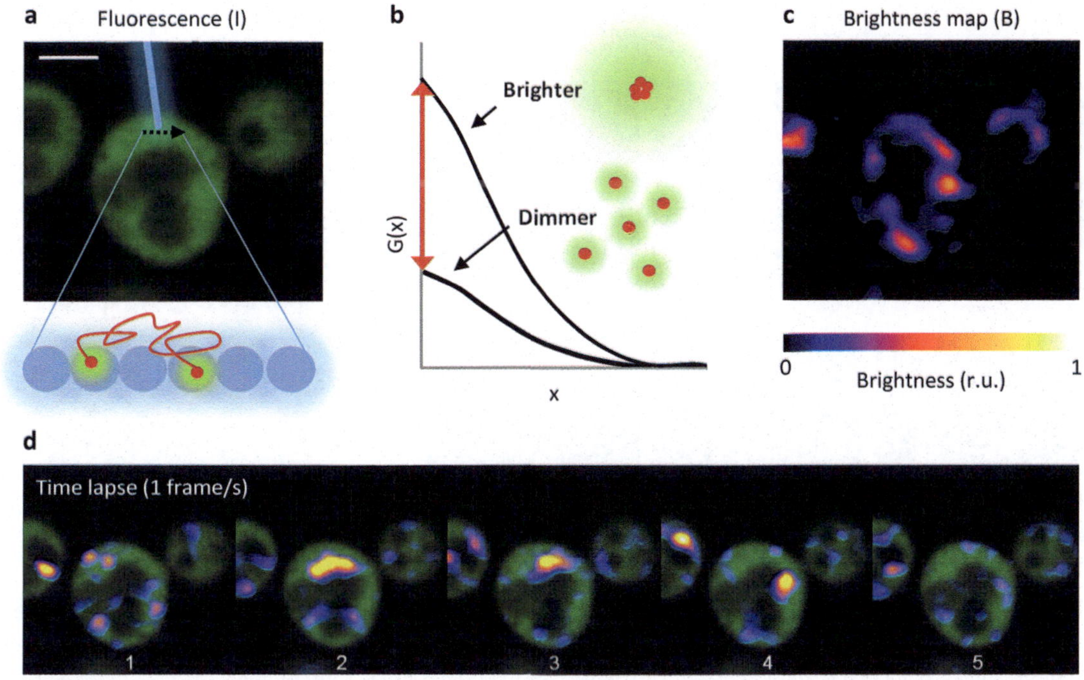

Fig. 1 RICS essentials and brightness maps. (**a**) In image acquisition by laser confocal microscopy, if scanning speed and pixel size are properly adjusted, a moving fluorescent particle or molecule can be recorded in different pixels during the same raster scan. The image corresponds to a budding yeast cell expressing a GFP fusion to Ssa1, an abundant Hsp70 chaperone. Bar is 2 μm. (**b**) Schematic showing autocorrelation curves corresponding to particles with one (dimmer) or five (brighter) fluorescent molecules. Note that the total intensity of fluorescence is the same in the two cases. (**c**) Brightness map obtained with CoinRICSJ from cell in panel **a**. Relative brightness levels are plotted using the *ImageJ* fire palette shown at the bottom. (**d**) Merged fluorescence (green) and brightness (*ImageJ* fire palette) images from cell in panel **a** obtained at the indicated times from a time-lapse experiment

counting detectors are preferable, this coincidence analysis does not require specific microscopy equipment and makes correlation microscopy available to any researcher interested in the dynamic analysis of functional and physical interactions of proteins in live cells.

2 Materials

1. Yeast strains and plasmids (*see* **Note 1**):
   (a) MAG261 (*MAT***a** *YDJ1-GFP-FS::HIS3*).
   (b) MAG676 (*MAT***a** *SSA1-GFP::HIS3*).
   (c) MAG1078 (*MAT***a** *YDJ1-GFP-FS::HIS3 SSA1-mCherry::HYG*).

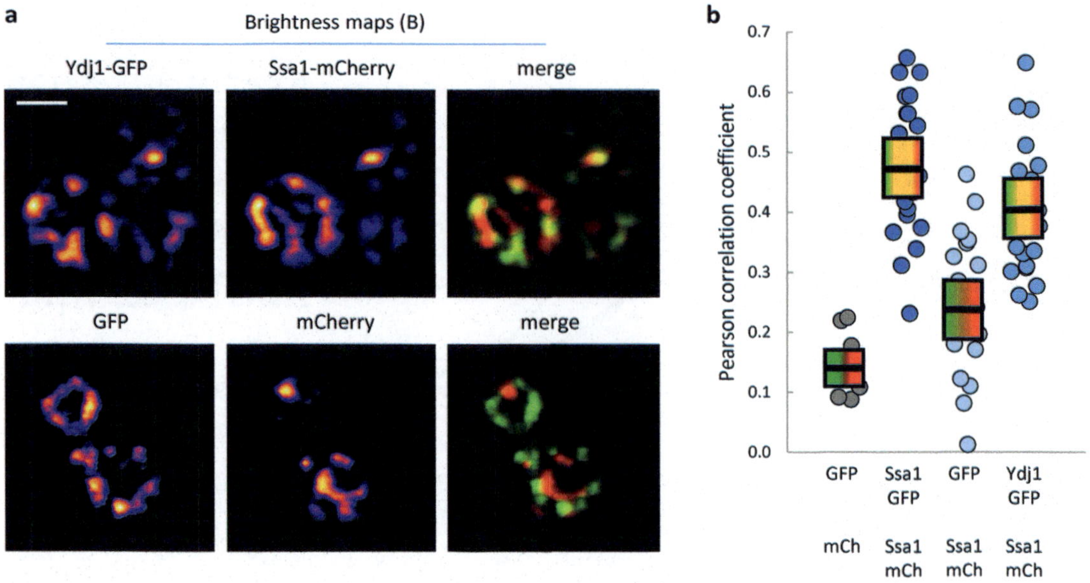

Fig. 2 Coincidence analysis of Ssa1-mCherry and Ydj1-GFP. (**a**) Brightness maps obtained with CoinRICSJ from yeast cells expressing either Ssa1-mCherry and Ydj1-GFP (top row) or mCherry and GFP (bottom row) as control. The corresponding merged images are shown at the right. Bar is 2 μm. (**b**) Coincidence analysis of the indicated proteins co-expressed in yeast cells as in panel **a**. Pearson correlation coefficients were obtained by using image mean values as threshold from individual cells (closed circles), and mean and confidence limits ($\alpha = 0.05$) for the mean of Pearson correlation coefficients are also plotted (boxes) for each pair of proteins

(d) MAG1512 (*MAT***a** *NAT::TEF1p-mCherry*) *pMAG1228* (*ARS-CEN URA3 TE1Fp-GFP*).

(e) MAG1802 (*MAT***a**/**α** *SSA1-GFP::HIS3 SSA1-mCherry::GEN*).

(f) MAG1803 (*MAT***a**/**α** *NAT::TEF1p-GFP SSA1-mCherry::GEN*).

2. SC + Glu medium (synthetic complete + glucose): 1.7 g/L yeast nitrogen base (w/o amino acids, w/o NH_4SO_4), 5 g/L NH_4SO_4, 20 mL/L 50× AA mix (10 g/L threonine, 5 g/L lysine, 5 g/L leucine, 5 g/L tryptophan, 5 g/L phenylalanine, 3 g/L isoleucine, 3 g/L methionine, 2.5 g/L histidine, 2.5 g/L adenine, 2.5 g/L uracil, 2 g/L arginine), and 2% glucose (*see* **Note 2**).
3. Glass-bottom 35-mm dish (Ibidi, #1.5).
4. Concanavalin A (type V, 0.2 mg/mL).
5. Zeiss LSM-780 confocal microscope with a 32 PMT GaAsP array (*see* **Note 3**).
6. Desktop computer with *ImageJ* 1.50i or higher (Wayne Rasband, NIH) on *Java* 1.8 and *CoinRICSJ* 1.0. The latest version of the plug-in and a sample file are available at www.ibmb.csic.es/groups/spatial-control-of-cell-cycle-entry.

3 Methods

3.1 Sample Preparation

1. Grow your yeast cells in liquid SC + Glu for 5–6 generations until OD_{600} 0.3–0.5.
2. Coat a glass-bottom 35-mm dish with a droplet of Concanavalin A solution for 20–30 min, and wash twice with water and once with SC + Glu.
3. Take 200 μL of the yeast culture into a 1.5-mL tube, pipette up and down several times to dislodge cell clumps, and place 50 μL on the dish for 2–5 min, depending on cell density. Remove non-attached cells and fill the dish with 400 μL of SC + Glu.

3.2 Microscope Settings and Acquisition Procedures

1. Place the dish in the microscope using a water-immersion 63×/1.3 NA objective at room temperature (*see* **Note 4**).
2. Locate a sample cell and focus using the bright-field channel to obtain the highest contrast at cell boundary.
3. Set pinhole at 1 AU (*see* **Note 5**), adjust zoom to obtain 0.03–0.05 μm/pixel (*see* **Note 6**), and set scanning speed to obtain a dwell time of 10–30 μs/pixel. The scanning rate can be adjusted to fit the specific dynamics of the target protein (*see* **Note** 7).
4. Set image size to fit one cell with as little background area as possible (*see* **Note 8**).
5. Adjust laser intensity to minimize photobleaching and avoid saturation of the photon-counting detector used (but *see* **Note 3** if analog detectors are used).
6. Use simultaneous recording mode to scan the two fluorescent proteins with excitation and emission settings that minimize fluorescence cross talk (*see* **Note 9**).
7. Obtain a time-lapse stack at 1–2 s/frame with no interleave and save (or convert) data to TIF files. At least 20 frames per stack are needed to allow removal of the immobile fraction of fluorescent molecules (*see* next section), but 100 frames are preferable to obtain more robust ACF data (*see* **Note 10**).

3.3 Image Analysis by CoinRICSJ

1. Install *CoinRICSJ* by copying the *CoinRICSJ_.class* file to the *ImageJ* plug-in folder.
2. Run *ImageJ* and execute *CoinRICSJ* from the *Plugins* pulldown menu.
3. Select the files to be analyzed in batch mode.
4. In the dialog box (Fig. 3), set the *channels* to be analyzed, the *pixel range* (16 pixels by default) used to calculate the *autocorrelation function (ACF)*, the *number of pixels to be excluded* (1 by default, but *see* **Note 3** if using analog detectors), and the

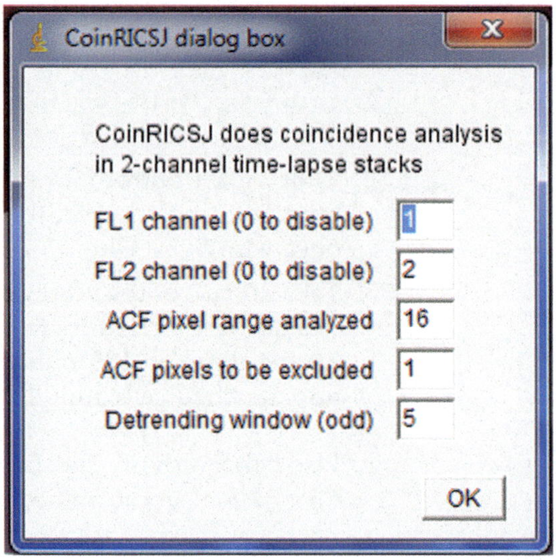

Fig. 3 CoinRICSJ settings. A dialog box allows to determine the channels to be analyzed, the pixel range used to calculate the autocorrelation function, the number of pixels to be excluded, and the frame range used for image detrending. *See* text and **Notes 3** and **11** for a detailed explanation

frame range (an odd number, 5 frames by default) used for image detrending (*see* **Note 11**).

5. Press *OK* to start the analysis. As shown in Fig. 4, a log window shows the progress of the analysis until it is finished and results are displayed. Then, B maps for the two channels are plotted in a new image window, and two additional windows display the global *ACF* data for each stack and the coincidence analysis of the obtained B maps, with *Pearson (PCC)* and *Manders (MCC)* correlation coefficients, setting the threshold as the mean intensity value or using the *Otsu* method. These data can be readily transferred to a spreadsheet for further analysis. Please *see* **Note 12** on the details of pixel autocorrelation and B map correlation analysis.

4 Notes

1. Any cells expressing compatible fluorescent protein fusions can be used, and, as long as they are readily detected by confocal microscopy with settings that minimize photobleaching, there are no specific requirements for expression levels or subcellular localization of the proteins of interest.
2. Use SC + Glu medium for better signal-to-noise ratio. Rich medium (YPD) is highly autofluorescent.

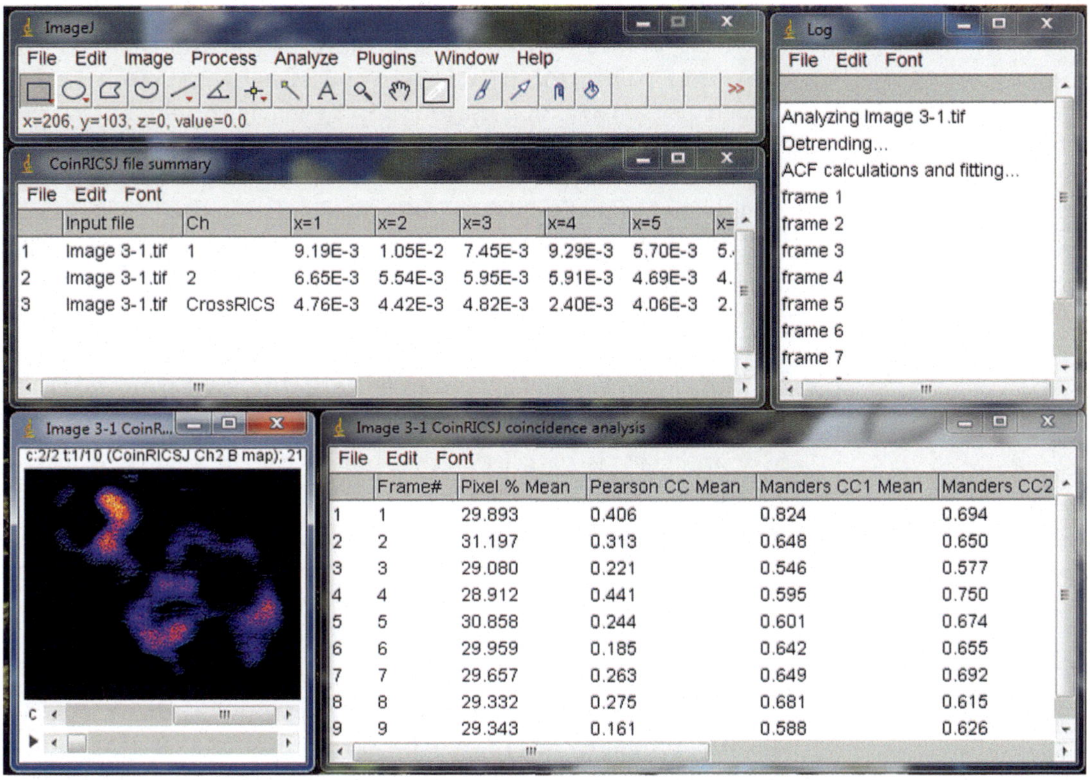

Fig. 4 CoinRICSJ sample output. A log window (right panel) shows the progress of the analysis until it is finished. The plug-in outputs a new image window with B maps for the two channels and two tables with the global ACF data for each stack (left table) and the coincidence analysis of the obtained B maps (bottom right table), with Pearson (PCC) and Manders (MCC) correlation coefficients setting the threshold as the mean intensity value or using the Otsu method

3. Photon-counting detectors are usually preferred in correlation microscopy, but analog detectors may also be used if acquisition parameters are carefully chosen [8, 9], particularly scan speed and detector offset. First, scanning speeds should be decreased as much as possible to minimize the effects of shot (or dark current) noise. Also, at very high scanning speeds, there is not enough time for the analog detection to reset itself before collecting data for the next pixel, which leaves a residual signal from the previous pixel and originates a basal correlation called bleed-through noise. Effects of both shot and bleed-through noise in analog systems can be tested and corrected as follows. Take images at different scan speeds with the microscope set with the laser off so that no light reaches the detector. Then, use *CoinRICSJ* to obtain the corresponding *ACFs* and choose the fastest scan speed that still produces almost no correlations at the closest pixels. Otherwise, if scan speed cannot be decreased further, the *CoinRICSJ* plug-in allows the user to remove additional pixels from the *ACF* before the

fitting procedure. Note that the autocorrelation procedure always excludes pixel self-correlations from the analysis. The second parameter to take into account when using analog detectors is the acquisition offset, which should be set just below the specimen intensity threshold to minimize background-dependent noise. It is important to note that, although *CoinRICSJ* provides with relative brightness values (*see* below in **Note 11**), B maps obtained with different offsets or gains cannot be directly compared.

4. A water-immersion objective is preferred to match the medium index of refraction and reduce chromatic aberration. Otherwise, images should be tested (and corrected if needed) for pixel displacement in the two channels.
5. If intensity is very low, the pinhole can be increased at the expense of losing axial resolution.
6. The pixel size needs to be small enough to obtain a minimum of ca. 10 data points within the spatial-temporal decay of the *ACF* to allow accurate fitting of the data, and, for highly dynamic cytosolic proteins, the required pixel size is $\leq$0.05 μm.
7. A pixel dwell time of 20 μs is ideal for measuring cytosolic protein movements if pixel size is set in the range of 0.03–0.05 μm/pixel, while for slower motions such as binding events, diffusion of large protein aggregates, or membrane-associated proteins, larger pixel sizes and dwell times should be tested in pilot acquisitions.
8. The selected frame for acquisition should only contain the cell or cell areas to be analyzed. Otherwise, images should be properly cropped before analysis.
9. Strains lacking one of the two fluorescent proteins should be used to test whether emission/excitation settings are able to minimize cross talk during acquisition in simultaneous mode.
10. In order to achieve higher accuracy in the whole-stack average *ACF* curve, there is a need to enlarge the statistics by using time-lapse series of at least 100 frames [10, 11].
11. The detrending procedure to remove the immobile fraction of molecules has been described [12]. Briefly, for each pixel within a frame, mean pixel intensity values are obtained from the specified number of frames above and below the frame analyzed. Then, the pixel is subtracted the obtained mean pixel value, and, to avoid having negative values, the average intensity of the frame is added to each pixel. Mainly for this reason, RICS only provides with relative, not absolute, brightness values.
12. Pixel autocorrelation and B map correlation analysis. After filtering out the immobile fraction, *CoinRICSJ* calculates for

each pixel the autocorrelation data **only** in the raster direction as follows:

$$G(\xi) = \frac{\delta I(x, y) \cdot \delta I(x + \xi, y)}{I^2}$$

where $G(\xi)$ is the autocorrelation variable as a function of ξ, the pixel distance in the scanning direction, and $\delta I(x,y)$ is the difference between the intensity of pixel at x,y and the mean frame intensity $\langle I \rangle$. Autocorrelation data for each pixel are obtained as the *ACF* average within a small square region defined by the same pixel range used to limit the *ACF*. Finally, no specific model of diffusion is assumed, and a simple linear fit is used to obtain the ordinate at the origin $G(0)$ from the *ACF* of each pixel. These $G(0)$ values are then used to calculate relative B values by $B = I{\cdot}G(0)$ and the corresponding maps for each frame. Finally, correlation between B maps is analyzed, and *Pearson (PCC)* and *Manders (MCC)* correlation coefficients are obtained, setting the threshold as the mean intensity value or using the *Otsu* method, which may contribute to better select a low number of pixels with high B values. These correlation coefficients assess the degree of spatiotemporal coincidence of moving particles of the two proteins as a function of the number of fluorescent molecules per particle.

Acknowledgments

This work was supported by grant BFU2016-80234-R of the Spanish Agency of Investigation to M.A. D.F.M. received a fellowship from the *Generalitat de Catalunya*.

References

1. Hess ST, Huang S, Heikal AA, Webb WW (2002) Biological and chemical applications of fluorescence correlation spectroscopy: a review. Biochemistry 41:697–705
2. Ries J, Schwille P (2012) Fluorescence correlation spectroscopy. BioEssays 34:361–368
3. Singh AP, Wohland T (2014) Applications of imaging fluorescence correlation spectroscopy. Curr Opin Chem Biol 20:29–35
4. Wiseman PW (2015) Image correlation spectroscopy: principles and applications. Cold Spring Harb Protoc 2015:336–448
5. Digman MA, Gratton E (2012) Scanning image correlation spectroscopy. BioEssays 34:377–385
6. Digman MA, Stakic M, Gratton E (2013) Raster image correlation spectroscopy and number and brightness analysis. Methods Enzymol 518:121–144
7. Finka A, Sharma SK, Goloubinoff P (2015) Multi-layered molecular mechanisms of polypeptide holding, unfolding and disaggregation by HSP70/HSP110 chaperones. Front Mol Biosci 2:29
8. Brown CM, Dalal RB, Hebert B, Digman MA, Horwitz AR, Gratton E (2008) Raster image correlation spectroscopy (RICS) for measuring fast protein dynamics and concentrations with a commercial laser scanning confocal microscope. J Microsc 229:78–91
9. Moens P (2011) Fluorescence correlation spectroscopy, raster image correlation spectroscopy, and number and brightness on a commercial confocal laser scanning microscope

with analog detectors (Nikon C1). Microsc Res Tech 74:377–388

10. Hendrix J, Baumgärtel V, Schrimpf W, Ivanchenko S, Digman MA, Gratton E et al (2015) Live-cell observation of cytosolic HIV-1 assembly onset reveals RNA-interacting Gag oligomers. J Cell Biol 210:629–646

11. M a D, Gratton E (2009) Analysis of diffusion and binding in cells using the RICS approach. Microsc Res Tech 72:323–332

12. Digman MA, Brown CM, Sengupta P, Wiseman PW, Horwitz AR, Gratton E (2005) Measuring fast dynamics in solutions and cells with a laser scanning microscope. Biophys J 89:1317–1327

Chapter 18

3D Tracking of Migrating Cells from Live Microscopy Time-Lapses

Sébastien Tosi and Kyra Campbell

Abstract

With rapidly advancing microscopy techniques for live cell imaging, we are now able to image groups of migrating cells in many different in vivo contexts. However, as the resulting data sets become larger and more complex, following the behavior of these cells and extracting accurate quantitative data become increasingly challenging. Here we present a protocol for carrying out accurate automated tracking of cells moving over time in 3D, implemented as custom-built macro scripts for *ImageJ*. As opposed to many generic tracking workflows, the workflow we propose here accounts for the overall movement of the embryo, allows the selection of subgroups of cells, and includes a step for the complete assisted review of all 3D tracks. Furthermore, it is easy to add new custom track measurement to the code provided. Together, these present a reliable method for the precise tracking of cells, from which distinct subsets of cells can be selected from within a population.

Key words 3D cell tracking, Cell migration, Live in vivo imaging, Endoderm migration, *Drosophila melanogaster*, Tissue morphogenesis, Directional persistence, Image stabilization, Results validation

1 Introduction

Cell tracking is a powerful tool for studying the behavior of migrating cells. While many tools have been developed for following cells moving in 2D, reliably tracking cells over time in 3D has remained extremely challenging. These difficulties are further compounded when tracking cells in in vivo contexts such as in developing embryos, where the signal can often be weak and the surrounding environment in continual flux. Furthermore, as cells migrate during embryogenesis, embryos often undergo major reshaping and large overall movements, and it can be difficult to separate one from the other.

As a consequence, the development of many tissues has remained poorly characterized, also due to their depth and inaccessibility during embryogenesis. An example of this is the endoderm germ layer, which comprises one of the most fundamental cell types

Elena Rebollo and Manel Bosch (eds.), *Computer Optimized Microscopy: Methods and Protocols*, Methods in Molecular Biology, vol. 2040, https://doi.org/10.1007/978-1-4939-9686-5_18, © Springer Science+Business Media, LLC, part of Springer Nature 2019

in multicellular organisms. The endoderm gives rise to the entire digestive tract, as well as other vital organs such as the lung, thyroid, pancreas, and liver. Past studies on endoderm have been hindered in comparison to other tissues such as the ectoderm and mesoderm, as it is less accessible and more difficult to image during normal and perturbed development.

Over the past few years, the first live cell imaging studies of endoderm migration in *Drosophila melanogaster* embryos were published [1, 2]. These studies have been aided by the identification of molecular markers and development of fluorescent reporters for endoderm, as well as advanced live cell and deep tissue imaging techniques [2–4]. It is not possible to image these cells using standard confocal microscopy due to their depth within the embryo (>40 μm beneath the surface of the embryo) and their proximity to the highly fluorescent yolk sac. The use of two-photon microscopy can partially overcome these difficulties; migrating endoderm cells are then imaged using a nuclear marker and different cell types within the migrating endoderm being distinguishable by nuclear size (Fig. 1).

However, analyzing these movies still presents numerous challenges: the nuclei are often heavily clustered, their contours are somewhat fuzzy, and the temporal resolution is limited by low fluorescence emission. Also, the tissues surrounding the migrating cells often undergo a strong directed movement, making nuclei harder to follow. To address these issues and allow a semiautomated tracking of the nuclei, some tools were developed at the Advanced Digital Microscopy Core Facility at IRB Barcelona; they are described in detail in this protocol. The workflow includes preprocessing movies prior to tracking to compensate for gross overall

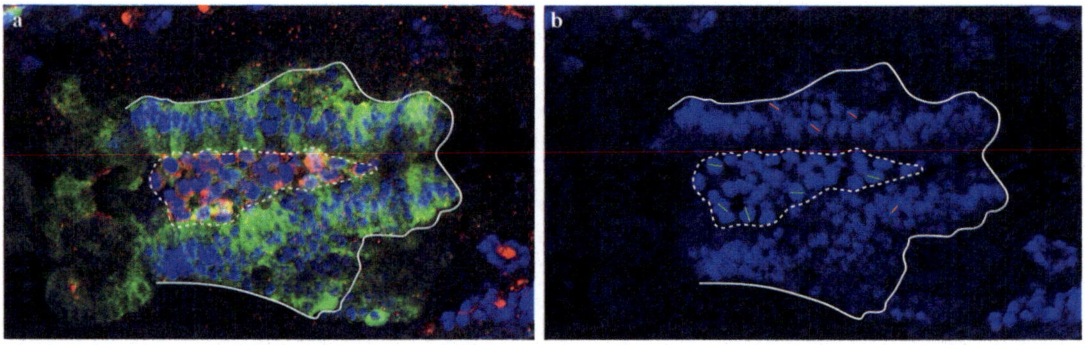

Fig. 1 Two of the main cell types in migrating endoderm cells in *Drosophila* embryos can be distinguished by their nuclear size. (**a**) A stage 12 *Drosophila* embryo in which all endoderm cells have been labeled by a nuclear marker, hindsight (hnt) (blue), GFP driven by the huckebein-Gal4 (green), and a subpopulation of cells, the interstitial cell precursors (ICPs) by the inscuteable antibody (Insc) (red). The white line outlines the principle midgut epithelial cells (PMECs), which are Insc negative, and the dotted white line the ICPs. (**b**) These two cell populations have distinct nuclear sizes. PMECs have a nuclear diameter (red lines) of 3.3 μm ± 0.4 ($n = 117$), whereas ICPs have a nuclear diameter (green lines) of 5.4 μm ± 1.2 ($n = 112$)

movements of the surrounding embryo, assisted track validation which ensures that statistics only include correctly tracked nuclei, and spatial selection of nuclei subpopulations, based on hand-drawn region of interests (ROIs). This last feature makes this method particularly suited to following the behavior of distinct cell types within a heterogeneous population. Assisted track validation is critical given the complexity of the images, and overall movement compensation alleviates the strain put on conventional object trackers by limiting object's excursion (we used *TrackMate* [5] in this protocol). Importantly, even though the movement is compensated, the protocol presented here quantifies absolute nuclei movements, the original movement being restored during the last step of the workflow. This means that no assumption is made about a reference frame to estimate nuclei displacements. These recently developed tools should be generally applicable or adaptable to the tracking of multiple cell populations in 3D over time imaged by any suitable microscopy modality.

2 Materials

2.1 Movies to Be Analyzed

Sample 3D microscopy time-lapses were acquired as described in detail in Ref. [2]. In general, any 3D time-lapses where objects appear as spots or blobs, such as cells labeled with a nuclear marker, can be put through this workflow. A sample movie can be downloaded at [6]. In this movie the 48Y-Gal4 driver was used to drive the expression of a nuclear GFP (stinger-GFP) in the embryonic *Drosophila* midgut. This embryo was imaged using an inverted Leica SP5 two-photon microscope with excitation at 890-nm wavelength and a 63× oil immersion lens (NA = 1.4); 15-μm-thick image stacks (1.5-μm step) were acquired at 2-min intervals for a period of 20 min.

2.2 Analysis Software

The macros and plugins presented here are for use with *ImageJ*, the open-source image processing and analysis software originally developed at the National Institutes of Health (Bethesda, Maryland, USA). To ensure compatibility, we recommend using *Fiji Lifeline 2014 June 02*, an *ImageJ* distribution for Life Sciences that you can find in "Other Downloads" section from *Fiji* download webpage [7]. *Fiji* automatically comes with *TrackMate* [5], a plugin which is used in this protocol. In principle *Fiji* should run on all major computer platforms, including Microsoft Windows, Mac OS, and Linux (*see Fiji* website for detailed requirements).

Additional software components required for this protocol are briefly described below and can be downloaded at [8]. The readers interested in understanding or modifying the code of the two *ImageJ* macros can refer to the document "*ImageJ* macros code description," also downloadable from this repository.

2.3 TrackMate-SpotDistanceFilter.jar ImageJ Plugin

This plugin is a custom *TrackMate* extension enabling the elimination of blob detections that are abnormally close; it can therefore improve the overall accuracy of nuclei detection, especially when nuclei are heavily clustered. To install the plugin, simply copy the file *TrackMateSpotDistanceFilter.jar* to the "plugins" subfolder of *Fiji* installation folder.

2.4 Preprocess "TimeLapse_batch.ijm" ImageJ *Macro*

This *ImageJ* macro applies a linearly increasing translation opposing the overall embryo movement to stabilize it in the time-lapse. It can be copied anywhere, drag, and drop the file to *Fiji* to use it.

2.5 "Review-MeasureTracks.ijm" ImageJ Macro

This *ImageJ* macro enables the user to select subgroups of cells; it then interactively displays the corresponding tracks detected by *TrackMate* for review and computes average speed and directional persistence from validated tracks. It can be copied anywhere; drag and drop the file to *Fiji* to use it.

3 Methods

In this section we describe the workflow to perform assisted 3D tracking of cell nuclei; it is subdivided into three steps. Firstly, movies are preprocessed to compensate for the overall movement of the surrounding environment. Next, nuclei are automatically tracked in 3D in the compensated movie by using *TrackMate*. Then, tracks starting bounding boxes can be hand drawn to select specific subpopulations of cells; the movement compensation of the first step is numerically canceled out in *TrackMate* tracks, and they are interactively overlaid on the original movie for user inspection. For each new validated track, velocity and persistence of movement are automatically calculated and accumulated to a table. This table can be easily exported for further analysis.

3.1 Compensation of Overall Directed Movement

1. Decide if the time-lapse image series needs compensation of overall movement (*see* **Note 1**). If not skip to Subheading 3.2.
2. Open time-lapse image series (drag and drop TIFF file to *Fiji*).
3. Estimate gross X/Y frame shifts manually by measuring the total displacement of several reference structures affected by the overall movement (from movie start to end), averaging them, and dividing by the number of time steps. For instance, for the time-lapse depicted in Fig. 2 ("sample_movie.tif"), we estimated XShift = −10 pix/frame and YShift = 0 pix/frame.
4. Open macro "PreProcessTimeLapse_batch.ijm" (drag and drop to *Fiji*).
5. Hit *Run* in the macro editor window.

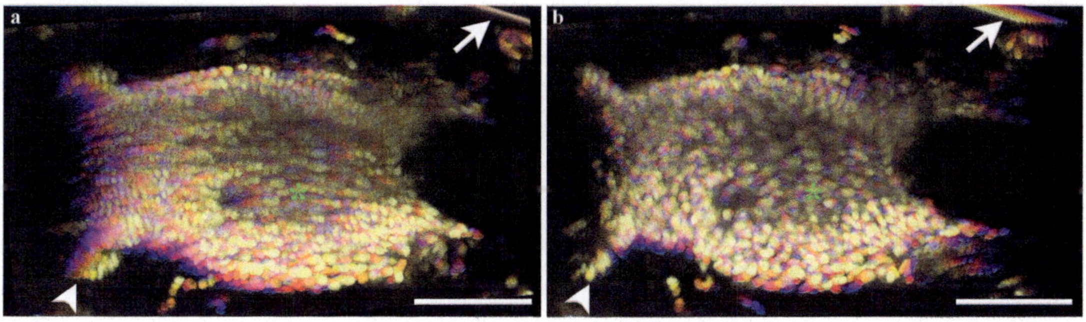

Fig. 2 Preprocessing is used to compensate for large uniform directed movements prior to tracking. A temporal color coded projection has been applied to the time series of migrating endoderm nuclei in *Drosophila*, which enables visualization of the movement over time from a single image. (**a**) shows a time series that has not been compensated. Static objects appear as a single color (arrow), fast-moving objects moving with uniform movement appear as parallel colored smears (arrowhead), and objects moving with less uniform movement appear with mixed blurs of color (asterisks). (**b**) shows the same time series as in (**a**). After compensation with *Xshift*, set to −10 pix/frame, and *Yshift* set to 0 pix/frame. Static objects such as the external membrane of the embryo now appear as parallel colored smears (arrow), fast-moving objects with the uniform movement that the image has been compensated for appear as a single color (arrowhead), and objects with a movement that is distinct from the compensated movement appear as fuzzy colored regions (asterisks). Note that the nuclei in (**b**) appear more distinct than in (**a**) as a result of the stabilization. Scale bar = 50 μm

6. Enter frame shifts *Xshift* and *Yshift* to apply. For the sample movie, −10 pix/frame and 0 pix/frame, respectively.
7. Press *OK*.
8. Save stabilized time series as a new TIFF file ([File > Save as Tiff...]). The shifts applied are automatically appended to filename for reference.

3.2 Nuclei 3D Tracking with *TrackMate*

1. Open the time-lapse image series to be tracked in *Fiji*. If the movie has been compensated for overall movement, open the compensated movie.
2. Launch *TrackMate ImageJ* plugin ([Plugins > Tracking > *TrackMate*]) (*see* **Note 2** for details on *TrackMate*). Ensure that spatial and temporal calibrations are correct at the first step (if not, update the values from the dialog box), and press *Next*.
3. Select *TrackMate LoG detector* (*see* **Note 3**) and press *Next*.
4. We recommend testing varying *Estimated blob diameter* and *Threshold*, with the aim of finding a balance between maximal nuclei detected and minimal false detections in the background. For this tick *Preview* and start tweaking the values, preferably starting with a large threshold and going down to smaller values not to trigger too many spot detections. *See* Fig. 3 for the effects of varying *Estimated blob diameter* and

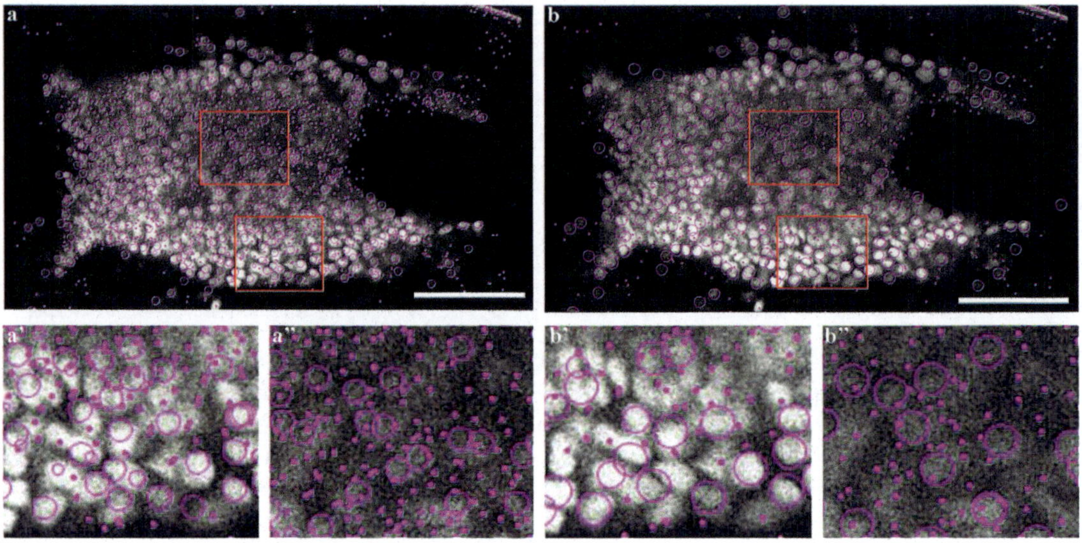

Fig. 3 LoG detector "Estimated blob diameter" cannot be used to distinguish between cell populations based on small differences in nuclear diameter. (**a**, **b**) Single z-slices from a time series of migrating endoderm nuclei in *Drosophila*; nuclei are in white, magenta circles show "blobs" detected by *TrackMate*, and dots also represent "blobs" detected but in a different z-slice. (**a**) Estimated blob diameter of 3.5 μm and threshold set to 30 lead to smaller nuclei (average diameter 3.3 μm) as well as larger nuclei (average diameter of 5.4 μm) being detected, but many nuclei are detected multiple times, and there is a lot of background noise (*see* **a'** and **a"**). (**b**) Using an *Estimated blob diameter* of 5 μm and a *Threshold* set to 10 allows a more robust detection of both cell types and decreases multiple nuclei detection and background noise (*see* **b'** and **b"**). Scale bars = 50 μm. Insets are depicted by red boxes

Threshold on a movie of migrating *Drosophila* endoderm cells, which displays a mixture of nuclei about 3 μm and 5 μm in diameter (*see* **Note 4**). For the sample movie, 4-μm estimated blob diameter and a threshold of 10 is a good trade-off. Applying median filter is not necessary for this movie since there is no visible bright isolated outlier pixel in the images; *enable sub-pixel localization* for increased localization accuracy and leave all other settings to default before pressing *Next*. Processing should take only a few seconds on a modern computer; upon completion press *Next* again.

5. Since we have yet no visual cue on the quality of spot detection, at initial thresholding, leave setting to default, and press *Next*.
6. At *Select a view*, pick *Hyperstack Displayer* since we want to check spot detection and tracks inside *ImageJ* native hyperstack viewer.
7. Multiple spots can typically be detected within one nucleus, especially when the signal is faint (*see* **Note 5**). To mitigate multiple detections, we designed a custom "*TrackMate* spot filter extension" called *Closest Distance*, which ensures that only the brightest spot among close by detections is kept. To apply

it, at step *Set filters on spots*, press + green button, select *Closest Distance*, and adjust minimum distance by dragging and dropping the left edge of the blue region. For the sample movie, we used 4 μm (the typical nuclei diameter) to eliminate multiple detections inside single nuclei. If you do not see the filter *Closest Distance* in the drop-down list, make sure that you completed Subheading 2.3 (*ImageJ* restart is required in case you did not complete this step). Press *Next*.

8. Use *Simple LAP Tracker* to link from frame to frame the nuclei that have been detected by *TrackMate* (*see* **Note 6**). Press *Next*.
9. Adjust parameter *Linking max distance* to the maximum distance the nuclei are expected to move from one frame to the next. This limits the permitted spot displacements and should reduce tracking errors, as well as speed up calculations. For the sample movie, we estimated this maximum displacement to about 4 μm.
10. In cases where the nuclei signal is steady throughout, we recommend not allowing any track gap closing (both gap closing-related parameters set to 0). This enforces that tracks hold a detected spot at every time position and discards tracks not fulfilling this condition. Press *Next*. Wait for processing to complete and press *Next* again.
11. The next step ensures that only tracks fulfilling specific user-defined criteria are kept. We recommend only keeping tracks spanning the duration of the whole time-lapse. To discard shorter tracks, discard any that include less spots than the number of time frames of the time-lapse. For example, for eight time frames, *Set filters on tracks* by pressing + green button, select *Number of spots in track*, and adjust the left edge of the blue region slightly above 7. For longer movies, or if you consider that too many tracks are discarded, you can slightly decrease this value; we used a minimum number of six spots for the sample movie.
12. At this step you can visualize tracks with different display options to get a quick impression about the quality of the tracking. If you notice that something went really bad, go back to the previous steps and start to troubleshoot. If everything seems fine, export tracks by pressing *Analysis* and saving the *Spots in tracks statistics* table to an *.xls* file (from table menu); this window is at first hidden by other windows that you can safely close.

3.3 Track Selection, Visualization, and Validation

1. Close *TrackMate*, all windows, and all images, and open the original time-lapse image series (not compensated).
2. Import the table *Spots in tracks statistics* to results table (drag and drop .xls file to *ImageJ*). In case you could not complete the previous section successfully, this file is provided for the

sample movie in the repository (compensation: Xshift −10 pix/frame, Yshift 0 pix/frame).

3. Drag and drop macro "ReviewMeasureTracks.ijm" to *Fiji* and run it (*run* button in *ImageJ* macro editor).
4. Adjust *X* and *Y shifts* to the same values that were used in **step 3** of Subheading 3.1. If the movie was not compensated, then enter 0 and 0. Leave *Manual check* ticked if you want to review the tracks and press *OK*.
5. Draw cell start bounding boxes around the cells that are to be reviewed/tracked in the first time frame, each time press *t* to add a bounding box to the *ROI Manager*, and then press *OK*. If no bounding box is defined, then the whole movie is processed.
6. Review the first track by selecting the window of the ROI manager and pressing the "up" and "down" arrows to scroll through the track. You should see a moving spot drawn over the nucleus that is currently inspected; the display is automatically adjusted in time and Z position to aid the validation. A valid track should obviously follow the same nucleus the whole way through.
7. Press *OK* after the track has been reviewed.
8. Keep/discard the track when asked.
9. Process all tracks in the same way (*see* **Note 7**).
10. Once all tracks selected in the initial bounding boxes have been reviewed, the tracks that have been "Kept" will be shown overlaid on a z-slice from the first time frame of the series (Fig. 4). The table *Tracks Statistics* stores quantitative data on the "Kept" tracks including total track length, total displacement (line distance from start to end), mean speed, and directional persistence (Fig. 5). Measurements can easily be exported by saving the table to *.xls* file.

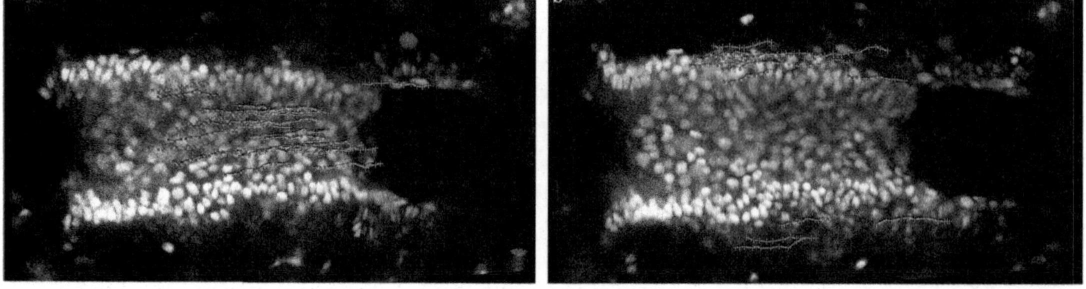

Fig. 4 Manually selected tracks are overlaid on a z-slice from the first time frame from the time-lapse series. (**a**) An example of the ICP cells tracked using a single ROI as a starting point. (**b**) An example of PMEC cells tracked using multiple ROIs

Results

	Track ID	Length (micron)	Total Disp (micron)	Mean speed (micron/sec)	Persistence
1	211	40.320	38.530	0.034	0.960
2	354	29.130	27.740	0.024	0.950
3	375	30.560	26.860	0.025	0.880
4	586	29.050	27.990	0.024	0.960
5	588	28.990	26.950	0.024	0.930
6	589	31.230	28.900	0.026	0.930
7	684	32.200	29.880	0.027	0.930
8	772	31.360	30.140	0.026	0.960
9	773	27.630	25.660	0.023	0.930
10	925	27.380	25.490	0.023	0.930
11	946	27.220	26.060	0.023	0.960
12	211	40.320	38.530	0.034	0.960
13	675	31.350	28.130	0.026	0.900
14	743	29.580	26.300	0.025	0.890
15	837	30.730	23.460	0.026	0.760
16	211	40.320	38.530	0.034	0.960
17	349	34.440	31.170	0.029	0.910
18	354	29.130	27.740	0.024	0.950
19	375	30.560	26.860	0.025	0.880
20	402	25.480	23.860	0.021	0.940
21	581	34.070	32.650	0.028	0.960
22	586	29.050	27.990	0.024	0.960
23	588	28.990	26.950	0.024	0.930
24	675	31.350	28.130	0.026	0.900
25	683	27.140	25.480	0.023	0.940
26	684	32.200	29.880	0.027	0.930
27	739	26.220	22.510	0.022	0.860
28	740	35.180	33.010	0.029	0.940
29	772	31.360	30.140	0.026	0.960
30	773	27.630	25.660	0.023	0.930
31	925	27.380	25.490	0.023	0.930
32	946	27.220	26.060	0.023	0.960

Fig. 5 The results table summarizes measurements from selected tracks: *track length*, *displacement* (line distance from start to end), *mean speed*, and *directional persistence* (ratio between track displacement and track length)

4 Notes

1. If the cells under study exhibit a strong, directed movement (e.g., cell migration or embryo movement), generic spot trackers such as *TrackMate* are typically inaccurate. Indeed, when linking spots across contiguous frames, trackers typically assume small displacements, and there is no consideration for larger movements or for underlying external movements. This can sometimes be compensated for automatically by a technique called registration [9]. Alternatively, assuming that the

strong directed movement is constant across the movie and mostly occurs in *X* and *Y* dimensions, movies can be preprocessed by incrementally shifting frames by a fixed offset using a custom *ImageJ* macro (Subheading 3.1; *see* Fig. 2). The original movement is restored after tracking to compute measurements from the absolute movements.

2. *TrackMate* performs both spot/blob detection and tracking (linking detected blobs across frames to build tracks). *TrackMate* offers several blob detection and tracking algorithms [5] both with sets of adjustable parameters.
3. *Laplacian of Gaussian (LoG)*-based blob detection is a robust technique for blob detection of a given characteristic size. It implements an adjustable *3D Laplacian of Gaussian filter* [10], followed by gated local intensity minima detection with user-defined intensity threshold. *LoG* response is negative inside bright blobs with intensity minima expected to be located close to blob centers. *DoG* and especially *downsampled LoG* can be significantly faster than *LoG*, but their increased speed can come at the cost of slight accuracy loss.
4. We attempted to distinguish between two of the main cell types in the endoderm at the *LoG* detector *Estimated blob diameter* stage, as they have distinct nuclear diameters; interstitial cell precursors (ICPs) average 5.4 μm in diameter, whereas the principle midgut epithelial cells (PMECs) average 3.3 μm (*see* Fig. 1). However, we found that using the smaller estimated blob diameter of 3.5 never allowed us to distinguish between ICPs and PMECs, despite putting a high threshold value such as 30 (Fig. 3a). Furthermore this also led to a lot of background and multiple detections of single nuclei (Fig. 3a). We found that increasing the diameter to 4 or 5 μm and decreasing the threshold to 10 led to the most reliable detection of both cell types, with minimal background (Fig. 3b). For this reason, we decided the best method was to track all cells together, but then include a step to manually select cell subpopulations.
5. In practice, it turned out to be difficult to achieve satisfying detection accuracy solely by adjusting detection parameters (*LoG scale* and *intensity threshold*). Typically, at the sensitivity required to detect even the faintest nuclei, several spots were detected, especially inside inhomogeneous, large, or elongated nuclei. To discard most of these spurious detections, we implemented a custom *TrackMate* extension (*Closest Distance*) keeping only the brightest spot among close by detections (user-defined range)—Subheading 3.2, **step** 7.
6. Once blobs are detected, *TrackMate* attempts to link them by minimizing centroid displacements across time frames. The simplest algorithm consists in linking spots to their closest

neighbor in the next frame. This is not necessarily optimal when considering all centroid displacements, especially for false or missed detections. A notoriously well-achieving algorithm is the *Hungarian linear assignment* [11]; it minimizes the summed displacements of all centroids between two time frames. In *TrackMate*, this algorithm is called *Simple LAP Tracker*, and this is the one we recommend using.

7. You can interrupt the macro at any time by pressing *Cancel* when asked if a track should be kept. The measurements of the "Kept" tracks are safely accumulated to *Tracks Statistics table*; to display the selected tracks, tick *Show All* from the window of the *ROI Manager*.

Acknowledgments

This work was supported by the Institute for Research in Biomedicine Barcelona and by a Wellcome Trust/Royal Society Sir Henry Dale Award to KC (Grant number R/148777-11-1).

References

1. Campbell K, Whissell G, Franch-Marro X, Batlle E, Casanova J (2011) Specific GATA factors act as conserved inducers of an endodermal-EMT. Dev Cell 21 (6):1051–1061. https://doi.org/10.1016/j.devcel.2011.10.005
2. Campbell K, Casanova J (2015) A role for E-cadherin in ensuring cohesive migration of a heterogeneous population of non-epithelial cells. Nat Commun 6:7998. https://doi.org/10.1038/ncomms8998
3. Wu T, Hadjantonakis AK, Nowotschin S (2017) Visualizing endoderm cell populations and their dynamics in the mouse embryo with a Hex-tdTomato reporter. Biol Open 6 (5):678–687. https://doi.org/10.1242/bio.024638
4. Wen JW, Winklbauer R (2017) Ingression-type cell migration drives vegetal endoderm internalisation in the Xenopus gastrula. elife 6: e27190. https://doi.org/10.7554/eLife.27190
5. Tinevez JY, Perry N, Schindelin J, Hoopes GM, Reynolds GD, Laplantine E, Bednarek SY, Shorte SL, Eliceiri KW (2017) *TrackMate*: an open and extensible platform for single-particle tracking. Methods 115:80–90. https://doi.org/10.1016/j.ymeth.2016.09.016
6. Sample movie. https://goo.gl/aZBZk8
7. *Fiji*. https://*ImageJ*.net/*Fiji*/Downloads
8. Assisted Nuclei 3D Tracking. https://github.com/SebastienTs/AssistedNuclei3DTracking
9. Thévenaz P, Ruttimann UE, Unser M (1998) A pyramid approach to subpixel registration based on intensity. IEEE Trans Image Process 7(1):27–41
10. Lindeberg T (1993) Detecting salient blob-like image structures and their scales with a scale-space primal sketch: a method for focus-of-attention. Int J Comput Vis 11(3):283–318. https://doi.org/10.1007/BF01469346
11. Kuhn H (1955) The Hungarian Method for the assignment problem. Naval Res Logistics Quarterly 2:83–97

Part IV

Methods Based on Machine Learning

Chapter 19

A Cell Segmentation/Tracking Tool Based on Machine Learning

Heather S. Deter, Marta Dies, Courtney C. Cameron, Nicholas C. Butzin, and Javier Buceta

Abstract

The ability to gain quantifiable, single-cell data from time-lapse microscopy images is dependent upon cell segmentation and tracking. Here, we present a detailed protocol for obtaining quality time-lapse movies and introduce a method to identify (segment) and track cells based on machine learning techniques (*Fiji's Trainable Weka Segmentation*) and custom, open-source *Python* scripts. To provide a hands-on experience, we provide datasets obtained using the aforementioned protocol.

Key words Computational image analysis, Single-cell quantification, Cell lineage analysis, Cell segmentation, Cell tracking, Machine learning, Fluorescence microscopy, Bacterial growth

1 Introduction

During the last decade, there has been a transition in the analysis of cellular physiology from the batch level (population-scale) to the single-cell level. This transition has been stimulated by the development of quantitative and high-throughput techniques that require computer-aided methods to extract information at the single-cell level. Using these approaches researchers have been able to show, for example, the relevance of stochastic sources in gene expression [1] or the logic underlying cell size homeostasis [2]. Thus, single-cell analysis is less subject to averaging effects (Fig. 1) and offers a level of discrete detection that is unobtainable with traditional techniques [3–6]. In this context, the adoption of single-cell microscopy techniques has been limited because identifying, tracking, and quantifying single cells within a population of cells are usually difficult, time-consuming, and prone to errors that

Electronic supplementary material: The online version of this chapter (https://doi.org/10.1007/978-1-4939-9686-5_19) contains supplementary material, which is available to authorized users.

Elena Rebollo and Manel Bosch (eds.), *Computer Optimized Microscopy: Methods and Protocols*, Methods in Molecular Biology, vol. 2040, https://doi.org/10.1007/978-1-4939-9686-5_19, © Springer Science+Business Media, LLC, part of Springer Nature 2019

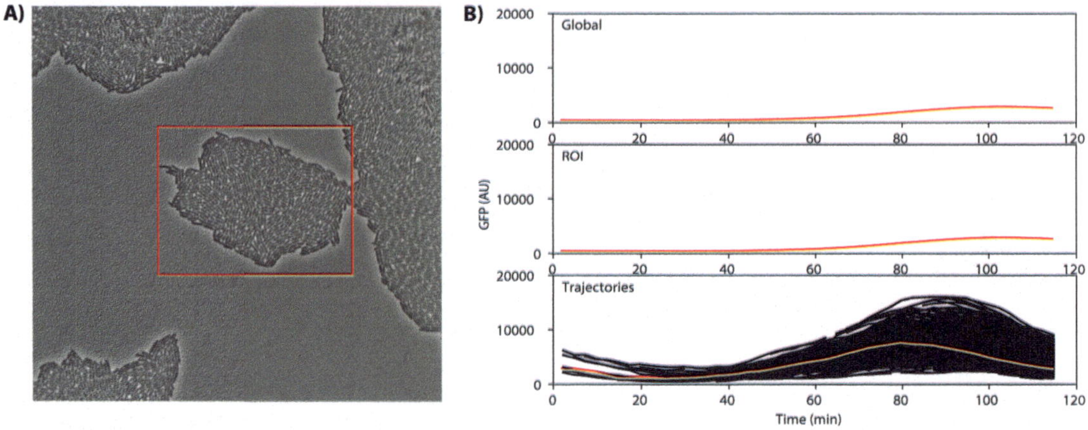

Fig. 1 (**a**) The final frame of the image dataset. The region of interest (ROI) is outlined in red. (**b**) Median fluorescence for given selections over time (below). Global: the median fluorescence over time for the whole image. ROI: the median fluorescence over time for the ROI is outlined in red in (**a**). Trajectories: the fluorescence over time for each trajectory (black) and the median fluorescence for all cells (red)

require manual corrections. Indeed, the identification stage implies a methodology able to recognize and outline the domain of individual entities (segmentation) and is particularly critical since tracking and quantification depend on it.

Traditional cell segmentation algorithms are based on image-processing techniques that ultimately compute gradients and use thresholding to measure the intensity and spatial relationships of pixels in order to detect cell boundaries. The latter is especially challenging in dense cell populations, e.g., bacterial colonies, and, while some edge detectors have been proven to be more effective than others, e.g., *Marr-Hildreth* vs. *Canny* [7], small changes in the microscopy illumination conditions require, more often than not, nontrivial adjustments of the segmentation parameters. In that regard, during the last years, a number of segmentation/tracking software suites have been publicly released [8–10]. Here we highlight three examples that, while essentially based on, and consequently constrained by, the aforementioned methodology, stand out because of their additional features and reliability. *MicrobeJ* [11, 12] is a plugin available through *Fiji/ImageJ* [13] that has a wide variety of tools available to analyze cell morphology and track cells in their user-friendly interface. *Oufti* [14] offers a user-friendly interface and a number of functionalities for quantitative analysis that include subpixel resolution for "reading" fluorescent signals within single cells. On the other hand, *CellX* [15] uses a novel approach for cell segmentation based on membrane patterning that is versatile in terms of cell shapes and robust to image noise.

More recently, the advent of artificial intelligence and machine learning techniques into the field has made possible the development of segmentation/tracking tools able to learn from training datasets and improve from experience without the need of explicit programming or parameter tweaking, e.g., *CellProfiler* [16] or more recently *SuperSegger* [17]. Here, following these ideas, we present a detailed protocol that utilizes an open-source *Fiji/ImageJ* plugin [13], the *Trainable Weka Segmentation tool* [18], complemented by custom-made open-source *Python* scripts. The computational methods herein can be used to count and track objects in any series of 16-bit tiff images. We have used these methods to count colonies on agar plates and track cells in microscope images. Here we detail one method of obtaining microscope images, which aims to reduce the training queue and improve the segmentation/ tracking process. To give a hands-on experience, we provide a dataset in the context of bacterial growth that was obtained using this method [19]. We have also recorded a video tutorial (Video 1) available on YouTube [20] and in the data repository to assist the users.

2 Materials

2.1 Reagents

1. 5× A Salts (composition per 100 mL): 0.046 g $(NH_4)_2SO_4$, 2.25 g KH_2PO_4, 5.25 g K_2HPO_4, 0.25 g sodium citrate tribasic·$2H_2O$, and 100 mL sterile deionized water [21]. Proceed to filter sterilize.
2. A minimal medium (composition per 100 mL): 20 mL (5×) A Salts, 80 mL sterile deionized water, 100 μL (1 M) $MgSO_4 \cdot 7H_2O$, 250 μL (80%) glycerol, 0.4% (w/v) glucose, and 1 mL (10%) casamino acids [21].
3. Lysogeny Broth Miller (LB; *see* list of abbreviations in Supplementary Table S1).
4. 100 mM Isopropyl β-D-1-thiogalactopyranoside (IPTG).
5. 50 mg/mL Kanamycin.
6. Silica gel.
7. Low melting agarose (2-hydroxyethyl agarose, Sigma).

2.2 Equipment

1. Two coverslips 24 × 60 mm, thickness No. 1.
2. One biopsy punch 6 mm diameter.
3. Desiccator (or an airtight container that can be used as a desiccator).
4. Microwave.
5. Parafilm.
6. A 96 well plate lid or an equivalent glass surface.
7. A 0.2 μm pore-size filter.

8. A 10 mL syringe (in which the 0.2 μm pore-size filter can be coupled).
9. Scalpel.
10. IBIDI μ-dish 35 mm low (bottom: polymer coverslip No. 1.5, thickness: 180 μm).
11. Microscope/imaging equipment: we used an inverted microscope (Leica DMi8) equipped with a 100×/1.40 oil objective (HC PL APO, Leica), Köhler illumination, a CMOS camera (Hamamatsu ORCA-Flash4.0 camera, V2), and green fluorescent protein (GFP; Ex: 470/40 nm, Em: 525/50 nm) filter. Excitation was performed using a led lamp (Lumencor light engine SOLA SE 5-LCR-VA). Leica Application Suite X (LAS X) software was used to control microscope and acquire images using Live Data Mode.
12. Temperature control system: IBIDI heating system. All experiments are performed at 37 °C.

2.3 Operating System, Software, and Data Repository

All of the following are open-source, and downloads are available online (*see* Tables 1, 2, and Supplementary Table S2).

1. *Ubuntu 16.04 LTS*, a Linux operating system. Alternatively, Mac OS X can also be used to run the pipeline. To run the pipeline on other operating systems, a virtual machine [22] can be installed to use *Ubuntu* (*see* **Note 1**).
2. *Fiji/ImageJ*, an open-source *Java* image-processing program [13].
3. *Anaconda 2.7* is an open-source distribution of *Python*, a programming language that has a wide range of tools and libraries for image analysis, including *SciPy* and *NumPy* (*see* **Note 2**). Our scripts have exclusively been tested with *Python* 2.7.
4. *OpenCV* is a *Python* package required for the pipeline that is not included in the initial *Anaconda* download (Table 2). Only *OpenCV* downloaded through *Anaconda*, using the command *conda install opencv* in the terminal, has been tested to work with our scripts (*see* **Note 3**).
5. *Avconv*, a library for video and audio conversion. This library can be installed in *Ubuntu* using the following command in the terminal: *sudo apt install libav-tools*. To install *Avconv* for Mac OS X, the command is *brew install libav* (*Homebrew* must be installed prior to installing *Avconv*).
6. A list of scripts is available in Table 2. Scripts must be run in an *Anaconda Python* 2.7 environment with *OpenCV* installed to function as designed. The Master Script, *SegmentandTrack.py*, calls the remaining scripts to run the entire analysis pipeline based on user input. All scripts and imaging datasets are available at the OSF public repository [19]. These images can be

Table 1
List of all software and operating systems

Software	Function	Website
Ubuntu 16.04 LTS	Linux operating system	https://www.ubuntu.com/download [28]
VirtualBox (optional)	Virtual machine to run Linux environment	https://www.virtualbox.org/wiki/Downloads [22]
Fiji ImageJ (includes Weka segmentation tool)	Classification of images using machine learning	https://imagej.net/Fiji/Downloads [13, 23]
Anaconda (Python 2.7)	Open-source distribution of Python and related packages (including NumPy and SciPy)	https://www.anaconda.com/download/ [29]
OpenCV	A Python package with tools for image analysis (not included in Anaconda). On a Linux machine, install through Anaconda using the command "conda install opencv" in the terminal	https://pypi.python.org/pypi/opencv-python [30]
Avconv	Software package for handling videos	https://libav.org/avconv.html#Description [31]

used to recreate Videos 2–4. Figure 2 demonstrates how an image was segmented prior to cell tracking.

7. 16-bit tiff images. Sample image datasets are available in the data repository [19]. These datasets contain 467 frames and 20 frames.
8. The recommended hardware to process the complete imaging dataset (467 frames) is 8 GB or more of RAM and a modern processor. An image subset (20 frames) requiring a less demanding hardware configuration is also available in the data repository. Adjustments in *Fiji* memory settings may be required for processing the dataset (the *Fiji* software requires at least 1 GB RAM for its processes [23]).

3 Methods

Here we present one method of obtaining images, followed by an image-processing pipeline that can be applied to a variety of image datasets. The processing pipeline consists of a series of steps that were designed to analyze a swath of single-cell datasets (Table 2). These scripts are designed to work on 16-bit tiff images that use a file naming system containing the letters "p" or "g" to indicate phase or fluorescence, respectively, followed by three numbers to

Table 2
List of custom scripts

Script	Subheadings	Language	Brief description	CSV data
SegmentandTrack.py	3	Python	The master script to run the pipeline based on user input	N/A[a]*
Image_alignment.py	3.4	Python	Aligns images based on differences calculated through FFT	N/A
Segmentation.ijm	3.5	Fiji macro (ijm)	Calls Trainable Weka Segmentation tool and can be used to train or apply classifiers	N/A
Batch_segment.bsh	3.5	BeanShell	Called by RunWeka.py to segment a batch of images	N/A
RunWeka.py	3.5	Python	Calls Segmentation.ijm and Batch_segment.bsh	N/A
TrackCellLineages.pyCellLineages	3.6	Python	Labels a binary mask and calculates the differences between a given cell and cells within a given area in the previous image. Automatically saves single-cell data. Uses the calculated differences to find trajectories and identify cell lineages. Labels lineages from the first frame and outputs lineage data.	Single-cell data and lineage data summary
Lineage_analysis.py	3.7	Python	Outputs csv files with frame-by-frame data for lineages tracked in TrackCellLineages.py	Lineage data for individual lineages
Image_analysis.py	3.7 and 3.8	Python	Analyzes global or ROI data	Global and ROI data

All scripts can be modified and include comments to facilitate modification. Scripts can be downloaded at GitHub https://github.com/hdeter/CellTracking [32] or at the public repository http://osf.io/gdxen/ [19]

[a]CSV files output by other scripts will also be output when running SegmentandTrack.py

indicate frame and then the extension ".tif" (e.g., 20171212_book-**p-001.tif**, indicating **p**hase contrast channel and frame number **001**). Phase and fluorescence images must be separate (not stacked). Modification of the scripts is required to use alternative naming systems, and some modification is required for datasets containing two or more fluorescence channels (*see* **Note 4**).

3.1 Sample Preparation

The *E. coli* strain used for imaging was MG1655 transformed with a ColE1 plasmid containing resistance to Kanamycin and the inducible combinatorial promoter $P_{lac/ara\text{-}1}$ [24] controlling the expression of GFP marked with an LAA degradation tag. We broke the sample preparation down to three major steps:

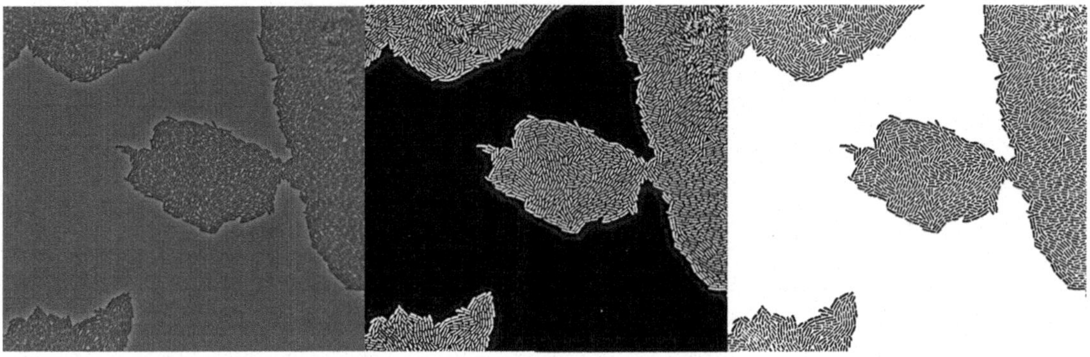

Fig. 2 An example of cell segmentation using our method. Left: phase image. Center: Mask 1, a probability mask based on the classification of the phase image. Right: Mask 2, a binary mask based on the classification of Mask 1

1. Grow cells overnight from a glycerol stock in 10 mL LB with 10 μL (50 mg/μL) Kanamycin at 37 °C, shaking at 200 rpm.
2. Dilute cells to a final OD_{600} ~0.01 in 5 mL A minimal media (*see* **Note 5**) with 1.5 μL (100 mM) IPTG, to achieve a final concentration of inductors of 0.03 mM IPTG.
3. Let this culture grow at 37 °C and 200 rpm until OD_{600} is between 0.2 and 0.3 (~3 h). While cells are growing, prepare agarose pads (**steps 4–8**).
4. Cover a 96 well plate lid with parafilm, extending the parafilm over its surface. Place one coverslip in the middle of the parafilmed area, and gently press the cover glass' four corners against the parafilm to "fix" it.
5. Mix 0.2 g low melting agarose (*see* **Note 6**) in 10 mL A minimal medium, and go through three to four cycles of heating the agarose solution in a microwave until it starts to boil and vortexing the mixture.
6. Once a homogeneous agarose solution is ready and it has cooled down to around 50 °C, add the necessary inducers/antibiotics. In our case, we added 3 μL (100 mM) IPTG. Vortex the mixture gently, and filter it immediately with a 0.2 μm pore-size filter using the 10 mL syringe (*see* **Note 7**).
7. Allow the mixture to sit (for a few minutes) so that bubbles accumulate by the surface (*see* **Note 8**). Pipette around 3 mL of the agarose mixture from the bottom part of the tube (bubble-free) onto the cover glass slide. Immediately place another coverslip on top of the poured agarose to create an agarose *sandwich*. Take special care to minimize introduction bubbles while preparing the sandwich (first make a gentle contact between the upper cover glass and the agarose in one of the extremes, and then proceed to lower down the rest of the cover glass).

8. Let the pad dry at room temperature for at least 1 h before spotting cells onto it (*see* Subheading 3.1, **step 13**). Note that if the pad is left to dry "too much," it will shrink and will not be useful/reliable. To prevent excessive evaporation from edges, cover the agarose sandwich with a lid. Additional details can be found here [25].
9. Once cells have grown between OD_{600} 0.2 and 0.3, dilute them to a final concentration of ~0.01 OD_{600} using A minimal media with 0.03 mM IPTG to a final volume of 1 mL.
10. Carefully slide/detach the upper cover glass from the agarose sandwich, and cut a smaller pad using a 6 mm diameter biopsy punch (*see* **Note 9**).
11. Spot 2 μL of the diluted culture ("seed the pad") from **step 2** onto this smaller pad.
12. Incubate the pad at 37 °C for 15 min, and then place it in a desiccator (containing silica gel) at room temperature for another 15 min. This drying process of the seeded pad should avoid any drift in the imaging (*see* **Note 10**).
13. After drying, place the prepared pad in a cover glass bottom dish so that the seeded surface is in direct contact with the bottom of the dish (use a scalpel to do this, and avoid touching the seeded surface of the pad).
14. Seal the dish with parafilm to prevent pad shrinkage due to evaporation. The sample is now ready for imaging.

3.2 Imaging

1. Set up the microscope for Köhler illumination. A precise alignment of the optical components in the optical path (including the phase contrast ring) is a critical step for getting quality images and, as a consequence, a reliable segmentation. The procedure to do this is outside the scope of this manuscript, and there are excellent interactive tutorials on the web showing how to achieve alignment, including one available from Nikon [26].
2. To generate our image dataset, we used the following time-lapse sampling times: phase images were taken every 30 s and fluorescence images of GFP were taken every 5 min (*see* **Note 11**). We set 15 and 30 ms exposure time (using the maximum intensity of the lamp) for phase contrast and GFP channels, respectively. These conditions are experiment and equipment dependent.

3.3 Running the Microscope Image Analysis Pipeline

We have developed specific custom scripts that utilize open-source software (*see* Subheading 2.3) for cell segmentation and lineage tracking. To facilitate use, we provide *SegmentandTrack.py*, a Master Script, to run the entire pipeline based on user input (Tables 2 and 3). The pipeline (*see* Table 2) has been tested using an *Ubuntu*

Table 3
List of prompts for user input by SegmentandTrack.py

Prompt	Description	Subheadings
Do you wish to align images? (Y/N)	Answer "y" to align images	3.5
Do you wish to train and/or apply a classifier? (Y/N):	Answer "y" to segment images	3.6
Do you wish to track cells? (Y/N)	Answer "y" to track cells	3.7
Do you wish to output csv files detailing data for individual lineages? (Y/N)	Answer "y" to output frame-by-frame data for individual lineages as a csv file	3.7
Do you wish to analyze images (needed to get whole image fluorescence or render videos)? (Y/N)	Answer "y" to analyze the entire image and render videos	3.8
Do you wish to render videos? (Y/N)	Answer "y" to output videos	3.9
Enter the name of the image directory, relative to the working directory (e.g., Practice)	Enter the name of the directory containing your images (must be in your working directory)	3.3
Enter the number of the first frame in the dataset (e.g., 448)	Enter a whole digit integer corresponding to the first image in the dataset	3
Enter the number of the last frame in the dataset (e.g., 467)	Enter a whole digit integer corresponding to the last image in the dataset	3
Enter the time per frame in minutes (e.g., 0.5)	Enter the time in minutes (e.g., 30 s is 0.5 min)	3.2
Enter the number of the first frame with a fluorescence image (e.g., 449; for no fluorescence enter 0)	Enter a whole digit integer corresponding to *the first fluorescence image* in the dataset	3
Enter the number of frames between fluorescence images (i.e., every nth image; for no fluorescence enter 0)	Enter how often fluorescence images occur (e.g., every nth frame)	3.2
Enter the images' file name preceding the channel and file number (e.g., 20171212_book)	Enter the portion of the file name that does not change (i.e., precedes channel and file number)	3
Enter the name of fluorescence channel 1 (e.g., GFP)	Enter the name of the fluorescence channel	3.2
Do you have an ROI file for a stationary area? (Y/N)	Answer "y" if using an ROI for image alignment	3.4 and 3.5
Enter the path to the csv file, relative to the working directory (e.g., Align_roi.csv)	Enter the path to the csv file containing the ROI to use for image alignment	3.4 and 3.5
Enter the name of the directory to output images into (e.g., Aligned)	Enter a name to call the directory to which images will be output	3.5
Do you have masks for the images? (Y/N)	If not segmenting images, answer whether or not you have binary masks	3.6

(continued)

Table 3
(continued)

Prompt	Description	Subheadings
Enter the name of the directory containing masks, relative to [image directory]	If you have binary masks, enter the directory containing them, which must be within your image directory	3.6
Enter the absolute path to the Fiji executable file (e.g., /home/user/Downloads/Fiji.app/ImageJ-linux64)	Enter the full path to the executable file, located in the Fiji.app folder on your computer	3.6
Is there an open instance of Fiji? (Y/N)	Answer "y" once Fiji is open	3.6
How many rounds of classification are you running? (1 or 2)	Determines how many rounds of segmentation to run; we recommend two rounds	3.6
Would you like to batch classify the images in the background? (Y/N)	Answer "y" for background classification (only available on Linux), which allows multiple images to be segmented simultaneously	3.6
Enter how many processes are available to use for multiprocessing; set to 1 for no multiprocessing:	If classifying images in the background, enter the number of processes (image classifications) that can be run simultaneously. Check the number of threads available on your specific computer	3.6
Do you have a trained classifier? (Y/N)	Answer "n" if you need to train a classifier in Fiji. If you already have a model file (classifier) saved, answer no	3.6
Enter the path to the classifier, relative to the working directory (e.g., Aligned/classifier.model)	Enter the path to the classifier relative to your working directory (it will likely be saved within your image directory)	3.6
Enter the name of the directory within [image directory] to output masks into (e.g., Mask 1)	Enter a name to call the directory to which images will be output	3.6
Enter the name of the directory containing images to classify, relative to the working directory (e.g., Aligned/Mask 1)	Enter the name of the image directory with the images you wish to segment	3.6
Enter the minimum cell area for tracking (e.g., 100)	Enter a minimum size for cell tracking (area in pixels)	3.7
Enter the maximum cell area for tracking (e.g., 2500)	Enter a maximum size for cell tracking (area in pixels)	3.7
Enter the minimum number of frames to track cells through (e.g., 15)	Enter the minimum number of frames to track cells. Trajectories with a shorter length will be excluded	3.7
Enter the name of the directory to output csv files into, relative to the working directory	Enter a name to call the directory to which the files will be output	3.7

(continued)

Table 3 (continued)

Prompt	Description	Subheadings
Do you wish to get data for all of the lineages? (To only analyze select lineages based on lineage name answer no; Y/N)	Answer "y" to output a csv file for each lineage. Answer "n" to input the names of specific lineages for which you wish to output a csv file	3.7
Enter the number of lineages you wish to analyze	Enter how many lineages you wish to analyze as a whole digit integer (*see* **Note 28**)	3.7
Input the full name for each lineage (e.g., 0001, 0001–2, 10001) Lineage #	For each lineage you wish to analyze, input the name of the lineage and hit enter (*see* **Note 28**)	3.7
Do you wish to analyze a region of interest? (Y/N)	Answer "y" to get fluorescence for an entire region of interest (must have ROI file)	3.8
Do you wish to crop the images based on an ROI? (Y/N)	Answer "y" to crop the images (and final video) to a region of interest (must have ROI file)	3.9
Enter the path to the csv file for the ROI to analyze, relative to the working directory (e.g., ROI.csv)	If analyzing or cropping to an ROI, input the path (relative to the working directory) to the csv file	3.8 and 3.9
Do you want to number the cells in the images based on lineage tracking? (Y/N)	Answer "y" to write cell labels into the final video	3.9
Do you want to contour cells based on masks? (Y/N)	Answer "y" to contour (outline) cells in the final video	3.9

16.04 LTS operating system (recommended) and Mac OS X (*see* **Note 1**). For convenience, place the scripts in a folder that also contains the images in a subfolder. It is essential to read the entire protocol before running the analysis pipeline for optimal operation.

To run *SegmentandTrack.py* (*see* Video 1 for a video tutorial):

1. Open a terminal. Keep in mind that the terminal is case-sensitive.
2. Change directories to the folder that contains the scripts and the images that are to be analyzed (Fig. 3a). Note that directories will differ based on individual machine and file locations (*see* **Note 12**).
3. Type "Python SegmentandTrack.py" (Fig. 3a).
4. Answer the prompts (Table 3; please *see* Subheadings 3.4–3.9 for further explanation). For yes or no (Y/N) questions, answer "y" for yes or "n" for no.

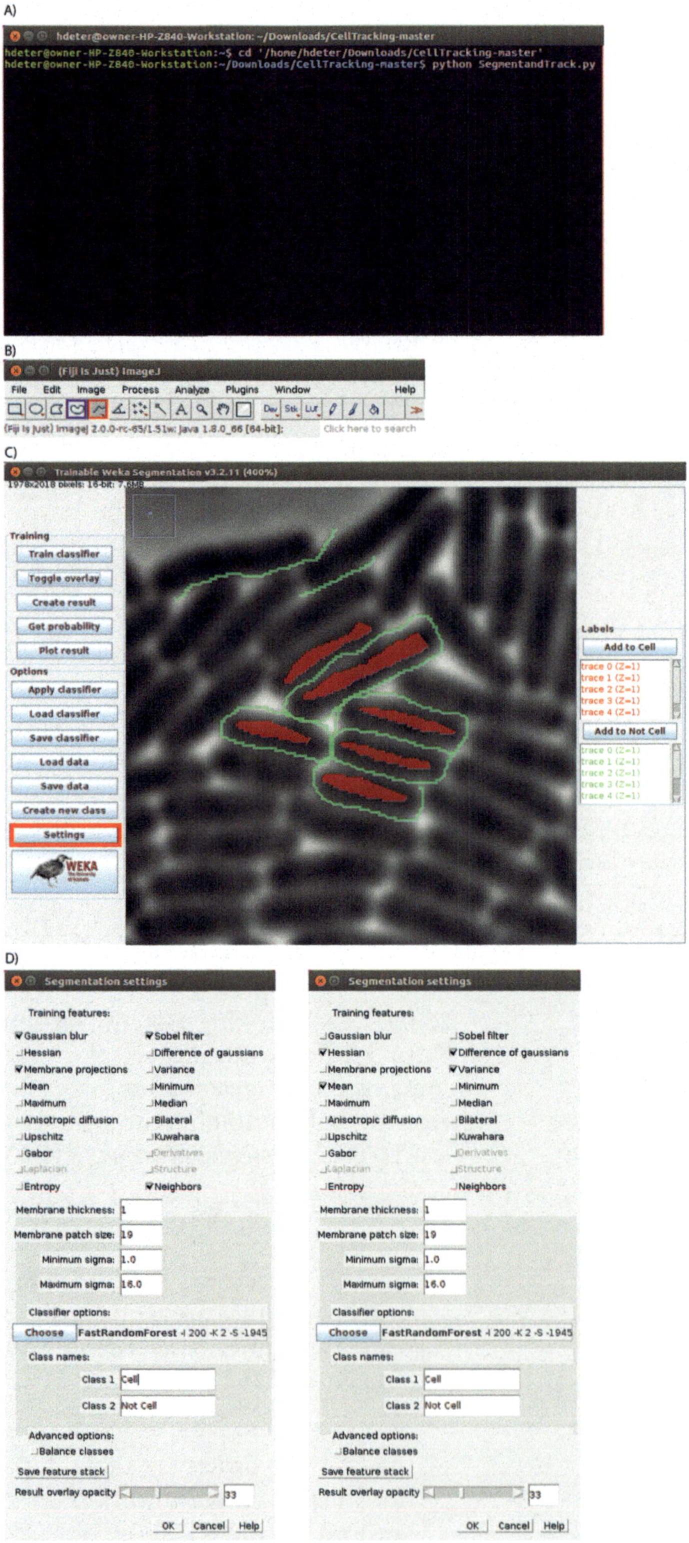

Fig. 3 The *Weka* settings for our classifiers. Here we highlight default settings that have worked well in the past, but these parameters can be altered depending on the user requirements. (**a**) An example of commands to run a

3.4 Selecting a Region of Interest (ROI) (Optional)

Regions of interest (ROIs) can be used to crop images or select a specific region for a particular type of analysis. SegmentandTrack.py offers two options that use an ROI (*see* Subheadings 3.5 and 3.9), which relies on a csv (comma-separated values) file created using the *Fiji* measure tool as follows (*see* Video 1 for a video tutorial):

1. Open the desired image in *Fiji*. To observe the same region throughout multiple images, import an image sequence using [File > Import > Image Sequence. . .].
2. Remove any scale associated with the images using [Analyze > Set Scale. . .] (*see* **Note 13**).
3. Go to [Analyze > Set measurements. . .], and choose *Bounding rectangle* as the only checked box; use 0 decimal places (the results need to be whole, even numbers; *see* **Note 14**).
4. Select the desired region with the rectangular selection tool and use the *measure* tool, located at [Analyze > Measure] (Fig. 3b). Save the results as a *csv* file (e.g., filename.csv) in the same directory as the scripts.

3.5 Image Alignment (Optional)

Image alignment is an optional image pre-processing step, yet it is essential if there are significant shifts between phase images, because cell tracking is accomplished by comparing cell locations between consecutive frames. Alignment is not required if the image registration does not shift significantly. If running the alignment on an image with a large number of moving objects, we suggest using an ROI during the alignment (*see* **Note 15**). Align the images as follows (*see* Video 1 for a video tutorial):

1. Answer "y" when asked, "Do you wish to align images?"
2. If using an ROI to base the alignment upon (optional), make an ROI file (*see* Subheading 3.4) that indicates a stationary area of the image (the largest possible background region that contains a minimal number of moving objects throughout the images of the image set; *see* **Note 15**).

Fig. 3 (continued) Python script in the terminal. Note that the directory the script is running from can be different. (**b**) The *freehand* selection tool (purple) is useful for selecting cells for cell segmentation. The *line* tool (red) is useful for outlining cells and dividing cells to add to the "Not cell" label (*see* **Note 22**). (**c**) An example image loaded in *Fiji* for training in *Weka* by going to [Plugins > Segmentation > Trainable *Weka* Segmentation]. To adjust the parameter and to make labels, click on *Settings* (red). The image loaded into *Weka* has magnified to demonstrate cell selection (*see* **Note 13**). (**d**) Left: the settings used to classify the phase images to a probability mask. Right: the settings used to classify the phase images to a binary mask (*see* **Note 20**). To follow our example, we change the *Weka* default names from "Class 1" and "Class 2" to "Cell" and "Not cell" (*see* **Note 21**)

Table 4
Fiji and Weka terminology

Term	Definition
Classifier	A machine learning algorithm that categorizes input data
Binary mask	A black and white image in which pixels identified as a given label by the classifier are black, and the remaining image is white
Probability mask	A grayscale image based on the probability calculated by the classifier that each pixel is a given label

3. If using an ROI, answer "y" when asked, "Do you have an ROI file for a stationary area?" If aligning based on the whole image, answer "n" for the same prompt.
4. If using an ROI, enter the path to the *csv* file, relative to the working directory.
5. Enter the name of the directory into which images will be saved.

3.6 Cell Segmentation

A single-cell analysis is primarily dependent upon cell segmentation; we use the *Trainable Weka Segmentation* tool in *Fiji*. Weka uses machine learning to train a classifier based on training data selected by the user [27] (Table 4). Subheading 3.6, **steps 1** and **2**, covers how to use Weka to classify images, and we have developed our custom scripts to make the process of training and applying a classifier faster and more direct (*see* **Note 16**; *see* Video 1 for a video tutorial).

1. Subheading 3.6, **step 1**, details how to use *Trainable Weka Segmentation* to classify images. We have had success in the past using two rounds of classification because this method improves segmentation of neighboring cells (*see* **Note 17**). Answer "y" when asked, "Do you wish to train and/or apply a classifier?"
2. Input the full path to the *Fiji* executable file located in the Fiji. app directory when prompted (*see* **Note 12**).
3. Enter how many rounds of classification you are running; we recommend two rounds (*see* **Note 17**).
4. If using a *Linux* machine, answer "y" if you would like to classify the images in the background (it is faster). Then enter how many processes are available for multiprocessing. The number of threads available for these processes will depend on the number of processors available to the individual machine (less if using a virtual machine). Do not use more than half of the available threads for multiprocessing.

5. To train a classifier on a subset of phase images (*see* tutorial Video 1), open an instance of *Fiji*.
6. Answer "y" when asked by *SegmentandTrack.py*, "Is there an open instance of *Fiji*?" (*see* **Note 16**).
7. Answer the *Fiji* prompts. If you click "Yes" in response to "Are you classifying a phase image?" the script uses a bandpass filter to subtract background and outputs a probability mask (first round of classification; *see* **Note 18**). When you click "No" (i.e., when classifying the probability mask for the second round of classification), the script does not run a bandpass filter on the image and outputs a binary mask (which the pipeline uses for cell tracking).
8. At the *Trainable Weka Segmentation* window (also located at [Plugins > Segmentation > Trainable Weka Segmentation]), change the settings as desired (Fig. 3c). Although there are a few different classifiers available, we have had success in the past using the *Fast Random Forest* classifier, which is the Weka default setting. The training features in Weka should be adjusted based on the image dataset to improve classification (*see* **Note 19**). Figure 3d shows the settings we used with this dataset (*see* **Note 20**).
9. Select regions of your image and add them to the appropriate label (e.g., "Cell" or "Not cell"), and then train the classifier (*see* **Note 21**). For cell segmentation, the *freehand* selection tool (Fig. 3b) is useful for selecting cells. To indicate the separation of cells that are recently divided, draw a line using the *freehand* line tool, and add the line to the "Not cell" label (Fig. 3c; *see* **Note 22**).
10. Repeat **step 10**, selecting data based on the results of the classifier (*see* **Note 22**).
11. Save the data and classifier (*see* **Note 22**).
12. Press *OK* in the dialog box. This causes the script to prompt the user for filenames, and then open another image in *Trainable Weka Segmentation*, and load the previously saved classifier and data.
13. Continue to train the classifier on multiple images by repeating **steps 9–12** (*see* **Note 22**). Be sure to save (**step 11**) both the classifier and data after every round of training.
14. To finish training the classifier, select *no* when asked, *Do you wish to continue training the classifier?* This will trigger the script to close the open windows in *ImageJ* and continue in the terminal.
15. In the terminal, enter the path to the classifier, relative to the working directory.

16. Enter the name of the directory within your image directory into which the masks will be saved.
17. Based on this input, the script segments the images. If using two rounds of classification, you will need to repeat **steps 5–16** using the probability masks. The first round of classification will produce a probability mask, and the second will produce the binary mask (Fig. 2, *see* **Note 17**).

3.7 Obtaining Single-Cell Data and Cell Lineages

1. Answer "y" when asked, "Do you wish to track cells?"
2. Enter the minimum and maximum cell areas (in pixels) for tracking. If you would like to determine values that work well for your dataset, you can measure the area of the cells, background artifacts, or features with the measure tool in *Fiji* (*see* **Note 23**).
3. Enter the minimum number of frames through which a trajectory must be tracked to be included in the analysis (*see* **Note 24**).
4. *SegmentandTrack.py* will then run the analysis. This analysis uses the binary masks created during cell segmentation (*see* Subheading 3.6; **Note 25**) and outputs csv and pickle files containing single-cell and lineage data (Tables 2 and 5; *see* **Note 26**). The lineage data includes the mean cell doubling time for a given lineage (*see* **Note 27**). Each lineage is numbered, with dashes to indicate a branch (e.g., Lineage 0001–1 divided from lineage 0001, Fig. 4; *see* Videos 2 and 3).
5. If desired, our script can also output *csv* files with individual lineage information (Table 5). Enter "y" when asked, "Do you wish to output csv files detailing data for individual lineages?" to obtain frame-by-frame data. Then either answer "y" when asked, "Do you wish to get data for all of the lineages?" or

Table 5
CSV files output by the pipeline

Filename	Data within	Subheadings
Data_[Name]_xy1_c2.pkl	Time, filtered, and unfiltered median fluorescence of frame and ROI (*see* **Note 29**)	3.8
Global_cell_statistics.csv	Time, cell count, mean, median and standard deviation of cell fluorescence	3.7 and 3.8
Lineagedata.csv	Trajectory name, last time tracked, initial time tracked, mean, and standard deviation of the area over the course of the lineage, mean and standard deviation of lineage fluorescence, and doubling time	3.7 and 3.8
[Trajectory number].csv	Frame, time, cell area, cell *x* and *y* positions, and cell fluorescence	3.7 and 3.8

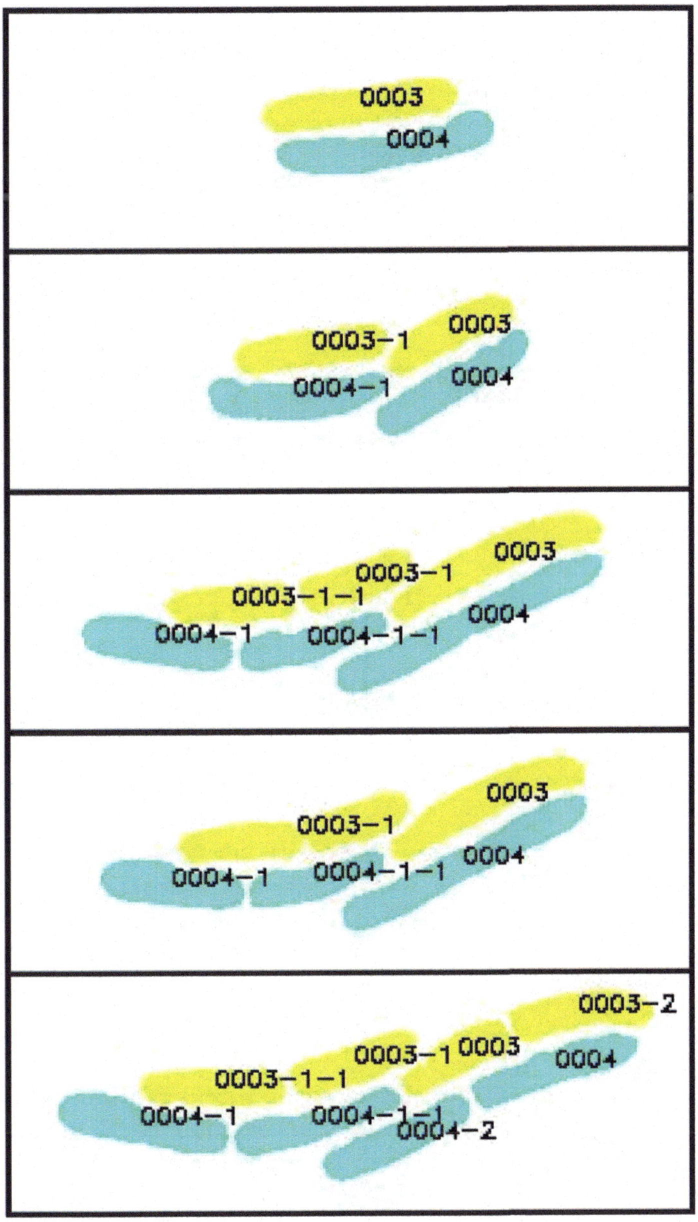

Fig. 4 Cell lineages are kept track of by renaming trajectories that overlap with another trajectory and therefore have a common ancestor (*see* Video 2). Cells in the top images were tracked and labeled, and the output quantified as the cells divide (bottom images) using this method

answer "n" and you will be prompted to input the number of lineages you wish to analyze, and then input each lineage name separately (*see* **Note 28**).

6. Advanced users can further analyze the data using the pickle files (*see* **Note 26**).

3.8 Measuring Fluorescence

Fluorescence is reported in arbitrary units (AU) and based on the mean or the median pixel intensity of the fluorescence image over the area. Pixel intensity of 16-bit images can be measured using *Fiji* (the *measure* tool located under the *Analyze* tab; shortcut key: Ctrl/Cmd + M) or by converting the image to a *NumPy* array and getting the value of the desired pixels. Our custom scripts utilize the latter and report raw fluorescence data (no background subtraction; *see* **Note 29** and Table 5). Single-cell fluorescence is analyzed when tracking cells (*see* Subheading 3.7).

To analyze the whole image and ROI fluorescence:

1. Answer "y" when asked, "Do you wish to analyze images?"
2. If you wish to analyze an ROI, answer "y" when asked, "Do you wish to analyze a region of interest?"
3. Enter the path to the ROI file to analyze (*see* Subheading 3.4), relative to the working directory. A different ROI than the one used for image alignment is required.
4. Fluorescence data will be output as a csv file into the working directory.

3.9 Video Rendering

The pipeline run by *SegmentandTrack.py* uses *Avconv* to render videos from an image directory (*see* **Note 30**).

1. Answer "y" when asked, "Do you wish and analyze images?" and "Do you wish to render videos?" to combine the phase contrast and fluorescent channels (*see* **Note 31**).
2. To crop the video to an ROI, answer "y" when asked, "Do you want to crop the images based on an ROI?" Enter the path to the ROI file to analyze (*see* Subheading 3.4), relative to the working. A different ROI than the one used for image alignment is required, but the same ROI must be used for analyzing fluorescence of an ROI (*see* Subheading 3.8).
3. Cells can be numbered based on lineage (*see* Video 2) if you answer "y" when asked, "Do you want to number the cells in the images based on lineage tracking?"
4. Binary masks can also be utilized to contour (outline) the cells (Video 4) if you answer "y" when asked, "Do you want to outline cells based on masks?"

4 Notes

1. We designed our scripts to function on *Ubuntu* because it is a free and open-source software operating system that can be utilized by all users, and it is compatible with other common operating systems. All of our *Python* scripts have been tested and optimized to run on *Ubuntu 16.04 LTS* (Linux). The

scripts have also been tested on Mac (OS X), although segmentation cannot be run as a background process on a Mac. A virtual machine can be used to run the software in an *Ubuntu* environment by other operating systems (e.g., Windows, Mac), but this method requires more RAM (*see* Subheading 2.3).

2. To install the *Anaconda* shell file in Linux (*see* Table 1), type the following commands in the terminal and replace "path/to/script.sh" with the path to your shell file: (1) "sudo chmod +x path/to/script.sh, (2) /path/to/script.sh." Then you must add Anaconda to the path in Linux (so that conda commands can be run through the terminal), and enter the command "export PATH=~/anaconda2/bin:$PATH" into the terminal.
3. The scripts have only been tested to work using *OpenCV* (version 3.1.0) downloaded through *Anaconda*. Using a different version of *OpenCV* may result in an error reading "ImportError: No module named cv2."
4. The scripts are hardcoded to be specific to the file naming system herein described (*see* Subheading 3) and to analyze data without fluorescence or containing a single fluorescence channel. The code is capable of analyzing more than one fluorescence channel but requires some editing to output data.
5. To minimize background autofluorescence, we recommend the use of minimal media supplemented with a carbon source, and the addition of casamino acids can be added to have faster-growing cells. However, using minimal media for cell imaging results in slower cell growth, which facilitates cell tracking when processing time-lapse images. If the media has substantial autofluorescence, background subtraction may be useful in minimizing its interference (*see* Subheading 3.6, **step 1**; **Note 18**).
6. When preparing agarose pads for imaging bacteria, use only low melting agarose (regular agarose is not suitable for this application).
7. If heat-sensitive compounds need to be added to the agarose solution, make sure that the agarose mixture has cooled down to around 50 °C before the addition. Also, add the compound before filtering to ensure you are working with a known volume (during the filtering process part of the solution is lost). The filtering process helps to homogenize the solution.
8. A compromise needs to be reached between letting the filtered agarose mixture rest after the final vortex (so that the bubbles can migrate closer to the surface) and excessive cooling of the agarose solution. Bubbles trapped in the agarose solution can interfere with the quality of imaging. However, over-cooling the agarose mixture will result in a nonuniform pad (thus compromising the quality of imaging), and ultimately it will prevent pouring onto the coverslip.

9. Fixing the pad area and the final culture OD allows total control of the number of cells in the field of view when imaging, thus ensuring consistent cell density between experiments.
10. An extra pad can be cut and placed in the cover glass bottom dish together with the seeded pad to avoid possible drift in the imaging. The objective should then be positioned in the middle of this extra pad when resting between snapshots. In this way, the time that the seeded pad is in close contact with the heated objective (as the objective tends to heat along the imaging process) will be limited, thus preventing the seeded pad from melting.
11. When working with fluorescent samples, it is ideal to avoid photobleaching and phototoxicity. Sampling the fluorescence channel once every 2–3 min can prevent both, but that will depend on how many fluorescence channels are being considered, the exposure time for each one of the channels, the lamp intensity, and the organism being imaged.
12. To find the directory in which a file is located in *Ubuntu*, right click on the file and select *Properties*. The directory of the file can be found in *Location* under the *Basic* tab (e.g., /home/user/Downloads). You can copy and paste this path into the terminal using the mouse.
13. It is essential that the scale of the bounding rectangle be in pixels because the scripts import the values as pixel values and therefore any other scale would lead to the ROI being different from intended. In the case of image alignment (*see* Subheading 3.5), this could unintentionally result in a region that includes moving objects, which will hinder the alignment (*see* **Note 15**).
14. The *Python* scripts *Image_alignment.py* and *Image_analysis.py* require bounding box input to be whole number integers. Additionally, *Avconv* cannot render a video using images with an odd length or width, and therefore any ROI that is used to crop the images in *Image_analysis.py* must have an even length and width in pixels to output videos.
15. *Fast Fourier transform* (FFT) alignment works best on regions that have limited change between images, and therefore moving objects (such as cells) should be minimally included in the region to be aligned. When selecting an ROI in *Fiji* (*see* Subheading 3.4), use an image stack to scroll through multiple images and ensure that there is minimal movement within the selected area.
16. *SegmentandTrack.py* provides the option to run classification without training a new classifier, which allows a classifier to be used across multiple experiments. Furthermore, *Segmentation.ijm* can be run in *Fiji*, independently of the *Python* scripts, to call *Weka* and load images and classifiers for training.

17. The purpose of the segmentation is to create a binary mask that can then be used to label and identify single cells. However, an initial probability mask gives a more nuanced picture of the classification results than a binary mask and allows for a second round of classification (Fig. 2). We have empirically determined that two rounds of classification can result in a final segmentation that is more accurate and sometimes faster than when classifying with a single, larger classifier. However, one round of classification may be sufficient for certain datasets.
18. Subtracting background: In *Fiji*, we use the *bandpass filter* plugin, which removes high and low spatial frequencies, to regularize the image and subtract the background. The plugin is located under [Process > FFT > Bandpass Filter...]. We have had previous success with large structures filtered to 100 pixels and small structures filtered up to 0 pixels, no suppression, and a 5% tolerance. The *bandpass filter* is included in the custom *Fiji* scripts *Segmentation.ijm* and *Batch_segment.bsh* for the first round of classification.
19. *Weka* settings are saved with the data and therefore can only be set when training the classifier on the first image. Further information on *Weka* settings is on the *Fiji ImageJ* website [23, 27]. Additionally, the *Weka Explorer* is a tool provided by *Weka* to aid in determining which classifiers or training features should be used for a given dataset. It can be accessed by clicking on the *Weka* logo in *Trainable Weka Segmentation*.
20. The training features we used for our classification were different for each round of classification due to the differences between the images being classified. For the first classifier (phase to probability mask), we used *Gaussian blur*, *Sobel filter*, *Membrane projections*, and *Neighbors*. For the second classifier (probability mask to binary mask), we used *Hessian*, *Difference of Gaussians*, *Variance*, and *Mean* (Fig. 3d).
21. In the *Settings* of *Trainable Weka Segmentation*, the labels can be named to help guide the user. We call label 1 "Cell" and label 2 "Not cell," wherein "Not cell" includes anything that is not a cell, including background, features, etc. (Fig. 3c, d). The same method can be used for identifying other objects in an image; for example, labeled organelles could be identified and tracked in eukaryotes.
22. For good-quality segmentation results, it helps to outline just outside the edges of a cell and add it to the "Not cell" label (Fig. 3c). When training the classifier, it is important not to overfit the data (i.e., when the classifier matches a training set closely but is no longer applicable to the more extensive dataset). Furthermore, extensively training the classifier can slow the classification process with minimal returns on efficacy. To

Table 6
Common Fiji and Weka shortcut keys

Shortcut key	Function	Subheadings
Ctrl/Cmd + Shift+X	The crop tool is located under the image tab	3.5
Ctrl/Cmd + Shift+I	Invert the image using the invert tool under the edit tab	3.6, **step 2**
Ctrl/Cmd + M	The measure tool located under the analyze tab	3.8

The control (Ctrl) and Command (Cmd) keys are generally associated with Windows and Mac (Apple) computers. The shortcut keys in this table should function in most cases

prevent this, it is important to save classifiers and data intermittently (with sequential names), in case a later classifier results in a decrease in efficacy. Our classifiers were trained on 5–15 cells in every 20–30 images.

23. To measure an area in *Fiji*, first go to [Analyze > Set Measurements. . .], then select ***area***, and hit *OK*; to measure it go to [Analyze > Measure] (or Ctrl+"M") (*see* Table 6). Our results were generated using a minimum value of 100 and a maximum value of 2500 (Videos 1 and 2).
24. The maximum number of frames is two less than the total number of frames (the first and last frames are not included in the analysis). Our results were generated by tracking cells through a minimum of 25 frames and a maximum of 465 (Videos 1 and 2). Currently, the script is hardcoded so that the overlap requirement to track cells between frames is at least half and to end a trajectory if the cell decreases in area by more than 40%. Changing these values (*see* comments in *TrackCellLineages.py*) can increase or decrease the fidelity of the tracking.
25. We track cells from the last frame to the first frame, because the process of cells merging is more apparent than cells dividing. Currently, the distance radius to test for cell overlap (threshold in *TrackCellLineages.py*) is set to 150, but this can be adjusted based on the dataset by editing the script. A smaller radius can speed up the process, while a larger one may be necessary for larger or faster moving cells.
26. The *csv* files can be opened using *Microsoft Excel*, *LibreOffice Calc*, and most other spreadsheet software. *Pickle* files are used to pack and unpack *Python* objects. The *pickle* files produced by *TrackCellLineages.py* ("lineagetracking.pkl" and "lineagetrackingsummary.pkl") can be unpacked and used for further analysis by advanced users; *see Lineage_analysis.py* for an example.
27. Our method determines doubling time based on two observed divisions, and therefore a lineage must divide at least twice in the course of the experiment to determine doubling time. The doubling time output by *TrackCellLineagse.py* is the mean of

the amount of time between each division over the entire lineage.

28. Lineage names are output in "lineagedata.csv," and visual output (masks with cells colored and labeled according to their lineage) is available in the Lineages subfolder, created within the folder containing the analyzed images (Video 2). Outputting all of the files at once (answering "y" when asked, "Do you wish to get data for all of the lineages?") is often faster than typing in multiple individual names.
29. There are a few options for filtering the data and subtracting background available within the script (*see* **Note 30**), but modifications are required to use them for single-cell tracking. *Global fluorescence data* reports both unfiltered and filtered data, wherein filtered refers to data that has been normalized so that the maximum fluorescence measurement within the frame is 1 and the minimum is 0.
30. Any processing of the images (e.g., background subtraction) must be done before rendering the videos when using these methods. Background subtraction is included in the "filtered" results when analyzing whole image fluorescence (*see* **Note 29**). There are a few different filters to choose from in *Image_analysis.py* to subtract the background. These can be changed by adjusting *plot_filterIndex.*
31. *Image_analysis.py* includes functions for adjusting image brightness, scaling the image values, and adjusting color. The parameters for these adjustments can be modified based on the dataset as described by the script comments.

References

1. Rosenfeld N, Young JW, Alon U, Swain PS, Elowitz MB (2005) Gene regulation at the single-cell level. Science 307:1962–1965. https://doi.org/10.1126/science.1106914
2. Campos M, Surovtsev IV, Kato S, Paintdakhi A, Beltran B, Ebmeier SE, Jacobs-Wagner C (2014) A constant size extension drives bacterial cell size homeostasis. Cell 159:1433–1446. https://doi.org/10.1016/j.cell.2014.11.022
3. Brehm-Stecher BF, Johnson EA (2004) Single-cell microbiology: tools, technologies, and applications. Microbiol Mol Biol Rev 68:538–559, table of contents. https://doi.org/10.1128/MMBR.68.3.538-559.2004
4. Butzin NC, Mather WH (2015) Synthetic genetic oscillators. Rev Cell Biol Mol Med 2:100–125
5. Ferry MS, Razinkov IA, Hasty J (2011) Microfluidics for synthetic biology: from design to execution. Methods Enzymol 497:295–372. https://doi.org/10.1016/B978-0-12-385075-1.00014-7
6. Nketia TA, Sailem H, Rohde G, Machiraju R, Rittscher J (2017) Analysis of live cell images: Methods, tools and opportunities. Methods 115:65–79. https://doi.org/10.1016/j.ymeth.2017.02.007
7. Vallotton P, Turnbull L, Whitchurch CB, Milli L (2010) Segmentation of dense 2D bacilli populations. 2010 International Conference on Digital Image Computing: Techniques and Applications, IEEE. p 82–86
8. Chowdhury S, Kandhavelu M, Yli-Harja O, Ribeiro AS (2013) Cell segmentation by multi-resolution analysis and maximum likelihood estimation (MAMLE). BMC Bioinformatics 14(Suppl 10):S8. https://doi.org/10.1186/1471-2105-14-S10-S8

9. Sadanandan SK et al (2016) Segmentation and track-analysis in time-lapse imaging of bacteria. IEEE J Sel Top Signal Process 10:174–184. https://doi.org/10.1109/Jstsp.2015.2491304
10. Hu Y, Wang S, Ma N, Hingley-Wilson SM, Rocco A, McFadden J, Tang HL (2017) Trajectory energy minimization for cell growth tracking and genealogy analysis. R Soc Open Sci 4:170207. https://doi.org/10.1098/rsos.170207
11. Ducret A, Quardokus EM, Brun YV (2016) MicrobeJ, a tool for high throughput bacterial cell detection and quantitative analysis. Nat Microbiol 1:16077. https://doi.org/10.1038/nmicrobiol.2016.77
12. Boyle EA, Li YI, Pritchard JK (2017) An expanded view of complex traits: from polygenic to omnigenic. Cell 169:1177–1186. https://doi.org/10.1016/j.cell.2017.05.038
13. Schindelin J et al (2012) *Fiji*: an open-source platform for biological-image analysis. Nat Methods 9:676–682. https://doi.org/10.1038/nmeth.2019
14. Paintdakhi A, Parry B, Campos M, Irnov I, Elf J, Surovtsev I, Jacobs-Wagner C (2016) Oufti: an integrated software package for high-accuracy, high-throughput quantitative microscopy analysis. Mol Microbiol 99:767–777. https://doi.org/10.1111/mmi.13264
15. Dimopoulos S, Mayer CE, Rudolf F, Stelling J (2014) Accurate cell segmentation in microscopy images using membrane patterns. Bioinformatics 30:2644–2651. https://doi.org/10.1093/bioinformatics/btu302
16. Kamentsky L et al (2011) Improved structure, function and compatibility for CellProfiler: modular high-throughput image analysis software. Bioinformatics 27:1179–1180. https://doi.org/10.1093/bioinformatics/btr095
17. Stylianidou S, Brennan C, Nissen SB, Kuwada NJ, Wiggins PA (2016) SuperSegger: robust image segmentation, analysis and lineage tracking of bacterial cells. Mol Microbiol 102:690–700. https://doi.org/10.1111/mmi.13486
18. Arganda-Carreras I, Kaynig V, Rueden C, Eliceiri KW, Schindelin J, Cardona A, Sebastian Seung H (2017) Trainable *Weka* segmentation: a machine learning tool for microscopy pixel classification. Bioinformatics 33:2424–2426. https://doi.org/10.1093/bioinformatics/btx180
19. Deter HS, Dies M, Cameron CC, Butzin NC, Buceta J (2018) A bacteria segmentation/tracking tool based on machine learning. http://osf.io/gdxen/. doi: https://doi.org/10.17605/osf.io/gdxen
20. Deter HS (2018) Cell segmentation and tracking tools. YouTube. https://goo.gl/uJ3j8A
21. Green MR, Sambrook J, Sambrook J (2012) Molecular cloning: a laboratory manual, 4th edn. Cold Spring Harbor Laboratory Press, Cold Spring Harbor, NY
22. (2018) https://www.virtualbox.org/
23. Rueden CT, Schindelin J, Hiner MC, DeZonia BE, Walter AE, Arena ET, Eliceiri KW (2017) ImageJ2: ImageJ for the next generation of scientific image data. BMC Bioinformatics 18:529. https://doi.org/10.1186/s12859-017-1934-z
24. Lutz R, Bujard H (1997) Independent and tight regulation of transcriptional units in Escherichia coli via the LacR/O, the TetR/O and AraC/I1-I2 regulatory elements. Nucleic Acids Res 25:1203–1210
25. Young JW et al (2011) Measuring single-cell gene expression dynamics in bacteria using fluorescence time-lapse microscopy. Nat Protoc 7:80–88. https://doi.org/10.1038/nprot.2011.432
26. Parry-Hill M, Sutter RT, Davidson MW (2018) Microscope alignment for Köhler illumination. Nikon. https://wwwmicroscopyucom/tutorials/kohler. Accessed 7 July 2018
27. Arganda-Carreras I, Kaynig V, Rueden C, Eliceiri KW, Schindelin J, Cardona A, Seung HS (2017) Trainable *Weka* segmentation: a machine learning tool for microscopy pixel classification. Bioinformatics 33:2424–2426. https://doi.org/10.1093/bioinformatics/btx180
28. (2018) https://www.ubuntu.com
29. (2018) https://www.anaconda.com
30. (2018) *Python* Software Foundation. https://pypi.org
31. Libav (2018). https://libav.org
32. Deter HS (2018) CellTracking. https://github.com/hdeter/CellTracking
33. (2018) https://opencv.org/
34. Beanshell (2018). http://www.beanshell.org
35. (2018) https://docs.scipy.org

Chapter 20

2D + Time Object Tracking Using *Fiji* and *ilastik*

Andrea Urru, Miguel Angel González Ballester, and Chong Zhang

Abstract

Tracking cells is one of the main challenges in biology, as it often requires time-consuming annotations and the images can have a low signal-to-noise ratio while containing a large number of cells. Here we present two methods for detecting and tracking cells using the open-source *Fiji* and *ilastik* frameworks. A straightforward approach is described using *Fiji*, consisting of a pre-processing and segmentation phase followed by a tracking phase, based on the overlapping of objects along the image sequence. Using *ilastik*, a classifier is trained through manual annotations to both detect cells over the background and be able to recognize false detections and merging cells. We describe these two methods in a step-by-step fashion, using as example a time-lapse microscopy movie of HeLa cells.

Key words Cell tracking, Segmentation, Classification, Fiji, ilastik

1 Introduction

Nowadays, one of the main challenges for biologists is to be able to accurately and effectively monitor and follow cell growth and development through time. Manually tracking nuclei in microscopy images is highly time-consuming, as the user needs to take into account several factors such as frame rate, size, shape, etc. to distinguish cells in possibly low-quality images that may contain hundreds to thousands of cells. To allow for high-throughput tracking experiments, there is a strong need for automated analysis methods that require minimal annotations from the user. This is not an easy goal to reach, especially in highly populated images, due to the complexity of performing a robust and accurate detection of the objects. As a matter of fact, tracking a cell usually needs a robust segmentation, since the consequent trajectory linking stage between consecutive frames will become a one-to-one matching problem or a one-to-two in case divisions occur.

Electronic supplementary material: The online version of this chapter (https://doi.org/10.1007/978-1-4939-9686-5_20) contains supplementary material, which is available to authorized users.

Elena Rebollo and Manel Bosch (eds.), *Computer Optimized Microscopy: Methods and Protocols*, Methods in Molecular Biology, vol. 2040, https://doi.org/10.1007/978-1-4939-9686-5_20,

Many solutions have been developed through the years, and most of them have been included in freely available and user-friendly software packages. Many of these tools (e.g., *MTrack2* [1], *TrackMate* [2, 3], *Particle Tracker* [4, 5]) are implemented as built-in or downloadable plugins in the general purpose image analysis platform *Fiji* [6], but *Fiji* itself also offers tools (e.g., *Analyze Particles* and *3D Objects Counter* functions) for segmenting and tracking objects in 2D and 3D. Software packages like *CellProfiler* [7] and *The Tracking Tool* [8] are able to handle large amounts of multidimensional image data and allow to segment and track single cells, while others are specialized in embryogenesis tracking, such as *StarryNite* [9]. Finally, a tool like *ilastik* [10] allows to segment and track objects without any previous experience in image processing. Since an extensive literature review is beyond the content of this chapter, we refer interested readers to a review [11] comparing cost, working platform, and main features for cell and particle tracking tools. In [12] more complex particle tracking approaches are compared by applying them to the same datasets.

In this chapter, two 2D object tracking approaches are described in a step-by-step fashion while applied to an example image of a mitotic nuclei 2D time series from the *MitoCheck* project [13]. Both approaches consist of three main steps: nuclear segmentation, tracking, and trajectory plotting. The first approach uses *Fiji* functions for the segmentation and tracking and a macro for the trajectory plotting. In particular, tracking is based on the overlap of 2D object positions along time. Thus, this approach is ideal for cells in monolayered cultures with frame rate high enough to catch a "continuous" cell motion, but not suitable for otherwise. The second approach uses an *ilastik* tracking workflow, which is not limited by the objects' overlap but is rather capable of tracking them in complex scenarios, such as the presence of dividing and merging events. We will demonstrate over the chapter how the workflow is able to learn how to segment and track dividing objects using minor user's annotations. Another example using *ilastik* in a *Drosophila* embryo 3D time series has been described in detail elsewhere [14].

2 Materials

1. The dataset provided with this chapter ("mitocheck_small.tif") can be downloaded from the Springer website and has been cropped, both in space and time, from its original image, a HeLa cells time-lapse publicly available at the *MitoCheck* project.

2. *Fiji* macros for tracking and plotting trajectories can be downloaded from the Springer website. In particular, "Fiji_solution.ijm" describes the approach using *Fiji* for both tracking and plotting trajectories, while "Fiji_ilastik_solution.ijm" describes the one involving *ilastik* for tracking cells and *Fiji* for plotting trajectories.
3. *Fiji* software version 2.0.0-rc-68/1.52e has been used for macros' development. *Fiji*'s latest stable version installer can be downloaded from [15] and has already several built-in plugins. More plugins can be easily installed by copying them into the plugin folder.
4. *ilastik* software version 1.2.2 has been used for tracking. Details on *ilastik* installation can be found in [16]. This software could require the installation of a commercial solver: *ilastik* tracking workflow described in Subheading 3.2.2 doesn't require the installation of the commercial solver CPLEX since version 1.2.0. For older versions of *ilastik*, *CPLEX* (or *Gurobi* [17]) may still need to be installed.

3 Methods

In this section we will perform 2D + time tracking on objects of interest using two different methods: one involving *Fiji* (Subheading 3.1) and the other involving *ilastik* (Subheading 3.2). For both methods, we will perform a tracking analysis of the results on *Fiji*, which compared to *ilastik* gives us the possibility to track the trajectories of the cells and plot their displacement and velocity.

The following subsections describe in detail both methods plus the final analysis; for each subsection an overview of the main steps that compose each methodology is presented, followed by the step-by-step corresponding procedure.

3.1 *Fiji* Solution

We will start by loading the image of interest and, after a brief pre-processing, we will move forward to the thresholding and the segmentation steps using the watershed transform. Finally, we will track the segmented objects through time. Before starting, we will click on [Plugins > Macros > Record...], a command which allows us to record all the subsequent operations and then turn them into a macro script. Once done, we will load the 2D + time image, in this case the multipage TIFF file "mitocheck_small.tif" (*see* **Note 1**), using [File > Open...] and browsing to the file location.

3.1.1 Object Segmentation

The first main step of the tracking is to segment, or identify, objects (cell nuclei in the matter at hand) in each time frame.

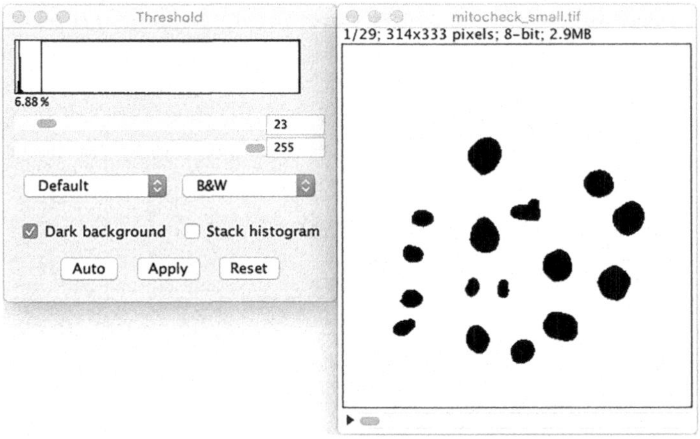

Fig. 1 Threshold window. The input settings (left) and the resulting thresholded image (right) are shown

1. Segmentation needs a pre-processing phase in which the stack can be smoothed in order to homogenize the intensity within each object using [Process > Filters > Gaussian Blur. . .] with a *Sigma* value of, e.g., 2.0 (this value allows to filter out high-frequency noise components without blurring the image too much) and hitting *OK*. By hitting *Yes* in the following pop-up window, this 2D filter will be applied to each slice or time frame independently, being the whole stack smoothed in a single run.
2. Since the background is quite clean, a simple thresholding is probably good enough to segment the objects. One of the available thresholding methods can be selected using [Image > Adjust > Threshold. . .], e.g., the *Default* method, ticking the *Dark background* option with minimum and maximum cutoff values set to 23 and 255, respectively (Fig. 1), and finally hitting *Apply*. A pop-up window will be shown, in which we will untick the *Calculate threshold for each image* option and hit *OK*.
3. Usually, there will be merged objects in the resulting binary mask, so [Process > Binary > Watershed. . .] can be run to split the most obviously wrongly merged ones (*see* **Note 2**). Click *Yes* to apply the transformation to the whole stack.

3.1.2 Find Connected Objects Along Time

At this point, a binary stack containing the segmented nuclei in all time frames should have been obtained. Next, the same object will be tracked/linked in consecutive time frames. The final result will be a label stack where each object is identified by a unique label along time (indicated by a distinct color, i.e., by a unique gray-level intensity value for each object). The strategy employed here is based on the overlap of the same object in temporal space between two consecutive time frames. If time is considered as the third

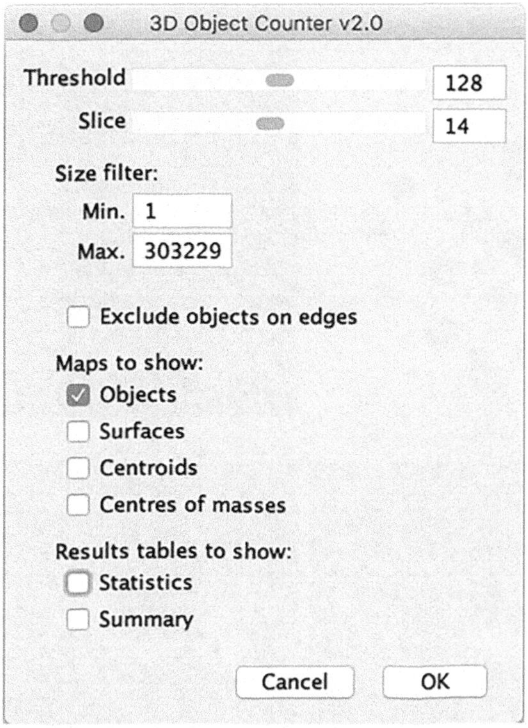

Fig. 2 Settings to perform the 3D object counting; the *Objects* option is ticked and the statistics unticked

dimension, the spatial overlap of the nuclei along time will translate into regions being connected in the third dimension, thus creating a pseudo-3D object.

1. *Fiji*'s function [Analyze > 3D Objects Counter] does the job of finding such connected components in 3D (*see* **Note 3**). This function works similarly as the 2D [Analyze > Analyze Particles...]. It can identify and count 3D objects in a stack, quantify each object's properties (*see* the full list in [Analyze > 3D OC Options]) in a table, and generate maps containing representations of specified results. When you select the function, a pop-up window will show up; in Fig. 2 you can find the settings to perform the counting. Once the 3D objects have been labeled, the *glasbey inverted* lookup table from [Image > Lookup Tables] can be applied to better visualize individual objects (*see* **Note 4**).
2. At this point we have two images, one containing a binary mask of the segmented objects and the other containing the labeled objects. We can, respectively, rename these images as "Mask" and "Labels" by selecting their windows and hitting [Image > Rename]. In Fig. 3 the first five frames having the segmented objects (their identity labels shown in random color) are displayed (*see* **Note 5**). You will now find in the *Recorder* window

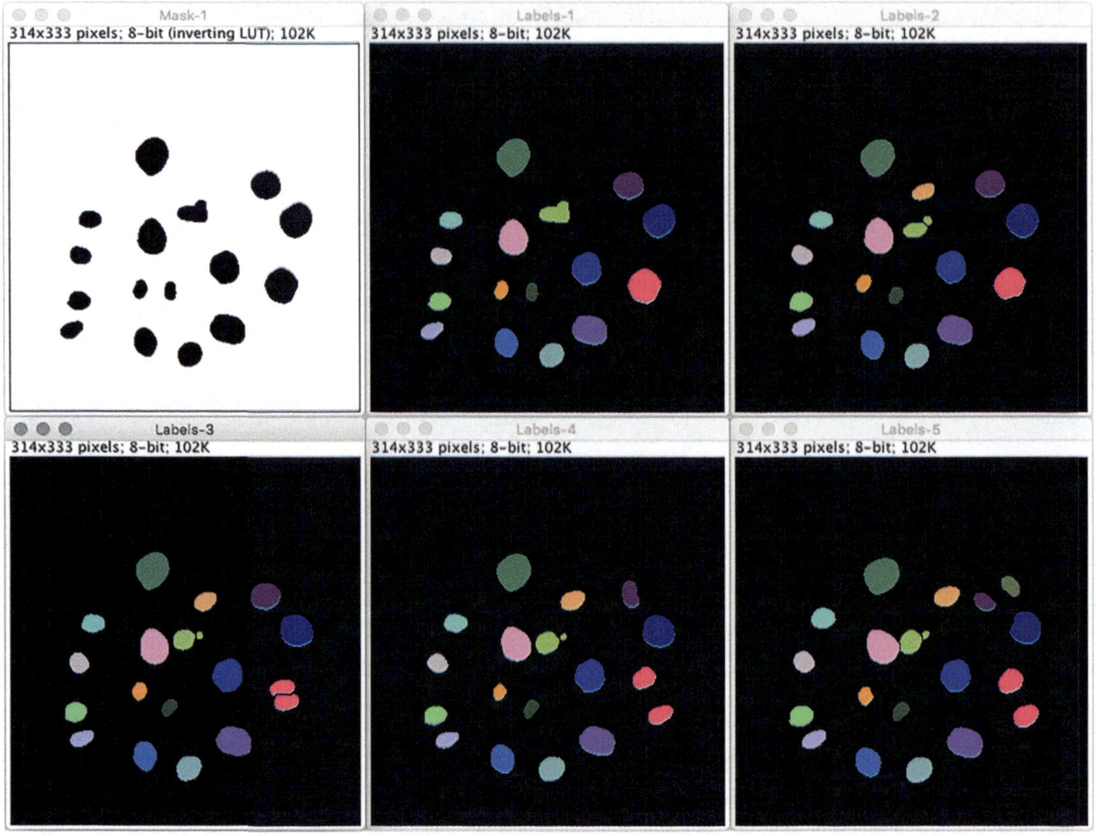

Fig. 3 Segmented first frame and objects counted and tracked through the first five frames, shown with labels represented by different random colors

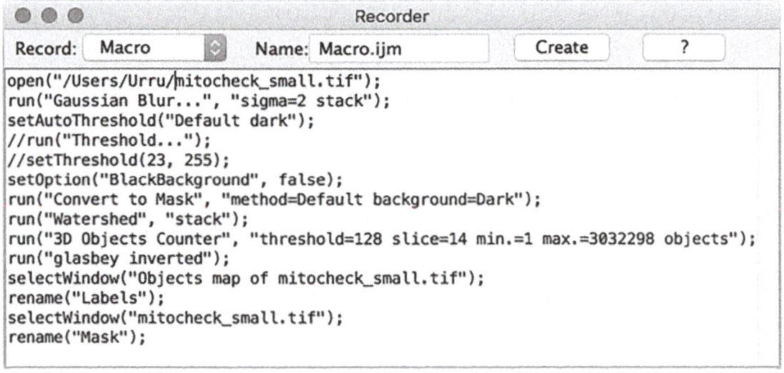


```
open("/Users/Urru/mitocheck_small.tif");
run("Gaussian Blur...", "sigma=2 stack");
setAutoThreshold("Default dark");
//run("Threshold...");
//setThreshold(23, 255);
setOption("BlackBackground", false);
run("Convert to Mask", "method=Default background=Dark");
run("Watershed", "stack");
run("3D Objects Counter", "threshold=128 slice=14 min.=1 max.=3032298 objects");
run("glasbey inverted");
selectWindow("Objects map of mitocheck_small.tif");
rename("Labels");
selectWindow("mitocheck_small.tif");
rename("Mask");
```

Fig. 4 Recorder window showing all the steps previously performed in macro code. We can simply copy and paste them to a new macro script

all the steps performed until this point, which compose the first part of the macro, as shown in Fig. 4. You can directly copy and paste the resulting code in a new macro script, which you can create using [Plugins > Macros > Startup Macros...] command. From now on, indeed, the analysis requires programming in macro language.

3.1.3 Tracking Analysis in Fiji

After correctly identifying and tracking the objects along time, the next step will be to extract trajectories and calculate displacements and velocities. We can do it directly on *Fiji*, by writing our own macro script. We will start from the two stacks obtained in the previous steps, the binary stack containing the thresholded objects (i.e., "Mask") and the objects' label stack (i.e., "Labels") obtained from the *3D Objects Counter* function. If we consider each object as a rigid body, we could simplify the problem and look at just one specific point of the object, e.g., the centroid (*see* **Note 6**). The piece of the source code reported in Fig. 5 (*see* **Note 7**) shows how to implement the steps described below for a selected cell.

1. First, store the size of the stack as a variable (called, e.g., "n") by means of the built-in macro function *nSlices.* We may want to use the command *waitForUser()* to have the chance to visually inspect the stack before deciding which object will be tracked. Then, store the ID (intensity value) of the object to be tracked in a new variable (e.g., "objID"), using the function *getNumber()*, that will allow the user to enter a number in the pop-up menu. Also, initialize an array (e.g., "objShowTime") of length "n" using the function *newArray()*. Every element of this array will be either given the value 1 or 2 based on how many cells with a certain "objID" have been found in the image (it will be 2 in case the cell has undergone mitosis). Initially, the array should be filled with zeros using the function *Array.fill()*. Finally, create two additional pairs of arrays (e.g., "objCenterX1," "objCenterY1"; "objCenterX2," "objCenterY2") to keep track of the coordinates of the daughter cell's centroids in case of mitosis, and fill them with the value −1. In order to obtain the centroid coordinates of an object, we will have to check the *Mean Gray Value* and *centroid* parameters in the [Analyze > Set Measurements...] option window.
2. A *for* loop will allow to go through the "n" slices. To synchronize the correct slice to the time index "i" inside the loop, use the built-in macro functions *selectImage()* and *setSlice()*, to be applied on the "Mask" stack. Then, use the [Analyze > Analyze Particles...] command to identify the objects, ticking the *Add to ROI Manager* box. By selecting the stack "Labels" together with the corresponding slice, ticking *Show All*, and clicking *Measure* in the *ROI Manager* window, the previously selected measurements will be displayed in the *Results* window.
3. Another *for* loop will allow to go through all the objects in the current frame. The number of objects can be obtained using the built-in function *nResults.* Moreover, the function *getResult()* returns a measurement, e.g., *getResult("Mean", 16)*, where number refers to the mean value of the 16th object. Since we have identified the object to track by its "objID" (i.e.,

```
n = nSlices;
// Select the object of interest, specified by "objID"
waitForUser("Click OK after deciding which object you would like to track (ID is given by index value in Labels image)...");
objID = getNumber("Insert Object ID:", 1);
// Define and initialize an array with a time flag for each frame, that will be set to 1 if the object "objID" is found
objShowTime = newArray(n); Array.fill(objShowTime,0);
// Define and initialize two couples of arrays to store the centroid coordinates in X and Y axes (taking into account possible mitosis)
objCenterX1 = newArray(n); Array.fill(objCenterX1,-1);
objCenterY1 = newArray(n); Array.fill(objCenterY1,-1);
objCenterX2 = newArray(n); Array.fill(objCenterX2,-1);
objCenterY2 = newArray(n); Array.fill(objCenterY2,-1);
// Check the measurements options, i.e. mean gray value and centroid
run("Set Measurements...", "mean centroid redirect=None decimal=2");
for(i = 1; i <= n; i++) {
    // Select the binary Mask stack image and get the slice i
    selectImage("Mask");
    setSlice(i);
    // Analyze the regions of each object in slice i, and add the regions to roiManager, to obtain ROI selection
    run("Analyze Particles...", "size=0-Infinity circularity=0.00-1.00 show=Nothing add");
    // We would like to measure the mean gray value in the object label image so as to get object with objID
    selectImage("Labels");
    setSlice(i);
    roiManager("Show All");
    roiManager("Measure");
    // Go through all the objects identified in the current slice to find the object with objID. If we don't find any object with ObjID, we copy coordinates from the previous frame.
    for(j = 0; j < nResults; j++) {
        current_objID = getResult("Mean", j);
        if (current_objID == objID) {
            // We increase the counter of a cell by 1 everytime an
            object with that ID is found
            objShowTime[i-1] = objShowTime[i-1] + 1;
            // We save the results in objCenter accordingly
            if (objShowTime[i-1] == 1) {
                objCenterX1[i-1] = getResult("X", j);
                objCenterY1[i-1] = getResult("Y", j);
            } elseif (objShowTime[i-1]==2) {
                objCenterX2[i-1] = getResult("X", j);
                objCenterY2[i-1] = getResult("Y", j);
            }
        }
        if (j == nResults-1) {
            // If we don't find any corresponding object, we assign the objects the position in the previous frame
            if (objShowTime[i-1] == 0) {
                objCenterX1[i-1] = objCenterX1[i-2];
                objCenterY1[i-1] = objCenterY1[i-2];
                objCenterX2[i-1] = objCenterX2[i-2];
                objCenterY2[i-1] = objCenterY2[i-2];
            }
            // If there is no mitosis, we assign the 2nd object the same coordinates of the 1st one (to later plot displacements)
            if (objShowTime[i-1] == 1) {
                objCenterX2[i-1] = objCenterX1[i-1];
                objCenterY2[i-1] = objCenterY1[i-1];
            }
        }
    }
    // Clean up, for each slice, Results and ROI Manager tables
    run("Clear Results");
    selectImage("Labels");
    roiManager("Show None");
    selectImage("Mask");
    roiManager("Show None");
    selectWindow("ROI Manager");
    run("Close");
}
```

Fig. 5 Macro code block for tracking analysis. The comments to the code are shown in green. The mentioned built-in functions are highlighted in brown color. Please note that we clean up the Results table and the ROI Manager after finishing checking each slice its intensity value in the label image), we can now go through the objects in the current slice and look for each of them using an *if* sentence, thus finding the objects whose mean value corresponds to a given "objID." If we find more than one object with the same "objID," it will mean that a mitosis has occurred, and we will save their coordinates in "objCenterX1" and "objCenterY1" or in "objCenterX2" and "objCenterY2" accordingly. A second *if* statement can also be implemented so that when an object is not found, its centroid is given the coordinates obtained in the previous frame. If mitosis does not occur, i.e., we have just one object with "objID," we assign "objCenterX2" and "objCenterY2" the same value of "objCenterX1" and "objCenterY1," respectively. The centroid coordinates can be obtained by getting the columns X and Y from *Results* again using *getResult()*.

4. Finally, to avoid overlapping between different trajectories, clear the results table and untick *Show All* before closing the *ROI Manager*.

3.1.4 Plotting Trajectories in Fiji

1. Create a duplicate of the Mask stack using [Image > Duplicate...] and rename it, as we will use this duplicate to visualize the cell's trajectory.
2. Create a *for* loop for going through all the "n" slices; depending on the value of "objShowTime" for that slice, plot a single line (if "objShowTime" is equal to 1) or two lines (if "objShowTime" is equal to 2). The built-in function *makeline()* can be used in order to plot the trajectory of the centroid positions overlaid on the image. Since this function only takes integer pixel positions, the function *floor()* is used to always take the integer part of the coordinates and omit the decimal part. The code in Fig. 6 draws the trajectory we want to plot using *Fiji*'s built-in function [Image > Overlay > Add Selection...] (*see* **Note 8**). The properties command configures the line appearance. We also use a variable plot and initialize it to 1 or 2 depending on whether a mitosis has taken place or not. Moreover, we save the slice where mitosis is detected in the variable mitosis. Figure 7 shows an example of trajectory overlaid on the mask image at four different frames.

3.1.5 Plotting Displacements or Velocities in Fiji

Examining the object displacements or velocity changes is another interesting measurement for tracking studies. Since we have already obtained the centroid positions at each time frame, the displacement of the centroid positions from one time point to the next will deliver the distance between the two positions. In Fig. 8 you can find the source code, including the steps below, to calculate and plot displacements, while an example of displacement plotted for two daughter cells is shown in Fig. 9.

```
// We create an empty stack in which we copy the segmented frames with the progressive
trajectory of the cell
selectImage("Mask");
run("Duplicate..." , "duplicate");
rename("Trajectory" + objID);

mitosis = newArray(1);

for (i = 1; i <= n-1; i++) {
    selectImage("Trajectory" + objID);
    // If there is mitosis, we draw two lines, while for a single-cell trajectory we
    use a single red line
    if (objShowTime[i] == 1) {
        plot = 1;
        makeLine(floor(objCenterX1[i-1]),floor(objCenterY1[i-
        1]),floor(objCenterX1[i]),floor(objCenterY1[i]));
        for (j = i; j <= n-1; j++) {
            setSlice(j+1);
            run("Properties... ", "  stroke=red width=1.5");
            run("Add Selection...");
        }
    } else {
        plot=2;
        // Record the mitosis frame for plotting it later on
        if (objShowTime[i-1]==1) { mitosis[0]=i-1; }
        makeLine(floor(objCenterX1[i-1]),floor(objCenterY1[i-
        1]),floor(objCenterX1[i]),floor(objCenterY1[i]));
        for (j = i; j <= n-1; j++) {
            setSlice(j+1);
            run("Properties... ", "  stroke=blue width=1.5");
            run("Add Selection...");
        }
        makeLine(floor(objCenterX2[i-1]), floor(objCenterY2[i-
        1]), floor(objCenterX2[i]),floor(objCenterY2[i]));
        for (j = i; j <= n-1; j++) {
            setSlice(j+1);
            run("Properties... ", "  stroke=green width=1.5");
            run("Add Selection...");
        }
    }
}
```

Fig. 6 Macro code block describing how to plot trajectories on the segmented images using *makeline* and *Add Selection* functions

1. Initialize a vector containing the right time frame for the plot, e.g., "xCoord," and use an *if* sentence to switch between one and two plots, depending on whether a mitosis was detected or not in the previous steps. If there was no mitosis (i.e., plot is equal to 1), the displacement of the cell over time can be stored in an array, e.g., "disp." The user will also need to create a variable, e.g., "max_d," to store the maximum value in the "disp" vector. On the other hand, if there was a mitotic event (i.e., plot is equal to 2), the displacement for the daughter cells over time can be stored in two arrays, e.g., "disp1" and "disp2." Two more variables need to be created, e.g., "max_d1" and "max_d2," to store the maximum values in "disp1" and "disp2" vectors, respectively.
2. The displacement is the square root of the sum of squared difference between each coordinate of the points considered.

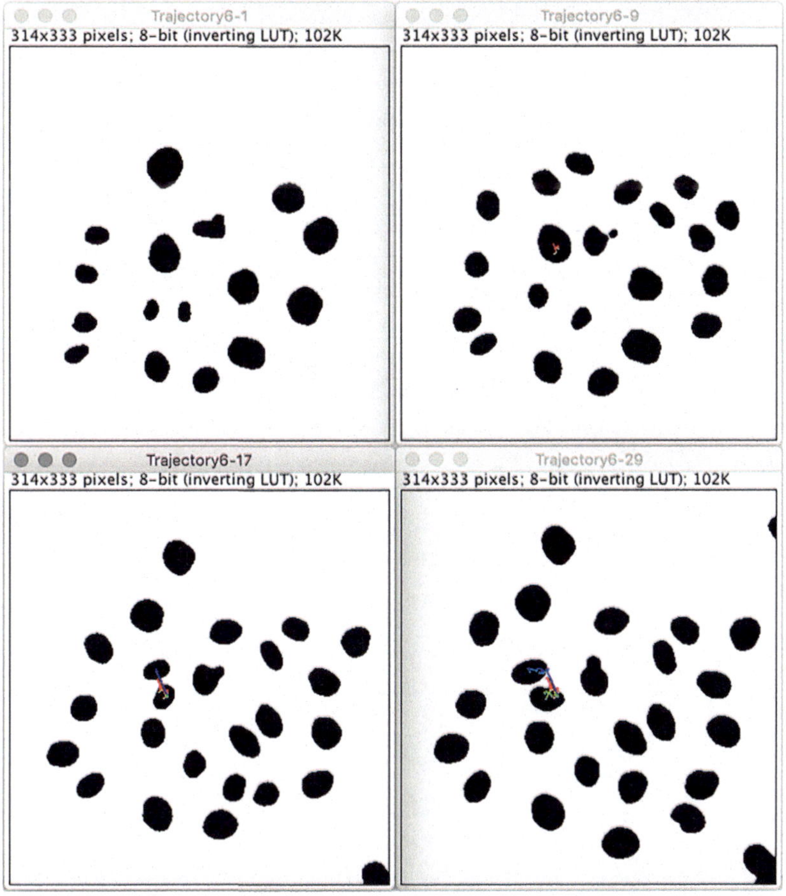

Fig. 7 An example of trajectory plotted on top of the image stack. Before mitosis the trajectory is plotted using a single red line, while after division the daughter cells are depicted by blue and green lines, respectively

The built-in function *sqrt()* calculates the square root of a given number. We also update the maximum values.

3. Finally, plot the displacements by using the built-in function *Plot()*. We can create the plot, set its axes length, set line color and width, and finally show it. The velocity can be easily obtained if the time interval is known as the displacement divided by the time interval.

4. To repeat the analysis for multiple cells, a *while* loop can be created to wrap the code relative to the tracking. A variable, e.g., "tracking," can be created and then initialized as the string "Yes." In this way, at the end of the analysis, using the built-in function *getString()*, we can decide whether repeating the analysis with another cell or not.

```
xCoord = Array.getSequence(n);
if (plot==1) {
    disp = newArray(n); Array.fill(disp,0);
    max_d = 0;
    for (i=1; i<=n-1; i++) {
        disp[i] = sqrt(pow(objCenterX1[i]-objCenterX1[i-1],2)+ pow(objCenterY1[i]-
        objCenterY1[i-1],2));
        if (disp[i] > max_d) { max_d = disp[i]; }
    }
    Plot.create("Displacement Cell " + objID, "Frame", "Displacement (pixel)");
    Plot.setLimits(0, nSlices, 0, max_d+1);
    Plot.setLineWidth(2);
    Plot.setColor("lightGray"); Plot.add("line", xCoord, disp);
    Plot.setColor("red"); Plot.add("circles", xCoord, disp);
    Plot.show();
} else {
    // Initialize parameters for the 2 plots
    disp1 = newArray(n); Array.fill(disp1,0);
    max_d1 = 0;
    disp2 = newArray(n); Array.fill(disp2,0);
    max_d2 = 0;
    for (i=1; i<=n-1; i++) {
        disp1[i] = sqrt(pow(objCenterX1[i]-objCenterX1[i-1],2)+pow(objCenterY1[i]-
        objCenterY1[i-1],2));
        if (disp1[i] > max_d1) { max_d1 = disp1[i]; }
        disp2[i] = sqrt(pow(objCenterX2[i]-objCenterX2[i-1],2)+pow(objCenterY2[i]-
        objCenterY2[i-1],2));
        if (disp2[i] > max_d2) { max_d2 = disp2[i]; }
    }
    if (max_d1 > max_d2) {
        max_d = max_d1;
    } else {
        max_d = max_d2;
    }
    m = mitosis[0];
    d_mitosis1 = newArray(1); d_mitosis1[0] = disp1[m];
    d_mitosis2 = newArray(1); d_mitosis2[0] = disp2[m];
    Plot.create("Displacement Cell " + objID + "-1", "Frame", "Displacement (pixel)");
    Plot.setLimits(0, n, 0, max_d+1);
    Plot.setLineWidth(2);
    Plot.setColor("lightGray"); Plot.add("line", xCoord, disp1);
    Plot.setColor("red"); Plot.add("circles", xCoord, disp1);
    Plot.setColor("green"); Plot.add("circles", mitosis, d_mitosis2);
    Plot.show();

    Plot.create("Displacement Cell " + objID + "-2", "Frame", "Displacement (pixel)");
    Plot.setLimits(0, n, 0, max_d+1);
    Plot.setLineWidth(2);
    Plot.setColor("lightGray"); Plot.add("line", xCoord, disp2);
    Plot.setColor("red"); Plot.add("circles", xCoord, disp2);
    Plot.setColor("green"); Plot.add("circles", mitosis, d_mitosis2);
    Plot.show();
}
```

Fig. 8 Macro code block showing the calculation and plotting of cell displacement

3.2 Ilastik Solution

The following subsections describe the tracking procedure using *ilastik*. In particular, the method is divided in two main parts represented by two different workflows: the *Pixel Classification* and the *Tracking* workflow. The final analysis of the tracking results will be performed in *Fiji*.

3.2.1 Pixel Classification Workflow

When using *Pixel Classification* workflow, the user segments foreground objects (e.g., cells) from background by defining two labels and providing examples through brush strokes.

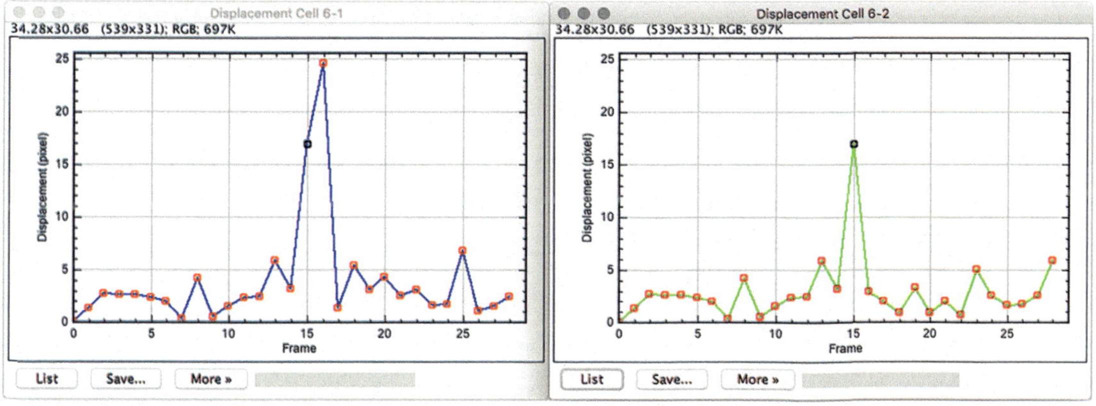

Fig. 9 An example of displacement plotted for two daughter cells, respectively, in blue and green. The moment of mitosis is highlighted with a black dot; note how the cells diverge after the event

1. Launch this workflow from the start screen of *ilastik* by creating a new project (Fig. 10).
2. In the *Input Data* applet, load data. The dataset we are using on the present example consists of a multipage TIFF file, so we have to select *Add separate image(s)...* (*see* **Note 9**).
3. After the data has been added to the project, it can be noticed that, in the *Shape* column of the top-right panel, the three dimensions of the stack will be shown, in particular (29, 333, 314) for the example data set, representing *time*, *y-axis*, and *x-axis*, respectively. This does not correspond to the *Axes* column, where *zyx*, rather than *tyx*, is shown (Fig. 11), because the software is considering the data as 3D rather than 2D + time. This issue can be corrected by double clicking on the dataset row, where the *Axes* field should be changed to *tyx*. The number of characters has to match the number of dimensions in the *Shape* field. Please note that *t* has to be present in this list, as tracking requires a time axis.
4. The next applet, on the left side of the window, is *Feature Selection*. Based on the features selected, *ilastik* will learn from the manual annotations to predict the different classes. In this example, all the features of scale σ equal to 1.6, 3.5, and 5.0 (*see* **Note 10**) can be selected.
5. In the *Training* applet, the user will have to manually brush strokes (Fig. 12), in particular identifying background pixels with Label 1 (red by default) and cell nucleus pixels with Label 2 (green by default). The *Live Update* toggle will show the result of the training while adding/removing annotations. In particular, we will see for every pixel its probability to belong to each of our classes (the most uncertain pixels will be those on the border of our cells, and they will have a very light shade of

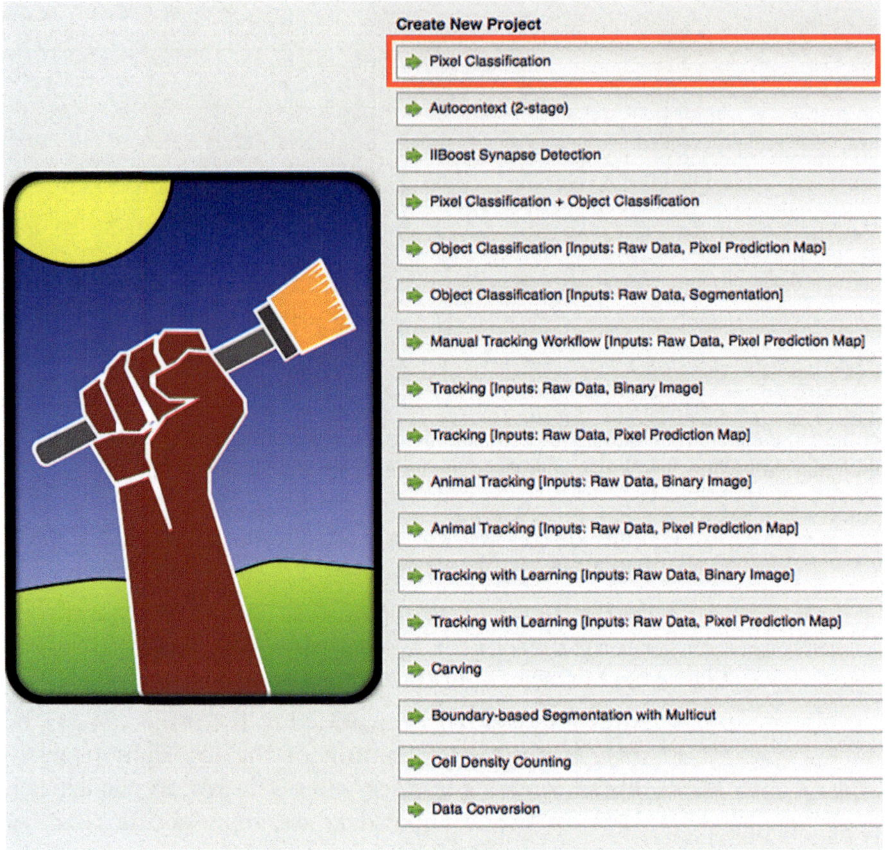

Fig. 10 *ilastik* start screen. A new project can be created selecting the *Pixel Classification* workflow (red box)

green or red). The segmentation resulting from this prediction map can be visualized by clicking on the eye of Segmentation (Label 1) or Segmentation (Label 2) in the *Group Visibility* menu.

6. If happy with the segmentation, the learned model can be applied to the entire dataset, and the prediction results can be exported for further use in tracking. In the *Prediction Export* applet, the location and file format can be changed using *Choose Export Image Settings* [18]. In this case, since the pixel classes' prediction will be used as intermediate results in another *ilastik* workflow, it is worth to save it as an HDF5 file (e.g., "prediction_map.h5"), an *ilastik*-friendly format. Now, the project can now be saved and closed before creating a new *Tracking* workflow project.

3.2.2 Tracking Workflow

1. Launch the *Tracking* workflow (*see* **Note 11**) from the start screen of *ilastik* by creating a new project (Fig. 13).

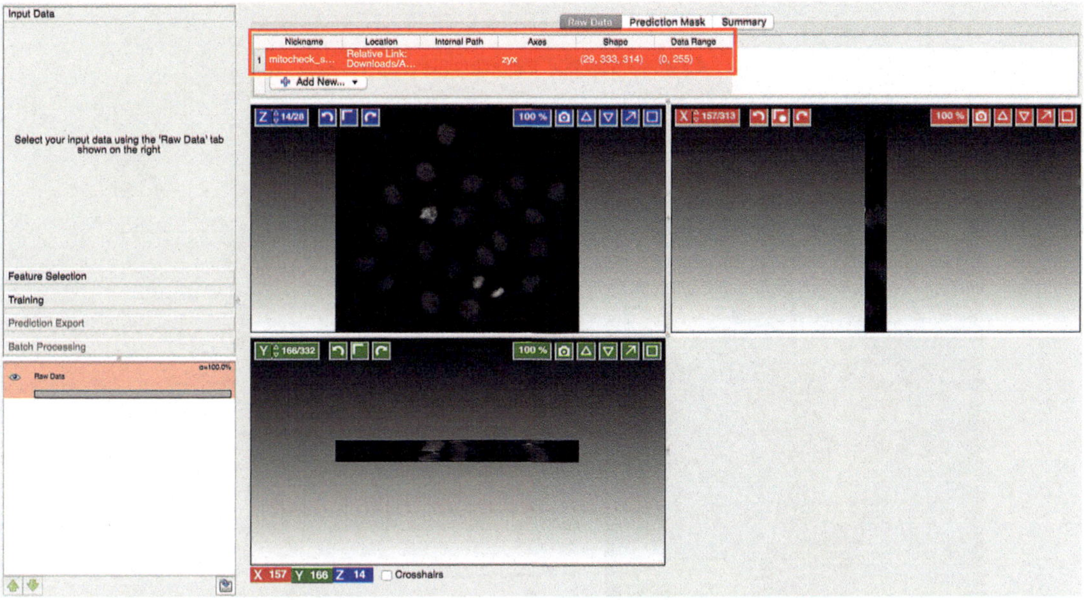

Fig. 11 Main *ilastik* menu snapshot. The red panel highlights the main characteristics of the uploaded file, before the Axes settings are modified

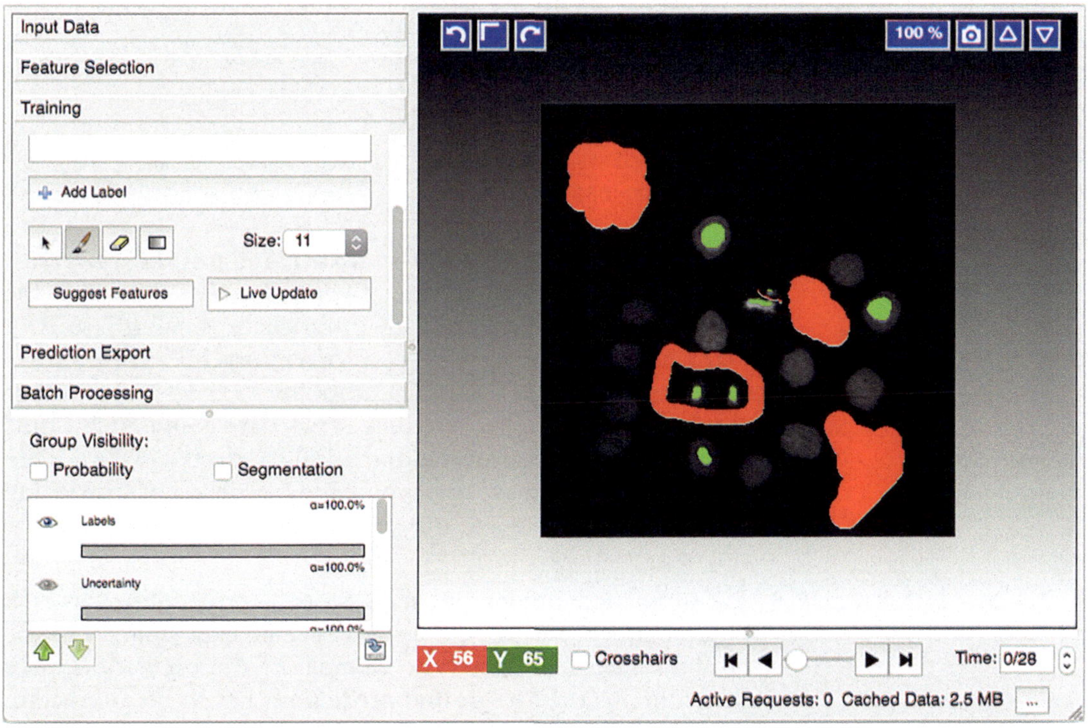

Fig. 12 An example of manual annotations useful to train the model. The annotations in red correspond to the background, while the ones in green identify the cells

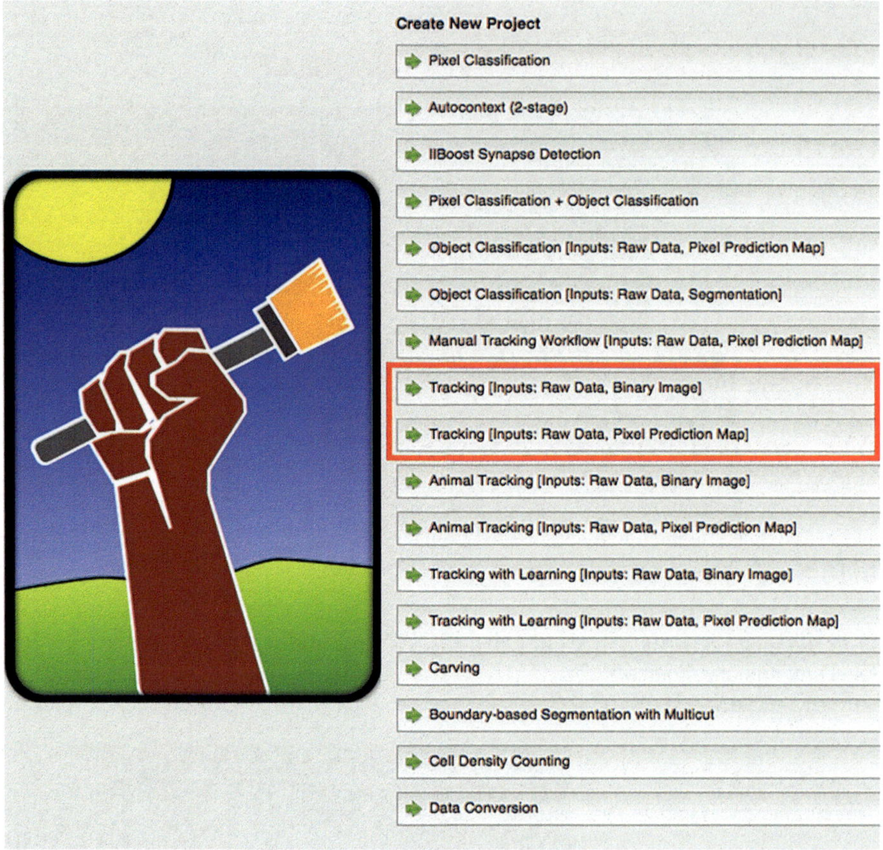

Fig. 13 *ilastik* start screen. Two tracking workflows are highlighted (red box): one uses the binary image and the other one—our case—the pixel prediction map

2. In the *Input Data* applet, the *Raw Data* and the *Prediction Maps* need to be specified in their respective tabs. For the particular example, add the file "mitocheck_small.tif" as *Raw Data*, and load the data set "prediction_map.h5" to *Prediction Maps*. Again, the *Tracking* workflow expects the image sequence to be loaded as a time series data containing a time axis; if the time axis is not automatically detected (as in HDF5 files), the axis tags have to be modified in a dialog when loading the data (e.g., the z-axis may be interpreted as t-axis by replacing z by t in this dialog).
3. The uploaded prediction maps store a probability for each single pixel/voxel to belong to a specific class defined in the pixel classification. First, the channel of the prediction maps that contains the foreground predictions has to be specified in the *Threshold and Size Filter* applet. For instance, if the user chose "Label 1" (red by default) to mark the foreground in the *Pixel Classification* workflow, *Input* will be 0; otherwise, if "Label 2" (green by default) was taken as the foreground

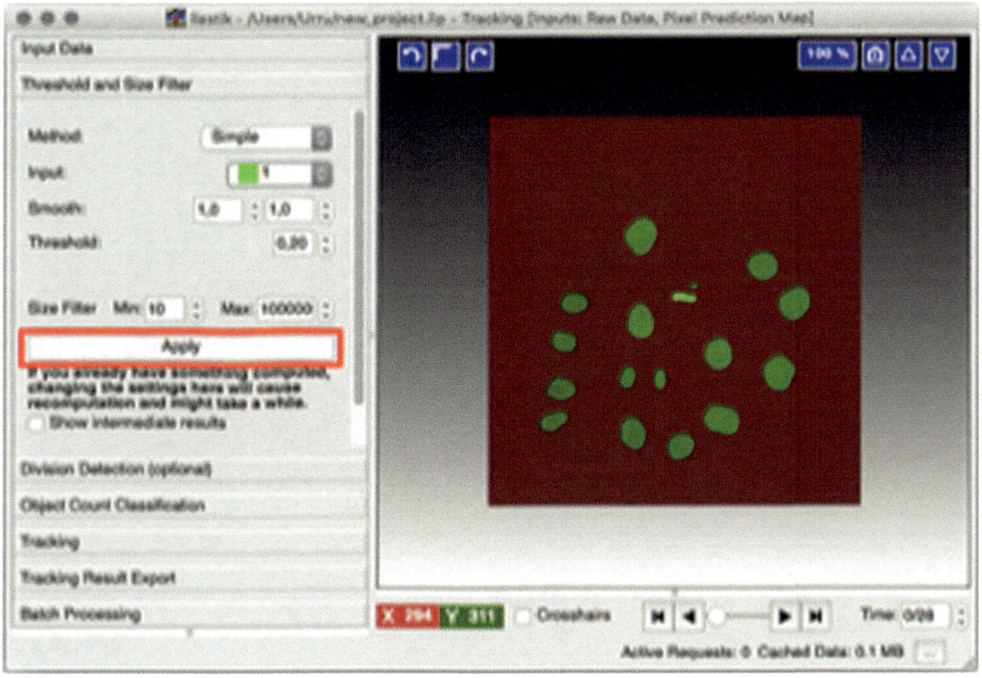

Fig. 14 Main *ilastik* screen displaying the *Threshold and Size Filter* applet. The foreground channel (i.e., Input) from the prediction map previously uploaded is selected; the image is then thresholded by clicking on the *Apply* toggle (highlighted in red)

label, then *Input* will take the value 1. Thus, we choose the *Input* to be 1 (Fig. 14). These probabilities can be smoothed over the neighboring probabilities with a *Gaussian* filter, specified by the *Sigma* values along the two axes (this is very useful when dealing with anisotropic data, i.e., data whose properties vary when measured along different directions). If we select *Simple thresholding*, the resulting probabilities are finally thresholded at the value specified: this thresholding method guarantees good performance in this case, but *ilastik* offers the option to use also different methods, which in some cases are more appropriate (*see* **Note 12**). The default values for the smoothing and thresholding should be fine in most of the cases (*see* **Note 13**). Finally, any objects outside the given *Size range* are not taken into account for the tracking. We could now proceed by pressing *Apply*. Please note that all of the subsequent computations and tracking will be invalid (and deleted) if parameters in this step are changed. This means we will need to repeat the following steps in order to update results according to the changes.

4. Next, go to *Division Detection*. The information in this applet is not needed for non-dividing object tracking applications, but it is useful when dealing with dividing objects, like in this case.

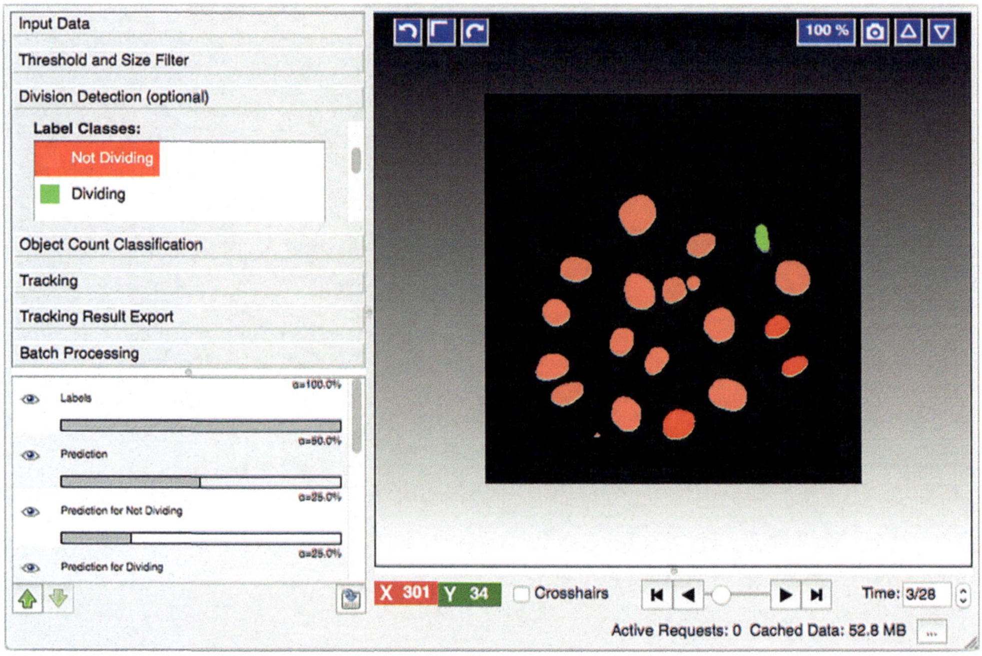

Fig. 15 Main *ilastik* screen displaying the *Division Detection* applet. A subset of nondividing cells should be selected in the following image frames. This can be done by selecting the class first and then going on the image and clicking on the cells to be classified. *Live Update* selection will give an idea of how the classification will work (cells will appear in light red or green based on their predicted class)

Here we will find two default classes, "Dividing" and "Not Dividing," and we will have to manually annotate/label a number of cells in the two categories, from which the algorithm will classify all the other cells. To place the labels, navigate to a frame where a parent cell is still a single detection but is split up into two children in the next frame, and mark it as a "Dividing" cell (Fig. 15). On the other hand, label as "Not Dividing" some of the other cells. By clicking *Live Update*, the user can see the prediction result and manually correct the misclassified cells or add new labels.

5. When happy with the results of the previous step, we can move to the next applet, the *Object Count Classification*, in which we can teach the algorithm to distinguish different cases, e.g., "False Detections" (*see* **Note 14**), "One Connected Object," "2 Connected Objects," and so on (Fig. 16). This classifier bases its decisions on features such as size, mean and variance of intensity, and shape information. The selection of these features can be adjusted by clicking the *Subset Features* toggle, but the default features usually perform well for a large variety of cell types. Also in this case, we can check *Live Update* to see the results of our manual annotations (*see* **Note 15**).

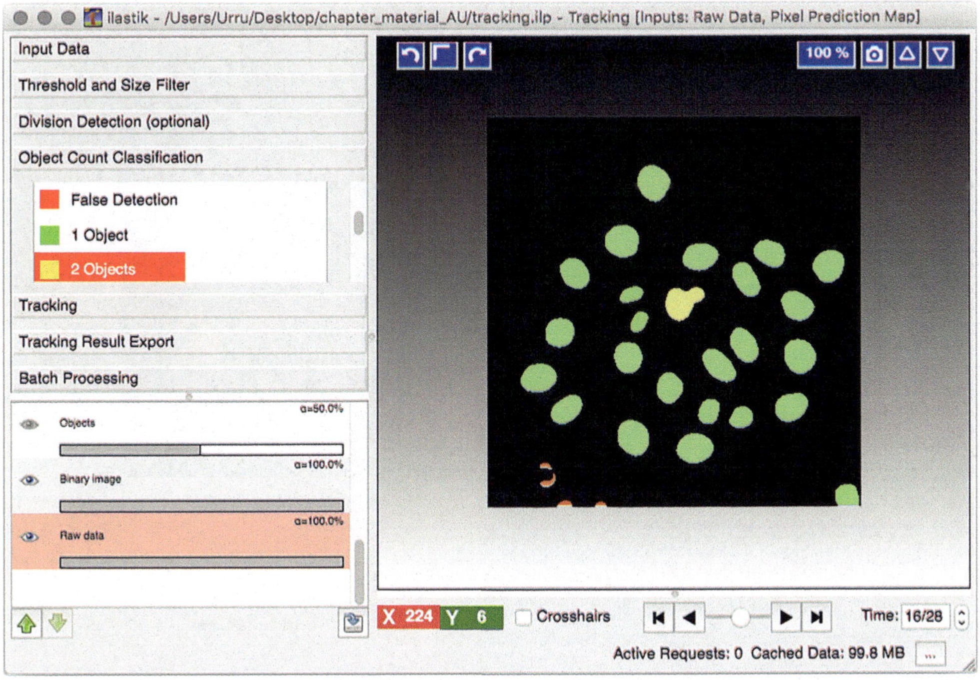

Fig. 16 Main *ilastik* screen displaying the *Object Count Classification* applet. During segmentation, subsets of *False Detections* should be detected: 1 connected objects and 2 connected objects (and so on if we have N-connected objects); also, a subset of features to classify the objects based on the training should be selected. By hitting *Live Update*, we can have an idea of how the classification will work

6. Next, the *Tracking* applet will be found, in which a number of parameters are shown, whose weights will determine the result of the optimization and thus the tracking results. To proceed with the actual tracking, click on the *Track!* button (Fig. 17). The default settings should work for most datasets, but we can tune them based on the results of the tracking. Finally, the tracking results can be exported as grayscale segmentation, where the background is black while every object has a different intensity value. We choose the source *Tracking-Result*, and each object is assigned a gray value corresponding to its lineage ID, but we have three more options for exporting (*see* **Note 16**). We can modify location and file format using *Choose Export Image Settings* [17], creating a file (e.g., objID.h5) that we can also import in *Fiji* for further analysis.

3.2.3 Tracking Analysis of ilastik Results in Fiji

The steps below describe a *Fiji* pipeline designed to analyze the results obtained in *ilastik*. This can be useful as in *ilastik* we are able to also track dividing cells.

1. Upload the HDF5 file from [File > Import > HDF5...], and browse for the corresponding folder (Fig. 18).

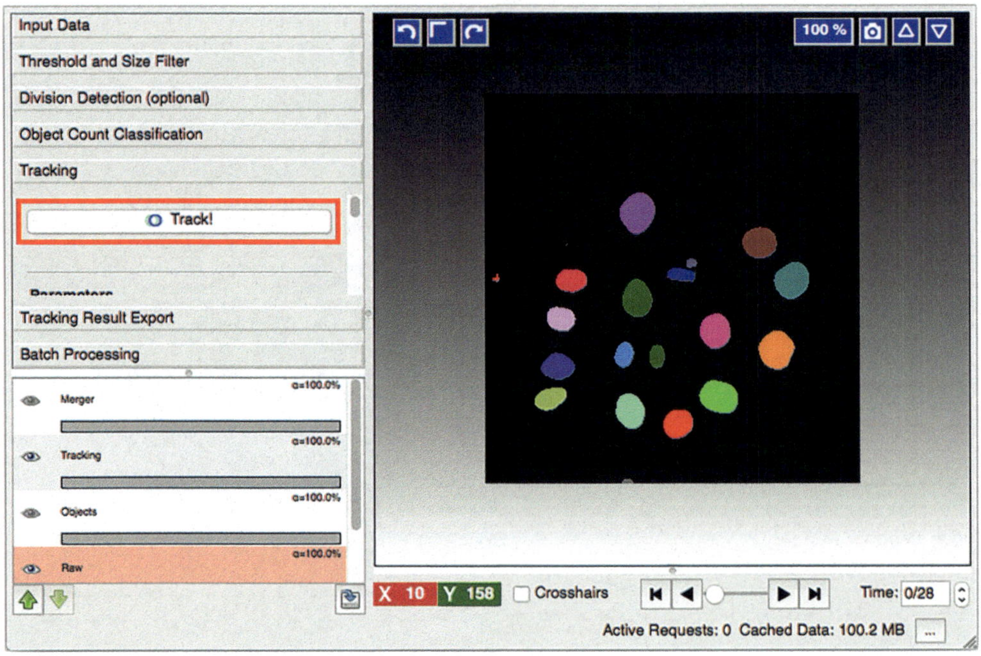

Fig. 17 Main *ilastik* screen showing the *Tracking* applet. We have to select the final parameters (place the mouse cursor on the parameter for an explanation of its function) and click on the *Track!* toggle to obtain the results of the tracking

2. The HDF5 image obtained using the *ilastik* protocol is the equivalent of the "Labels" image seen in the *Fiji* protocol. In this case, *ilastik* has identified the so-called False Detections with intensity equal to 1. To obtain the "Mask" image, duplicate the image and threshold it; use 2 as a threshold value to assign value 0 to the "False Detections," and assign the maximum value to all the other objects. Finally, rename the duplicated image as "Mask." Figure 19 shows all these steps implemented in a macro script.
3. For plotting trajectories, use the same approach described in Subheadings 3.1.3–3.1.5.

4 Notes

1. *Fiji* is able to import by default (i.e., simply dragging and dropping) many different bio-image formats (e.g., JPEG, TIFF, NRRD). You can find details on how to import image files and a link to the full list of bio-image formats supported in [19].
2. In some case, also a morphological operation of erosion [Process > Filters > Minimum. . .] with *Sigma* of 1.0 can be run in combination with watershed transform, to slightly shrink the

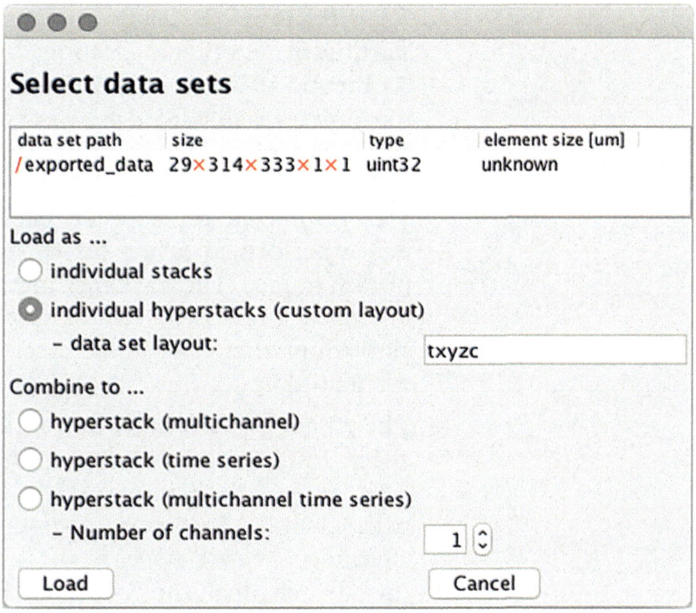

Fig. 18 Interactive panel to finally load the HDF5 file. We will need to select the dataset and insert the right labels for the axes

```
// Load image using HDF5 plugin (19)
run("HDF5...");
// The HDF5 plugin will need manual inputs so we wait for the user to go on with the analysis
waitForUser("Click OK after importing the HDF5 file as individual hyperstacks.");
filename = getTitle();
// We want to end up having the same initial situation as the analysis in Fiji, so one
Labels and one Mask stacks
selectImage(filename);
rename("ObjID");
run("Duplicate..." , "duplicate");
// We set the threshold to 2 as the cells with intensity 1 are the so-called false
detections found in ilastik
setAutoThreshold("Default dark");
setThreshold(2, 65535);
setOption("BlackBackground", false);
run("Convert to Mask", "method=Default background=Dark");
// We rename images to Labels (with glasbey inverted LUT) and Mask
selectImage("ObjID");
run("glasbey inverted");
rename("Labels");
selectImage("ObjID-1");
rename("Mask");
```

Fig. 19 Macro code needed to set up the HDF5 files obtained from the *ilastik* workflows, to be used for the tracking analysis in *Fiji*

obtained binary mask. This will prevent the merging of close neighboring objects in 3D in further steps and will also correct for any possible dilation of the objects due to the Gaussian filtering. Here again, click *Yes* to apply the filter to the whole stack.

3. An alternative to the *3D Object Counter* could be [Plugins > Process > Find Connected Regions]. It sometimes could be

faster than the *3D Object Counter* and has better options (e.g., starting from a point selection) but also lacks some flexibility. As mentioned in the introduction, there are other advanced tracking tools (included in *Fiji* or downloadable as *Fiji* plugins) available such as [Plugins > Tracking > TrackMate] and [Plugins > Tracking > Particle Tracker]. These trackers are however optimized for spotlike particle tracking; the linking is hence performed over a list of spot positions (spots detected in each frame). The tracking can be either straightforward (e.g., linking a spot to the closest spot in the next frame) or with algorithms that can handle cases when splitting and merging events occur.

4. The "Label" stack and the original stack can be merged to check results, using [Image > Colour > Merge Channels]. You can specify these two stacks in two of the available channels. Then, by unchecking the option *Ignore source LUT* and checking *Create Composite*, the *glasbey inverted* LUT is kept in the label channel, thus coloring IDs after merging.

5. Please note also that the method used here strongly relies on the single objects' spatial overlapping in temporal space between any two consecutive time frames and also on the fact that a single object at a later time point does not overlap multiple objects at an earlier time point. If there is no overlap or connection between two consecutive frames, then a new label will be assigned to the same object in the latter frame, since it is considered as another object just showing up. In conclusion, this method performs very efficiently in case of well-separated cells and with sufficiently high frame rate, while it can present issues in case of highly crowded image sequences.

6. For centroid we mean the arithmetic mean positions of all the points (in our case all the pixels) in the object.

7. As shown in the source code reported in the text, different colors will correspond to different items. By placing a double slash (i.e., //), we will create a comment line (colored in green), which won't be part of the actual code. It is good practice to add brief comments to make the code readable and understandable by other users. All the built-in functions will have a gold color, while the control structures (e.g., *for* cycle) will be in blue. Colors will help you to visually check if you typed correctly or if the functions exist in the first place.

8. An alternative to *makeLine* is the *makeSelection* built-in function with polyline type, which does not need to plot line segment by segment but takes the coordinates arrays (useful to plot trajectories all at once, and not the progressive trajectories). An alternative to [Image > Overlay > Add

Selection...] is [Analyze > Tools > ROI Manager] which allows keeping every line segment on display. A piece of code (alternative to the one displayed in Fig. 6) using this function is shown in Fig. 20.

9. If the data of interest consists of only one frame or is stored in a single file as either HDF5 volume or multipage TIFF, *Add separate image(s)...* should be selected; otherwise in the case of a stack with one file per time step or z-slice the, option *Add a single 3D/4D Volume from Sequence...* should be used. In the example used, the dataset is a multipage TIFF file, so the former has to be selected.
10. The user will find three main groups of features with different scales: (i) *Colour/Intensity*, useful when there is a difference in the color or brightness between foreground and background; (ii) *Edge*, useful when color gradients can be used to identify objects; and (iii) *Texture*, useful when foreground or background have different textures. The scales (in terms of pixels) of the features in these groups are very important as they represent the pixel diameter used to calculate the selected feature. In general, it is good to start with a large number of features and scales and then reduce them to find the best tradeoff between performance and computational time, as larger features scales lead to longer computational time.
11. There are two choices, one in case the prediction map is used (from the previously described *Pixel Classification* workflow) and the other one in case the user wants to use already segmented images from other sources or from the same *Pixel Classification* workflow.
12. We have the chance to choose between three different thresholding methods. In case of easily distinguishable objects, the *Simple* thresholding should be enough. In case of less clear, less uniform foreground, *Hysteresis* and *Graph Cut* could be useful. The former uses two thresholds (high (H) and low (L)): pixels higher than H are considered foreground, while the ones lower than L are labeled as background. The ones between H and L are considered foreground if connected to a foreground pixel. The latter uses a minimization function to find the cut in the graph formed by seeds selected by the user. This cut represents the edge between foreground object and background.
13. As mentioned before, although we have been using prediction maps as input files, nothing prevents the direct use of binary segmentation images instead in this tracking workflow. In this case, we do not want to apply the smoothing filter and thresholding to the binary images, so *Sigma* should be set to 0.0 to keep the original segmentation images and threshold to 0.5.

```
// We create an empty stack in which we copy the segmented frames with the progressive trajectory of the cell
selectImage("Mask");
setSlice(1);
run("Copy");
newImage("Trajectory" + objID, "RGB black", getWidth(), getHeight(), n);
setSlice(1);
run("Paste");

mitosis = newArray(1);

for (i = 1; i <= n-1; i++) {
    // If there is mitosis, we draw two green lines, while for a single-cell trajectory we use a red line
    if (objShowTime[i] == 1) {
        plot = 1;
        selectImage("Trajectory" + objID);
        setSlice(i+1);
        selectImage("Mask");
        setSlice(i+1);
        // Use makeline function to draw a line
        makeLine(floor(objCenterX1[i-1]),floor(objCenterY1[i-1]), floor(objCenterX1[i]),floor(objCenterY1[i]));
        run("Properties... ", " stroke=red width=1.5");
        roiManager("Add");
        roiManager("Show All");
        run("Flatten", "slice");
        run("Copy");
        selectImage("Trajectory" + objID);
        run("Paste");
        selectImage("Mask-1");
        close();
    } else {
        plot = 2;
        // Record the mitosis frame for plotting it later on
        if (objShowTime[i-1] == 1) { mitosis[0] = i-1; }
        selectImage("Trajectory" + objID);
        setSlice(i+1);
        selectImage("Mask");
        setSlice(i+1);
        // Use makeline function to draw a line
        makeLine(floor(objCenterX1[i-1]),floor(objCenterY1[i-1]), floor(objCenterX1[i]),floor(objCenterY1[i]));
        run("Properties... ", " stroke=blue width=1.5");
        roiManager("Add");
        roiManager("Show All");
        run("Flatten", "slice");
        run("Copy");
        selectImage("Trajectory" + objID);
        run("Paste");
        selectImage("Mask-1");
        close();

        selectImage("Trajectory" + objID);
        setSlice(i+1);
        selectImage("Mask");
        setSlice(i+1);
        // Use makeline function to draw a line
        makeLine(floor(objCenterX2[i-1]),floor(objCenterY2[i-1]), floor(objCenterX2[i]),floor(objCenterY2[i]));
        run("Properties... ", " stroke=green width=1.5");
        roiManager("Add");
        roiManager("Show All");
        run("Flatten", "slice");
        run("Copy");
        selectImage("Trajectory" + objID);
        run("Paste");
        selectImage("Mask-1");
        close();
    }
}
```

Fig. 20 An alternative piece of macro code describing how to plot trajectories on the segmented images using the *ROI Manager* built-in function

14. False Detections have to be limited as much as possible during the thresholding phase, leaving for the classifier just the most difficult ones to segment out. On the other hand, it is often not worth to spend too much time in the tuning of the parameters during the thresholding phase for few false detections, as probably the classifier will be able to recognize them after the training phase.
15. In both the *Division Detection* and the *Object Count Classification* applets, while one would think assigning more labels yields better results, it does not always work this way. It is good to give a sufficient number of labels to characterize the class, but giving too many of them might not generalize well the class itself. At the same time, the algorithm will try to reduce as much as possible the number of the wrong predictions with respect to the manually assigned label. Thus, whenever possible, try to be consistent while assigning labels between all the classes.
16. Choosing *Object-Identities*, objects intensity represents a unique ID in every frame. *Merger-Result* will export only the detections where the optimizer decided there were at least two objects merged. With the fourth choice, *Plugin*, you can export results in different formats. Exporting results in a CSV table, in combination with the *Object Identities* export, will allow you to uniquely identify which segment/frame a row of the table refers to. For visual inspection, *Tracking-Result* is usually the most helpful.

Acknowledgments

This work was financed by a fellowship (LCF/BQ/IN17/11620069) from "la Caixa" Foundation (ID 100010434) and by the Spanish Ministry of Economy and Competitiveness grant MDM-1025-0502 through the Maria de Maeztu Units of Excellence in R&D program. We also acknowledge support from the European Commission through the NEUBIAS network (COST action no. CA15124).

References

1. Stuurman N (2003) MTrack2 *Fiji* plugin. https://imagej.net/MTrack2
2. Tinevez J (2016) TrackMate *Fiji* plugin. http://imagej.net/TrackMate
3. Tinevez J, Perry N, Schindelin J et al (2017) TrackMate: an open and extensible platform for single-particle tracking. Methods 115:80–90
4. Sbalzarini I F (2006) Particle Tracker *Fiji* plugin. https://imagej.net/Particle_Tracker
5. Sbalzarini IF, Koumoutsakos P (2005) Feature point tracking and trajectory analysis for video imaging in cell biology. J Struct Biol 151 (2):182–195
6. Schindelin J, Arganda-Carreras I, Frise E (2012) *Fiji*: an open-source platform for

biological-image analysis. Nat Methods 9 (7):676–682. https://doi.org/10.1038/nmeth.2019
7. CellProfiler Project Website. http://cellprofiler.org
8. Hilsenbeck O et al (2016) Software tools for single-cell tracking and quantification of cellular and molecular properties. Nat Biotechnol 34(7):703–706. https://doi.org/10.1038/nbt.3626
9. StarryNite Project. http://starrynite.sourceforge.net
10. *ilastik* Toolkit. http://*ilastik*.org
11. Meijering E, Dzyubachyk O, Smal I (2012) Methods for cell and particle tracking. Methods Enzymol 504:183–200. https://doi.org/10.1016/B978-0-12-391857-4.00009-4
12. Chenouard N et al (2014) Objective comparison of particle tracking methods. Nat Methods 11(3):281–290. https://doi.org/10.1038/nmeth.2808
13. MitoCheck Project. http://www.mitocheck.org
14. Haubold C et al (2016) Segmenting and tracking multiple dividing targets using ilastik. Adv Anat Embryol Cell Biol 219:199–229
15. *Fiji* Homepage. https://*Fiji*.sc
16. *ilastik* documentation: installation guide. http://*ilastik*.org/documentation/basics/installation.html
17. Gurobi Optimizer. http://www.gurobi.com
18. *ilastik* documentation: exporting output. http://*ilastik*.org/documentation/basics/export
19. Importing image files—ImageJ. https://imagej.net/Importing_Image_Files

Chapter 21

Machine Learning: Advanced Image Segmentation Using *ilastik*

Anna Kreshuk and Chong Zhang

Abstract

Segmentation is one of the most ubiquitous problems in biological image analysis. Here we present a machine learning-based solution to it as implemented in the open source *ilastik* toolkit. We give a broad description of the underlying theory and demonstrate two workflows: *Pixel Classification* and *Autocontext*. We illustrate their use on a challenging problem in electron microscopy image segmentation. After following this walk-through, we expect the readers to be able to apply the necessary steps to their own data and segment their images by either workflow.

Key words Machine learning, *ilastik*, Semantic segmentation, Random forest

1 Introduction

Recent technical advances in imaging amount to a breakthrough: new microscopy techniques, automation, long-term high-throughput image, etc., allow to observe life at spatial and temporal scales that were previously inaccessible. Especially now that acquisition of multi-terabyte datasets is becoming routine, it is quantitative image analysis that is becoming the new bottleneck. Clearly, on these images traditional manual analysis is unfeasible, and computer-aided automatic image analysis is the order of the day [1, 2].

The large diversity of cell lines and microscopy imaging techniques requires the algorithms to perform robustly and equally well under different scenarios. Most of the existing tools or methods have been tailored for specific applications or projects. Applying to other assays or cell types typically requires parameter tuning or reprogramming of the software. Manual software adaptations are time-consuming and raise obstacles for life scientists due to their lack of expertise in software engineering.

Machine learning methods provide an effective way to automate the analysis, as they allow domain experts to inject their knowledge by training, i.e., by providing annotated examples

Elena Rebollo and Manel Bosch (eds.), *Computer Optimized Microscopy: Methods and Protocols*, Methods in Molecular Biology, vol. 2040, https://doi.org/10.1007/978-1-4939-9686-5_21,

such as malign vs. benign, foreground vs. background, etc. [3–5]. Since the data processing rules are constructed ("learned") from these examples and generic data properties, avoiding manual adjustments of the processing pipeline, machine learning methods are more flexible than conventional image processing techniques for solving complex multidimensional data analysis tasks. This chapter will demonstrate how one of the most common image analysis problems—segmentation—can be solved by using supervised machine learning through a simple and intuitive graphical user interface of the *ilastik* toolkit (6, *see* **Note 1**).

1.1 Machine Learning

A detailed explanation of the variety of categories of machine learning techniques goes beyond the scope of this chapter. Instead, we will explain the intuition behind the supervised learning approach on a segmentation problem example, which will include a tiny cutout from an electron microscopy image of a mouse brain, courtesy of Graham Knott, EPFL. Then we will introduce the basic terms and concepts used throughout the chapter.

The raw data is shown in Fig. 1a. The dark line visible in the image is a cell membrane, and the round objects are vesicles with neurotransmitters. In the following we will try to segment the cell membrane, but not the vesicles, which will be assigned to the same background class as the cytoplasm. First, let's try something simple: assign dark pixels to the membrane class and the light ones to the rest. Obviously, this will not fully work: while the cytoplasm pixels indeed have higher intensity than the membrane ones, vesicles are dark too. In Fig. 1b we labeled some of the membrane pixels in red and some of the non-membrane pixels in green. Figure 1c shows the histogram of the intensity values in the patch; in Fig. 1d one can see the same histogram, but the red and green bins additionally demonstrate where the labeled pixels have landed. We see that there is no way one could introduce a single threshold to separate the green and red classes by intensity. If we add another criterion, such as curvature, a two-dimensional scatterplot emerges (Fig. 1f). However, without labels there is still no clear way to separate the scatterplot into two clusters: the point cloud is uniform without a clear boundary. If we now plot the labels of Fig. 1b on top, a line can be drawn in such a way that most of the green points will be positioned on one side and most of the red points on the other. Roughly, this is the way supervised machine learning suggests to solve such problems: make the decision based on the examples provided by the user. This set of examples is referred to as the *training* set. The learning algorithms try to find settings at which the problem gets solved as well as possible for the training set. This is not the only objective of the learning algorithms; the other one is to generalize as well as possible to the unseen data. In this case, we are looking for a decision boundary that would separate the membrane and non-membrane pixels. The performance of the learned

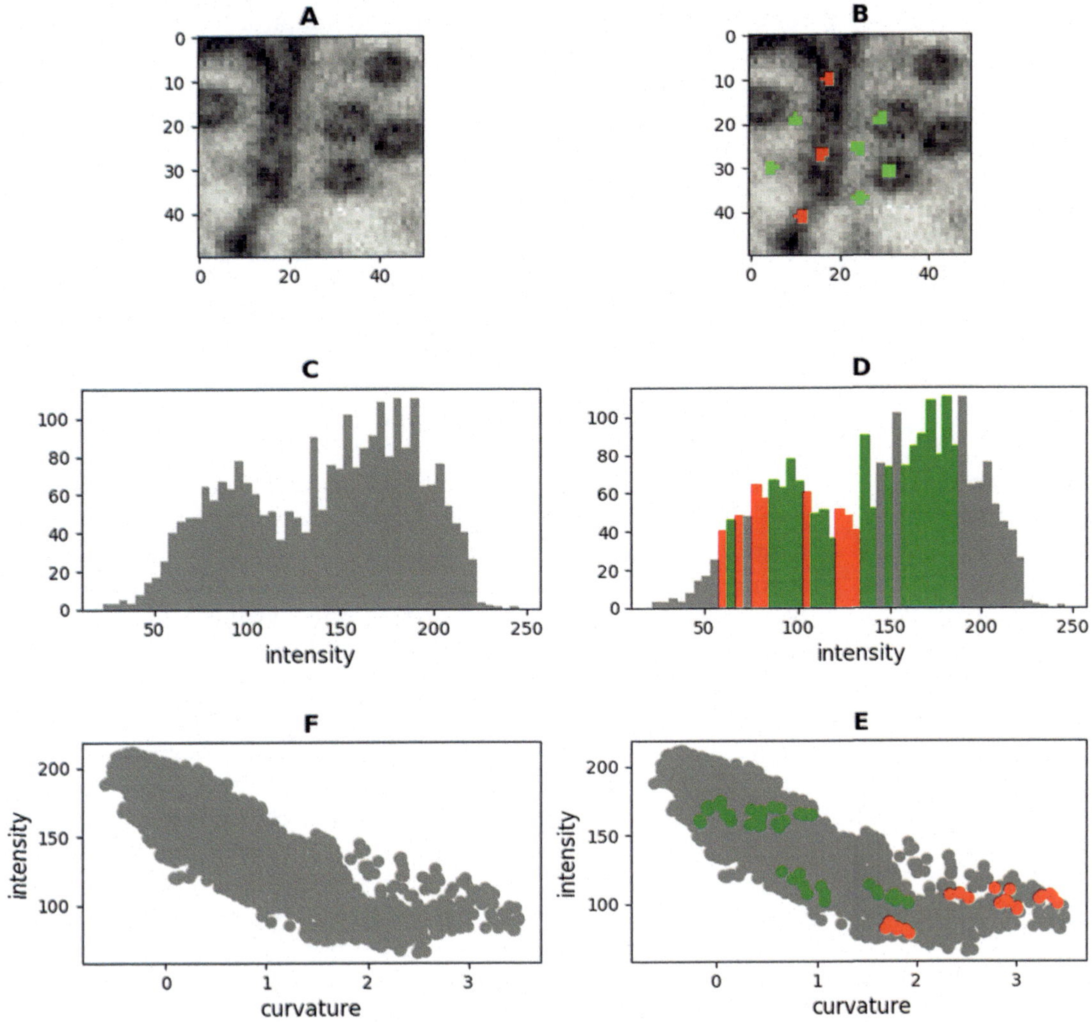

Fig. 1 (**a**) Raw data. (**b**) A few labeled pixels, red for cell membrane, green for the rest. (**c**) Overall intensity histogram. (**d**) Bins of the labeled pixels. (**e**) Intensity vs. second eigenvalue of the Hessian. (**f**) Dots for the labeled pixels. Note how the labeled pixels are not separable by a single threshold in the histogram (**d**) but separable in the scatterplot (**e**)

algorithm is measured on an independent *test* set, i.e., on the data that was not used in training. The criteria on which the algorithm bases its decisions are referred to as *features*. In our example, the features for each pixel consist of its intensity value and the value of the second eigenvalue of the Hessian matrix (representing curvature). The feature space is thus two-dimensional. In practice, many more features are usually used, and the decision surface separating the classes can also be nonlinear. For image analysis problems generic features describing primary elements of the image (edges, corners, texture, etc.) have been shown to be very powerful. Still, best results are achieved when the feature set is engineered exactly for the problem and data at hand. Omission of this requirement is one of the most attractive properties of deep learning-based classifiers: here the features are learned directly from the raw data, and the user only has to provide problem-specific labels [7]. However, since the representation of data has to be learned along with the decision boundary, a large amount of training annotations is required. While convolutional neural networks provide the best segmentation results right now, this chapter aims to introduce machine learning through tools that do not require computational expertise. It is not yet possible to train convolutional neural networks interactively from very sparse brushstroke annotations. We will thus refrain from describing them here and restrict ourselves to the "shallow" classifiers and generic image features.

1.2 *ilastik* and Its Segmentation Workflows

ilastik is a framework, graphical user interface (GUI), and suite of workflows for automated segmentation, classification, tracking, and counting in 2D and 3D multispectral images and videos. It casts segmentation and classification as interactive machine learning problems, allowing targeted annotation and making for significantly steeper machine learning curves. This reduces human effort. Once the training is completed, batch mode is set up to quantitatively analyze multi-terabyte datasets. In addition, *ilastik* provides semi-automated interactive workflows for the seeded segmentation of 2D and 3D objects as well as for tracking. Both are useful for the construction of structured ground truth. *ilastik* is free and open source, available online [8]. In this chapter we will briefly describe a subset of *ilastik* workflows that abstract from highly specialized use cases and can be combined to address a variety of fundamental (bio-)image segmentation problems. In particular, we will address the *Pixel Classification* workflow for semantic segmentation and its more advanced extension: the *Autocontext* workflow.

1.3 Overview of This Chapter

This chapter will give an overview of segmentation-related workflows in *ilastik* and demonstrate one of them—the *Autocontext* workflow—in a detailed step-by-step "walk-through," using 3D electron microscopy data (Subheading 2.2). We will show how the workflow can be trained from a few interactively added annotations, giving additional insight into training the *Pixel Classification* workflow as it constitutes the major part of *Autocontext*. After following this walk-through and experimenting with the interactive machine learning via a GUI, we expect that readers will be able to continue the work on their own data with this and other workflows. Subheading 3 describes all the necessary steps in detail.

2 Materials

2.1 Software

The *ilastik* toolkit is open source and can be downloaded freely from [8]. Binaries for all major platforms are available along with example projects for all workflows. In this chapter we will use *ilastik* version 1.3.0, but older versions such as 1.2.* can be used as well, although the user interface might prove to be slower in those. If large datasets are to be processed, it is important to use a computer with sufficient amount of RAM (*see also* **Note 2**). For *Autocontext* workflow in particular, we recommend at least 16 GBs of RAM for large 2D data and, if possible, even more for 3D data, to ensure smooth and interactive training experience.

2.2 Dataset

As example data we will use an electron microscopy image stack from mouse somatosensory cortex, provided by Graham Knott, EPF Lausanne. Imaging was performed on a FIB/SEM microscope with approximately isotropic resolution of $5 \times 5 \times 6$ nanometers. A small crop of this dataset can be found in the *ilastik* workflow examples [9]. The complete dataset used in the figures below can be found at [10]. We also provide the *Pixel Classification* and *Autocontext* workflows that we used to obtain the results. It has to be noted that a powerful machine is needed to run the *Autocontext* workflow in 3D. Users without access to a machine with more than 16 GB of RAM can still use the *Pixel Classification* workflow and try *Autocontext* on the small crop above. Overall, this dataset serves as a good illustration for the *Autocontext* workflow since different elements of ultrastructure are clearly visible. Example labels of the biological structures in this dataset can be seen at Fig. 6 (with their color codes in Fig. 5).

3 Methods

3.1 Pixel Classification Workflow

This workflow employs a classifier (by default, a *Random Forest* [11]) to estimate, for each pixel in a (multidimensional) image, the probability that it belongs to one of several classes, such as "cell" vs. "background" vs. "artifact," defined by the user. The decisions are based on generic image features, which can be computed at different scales. The features characterize intensity, edge proximity, and texture for every pixel of the original image and of the image smoothed with a Gaussian of the specified scales. For colored or multispectral data, features are computed separately for each channel. An important characteristic of *ilastik* is that the training set is created interactively in response to the predictions made with the current classifier. Such interactivity allows the user to concentrate on the difficult examples, giving the classifier more information on cases in which it still makes errors. Compared to "blind" labeling without interactive classifier feedback, this strategy makes steeper learning curves and requires substantially less annotations [12]. Additionally, image features can be visualized to help with appropriate selection of features and their scales. A detailed guide on this workflow is provided below as part of the *Autocontext* workflow description (Subheading 3.2, **steps 2–8**). The *Pixel Classification* workflow is not suited to distinguish extended objects by size, shape, or color distribution. However, if the objects themselves can be segmented from the background with the *Pixel Classification*, the *Object Classification* workflow can take over the task of separating them into different populations.

3.2 Autocontext Workflow

This workflow improves the *Pixel Classification* workflow by training a cascade of classifiers. The approach of *Autocontext* has been introduced by Tu and Bai in [12]. Briefly, the idea behind it is to perform segmentation in multiple stages and let each stage use the results of the previous one. Since the original image features are also present at each segmentation round, the results are guaranteed not to get worse, given a sufficient amount of training data. The original method of [12] samples the results of the previous segmentation stages on a stencil around each pixel. In contrast, we add the results of the previous stage to the input data as new channels. This approach enables the algorithm to not only see the predictions of the previous stage but also to compute features on them. It can then learn to interpret edges and texture variations in probability maps as well as in the raw data. While this approach does not allow for orientation-specific context, which can be important in, e.g., medical images registered to an atlas, we believe it is more appropriate for biological images, where the objects of interest do not maintain a preferred orientation. If more than two stages are used, the procedure is repeated. This workflow is particularly efficient in cases where the image data shows multiple distinct classes.

In the following we give a step-by-step description of the application of the *Autocontext* workflow to the dataset described in Subheading 3.2.

1. Choose the *Autocontext* (two-stage) workflow in the *ilastik* start-up window (Fig. 2). A new window will pop up, and we can save the project as "autocontext.ilp" (or any other name you prefer) in the *Save As* textbox, by clicking the *Save* button.
2. The empty workflow window is now loaded. We can load the data in, from a stack or from an individual file. Since our dataset is a single file (i.e., "smallFibStack.h5"), we will load it by selecting [Raw Data > Add New... > Add separate Image (s)...]. If the axes of the dataset are not interpreted correctly by default (e.g., the data gets loaded as a time sequence instead of a 3D stack), double-click on the line with the dataset name, and correct the axes attribution in the *Data Properties* dialog

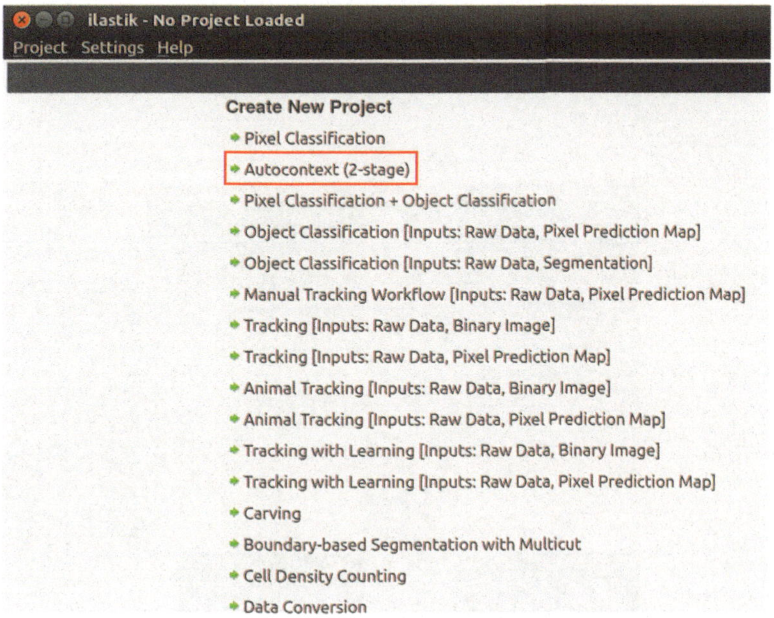

Fig. 2 *ilastik* start screen, with the *Autocontext* workflow highlighted. Note that the order of workflows can depend on the operating system

(*see also* **Note 3** for information on supported data formats). If one looks at the list of applets on the left, the structure of the workflow becomes very clear: this is just Pixel Classification, repeated twice (i.e., two times *Feature Selection* and *Training*, as shown in Fig. 3).

3. All *ilastik* workflows proceed from top to bottom. The next applet after *Input Data* is *Feature Selection*. After clicking the *Select Features...* button, the corresponding dialog allows one to select the features that will be used for classification. Simply selecting all features does not hurt the classification performance in general, but it makes the computation slower, especially for 3D datasets and features at large scales. As a compromise, one can select features in a checkerboard pattern, as shown in Fig. 4. The trick on what features and scales are to be checked or unchecked can be found in **Note 4**. Only a set of generic image features is offered in *ilastik*, but those should be effective for most images. However, customized features are also possible; please *see* **Note 5** for usage. We can repeat and modify the Feature Selection list as many times as we like.

4. By clicking the first *Training* applet, let's proceed to training the first classifier to distinguish various elements of the cell ultrastructure. By clicking *Add Label* eight times, we define eight classes, or *labels* as in the software, as shown in Fig. 5: cell membrane, cytoplasm, mitochondria, mitochondrial membrane, myelin sheath, synapse (including the cleft and the

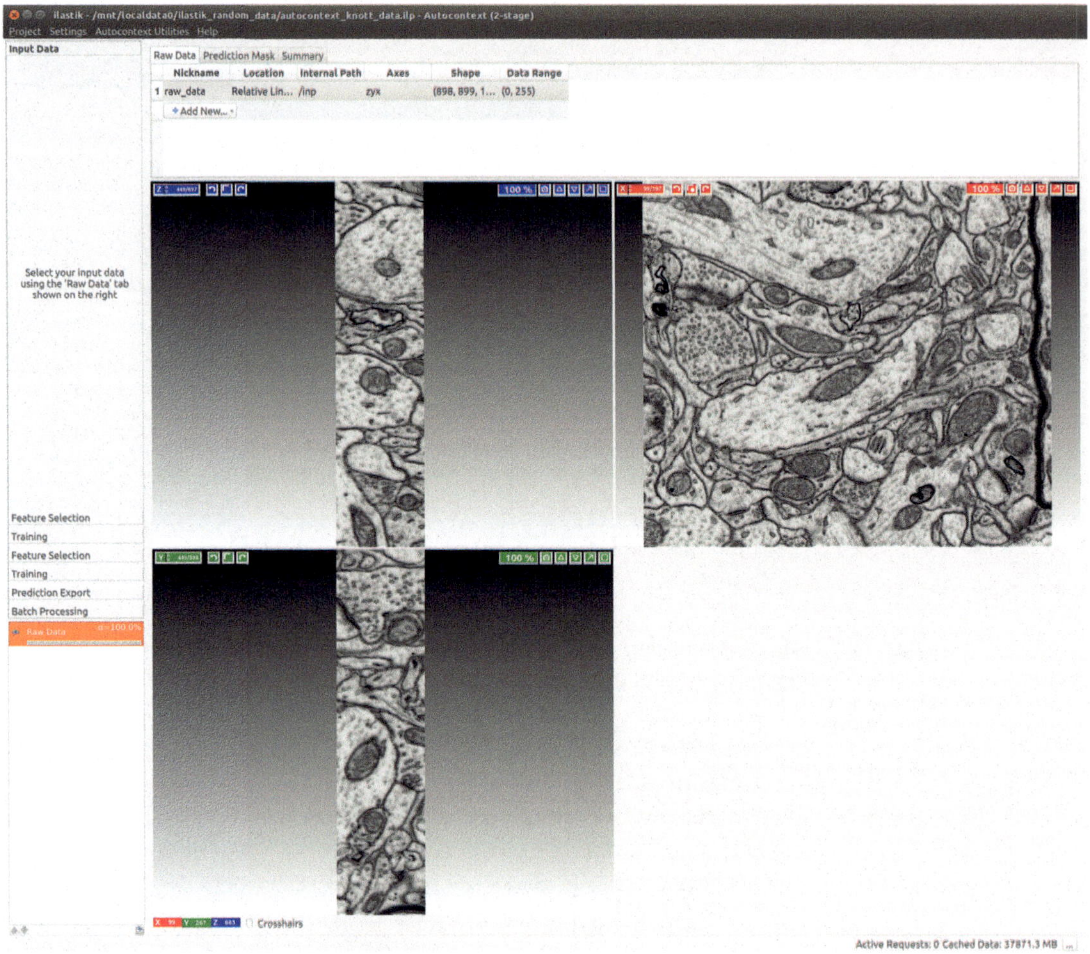

Fig. 3 *Input Data* applet with the loaded dataset. Note the list of applets on the left, *ilastik* workflows always proceed by executing them top to bottom

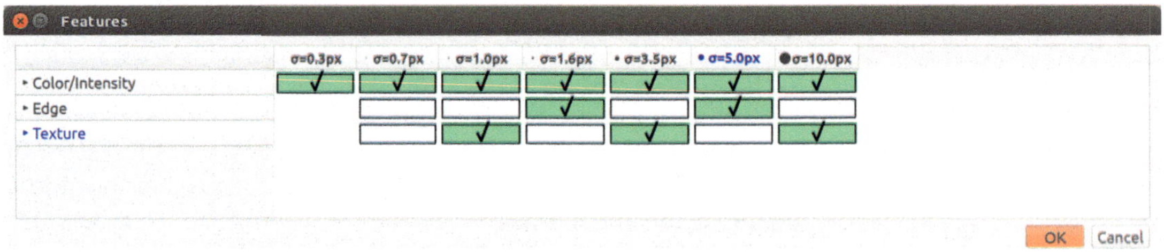

Fig. 4 *Feature Selection* dialog with the features we used. Other combinations of features can be used as well, including simply selecting all features. This, however, is fairly computationally intensive in 3D

postsynaptic density), vesicle, and membrane-bound organelles other than vesicles ("Inner"). By double-clicking on each label name, we can rename them or change their color (note that label color is changed from default for class "Vesicle" to make

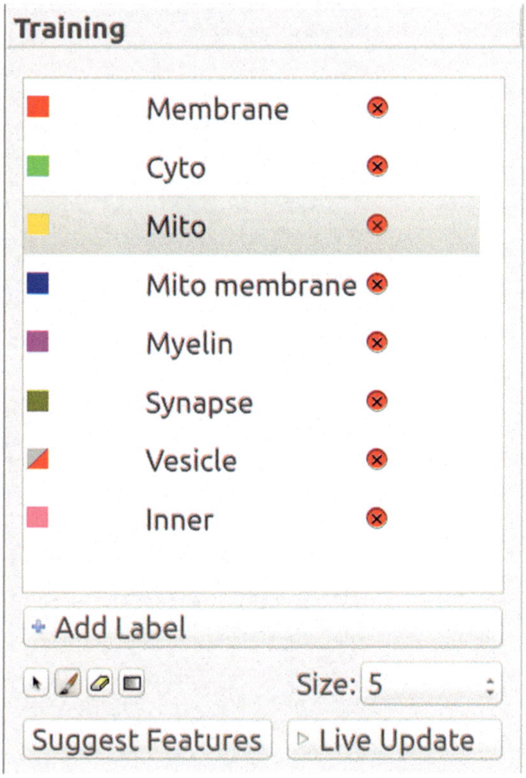

Fig. 5 Labels created to train the first stage of the workflow. You can change the label name, color, or prediction display color by double-clicking on it in this widget

its labels stand out better against those of class "Membrane"). Below the *Add Label* button, there are four cursors: navigation, brush, eraser, and gray-scale contrast enhancement. The most used ones are the first three. By changing the *Size* parameter on the right, the brush stroke width can be adjusted. Now we can manually introduce labels, or "paint," on the image example classes that we think belong to each of the classes. When a mistake is made, the *eraser* cursor can remove the strokes. Given the fact that these strokes serve as the training examples for the classifier to predict, one should always try to label as accurately as possible on the pixel level. Therefore, you may also want to use the mouse roller to zoom in or out the labeled local regions. Figure 6a shows the labels we introduced on one slice of the data.

5. If we now press the *Live Update* button, *ilastik* will proceed to compute features for the pixels in the field of view and to train a classifier, based on all the labels the user has provided. If more than one image has been labeled, the labels will be pooled together to train a single classifier. Once the classifier is trained,

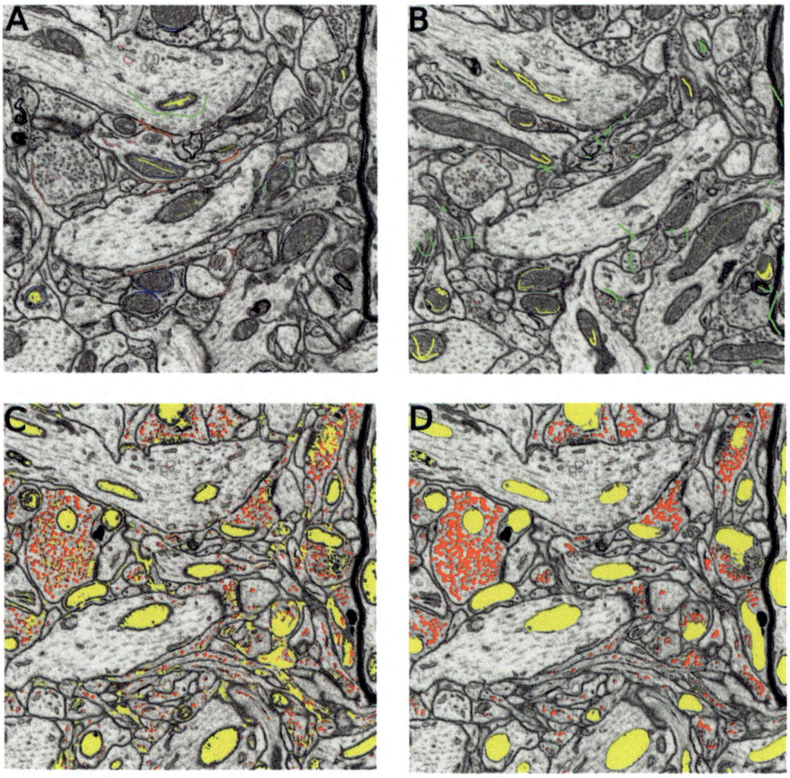

Fig. 6 (**a**) Labels for the first stage of the *Autocontext* workflow, with classes as defined in Fig. 5. (**b**) Labels for the second stage of *Autocontext*. No other labels were used in either classification step. (**c**) Prediction results for the first stage. (**d**) Prediction results for the second stage. Note especially how the mitochondrion class is not mixed into vesicle clouds anymore and how much smoother the predictions have become

it will be applied to the currently visible part of the image in order to give the user immediate feedback on the current classification accuracy. Note that the computed features are cached, so if more labels are added, the prediction results are updated much faster than the first computation time.

6. The first results are loaded. How would we decide where to add more labels? First of all, address the areas where the classifier is making wrong predictions, and label its errors into correct classes. If the *Live Update* button is pressed, the prediction results will get updated automatically. Besides errors, it is also recommended to keep an eye on the *uncertainty* layer (by clicking the *eye* icon on and off this layer), which highlights the areas the classifier is not certain about. Placing "uncertain" pixels as one class label or another will help the classifier to learn better. Areas very near to the boundary between classes will always be somewhat uncertain, but if there were high-

uncertainty areas that belong to a single class, the classifier would benefit from receiving more labels there.

7. Once the results on the current field of view are sufficiently good or stop improving, pan or scroll to a different area of the image or stack. Remember that the basic assumption of supervised machine learning is that the training set is representative of the data to be processed. It is thus helpful to give labels in visually different areas, especially if the data suffers from contrast or illumination changes (*see also* **Note 6**).
8. Once the classifier predictions stop improving, we have done what we could at the first classification stage. Figure 6c shows the first stage classifier performance on predicting classes of each pixel from a different, completely unlabeled slice of the stack. As we can see, the classifier may still have problems distinguishing vesicles from inner mitochondrial membranes (*see* **Note 7** for general limitations of Pixel Classification). While this classifier could benefit from seeing more labels as discussed in the previous point, we limited ourselves to just one slice for illustration purposes. Let's proceed to the second classification stage.
9. Open the next *Feature Selection* applet and select more features. Make sure the *Live Update* is unselected so that we could proceed further to the next *Feature Selection* step. Note how many channels these features now have: this is because we appended the predictions of the first stage to the raw data (*see also* **Note 8**).
10. Proceed to the second training applet. Choose a different slice or area of the data and add more labels. Now we will only define three classes, mitochondria, vesicles, and the rest (*see* **Note 9**), since segmenting vesicles and mitochondria was our original intention. Multiple classes introduced in the first stage help the classifier learn context relations between different areas in the image stack. Now that this is learned, we turn to the actual problem at hand. Figure 6b shows our annotations. Press the *Live Update* button and wait for new predictions. This is going to take longer than in the first stage, since *ilastik* first has to compute the first stage predictions at the labeled slice and then compute features out of those (*see* **Note 10**).
11. Verify the results on a different slice or area of the data. Figure 6d shows the segmentation we achieve at the second stage. This is clearly better than the first stage: there are almost no pixels attributed to mitochondria in the vesicle clouds, and the predictions are in general much cleaner.
12. We can now proceed to export the results in the *Prediction Export* applet. Intermediate computations (features) and results of both stages are available for export here. For results

the options include probabilities—a pixelwise map of probabilities for each class—and simple segmentation, a winner-takes-all assignment of the most probable class for each pixel. A click on the *Choose Export Image Settings...* button will bring up a dialog with export options. When exporting probability maps, we recommend converting them from type *floating 32-bit* to *unsigned 8-bit*. The *Random Forest* classifier in *ilastik* uses 100 trees, so the resulting probability maps can safely be stored in an 8-bit image.

3.3 What's Next?

In *ilastik* itself, the results of the *Pixel Classification* or *Autocontext* workflows can serve as input for other workflows. The *Object Classification* workflow can utilize object-level features, such as shape, intensity statistics inside the object, or neighborhood appearance to separate segmented objects into different populations. Given pixel and object classification, the *Tracking* workflow [13] can then connect objects over time and division events.

A different application can be found in the problem of boundary-based segmentation: sometimes we find it necessary to extract one or more regions from a 2D or a volumetric image, where the regions of interest are demarcated only by a more or less pronounced boundary, while the actual region appearance is uninformative. Electron microscopy or phase contrast microscopy images can serve as examples of this type of data. *ilastik* contains two workflows for boundary-based segmentation: *Carving* and *Multicut* (*boundary-based segmentation with Multicut* is the title on the start-up screen). Both of them can use Pixel Classification results as boundary information. *Carving* would then allow to segment the regions one by one based on user seeds [14], while the *Multicut* workflow produces a dense segmentation of all regions in the image [15].

Outside of *ilastik*, *Pixel Classification* or *Autocontext* results can be used for a variety of applications, from co-localization to counting to tracking.

Of course, *ilastik* is not the only solution to save the world; please *see* **Note 11** for other options of software to tackle similarly image segmentation problems.

4 Notes

1. Apart from the image segmentation, which our chapter belongs to, many image-based biological studies can benefit from them, to name just a few: object detection, counting, stereo reconstruction, tracking, sequence analysis, etc.
2. What if the data is bigger than the RAM of my computer? Do not worry. *ilastik* can achieve interactive machine learning on

datasets that do not fit into RAM. This is ensured by its lazy flow architecture, which computes only what is strictly necessary to produce results for the currently displayed field of view or for the output image (or volume) requested in batch processing.

3. While *ilastik* can read most common image formats, the preferred data format for *ilastik* is hdf5. This is especially important for large data, which hdf5 chunked storage allows to load partially. The simplest way to convert data to appropriate hdf5 is to use the *ilastik* "Import Export" plugin in *Fiji* or the *ilastik Data Conversion* workflow.
4. In *ilastik*, features (responses) can be visualized as images, such that one can see and select the features and scales that can help distinguish most different classes. By simply clicking items in the *Features* applet, located at the bottom-left part of the window, the corresponding feature (response) images will be shown on the right.
5. If you would like to try segmenting with your own pixel features computed outside *ilastik*, add them as new channels to the raw data, and only select the intensity feature with $sigma = 0.3px$ from the *ilastik* ones.
6. Since the classification algorithm assumes that training and test data are sampled from the same distribution, it is beneficial to make the dataset as homogenous as possible before processing it with *ilastik*. If this is not possible, introduce labels in all visually different parts of the dataset.
7. All pixel-level features in *ilastik* are computed in a circular (spherical) neighborhood of a given pixel. The classifier thus cannot take into account any information from beyond this neighborhood: if a human cannot determine the class of the central pixel from only looking at a 60-pixel circle around it, *ilastik* will likely not be able to do it either.
8. *Autocontext* workflow needs a lot of computational resources. The main reason lies in the sheer number of features that need to be computed at the second stage. Since all these features are needed to make a prediction, a large amount of RAM is consumed. *ilastik* has a balancing mechanism which frees up the least-used cached features, but that, in turn, forces a recomputation of these features when they are needed again. To reduce RAM requirements, the workflow can be broken into two separate stages: perform Pixel Classification as usual first, and export the probability maps (as unsigned 8-bit type). Now use *ilastik*'s option to stack input data across channel: open another Pixel Classification project, and stack raw data with the predictions of the first stage. Proceed to label as described for the second stage. While this approach is less resource-intensive, it

requires a noninteractive phase where the complete dataset is predicted with the classifier of the first stage.

9. The entities that are assigned to different classes in the first stage need not represent biologically different classes. Another option is to label different textures or out-of-focus areas as different classes in the first stage and only introduce semantic classes in the second stage.

10. Recall that *ilastik* uses lazy computation but strives to provide interactive feedback in the current field of view. To get a faster turnaround, it is thus recommended to zoom in and, for 3D data, maximize one of the orthogonal views when the *Live Predict* mode is enabled.

11. Are there other learning-based tools for segmentation? *Fiji* offers a *Trainable Weka Segmentation* plugin [16] that, however, cannot compute features in 3D, cannot process data larger than RAM, and, in general, is slower than *ilastik*. "Vaa3D" has a plugin for interactive cell segmentation [17]. *Cytomine* also offers a data mining module [18], which is especially well suited for 2D histopathology images, but the necessary learning effort is relatively high. A recently published tool *fastER* [19] can take fuzzy scribbles as annotations for training, but it only works on 2D images and does not perform well on cells that are not homogenous in intensity.

References

1. Myers G (2012) Why bioimage informatics matters. Nat Methods 9(7):659–660
2. Meijering E, Carpenter AE, Peng H et al (2016) Imagining the future of bioimage analysis. Nat Biotechnol 34(12):1250–1255
3. Coelho L, Glory-Afshar E, Kangas J et al (2010) Principles of bioimage informatics: focus on machine learning of cell patterns. In: Blaschke C, Shatkay H (eds) ISBM/ECCB, 2010. Lecture notes in bioinformatics, vol 6004, pp 8–18
4. Sommer C, Gerlich D (2013) Machine learning in cell biology-teaching computers to recognize phenotypes. J Cell Sci 126 (24):5529–5539
5. Kan A (2017) Machine learning applications in cell image analysis. Immunol Cell Biol 95:525–530
6. Sommer C, Straele C, Koethe U et al (2011) *ilastik*: interactive learning and segmentation toolkit. 8th IEEE International Symposium on Biomedical Imaging (ISBI). Proceedings, p 230–233
7. LeCun Y, Bengio Y, Hinton G (2015) Deep learning. Nature 521(7553):436–444
8. *ilastik* website (2018). www.*ilastik*.org/download
9. http://data.*ilastik*.org/smallFibStack.zip
10. http://data.*ilastik*.org/*ilastik*_data_and_autocontext_project.zip
11. Breiman L (2001) Random forests. Mach Learn 45:5–32
12. Tu Z, Bai X (2009) Auto-context and its application to high-level vision tasks and 3D brain image segmentation. Trans Pattern Anal Mach Intelligence 32(10):1744–1757
13. Haubold C, Schiegg M, Kreshuk A et al (2016) Segmenting and tracking multiple dividing targets using *ilastik*. Focus on bio-image informatics, p 199–229
14. Straehle CN, Köthe U, Knott G et al (2011) Carving: scalable interactive segmentation of neural volume electron microscopy images. In: Fichtinger G, Martel A, Peters T (eds) Medical Image Computing and Computer-Assisted Intervention (MICCAI), 2011. Lecture notes in computer science, vol 6891. Springer, Berlin Heidelberg, pp 653–660

15. Beier T, Pape C, Rahaman N et al (2017) Multicut brings automated neurite segmentation closer to human performance. Nat Methods 14(2):101–102
16. Arganda-Carreras I, Kaynig V, Rueden C et al (2017) Trainable Weka segmentation: a machine learning tool for microscopy pixel classification. Bioinformatics 33 (15):2424–2426
17. Li X, Zhou Z, Keller P et al (2015) Interactive exemplar-based segmentation toolkit for biomedical image analysis. In: IEEE 12th International Symposium on Biomedical Imaging, p 168–171
18. Marée R, Rollus L, Stévens B et al (2016) Collaborative analysis of multi-gigapixel imaging data using Cytomine. Bioinformatics 32 (9):1395–1401
19. Hilsenbeck O, Schwarzfischer M, Loeffler D et al (2017) fastER: a user-friendly tool for ultrafast and robust cell segmentation in large-scale microscopy. Bioinformatics 33 (13):2020–2028

Index

A

Adjust B&C 286
Algorithm 4, 23, 75, 109, 135, 157, 196, 225, 279, 304, 338, 360, 394, 440, 449
Analyze particles 16, 44, 46, 56, 77, 80, 83, 84, 93, 107, 109, 124, 161, 171, 316, 317, 320–322, 328, 424, 427
Arrays 53, 57, 59–62, 64, 65, 68, 89, 90, 121, 147, 165, 167, 172, 220, 230, 307, 310, 312, 313, 320, 322, 323, 378, 415, 427, 432, 444
Aspect ratio (AR) 100, 109–111, 113, 114
Autocontext workflow 451, 453–455, 457–461

B

Background 5, 26, 43, 73, 105, 124, 144, 156, 181, 219, 245, 285, 304, 350, 360, 379, 389, 411, 426, 450
Batch processing 31, 59, 461
BG subtraction from ROI 219, 221, 230
Binary image/mask 6–9, 12–19, 44–46, 67, 68, 75, 77, 79–82, 84, 93, 105, 109, 124, 126, 157, 160, 161, 166, 171, 172, 204, 216, 217, 219, 223–226, 231, 232, 301, 304, 310, 313, 315–317, 320, 321, 328, 350, 360, 405, 408, 411–415, 419, 426, 427, 438, 443, 445
Binary operations 126, 223, 304, 316, 360, 361, 367
Biofilm Intensity and Architecture Measurement (BIAM) 118–120, 131
Bio-Formats 33, 67, 92, 143, 170
Bit depth 3, 16–18, 54, 230, 313, 321, 322
Bleach correction 141, 144, 150
Boundary-based segmentation 455
Built-in *ImageJ* macro function 29, 32, 76, 92, 158, 170

C

Calibration bar 187, 287, 289–291
Cell lineage analysis 413–415
Cell tracking 147, 149, 385, 399–421
Code annotations 55, 58
Co-distribution 178, 219
Co-expression 177, 216, 229
CoinRICSJ 377–382
Co-localization 72, 177, 179, 200, 206, 208, 215–232
  coefficients 72, 121, 127, 178, 201, 216, 218, 219, 226, 227, 229, 256, 378, 380, 381, 383
  indicator 200, 201, 206
  quantifier 200, 201, 203, 205, 206
Composite image 46, 65, 85, 88, 218, 220
Conditional statement 54, 62, 65, 67, 165, 310
Connectivity map 121, 122, 124, 127, 130
Convolution 9, 10, 124, 131, 185, 256, 360, 362, 364, 452
Co-occurrence 178, 228, 376
Correlation 110, 111, 114, 178, 182, 201–204, 210, 228, 335, 343, 352
Cytofluorogram 184, 185, 201, 208

D

3D 6, 30, 72, 117, 136, 156, 180, 263, 362, 385, 451
Deconvolution 10, 195, 197, 198
DeconvolutionLab 181, 195
Densitometry parameters 11, 19
3D object 424, 427, 443, 451
3D reconstruction 6, 146, 151, 152
3D *suite* plugin 181, 190
Duplicate images 223
3D Viewer 146, 151, 152

E

Edge detection 124, 126, 131, 360, 361, 365
Enhanced Local Contrast (*CLAHE*) 159, 160, 162, 163, 170, 360, 362, 371

F

Fast Random Forest classifier 413
Feature Selection 343, 435, 455, 456, 459
Feret's diameter 17, 20, 151
File formats 33, 67, 92, 170, 337, 436, 441
FiloQuant 359–372
Filter/kernel 20, 105
Find Maxima 61, 68, 105, 113, 314
Fit Ellipse 109, 143, 145, 151, 161, 163

Elena Rebollo and Manel Bosch (eds.), *Computer Optimized Microscopy: Methods and Protocols* , Methods in Molecular Biology, vol. 2040, https://doi.org/10.1007/978-1-4939-9686-5, © Springer Science+Business Media, LLC, part of Springer Nature 2019

Flatten 145
Fluorescence correlation spectroscopy (FCS) 375, 376
Form factor (FF) 109–111, 113, 114
Förster Resonance Energy Transfer (FRET) 216, 235–272, 275–296
Functions 10, 29, 44, 52, 76, 99, 118, 147, 158, 187, 215, 249, 250, 276, 303, 360, 376, 402, 424

G

Gamma correction 112
Gaussian filter 11, 26, 79, 81, 93, 231, 304, 394, 439, 443
GitHub 29, 31, 56, 75, 119, 141, 281, 303, 404
Global thresholding 48, 94, 122, 124, 125, 157, 159, 161, 171

H

High content screening (HCS) 51, 101–103, 105, 111, 299–328, 331–354
Hyperstack 52, 143, 144, 150, 158, 163, 168, 169, 197, 198, 390

I

Ilastik 32, 33, 200, 207, 423–447, 449–462
Illumination co-registration 180
Image analysis 3–10, 13, 15, 17, 23–36, 41–49, 51, 52, 72, 73, 85, 100, 141, 155, 178, 300, 303, 310, 313, 319, 326, 331, 332, 336, 337, 349, 363–366, 371, 379–380, 402, 403, 405–409, 424, 449–451
Image analysis component (IAC) 26, 34
Image calculator 79, 84, 95, 105, 107, 195, 224, 226, 232, 250, 252, 267, 286
Image crop 150, 283, 285, 294, 296, 382, 409, 411, 415, 418
Image features 452–455
Image ID/image window 93, 288, 327, 380, 381
ImageJ/Fiji basic features 92, 180, 184, 185, 190, 195, 210
ImageJ, NIH Image, Fiji, ImageJ2 SCIFIO 29–31
Image masking 224
Image processing component (IPC) 26
Image resolution 8, 19, 370, 371
Image restoration 6, 181, 198
Image stabilization 388
Integrated density (*IntDen*) 85–88, 95, 226, 232
Interoperability 28, 31–33
Iteration statements 54, 55

J

JACoP 182, 184, 216, 219, 226, 227

L

Laplacian of Gaussian (*LOG*) 75, 77, 82–84, 93, 160, 161, 168, 171, 369, 389, 390, 394
Look-up table (LUT) 15, 16, 44, 83, 87, 124, 156, 160, 163, 187, 221, 222, 230, 256, 286, 289, 291, 317, 321, 444
Loop 54, 59–61, 64, 65, 67, 68, 89, 90, 121, 128, 147, 164, 165, 173, 307, 310, 312, 317, 320, 322, 427, 431, 433
Low-pass filter 194, 195, 231

M

Machine learning (ML) 31–33, 51, 198, 200, 332, 333, 341, 343, 352, 399–421, 449–462
Macroinstruction/macro 17, 32, 52, 72, 100, 119, 141, 157, 181, 218, 244, 279, 300, 424
Macro recorder 52, 53, 56–57, 60, 67, 77, 92, 147, 159, 170, 286
Maximum intensity projection 44
MetroloJ 180, 184, 186
MicrobeJ 400
Minimum/maximum local values 18
Morphometric parameters 9, 11, 19–20

O

Object centroid 6, 7, 427, 431, 444
Object classification 32, 454
Object displacements 431
Object overlap 109, 206, 424, 426, 444
Object/particle 5, 25, 43, 77, 113, 127, 138, 161, 181, 231, 262, 284, 301, 332, 360, 387, 401, 423, 450
Off-center illumination 19
Open source software (OSS) 23, 24, 26–29, 31–35, 332, 405, 415
Overlay 3, 73, 76, 111, 145, 200, 228, 291, 316, 431, 444

P

Python 30, 32, 34, 67, 401–404, 409, 411, 418, 420
Pipeline 8, 17, 30, 32, 34, 35, 52, 57, 58, 74, 77, 79, 85, 88, 89, 94, 95, 121, 122, 157, 159, 163, 171, 172, 218, 248, 286, 315, 402–409, 413–415, 438, 450
Pixel autocorrelation 380, 382

Pixel classification................................32, 434–436, 438, 445, 451, 453–455, 461
Plot profile................................170, 192, 197
Plot trajectories................................432, 444, 446
Plugins................................29, 52, 93, 118, 141, 170, 180, 216, 281, 320, 333, 362, 378, 387, 400, 424, 461
Prediction maps................................436, 438, 439, 445
Principal components analysis (PCA)................................341, 344
Programming/macros................................4, 31–34, 51–56, 60, 62–65, 67, 75, 76, 88, 92, 102, 130, 147, 158, 167, 170, 173, 244, 245, 286, 332, 387, 401, 402, 425, 428, 449
PSF Generator................................181, 195

R

Random Forest................................453, 460
Raster-image correlation spectroscopy................................375–383
Ratio................................20, 46, 100, 127, 144, 157, 216, 243, 278, 279, 360, 393
Ratio imaging................................267, 278–281, 283–291, 294, 295
Raw data................................17, 34, 35, 103, 120, 139, 149, 245, 281, 282, 285, 288, 295, 438, 450–452, 454, 459, 461
Region of interest (ROI)................................57, 72, 103, 126, 145, 159, 187, 219, 250, 252, 285, 317, 363, 392, 411
Registration................................144, 150, 180, 286, 295, 393, 411
Reslice................................137, 139, 145
Resolution analysis................................180
Results table................................57, 60–63, 68, 69, 77, 88, 90, 130, 302, 305, 307, 309, 314, 317, 322, 324, 366, 390, 393, 430, 431
Results validation................................23
Robust Automatic Threshold Selection (RATS)...122, 130
ROI Manager................................57, 59, 60, 62–64, 77, 84–86, 88, 94, 103, 107, 109, 126, 130, 131, 145, 161, 167, 171, 172, 190, 191, 221, 222, 230, 251, 285, 317, 392, 395, 427, 430, 431, 445, 446
Routine tasks................................34
Run time................................52–54, 65, 66

S

Salt and pepper noise................................105, 231
Script................................32, 52, 72, 101, 118, 149, 159, 207, 286, 300, 401, 425
Segmentation................................9, 32, 44, 73, 100, 121, 145, 157, 184, 228, 286, 301, 336, 370, 400, 423, 450
Semi-automatic macros................................65
Shape descriptors................................20, 109, 139, 145, 151
Signal to noise ratio (SNR)................................25, 144, 199, 200, 293, 314, 336, 372, 380
Skeletonize, Analyzeskeleton................................360, 361, 371
Spectral unmixing................................181, 194
Stack projection................................44, 139, 144, 150, 372
Substack................................121, 122, 127
Subtract background................................82, 94, 143, 144, 151, 159, 160, 163, 170, 221, 222, 231, 294, 314, 315, 327, 413
Supervised methods................................346

T

Thresholding................................6, 7, 13, 18, 26, 79, 93–96, 103, 105, 107, 122, 124–126, 131, 144, 151, 160, 171, 200, 219, 223, 225, 262, 304, 316, 317, 338, 360, 361, 363–365, 371, 390, 400, 425, 426, 439, 445, 447
Time-lapse image correlation analysis (ICS)................................376
TrackMate................................360, 367–369, 372, 387–391, 393–395, 424, 444
TrackmateSpotDistanceFilter................................388
Trainable Weka Segmentation tool................................401, 404, 412
Train classifier................................408, 412, 413, 418, 419, 455, 457

U

Unsupervised method................................344
User-defined function................................61, 63, 64, 67, 77, 78, 83, 89, 165, 220, 222, 223, 289–291, 303, 313–315, 317, 320, 322, 327, 328
User interface................................29, 31, 33, 119, 120, 128, 130, 332, 339, 400, 450, 451, 453

V

Variables................................52, 53, 57, 59–63, 65, 67–69, 78, 82, 88, 89, 128, 147, 165, 172, 220, 242, 243, 267, 286, 288, 301, 303, 305, 307, 333–335, 338, 341, 351, 383, 427, 431–433
Velocity changes................................431
Video rendering................................415
Volume................................118, 121, 139, 142, 146, 151, 152, 157, 189, 190, 205, 376, 406, 417, 445, 461

W

Wand tool................................145
Watershed................................14, 44, 46, 77, 83, 84, 93, 113, 304, 316, 328, 338, 425, 426, 442
Workflow................................4, 24, 73, 100, 120, 139, 179, 245, 301, 332, 361, 386, 424, 452

Printed by Printforce, the Netherlands